Atoms, Solids, and Plasmas
in Super-Intense Laser Fields

Atoms, Solids, and Plasmas in Super-Intense Laser Fields

Edited by

Dimitri Batani
University of Milano-Bicocca
Milan, Italy

Charles J. Joachain
Free University of Brussels
Brussels, Belgium

Sergio Martellucci
University of Rome "Tor Vergata"
Rome, Italy

and

Arthur N. Chester
Hughes Research Laboratories, Inc.
Malibu, California, USA

Kluwer Academic / Plenum Publishers
New York, Boston, Dordrecht, London, Moscow

Library of Congress Cataloging-in-Publication Data

Atoms, solids, and plasmas in super-intense laser fields/edited by Dimitri Batani ... [et al.].
    p.   cm.
  Includes bibliographical references and index.
  ISBN 0-306-46615-5
    1. Laser beams—Congresses.   2. Laser-plasma interactions—Congresses.   3. Photoionization—Congresses. 4. Multiphoton processes—Congresses. I. Batani, Dimitri.
II. Course of the International School of Quantum Electronics on Atoms, Solids, and Plasmas in Super-intense Laser Fields (2000: Erice, Italy)

QC689.5.L37 A76   2001
539.7'22—dc21

2001033810

The school on "Atoms, Solids, and Plasmas in Super-Intense Laser Fields" was organised in the framework of activity of FEMTO, a programme in the Physical and Engineering Science of the European Science Foundation. More information on this program can be obtained on the web site www.est.org/femto or by writing to the E.S.F. Contact Person Ms. Catherine Werner Cwerner@esf.org, 1 quai Lezay-Marnésia - F-67080 Strasbourg Cedex, tel. 33(0)388767128, fax. 33(0)388370532.

Proceedings of the 30th Course of the International School of Quantum Electronics on Atoms, Solids and Plasmas in Super-Intense Laser Fields, held July 8–14 July, 2000, in Erice, Sicily, Italy

ISBN: 0-306-46615-5

© 2001 Kluwer Academic/Plenum Publishers, New York
233 Spring Street, New York, New York 10013

http://www.wkap.nl/

10 9 8 7 6 5 4 3 2 1

A C.I.P. record for this book is available from the Library of Congress

All rights reserved

No part of this book may be reproduced, stored in a retrieval system, or transmitted in any form or by any means, electronic, mechanical, photocopying, microfilming, recording, or otherwise, without written permission from the Publisher

Printed in the United States of America

PREFACE

The recent developement of high power lasers, delivering femtosecond pulses of intensities up to 10^{20} W/cm^2, has led to the discovery of new phenomena in laser interactions with matter. At these enormous laser intensities, atoms, and molecules are exposed to extreme conditions and new phenomena occur, such as the very rapid multiphoton ionization of atomic systems, the emission by these systems of very high order harmonics of the exciting laser light, the Coulomb explosion of molecules, and the acceleration of electrons close to the velocity of light. These phenomena generate new behaviour of bulk matter in intense laser fields, with great potential for wide ranging applications which include the study of ultra-fast processes, the development of high-frequency lasers, and the investigation of the properties of plasmas and condensed matter under extreme conditions of temperature and pressure. In particular, the concept of the "fast ignitor" approach to inertial confinement fusion (ICF) has been proposed, which is based on the separation of the compression and the ignition phases in laser-driven ICF.

The aim of this course on "Atom, Solids and Plasmas in Super-Intense Laser fields" was to bring together senior researchers and students in atomic and molecular physics, laser physics, condensed matter and plasma physics, in order to review recent developments in high-intensity laser-matter interactions. The course was held at the Ettore Majorana International Centre for Scientific Culture in Erice from July 8 to July 14, 2000.

The lectures published in this book cover all the main aspects of laser interactions with atoms, solids, and plasmas at high intensities. We hope that they will provide scientists from various disciplines with a comprehensive survey of the subject, and a self-contained source from which to pursue a systematic study of the field.

This book is divided into three sections: The lectures, the special seminars on advanced topics given during the school, and a selection of works related to posters presented by participants during the school. The first two sections are divided into two parts of atomic physics and plasma physics, with a special introductory chapter on laser development written by Prof. G. Mourou.

We would like to express our gratitude to the European Science Foundation (E.S.F.), the Italian Ministry of Education, the Italian Ministry of University and Scientific Research, the Sicilian Parliament, the Italian Research Group on Quantum Electronics and Plasma Physics of the National Research Council (G.N.E.Q.P.), the University of Roma "Tor Vergata", the University of Milano-Bicocca and the Istituto Nazionale Fisica della Materia (I.N.F.M.) for their support. It is also a pleasure to thank all the lecturers who accepted our invitation to teach on this course. We would particularly like to thank the other members of the Advisory Committee of the course: Professors P. Knight, A. Maquet, W. Sandner, and H. Walther. We also wish to thank Professor A. Zichichi for his hospitality at the Ettore Majorana International Centre for Scientific Culture.

                                        D.Batani   A.N.Chester
                                        C.J.Joachain   S.Martellucci
                                      Directors of the Course   Directors of the School

                                                                                             January 2001

CONTENTS

PART 1. LECTURES

1. **ULTRAINTENSE LASERS AND THEIR APPLICATIONS** — 1
   G. Mourou

   Introduction — 2
   Intense and Ultraintense Laser Applications — 4

ATOMIC AND MOLECULAR PHYSICS

2. **THEORY OF MULTIPHOTON IONIZATION OF ATOMS** — 15
   N. J. Kylstra, C. J. Joachain, and M. Dörr

   Introduction — 15
   Theory: General Considerations — 19
   Theory: Basic Equations — 20
   Perturbation Theory — 22
   A Free Electron in a Laser Field — 23
   Floquet Theory — 24
   The Sturmian-Floquet Method — 24
   R-Matrix-Floquet Theory — 25
   Numerical Solution of the Time-Dependent Schrodinger Equation — 27
   Low Frequency Methods — 29
   The High Frequency, High Intensity Regime — 30
   Relativistic Effects — 31
   Outlook — 33

3. **EXPERIMENTS OF MULTIPHOTON DISSOCIATION AND IONIZATION OF MOLECULES** — 37
   H. Rottke

   Introduction — 37
   Ionization — 38
   Dissociation — 42
   Correlation in Molecular Fragmentation — 49
   Concluding Remarks — 55

4. **TWO-COLOR AND SINGLE-COLOR ABOVE THRESHOLD IONIZATION** — 59
   P. Agostini

   Part One: Two-Color Above Threshold Ionization and Applications — 59
   Introduction — 59
   Theoretical Approaches to Two-Color Above-Threshold Ionization — 60
   Experiments in Two-Color ATI — 63
   Conclusion — 71
   Part Two: Single-Color Above Threshold Ionization and Double Ionization — 71
   Introduction — 71
   High-Order Single Color ATI — 72

	Non-Sequential Double Ionization: Brief History and Recent Developments	79
	Conclusion	80
5.	**HIGH-ORDER HARMONIC GENERATION** P. Salières	**83**
	Introduction	83
	Physical Process of the Harmonic Emission	85
	Properties of the Harmonic Source	89
	Applications	93
	Conclusion	95
6.	**CLUSTERS IN INTENSE LASER FIELDS** J. W. G. Tisch	**99**
	Introduction	99
	Cluster Formation in Gas-Jets	100
	Experimental Details	101
	The Nanoplasma Model	104
	Review of Experimental Data	110
	Applications	116

PLASMAS

7.	**INTRODUCTION TO LASER-PLASMA INTERACTION AND ITS APPLICATIONS** S. Atzeni	**119**
	Introduction	119
	Laser-Plasma Interaction	121
	Applications of Plasmas Created by ns Laser Pulses	128
	Relativistic Laser-Plasmas	130
	Laser-Driven Inertial Fusion	132
	Outlook	141
8.	**EXPERIMENTAL STUDY OF PETAWATT LASER PRODUCED PLASMAS** M. Key	**145**
	Introduction	145
	PW Class Lasers	146
	Pre-Formed Plasma and Relativistic Self Focusing	148
	Relativistic Electrons, Bremsstrahlung and Positrons	149
	Photo-Nuclear Processes	149
	Generation of High-Energy Proton Beams and Proton Induced Nuclear Phenomena	152
	Fast Ignition	158
	Transport of Energy and Heating by Relativistic Electrons	160
	Conclusions	163
9.	**RELATIVISTIC LASER PLASMA INTERACTIONS** J. Meyer-ter-Vehn, A. Pukhov, and Zh. M. Sheng	**167**
	Introduction	167
	Relativistic Electron Motion in a Strong Laser Field	168
	The Cold Plasma Equations	172
	Relativistic Self-Focusing	173
	Direct Laser Acceleration of Electrons in Plasma Channels	177
	Plasma Waves and Wakefields	187
10.	**DENSE ULTRAFAST PLASMAS** J. C. Gauthier	**193**
	Introduction	193
	Dense Plasmas	194
	Basics of Short Pulse Laser Interaction with Solids	203
	Energy Transport by Photons, and Particles	213
	Atomic Physics and Spectroscopy of Dense Plasmas	217

X-Ray Source Applications of Dense Ultrafast Plasmas	222

11. MAGNETIC FIELDS AND SOLITONS IN RELATIVISTIC PLASMAS — 233
F. Pegoraro, S. Bulanov, F. Califano, T. Esirkepov, M. Lontano, N. Naumova, and V. Vshivkov

Introduction	233
Plasma Dynamics Constraints on the Magnetic Field Generation	234
Pressure Anisotropy, Repulsion of Opposite Currents and Magnetic Field Generation	236
Magnetic Interaction	239
Magnetic Vortices	240
Subcycle Relativistic Solitons	241
Higher Dimension Relativistic Solitons	242
Subcycle Soliton Generation and Interaction	243
Soliton Extraction and Bursts of Soliton Energy	244
Conclusions	245

PART 2. SEMINARS

ATOMIC AND MOLECULAR PHYSICS

12. R-MATRIX-FLOQUET THEORY OF TWO-ELECTRON ATOMS IN INTENSE LASER FIELDS — 249
M. Dörr

Introduction	249
Two-Electron Atoms	250
R-Matrix-Floquet Theory	251
Results	255
Outlook and Future Developments	257

13. INTENSE-FIELD MANY-BODY S-MATRIX THEORY: APPLICATION TO RECOIL-MOMENTUM DISTRIBUTIONS FOR LASER-INDUCED DOUBLE IONIZATION — 261
F. H. M. Faisal

Introduction	261
Intense-Field Many-Body S-Matrix Theory (IMST)	262
Recoil-Momentum Distribution in Double Ionization of He	265

14. ELECTRONS AND IONS IN RELATIVISTIC LASER FIELDS — 273
Y. I. Salamin, M. W. Walser, S. X. Hu, and C. H. Keitel

Introduction	273
Spin-Laser Interaction Effects	276
Electron Acceleration by Crossed Laser Beams	280

15. THE CLASSICAL AND THE QUANTUM FACE OF ABOVE-THRESHOLD IONIZATION — 285
G. G. Paulus and H. Walther

Introduction	285
The Classical Model of ATI	287
Quantum Interference Effects in ATI	293
Conclusion	300

16. RELATIVISTIC EFFECTS IN NON-LINEAR ATOM-LASER INTERACTIONS AT ULTRAHIGH INTENSITIES — 303
V. Véniard

Introduction	303
Two-Photon Bound Transitions in Hydrogenic Atoms	304
Laser- Assisted Mott Scattering	305
Relativistic Effects in Photoionization Spectra	307
Pair Creation in Ultra-Intense Laser Fields	310
Conclusions	312

PLASMAS

17. PLASMAS AT SOLID STATE DENSITY GENERATED BY ULTRA-SHORT LASER PULSES — 315
K. Eidmann

    Introduction — 315
    Isochoric Heating — 316
    Al-K-Shell Spectra — 320
    Conclusions — 324

18. SHOCK WAVE EXPERIMENTS AND EQUATION OF STATE OF DENSE MATTER — 327
M. Koenig

    Introduction — 327
    Hugoniot Relation, Diagnostics — 328
    EOS with Lasers: Some Key Issues — 330
    Relative EOS Measurements — 331
    Applications to Planetary Physics — 333
    Conclusions — 336

19. LASER PARTICLE ACCELERATION IN PLASMAS — 339
J. R. Marquès

    Introduction — 339
    Plasma Wave Properties — 340
    The Major Schemes of Laser Plasma Accelerators — 343
    Experiments on Laser Plasma Acceleration — 345
    Perspectives — 349

Part 3. POSTERS

20. ELLIPTIC DICHROISM IN ANGULAR DISTRIBUTIONS IN FREE-FREE TRANSITIONS IN HYDROGEN — 351
A. Cionga, F. Ehlotzky, and G. Zloh

    Introduction — 351
    Basic Equations — 352
    Weak Field Limit — 354
    Results and Discussion — 355

21. SHOCK ELECTROMAGNETIC WAVES RESULTING FROM HIGHER HARMONICS GENERATION IN TRANSPARENT SOLIDS — 357
V. E. Gruzdev and A. S. Gruzdeva

    Introduction — 357
    Formation of Electromagnetic Shocks — 358
    Interaction of Shew with Transparent Solids — 360
    Conclusions — 362

22. STUDY OF FAST ELECTRON PROPAGATION IN ULTRA-INTENSE LASER PULSE INTERACTION WITH SOLID TARGETS USING REAR SIDE OPTICAL SELF-RADIATION AND REFLECTIVITY - BASED DIAGNOSTICS — 363
J. J. Santos, E. Martinolli, F. Amiranoff, D. Batani, S. D. Baton, A. Bernardinello, L. Gremillet, T. Hall, M. Koenig, F. Pisani, M. Rabec Le Gloahec, and C. Rousseaux

    Introduction — 363
    Target Rear Side Optical Self-Radiation — 364
    Target Rear Side Reflectivity-Based Diagnostics — 368

23. DEMONSTRATION OF HYBRIDLY PUMPED SOFT X-RAY LASER — 375
F. Bortolotto

    Introduction — 375

Experimental Set-Up	376
Results	377
Conclusions and Outlooks	379

24. X-RAY EMISSION FROM LASER IRRADIATED STRUCTURED GOLD TARGETS 381
T. Desai, H. Daido, M. Suzuk, N. Sakaya, A. R. Guerrreiro, and K. Mima

Introduction	381
Experiments	381
Results	382
Discussion	383
Conclusion	387

25. MEASUREMENT OF SPECTRAL AND ANGULAR DISTRIBUTION OF HARD X-RAYS FROM LASER PRODUCED PLASMAS AND THEIR APPLICATION 389
S. Düsterer, H. Schwoerer, R. Behrens, C. Ziener, C. Reich, P. Gibbon, and R. Sauerbrey

Introduction	389
Experimental Set-Up and X-Ray Measurements	389
Discussion of the Results and the Theoretical Considerations	391
Application of the Mev X-Rays: Photoneutrons	392
Conclusion	393

26. EQUATION OF STATE OF WATER IN THE MEGABAR RANGE 395
E. Henry, D. Batani, M. Koenig, A. Benuzzi, I. Masclet, B. Marchet, M. Rebec, Ch. Reverdin, P. Celliers, L. Da Silva, R. Cauble, G. Collins, T. Hall, and C. Cavazzoni

Introduction	395
Experimental Set-Up	396
Experimental Results	397
Conclusions	399

27. XUV INTERFEROMETRY USING HIGH ORDER HARMONICS: APPLICATION TO PLASMA DIAGNOSTICS 401
J.-F. Hergott

INDEX 407

ULTRAINTENSE LASERS AND THEIR APPLICATIONS

Gérard A. Mourou

University of Michigan
Center for Ultrafast Optical Science
2200 Bonisteel Blvd., Rm. 6117/IST Bldg.
Ann Arbor, MI 48109-2099, USA

Abstract. Today laser optics stands on the threshold of a new frontier of unprecedented optical field strengths with single-cycle pulse durations. Traditional optics concerns physical phenomena in the eV regime. The new frontier will address GeV energy scales. In the last decade, lasers have undergone orders-of-magnitude jumps in peak power, with the invention of the technique of chirped pulse amplification (CPA) and refinement of femtosecond techniques. Modern CPA lasers can produce intensities greater than 10^{21} W/cm^2, one million times greater than previously possible. Above 10^{18} W/cm^2, an electron oscillating in the optical frequency laser field has a peak velocity approaching the speed of light, c. and the laser field exceeds many TV/m, fields greater than any other technology can muster. Under these conditions, relativistic nonlinearities emerge. The relativistic intensity regime is rich in novel physical effects, including acceleration of electrons and ions to multi-MeV energies, vacuum heating at surfaces, relativistic harmonic generation from free electrons, relativistic self-focusing and transparency. These ultra-intense lasers give researchers a tool to produce unprecedented pressures –Tbars, magnetic fields-Gigagauss-, temperatures – 10^{10} K- and accelerations- 10^{25}g-with applications to fusion energy, nuclear physics, high energy physics, astrophysics and cosmology

I. INTRODUCTION

Over the past fifteen years, we have seen a revolution in laser intensities [1]. This revolution stemmed from the technique of chirped pulse amplification (CPA), combined with recent progress in short-pulse generation and superior-energy-storage materials like Ti:sapphire, Nd:glass, and Yb:glass. The success of this technique was due to its very general concept, which fits small, university-type, tabletop-size systems as well as large existing laser chains built for laser fusion in national laboratories like CEA-Limeil in France; Lawrence Livermore National, Los Alamos National, and Naval Research Laboratories in the U.S.; Rutherford in the UK; Max Born Institute in Germany; and the Institute of Laser Engineering in Osaka, Japan. A record peak power of 1.5 petawatts (1 PW = 10^{15} W/cm^2) has been produced at LLNL. A number of laboratories are presently equipped with CPA ultrashort-pulse terawatt lasers, such as the Laboratoire d' Optique Appliquée in France, the University of Lund in Sweden, the Max-Planck Institute in Garching, the Jena University in Germany, the Japan Atomic Energy Research Institute (JAERI) in Kansai, Japan, and the NSF Center for Ultrafast Optical Science in the U.S. In this class of CPA lasers, up to 100 TW have been produced by JAERI. CPA lasers have made an new regime of intensities accessible, thus opening up a fundamentally new physical domain.

The large leap in intensities that we have experienced can be described in the following way. After a rapid increase in the 1960s with the invention of lasers, followed by the demonstration of Q-switching and mode-locking, the power of lasers stagnated due to the inability to amplify ultrashort pulses without causing unwanted nonlinear effects in the optical components. This difficulty was removed with the introduction of the technique of chirped pulse amplification [2], which took the power of tabletop lasers from the gigawatt to the terawatt—a jump of three to four orders of magnitude! This technique was first used with conventional laser amplifiers and more recently extended to Optical Parametric Amplifiers (OPCPA) [3]. The peak power in the 1990s reached the 100-TW level with the advance made in short-pulse generation down to the single-cycle regime [4].

More recently, deformable mirrors have been incorporated into CPA, making it possible with a low f# parabola to focus the laser power on a 1-µm spot size [5]. Present systems deliver focused intensities in the 10^{20}-W/cm^2 range. In the very near future, they will be able to produce intensities of the order of 10^{22} W/cm^2. We will certainly see a leveling off of laser intensity for tabletop-size systems at 10^{23} W/cm^2. This limit [6] is related, as we will explain later, to the saturation fluence F_{SAT} (energy per unit area) and damage threshold of the amplifying medium . Once this limit is reached, we will have accomplished a leap in intensities of eight orders of magnitude, and the only way to increase the focused intensity further will be by increasing the beam size, leaving the following questions to be answered: how could we go higher in intensity, and how much higher? To answer these questions, we need to evaluate the theoretical intensity limit.

The Theoretical Intensity Limit [6,7]

In a CPA system, the maximum energy per pulse obtainable is limited (1) by the damage threshold F_D of the stretched pulse—of the order of a few nanoseconds, and (2) by the saturation fluence $F_{SAT} = h\nu/\sigma$ of the amplifier. Where h is the Planck constant, ν is the laser frequency, and σ the cross section of the lasing transition at the lasing frequency ν, F_{SAT} corresponds to the energy per unit area, necessary to decrease the population inversion by a factor of two, in a time shorter than the excited state lifetime, T_f. For stretched pulses of the order of a nanosecond, F_D is of the order of 20 J/cm² and F_{SAT} = 1 J/cm² for Ti:sapphire. For Yb:glass, F_{SAT} = 40 J/cm² —significantly greater than the damage threshold of most optical components. In that regard, high-damage-threshold materials need to be developed in order to take full advantage of the excellent energy storage capability of this material. The minimum pulse duration is imposed by the Heisenberg uncertainty relation $\Delta\nu_a \cdot T = 1/\pi$, where $\Delta\nu_a$ is the transition bandwidth for a homogeneously broadened bandwidth. The maximum power per unit area is therefore given by

$$P_{th} = \pi \frac{h\nu}{\sigma} \Delta\nu_a \qquad (1)$$

The maximum focusable intensity I_{th} will be obtained when this power is focused on a spot size limited by the laser wavelength λ, leading to the expression

$$I_{th} = \pi \frac{h\nu^3}{\sigma} \cdot \frac{\Delta\nu_a}{c^2} \qquad (2)$$

Average Power[6]

The CPA technique made possible not only a jump in peak power by 4–5 orders of magnitude, but also an increase of three orders of magnitude in average power. For pre-CPA systems like dye and excimer lasers, the average power of a laser system is in the milliwatt range, whereas for CPA-type lasers the average power is three orders of magnitude higher, in the watt range. This is due to the fact that the size of the amplifying medium is much reduced, typically by a factor of one thousand. Also, material with good conductivity, like Ti:sapphire, can be used. The conductivity can also be enhanced by cooling the material to cryogenic temperature [8] or using the thermal lens induced in the crystal [9].

Ultraintense Laser Technologies

If we want to take full advantage of the extraordinarily large power available today, it is very important that it be focused on a diffraction-limited spot size and also that the pulse exhibits the highest intensity contrast. Considering that a plasma is created at intensities in the range of 10^{10}-10^{14} W/cm², according to the pulse duration and the type of material— dielectric or metal, we see that for peak power in the range of 10^{20} W/cm², a contrast between the peak and background of 10^6-10^{10} is necessary.

In modern CPA systems, the spatio-temporal control of the laser pulse is of extreme importance for reproducible and predictable results in laser-matter interaction and it dictates the current research in ultraintense laser technology. Here are some examples.

Temporal Control

In a CPA system, the pulse is subjected to an impressive amount of manipulation. Stretching by 10^4 times, amplification by 11 orders of magnitude, and compression by 10^4 leads to phase distortion that will prevent the pulse from being compressed to its initial value. Recently, two techniques have been demonstrated to minimize these effects. One, based on the acousto-optic effect [10], basically adjusts the optical path in an acousto-optic crystal of each frequency by means of an acoustic wave. A time-dependent frequency is chosen to impose a given optical path to each frequency. This is easily done in a birefringent acousto-optic crystal by controlling the pulse polarization state, i.e., ordinary/extraordinary polarization, according to the phase error to be corrected.

A second technique that has also recently been demonstrated uses a linear array of a segmented deformable mirror, DM. The DM is inserted in a zero-dispersion stretcher. The frequencies are spread over the mirror in one dimension and can experience a delay that can be adjusted.

Spatial Control [5, 12]

To reach the highest focused intensities, it is necessary to correct the wave front with a precision of the order of $\lambda/10$. Larger wave-front distortion will lead to less energy in the main peak and more in the wings. An uncorrected CPA laser has only between 5–30% of its energy in the central peak. The sources of distortions are both static and dynamic. Among the chief causes of static aberrations are thermal lensing and grating flatness, as well as the usual aberrations coming from the optical elements. The major source of dynamic aberrations is due to self-focusing in the optical elements, such as gain media, lenses, and Pockels cells. If we want to keep the dynamic aberrations to less than $\lambda/10$, we need to work with a total B value of 0.6. Let's recall that

$$B = \frac{2\pi}{\lambda} \int_0^l n_2 I(z) dz \qquad (3)$$

where n_2 is the nonlinear index of refraction, l is the material length traversed by the stretched pulse, and I(z) is the laser intensity of the stretched pulse. We have demonstrated [12] that the wave front could be corrected to better than $\lambda/10$ on large systems, as well as with kilohertz systems. A kilohertz, millijoule laser has been focused on a single-wavelength spot size, f#1 parabola, making possible the generation of relativistic intensities ($I > 10^{18}$ W/cm^2) at this repetition rate.

II. INTENSE AND ULTRAINTENSE LASER APPLICATIONS

The high-field applications are split into two regimes of intensities—the high intensities (<10^{18} W/cm^2) and the ultrahigh intensities (>10^{18} W/cm^2). The intensity level where the normalized vector potential

$$a = \frac{A}{mc^2} \qquad (4)$$

is equal to 1 is used as the dividing line. A is the laser vector potential, m the mass of the electron, and c the speed of light. A convenient expression to find the value of a is

$$I \cdot \lambda^2 = 1.37 \cdot 10^{18} \cdot a^2 W/cm^2 \qquad (5)$$

with λ in µm.

III. High-Intensity Applications ($<10^{18}$ W/cm^2)

III.1. Dielectric Breakdown, Micromachining, and Precision Surgery

At intensities of the order of 10^{14} W/cm^2, the field of the laser is large enough to ionize most dielectrics. What is remarkable about this regime is the deterministic character of the damage threshold, as opposed to the stochastic behavior that dominates for long (nanosecond) pulses. This was first demonstrated by D. Du et al. [13] and confirmed by Stuart et al. [14] and Lentzner, et al. [15]. These studies show that the damage threshold decreases like $\sqrt{T}$, where T is the pulse duration down to the 10-ps regime. This was demonstrated by the group of Bloembergen [16]. Below this value, the threshold decreases at a slower rate and in some cases increases (for single-shot experiments). More important are the damage threshold error bars or accuracy. The damage threshold is inaccurate for long pulses but Du et al. [13] discovered that it becomes accurate for pulses in the range and below 1 ps. This phenomenon can be explained in the following way.

A transparent dielectric has a low free-electrons density. For the light to be absorbed, it is necessary to raise the initial electron density ($\sim 10^8$/cc) to the plasma critical density value for the input light, i.e., $\sim 10^{21}$/cc. For a long pulse, this is done through the mechanism of impact ionization (avalanche), which is highly nonlinear for low fields, i.e., $<10^7$ V/cm. For short pulses, the seed electrons to the avalanche are not provided by the background electrons but by the multiphoton ionization (MPI) process, which is deterministic. Moreover, the impact ionization at high fields (short pulses) saturates, reinforcing the precise nature of the damage threshold.

This behavior gave rise to what is considered perhaps the most important applications of ultrashort-pulse lasers: micromachining [17] and precision surgery [18]. In micromachining, the applications range from drilling fine holes for fuel injectors, to marking diamonds, to repairing electronic masks, to making waveguides for optical communication applications [17].

Precision surgery has also been demonstrated for subsurface ophthalmic surgery [18]. Femtosecond laser-tissue interaction reveals that near-minimum fluence thresholds for photodisruption can be obtained. Shock-wave effects as well as cavitation bubble size are minimized, resulting in practically no collateral tissue damage. Femtosecond CPA lasers are used in photorefractive surgery and human corneal transplantation. Initial glaucoma experiments also show great potential [18].

III.2. High Harmonic Generation (Extreme Bound Electron Nonlinear Optics)

Coherent XUV can be produced in this intensity regime through harmonic generation in noble gases [19]. The electron driven by the laser field is brought back in the vicinity of its parent ion, where it produces a short wavelength pulse due to the very limited time spent near the nucleus [20]. The process is repeated for each cycle, giving rise to a discrete harmonic spectrum. It is important to work with short pulses to prevent the quick depopulation of the ground state. Harmonics of the laser up to the 330th have been observed. This region corresponds to the water window, 2.3–4.4 nm [21,22], between the oxygen and carbon K edges.

III.3. Isotope Enrichment

Recently a highly-efficient isotope enrichment process, based on laser ablation plumes generated from ultrafast laser pulses focused on a solid target, was discovered [23]. The technique works equally well for light boron as for heavy gallium elements. Enrichment factors of 2 or more above natural abundance have been observed. This technique has been used to deposit isotopically enriched films of boron nitride on silicon. This effect seems to be based on the strong axial magnetic field, 4–40 kG, taking place in the plasma. This magnetic field creates a miniature plasma centrifuge. From a practical standpoint, the direct deposition of engineered, isotopically enriched, thin films is a clear application area for this technique [24].

II.2. Ultrahigh-Intensity Applications (10^{18} W/cm^2)

This regime is clearly the most novel and was made possible by the availability of CPA-Tabletop Terawatt lasers. Today, we are on the threshold of a fundamentally new regime in nonlinear optics which could be as rich as the conventional bound electron nonlinear optics. A host of novel effects have been demonstrated: the generation of x-ray and γ-ray pulses [25–28], the production of high-energy electron [29–35] and ion [36–40] beams as well as neutron [41–42a], and the demonstration of relativistic harmonics from solids [43–44], self-focusing [45–50], and nonlinear Thomson scattering [51–53], as well as new ways to get into the regime of nonlinear quantum electrodynamics [54]. Let's note also the practical applications such as the fast ignitor [55] and compact laser accelerators for electrons, ions, and short-lifetime particles [56].

II.2.1. Ultrashort x-ray and γ-ray Pulses

Ultraintense pulses make possible the production of plasma with high energy density and highly transient and non-equilibrium states. To produce the shortest x-ray pulse, the plasma must have the highest temperature and density, while ionization and cooling must occur very rapidly. To reach such a state of matter, the laser radiation pressure must balance the plasma thermal pressure, thus minimizing the thermal expansion during the heating phase, keeping the density to the fully ionized, solid-density value. Higher density must also be achieved by keeping the foil thickness to a minimum. Using such a foil (500–1000 Å), subpicosecond thermal x ray (Al He$_\alpha$), 8Å was produced. The authors [57] think that this type of source could find direct applications in the study of molecular dynamics problems, and they are at the moment the best trade-off from 1–10 keV between brightness and pulse duration. Applications in precision radiography have been shown. The laser interaction with matter is only over a <10-μm diameter area. It will create a point source or a micro x-ray tube. The very small size of this source can produce high-resolution (μm) radiography, important for mammography [58–60].

II.2.2. Nonlinear Relativistic Optics: Harmonic from Solids

Harmonic generation on solid targets was first demonstrated with 10.6-μm CO_2 light [61]. The scaling of this concept in the visible, due to the large a^2 between 10 to 100, is a very attractive one. Simulations indicate that very high harmonic numbers >60 are expected, with conversion efficiency >10^6. Harmonics up to 75 were observed [43,44]. These harmonics were observed in spite of a prepulse causing a pre-plasma. Much like in the previous application, the radiation pressure plays an important role by steepening the gradient.

The interpretation relies on the moving relativistic mirror, where the electrons are pushed by the light pressure in the target. The more massive ions stay in place. In the process each electron feels the periodic potential of the ions. By moving in this potential back and forth, they are subjected to the Smith-Purcell effect, which occurs when free electrons travel along a periodic structure at a velocity near the speed of light. The spectrum has a plateau with equal amplitude harmonics with a cut-off frequency given by [33]

$$n_c = 8.5 \cdot 10^{-3} \frac{\lambda}{l_c} \left(\frac{I\lambda^2}{10^{14}\,Wcm^{-2}\mu m^2} \right)^{1/2} \quad (6)$$

It is important to keep the pulse very short, <100 fs, shorter than the collapse time of the lattice.

II.2.3. Nonlinear Relativistic Optics: Relativistic Self-Focusing

The intensity distribution across the beam will lead to a mass change proportional to the relativistic factor

$$\gamma = (1 + a^2) \quad (7)$$

across the beam. It will, therefore change the index of refraction according to

$$n(r) = \sqrt{1 - \frac{\omega^2_p(r)}{\omega^2}} \quad (8)$$

where ω is the laser frequency and

$$\omega_p = \left(\frac{4\pi n_o e^2}{m\gamma} \right)^{1/2}$$

is the plasma frequency. Very much as in classical nonlinear optics, self-focusing takes place at a critical power given by

$$P_c = 17.3 \cdot \left(\frac{\omega}{\omega_p}\right)^2 \text{ GW}$$

This effect was predicted by C. Max et al. [25] and was observed recently by a number of groups [26–30]. Because of the large intensities and intensity gradients taking place in the channel, relativistic nonlinear effects are being enhanced, giving rise for instance to large field gradients and electron acceleration. The very steep intensity gradient will lead to large ponderomotive potential that will have the tendency to exclude the electrons in the channel. In turn, the expelled electrons will drag the ions behind, increasing the index of refraction on the beam axis and enhancing the relativistic channeling [49].

II.2.4. Electron Acceleration

The very large ponderomotive pressures due to the colossal intensity gradients will displace the electrons with respect to the ions to create a large plasma wave. This charge displacement will produce a large, longitudinal electrostatic field, which can be as large as

$$E_s = m\omega_p c / e$$

Typical field values that have been produced by a number of groups are of the order of 100 GeV/m, or 10^4 times the fields produced by conventional technologies. It is worth noting that the electron beam generated is well collimated, with a divergence of a few degrees and with a low transverse emittance of 4 mm-mrad—as good as the best electron gun injector. The fact that the electrons are quickly accelerated to relativistic energies reduces the space-charge effects lowering the emittance. The number of electrons is also very copious of the order of 10^9 [30].

II.2.5. Ion Acceleration

Collimated beams of fast protons with energies as high as 55 MeV and a total number of the order of 10^9 have been produced. The cone angle is of the order of 40 degrees. These results have been generated by the petawatt system at Livermore and also by smaller systems at Rutherford and Michigan [36–40]. At Michigan, the protons originate from impurities at the front side and are accelerated over a region extending well into the target [17]. The acceleration gradients are of the order of 100 GeV/cm. The maximum proton energy is consistent with a model based on charge-separation electrostatic field based on vacuum heating. Rutherford has recently demonstrated heavy ions acceleration up to 430 MeV [62].

II.2.6. Pion Production and Acceleration

The generation of GeV protons interacting with a metallic target will produce pions. Pions have a lifetime of only 20 ns; for some high-energy applications like neutrino production, it is necessary to extend their lifetime by, say, 100-fold [63]. Therefore, it becomes necessary to increase their energy or their mass from 130 MeV to a value of ~10 GeV. This is not possible with conventional technology (~20 MeV/m), since the particle will have decayed before it has had the time to be accelerated to these values. Some simulations [64] have shown 10-fs pulses with intensities in the 10^{23}-W/cm^2 range in solid targets could produce GeV protons with <1 fs duration. The number of protons that are expected are in the range of 10^9. This well-synchronized proton pulse can produce a copious amount of pions that can be readily accelerated in an accelerating stage driven by a synchronized laser pulse.

II.2.7. Nuclear Reaction

Ultraintense lasers can drive nuclear reactions. Fusion neutrons have been produced using a deuterated target [41, 42] or deuterium clusters [42a].
Protons with mega-electron-volt energies are sufficient to produce transmutation. This has been demonstrated by a number of groups, Livermore[64], Rutherford [65], and the CEA Limeil [42]. Using only a compact Tabletop Terawatt system, the Garching group [41] was able to report a d(d,n) ^{3}He fusion reaction using a deuterated polyethylene target. Using a MeV deuteron beam produced by the interaction of an ultraintense laser in a polyethylene target, the Michigan group was able to produce a nuclear reaction in a ^{10}B target to produce ^{11}C [66].

II.2.8. Positron Production

In what could be considered the first high-energy experiment driven by a laser, the Livermore group with their petawatt laser [39] and the Garching group using their ATLAS laser reported the production of anti-matter positrons by ultraintense laser [67]. The Garching experiment is truly remarkable in the sense that it uses a true Tabletop Terawatt laser. They are reporting the production of 10^7 e$^+$ per second. This may be a new way to produce positronium for applications in materials science and to perform positronium spectroscopy. The positrons are produced by two basic concepts: (a) the direct or trident effect, where an electron with energy greater than 2 mc^2 is incident on a high-Z converter; or (b) the indirect pair production effect, where the high-energy electrons produce γ-photons with energy greater than 2 mc^2 via bremsstrahlung. These γ-photons will have enough energy to produce pairs.

It is important to note that in these two processes the binding energy is comparable the rest mass energy of the constituents e$^-$- e$^+$, which defines that it is truly a high-energy physics reaction.

II2.9. Nonlinear Quantum Electrodynamics

Pair production from vacuum. The direct production of pairs from vacuum will require an electric field of the order of the Schwinger field, given by

$$E_s = \frac{2mc^2}{e\lambda_c}$$

with the Compton wavelength

$$\lambda_c = \frac{\hbar}{mc}$$

E_s equals 10^{15} V/cm, corresponding to a laser intensity of 10^{30} W/cm^2. This is, of course, much larger, by many orders of magnitude, than what we can produce today or in the near future. This gargantuan intensity will demand a zettawatt (10^{21} W) or 1 MJ in 10 fs focused on 1-μm spot size.

One way to bridge this enormous intensity gap of ten orders of magnitude between what is needed and what can be produced today is by multiphoton pair production i.e. by colliding an ultraintense laser pulse (10^{18} W/cm^2) at the focus of the 50-GeV electron beam from SLAC. The electrons have a relativistic γ of 10^5 that will enhance the laser field by the same factor and the intensity by γ^2 or 10 orders of magnitude.

The multiphoton effect

$$\omega_\gamma + n\omega_0 \to e^+ + e^-$$

corresponding to n = 4, has been observed [68].

REFERENCES

1. G. A. Mourou, C. P. J. Barty, and M. D. Perry, "Ultrahigh-intensity lasers: physics of the extreme on a tabletop," Physics Today, Jan. 1998.
2. D. Strickland, G. Mourou, Opt. Commun. 56, 219 (1985); P. Maine, D. Strickland, P. Bado, M. Pessot, G. Mourou, IEEE J. Quantum Electron. 24, 398 (1988).
3. Dubeis et al., Opt. Comm. 88, 437 (1992).
4. C. P. Huang, H. C. Kapteyn, J. W. McIntoch, M. M. Murnane, Opt. Lett. 17, 139 (1992); F. Krausz, C. Spielmann, T. Brabec, E. Wintner, A. J. Schmidt, Opt. Lett. 17, 204 (1992); C. P. Huang, M. T. Asaki, S. Backus, M. M. Murnane, H. C. Kapteyn, H. Nathel, Opt. Lett. 17, 1289 (1992); B. Proctor, F. Wise, Opt. Lett. 17, 1295 (1992); B. E. Lemoff, C. P. J. Barty, Opt. Lett. 17, 1367 (1992); C. Spielmann, P. F. Curley, T. Brabec, F. Krausz, IEEE J. Quantum Electron. 30, 1100 (1994). A. Stingl, C. Spielmann, F. Krausz, and R. Szipöcs, Opt. Lett. 19, 204 (1994); I. D. Jung, F. X. Kätner, N. Matuschek, D. H. Sutter, F. Morier-Genoud, G. Zhang, U. Keller, V. Scheuer, M. Tilsch, T. Tschudi, Opt. Lett. 22, 1009 (1997).
4a. U. Morgner, F. X. Kätner, S. H. Cho, Y. Chen, A. H. Haus, J. G. Fujimoto, and E. P. Ippen, Opt. Lett. 24, 411 (1999).
5. F. Druon, G. Cheriaux, J. Faure, J. Nees, M. Nantel, A. Maksinchuk, and G. Mourou, Opt. Lett. 23, 1043-1045 (1998).
6. M. D. Perry, G. Mourou, Science 264, 917 (1994).
7. G. Mourou, "The Ultrahigh-Peak-Power Laser: Present and Future," Appl. Phys. B 65, 205-211 (1997).
8. S. Backus, C. Durfee, G. Mourou, H. C. Kapteyn, M. M. Murnane, "Pulse compression by use of deformable mirrors," Opt. Lett. 22, 1256 (1997).
9. F. Salin, and J. Squier, "Gain guiding in solid-state lasers," Opt. Lett. 17, 1352 (1992).
10. P. Tournois "Acousto-optic programmable dispersive filter for adaptive compensation of group-delay time dispersion in laser systems," Opt. Commun. 140, 245 (1997).
11. E. Zeek, K. Maginnis, S. Backus, U. Russek, M. Murnane, G. Mourou, H. Kapteyn, and G. Vdovin, Opt. Lett. 24, p. 493-5 (1999).
12. J. Queneuille, F. Druon, A. Maksimchuk, G. Cheriaux, G. Mourou, and K. Nemoto, Opt. Lett. 25, No. 7, April 2000.
13. D. Du, X. Liu, G. Korn, J. Squier, and G. Mourou, Appl. Phys. Lett. 64, 3071-3073, (994).
14. B. C. Stuart, et al., Phys. Rev. Lett. 74, 2248 (1995).

15. M. Lenzner, J. Krüger, S. Sartani, Z. Cheng, Ch. Spielmann, G. Mourou, W. Kautek, and F. Krausz, "Femtosecond optical breakdown in dielectrics," Phys. Rev. Lett. 80, 4076, (1998).
16. N. Bloembergen, IEEE J. Quantum Electron., QE-10, 375, 1974.
17. Balo (Spelling?)– Laser Focus - 2000
18. T. Juhasz, F. H. Loesel, R. M. Kurtz, C. Horvath, J. F. Bille, and G. Mourou, "Corneal refractive surgery with femtosecond lasers," IEEE J. Selected Topics in Quant. Electron. 5, 902 (1999).
19. G. Mainfray and C. Manus, "Multiphoton ionization in atoms," Rep. Prog. Phys. 54, 1333 (1991).
20. C. Joshi and P. Corkum, "Interactions of ultra-intense laser light with matter," Physics Today, p. 36, (1995).
21. C. Spielmann, C. Kan, N. H. Burnett, T. Brabec, M. Geissler, A. Scrinzi, Matthias Schürer, and F. Krausz, "Near-keV coherent x-ray generation with sub- 10-fs lasers," IEEE J. Selected Topics Quant. Electron., 4, (1998).
22. Z. Chang, A. Rundquist, H. Wang, M. Murnane, and H. C. Kapteyn, "Generation of coherent soft x-rays at 2.7 nm using high harmonics," Phys. Rev. Lett., 79, 2967 (1997).
23. P. P. Pronko, P. A. VanRompay, Z. Zhang, and J. A. Nees, "Isotope enrichment in laser-ablation plumes and commensurately deposited thin films," Phys. Rev. Lett., 83, 2596 (1999).
24. P. A. VanRompay, Z. Zhang, J. A. Nees, P. P. Pronko, "Isotope separation and enrichment by ultrafast laser ablation," SPIE Proceedings 3934, 43 (2000).
25. J. C. Kieffer, J. P. Matte, H. Pépin, M. Chaker, Y. Beaudoin, T. W. Johnston, C. Y. Chien, S. Coe, G. Mourou, and J. Dubau, "Electron distribution anistrophy in laser-produced plasmas from x-ray line polarization measurements," Phys. Rev. Lett., 68, 480 (1992).
26. J. D. Kmetec, C. L. Gordon III, J. J. Macklin, B. E. Lemoff, G. S. Brown, S. E. Harris, "MeV x-ray generation with a femtosecond laser," Phys. Rev. Lett. 68, 1527, (1992).
27. F. Beg, A. R. Bell, A. E. Dangor, C. Danson, A. P. Fews, M. E. Glinsky, B. A. Hammel, P. Lee, P. A. Norreys, M. Tatarakis, "A study of picosecond laser-solid interactions up to 10^{19} W/cm^2," Phys. of Plasmas 4, 447 (1997).
28. P. Norreys, M. Santala, E. Clark, M. Zeph, I. Watts, F. N. Beg, K. Krushelnick, M. Tatarakis, A. E. Dangor, X. Fang, P. Graham, T. McCanny, R. P. Singhal, K. W. D. Ledingham, A. Creswell, D. C. W. Sanderson, J. Magill, A. Machacek, J. S. Wark, R. Allott, B. Kennedy, D. Neely, "Observaton of a highly directional gamma-ray beam from ultrashort, ultraintense laser pulse interactions with solids," Phys. of Plasmas, 6, 2150 (1999).
29. C. E. Clayton, K. A. Marsh, A. Dyson, M. Everett, A. Lal, W. P. Leemans, R. Williams, and C. Joshi, "Ultrahigh-gradient acceleration of injected electrons by laser-excited relativistic electron plasma waves," Phys. Rev. Lett. 70, 37 (1993).
30. D. Umstadter, S. –Y. Chen, A. Maksimchuk, G. Mourou, and R. Wagner, "Nonlinear optics in relativistic plasmas and laser wake-field acceleration of electrons," Science 273, 472 (1996).
31. K. Nakajima, T. Kawakubo, H. Nakanishi, A. Ogata, Y. Kitagawa, R. Kodama, K. Mima, H. Shiraga, K. Suzuki, K. Yamakawa, T. Zhang, Y. Koto, D. Fisher, M. Downer, T. Tajima, Y. Sakawa, T. Shoji, N. Yugami, and Y. Nishida, "Proof-of-principle experiments of laser wakefield acceleration using a 1 ps 10 TW Nd:glass laser," Proc. AIP Conf. Advanced Accelerator Concepts, 335, P. S. Schoessow, Ed. New York: Amer. Inst. Phys., p. 145 (1995).
32. D. Gordon, K. C. Tzeng, C. E. Clayton, A. E. Dangor, V. Malka, K. A. March, A. Modena, W. B. Mori, P. Muggli, Z. Najmudin, D. Neely, C. Danson, C. Joshi, "Observation of electron energies beyong the linear dephasing limit from a laser-excited relativistic plasma wave," Phys. Rev. Lett. 80, 2133 (1998).
33. R. Wagner, S. –Y. Chen, a. Maksimchuk, and D. Umstadter, "Electron acceleration by a laser wakefield in relativistically self-guided channel," Phys. Rev. Lett. 78, 3125, (1997).
34. S. Y. Chen, M. Krishnan, A. Maksimchuk, R. Wagner, D. Umstadter, "Detailed dynamics of electron beams self-trapped and accelerated in a self-modulated laser wakefield," Phys. of Plasmas, 6, 4739 (1999).
35. C. Gahn, G. D. Tsakiris, A. Pukhov, J. Meyer-ler-Vehn, G. Pretzler, P. Thirolf, K. J. Witte, "Multi-MeV electron beam generation by direct laser acceleration in high-density plasma channels," Phys. Rev. Lett. 83, 4772 (1999).
36. G. Sarkisov, V. Yu. Bychenkov, V. N. Novikov, V. T. Tikhonchuk, A. Maksimchuk, S. –Y. Chen, R. Wagner, G. Mouru, D. Umstadter, "Self-focusing, channel formation, and high-energy ion generation in interaction of an intense short laser pulse with a He jet," Phys. Rev. E 59, 7042 (1999).
37. A. G. Zhidkov, A. Sasaki, T. Tajima, T. Aguste, P. D'Olivera, S. Hulin, P. Monot, A. Ya. Faenov, T. A. Pikuz, I. Yu. Skobelev, "Direct spectroscopic observation of multiple-charged-ion acceleration by an intense femtosecond-pulse laser," Phys. Rev. E 60, 3273 (1999).
38. T. Zh. Esirkepov, Y. Sentoku, K. Mima, K. Nishihara, F. Califano, F. Pegoraro, N. M. Naumova, S. V. Bulanov, Y. Ueshima, T. V. Liseikina, V. A. Vshivkov, and Y. Kato, "Ion acceleration by superintense laser pulses in plasmas," JETP Lett., 70, 82 (1999).

39. A. Maksimchuk, S. Gu, K. Flippo, D. Umstadter, V. Yu. Bychenkov, and W. Yu, "Forward ion acceleration in thin films driven by a high-intensity laser," Phys. Rev. Lett., 84, 4108, (2000).
40. E. L. Clark, K. Krushelnick, M. Zeph, F. N. Beg, M. Tatarakis, A. Machacek, M. I. Santala, I. Watts, P. A. Norreys, and A. E. Dangor, "Energetic heavy-ion and proton generation from ultraintense laser-plasma interactions with solids," Phys. Rev. Lett., 85, 1654 (2000).
41. G. Pretzler, A. Saemann, A. Pukhov, D. Rudolph, T. Schätz, U. Schramm, P. Thirolf, D. Habs, K. Eidmann, D. G. Tsakiris, J. Meyer-ler-Vehn, and K. J. Witte, "Neutron production by 200 mJ ultrashort laser pulses," Phys. Rev. E 58, 1165 (1998)
42. L. Disdier, J. P. Garconnet, G. Malka, J. L. Miquel, "Fast neutron emission from a high-energy ion beam produced by a high-intensity subpicosecond laser pulse," Phys. Rev. Lett. 82, 1454 (1999).
42a. T. Ditmire, J. Zweiback, V. P. Yanovsky, T. E. Cowan, G. Hays, and K. B. Wharton, "Nuclear fusion from explosions of femtosecond laser-heated deuterium clusters," Nature, 398, 489 (April 1999).
43. R. Lichter, J. Meyer-ler-Vehn, and A. Kukhov, "Short-pulse laser harmonics from oscillating plasma surfaces driven at relativistic intensity," Phys. Plasmas, 3, 3425 (1996).
44. M. Zeph, D. G. Tsakiris, G. Pretzler, I. Watts, D. M. Chambers, P. A. Norreys, U. Andiel, A. E. Dangor, K. Eidman, C. Gahn, A. Machacek, J. S. Wark, and K. Witte, "Role of plasma scale length in the harmonic generation from solid targets," Phys. Rev. E 58, R5253, (1998).
45. C. Max, J. Arons, and A. B. Langdon, "Self-modulatin and self-focusing of electromagnetic waves in plasmas," Phys. Rev. Lett., 33, 209 (1974).
46. P. Sprangle, C. M. Tang, and E. Esarey, "Relativisitic self-focusing of short-pulse radiation beams in plasmas," IEEE Trans. Plasma Sci., PS-15, (1987).
47. A. B. Borisov, A. V. Borovskiy, O. B. Shiryaev, V. V. Korobkin, A. M. Prokhorov, J. C. Solem, T. S. Luk, K. Boyer, and C. K. Rhodes, "Relativistic and charge-displacement of self-channeling of intense ultrashort laser pulses in plasmas," Phys. Rev. A, 45, 5830 (1992).
48. P. Gibbon, P. Monot, T. August, and G. Mainfray, "Measurable signatures of relativistic self-focusing in underdense plasmas," Phys. Plasmas, 2, 1304, (1995).
49. S.-Y. Chen, G. S. Sarkisov, A. Maksimchuk, R. Wagner, and D. Umstadter, "Evolution of a plasma waveguide created during relativisitic-ponderomotive self-channeling of an intense laser pulse," Phys. Rev. Lett., 80, 2610 (1998).
50. J. Fuchs, G. Malka, J. C. Adam, F. Amiranoff, S. D. Baton, N. Blanchot, A. Héron, G. Laval, J. L. Miquel, P. Mora, H. Pépin, and C. Rousseaux, "Dynamics of subpicosecond relativistic laser plasma self-channeling in an underdense preformed plasma," Phys. Rev. Lett., 80, 1658, (1998).
51. E. S. Sarachik, G. T. Schappert, "Classical theory of the scattering of intense laser ratiation by free electrons," Phys. Rev. D, 1, 2738 (1970).
52. S.-Y. Chen, A. Maksimchuk, and D. Umstadter, "Experimental observation of relativistic nonlinear Thomson scattering," Nature 396, 653, (1998).
53. S.-Y. Chen, A. Maksimchuk, E. Esarey, and D. Umstadter, "Observation of phase-matched relativistic harmonic generation," Phys. Rev. Lett., 84, 5528 (2000).
54. C. Bula et al., Phys. Rev. Lett. 76, 3116 (1996).
55. M. Tabak, J. Hammer, M. Glinsky, W. L. Kruer, S. C. Wilks, J. Woodworth, E. M. Campbell, and M. D. Perry, "Ignition and high gain with ultrapowerful lasers," Phys. Plasmas 1, 1626 (1994).
56. D. Habs – Private communication
57. P. Gallant, Z. Jiang, C. Y. Chien, P. Forget, F. Dorchies, J. C. Kieffer, H. Pépin, O. Peyrusse, G. Mourou and A. Krol, "Spectroscopy of solid density plasmas generated by irradiation of thin foils by a fs laser," J. of Quant. Spect. & Rad. Trans., 65(1-3), 243 (2000).
58. C. B. J. Barty, M. Ben-Nun, T. Guo, F. Ráksi, C. Rose-Petruck, J. Squier, K. Wilson, V. Yakovlev, P. M. Weber, Z. Jiang, A. Ikhlef, J.-C. Kieffer, "Ultrafast x-ray diffraction and absorption," Time-resolved Diffraction, eds. J. Helliwell and P. M. Rentzepis, New York, Oxford University Press, pp. 44, 1998.
59. S. Svanberg, "High-powered lasers and their applications," Adv. in Quant. Chem., 30, 209 (1998).
60. Kieffer – Private Communication
61. R. Carman, C. Rhodes, R. Benjamin, Phys. Rev. A 24, 2649 (1981).
62. E. L. Clark, K. Krushelnick, M. Zepf, F. N. Beg, M. Tatarakis, A. Machacek, M. I. Santala, I. Watts, P. A. Norreys, and A. E. Dangor, "Energetic heavy-ion and proton generation from ultraintense laser-plasma interactions with solids," Phys. Rev. Lett., 85, 1654, (2000).
63. D. Habs – Private Communication
64. T. E. Cowan, A. W. hunt, T. W. Phillips, S. C. Wilks, M. D. Perry, C. Brown, W. Fountain, S. Hatchett, J. Johnson, M. H. Key, T. Parnell, D. M. Pennington, R. A. Snavely, Y. Takahashi, "Photonuclear fission from high energy electrons from ultraintense laser-solid interaction," Phys. Rev. Lett., 84, 903 (2000).

65. K. W. Ledingham, I. Spenser, T. McCanny, R. P. Singhal, M. I. K. Santala, E. Clark, I. Watts, F. N. Beg, M. Zeph, K. Krushelnick, M. Tatarakis, A. E. Dangor, P. A. Norreys, R. Allott, D. Neely, R. J. Clark, A. C. Machacek, J. S. Wark, A. J. Cresswell, D. C. W. Sanderson, and J. Magill, "Photonuclear physics when a multiterawatt laser pulse interacts with solid targets," Phys. Rev. Lett., 84, 899 (2000).
66. K. Nemoto, A. Maksimchuk, V. Bychenkov, S. Banerjee, K. Flippo, D. Umstadter, and G. Mourou, "Laser-triggered 10-MeV ion forward acceleration and table top isotope production," submitted to Applied Physics Letters (2000).
67. C. Gahn, G. D. Tsakiris, G. Pretzler, K. J. Witte, C. Delfin, C. –G. Wahlström, D. Habs, "Generating positrons with femtosecond-laser pulses," Appl. Phys. Lett., 77, 2662 (2000).
68. C. Bula et al., Phys. Rev. Lett. 76, 3116 (1996).

THEORY OF MULTIPHOTON IONIZATION OF ATOMS

N. J. Kylstra[1], C. J. Joachain[2] and M. Dörr[3]

[1]Department of Physics
University of Durham
Durham DH1 3LE
United Kingdom
n.j.kylstra@durham.ac.uk

[2]Physique Théorique, CP 227
Université Libre de Bruxelles
B-1050 Brussels
Belgium
cjoacha@ulb.ac.be

[3]Max-Born-Institut
D-12489 Berlin
Germany
doerr@mbi-berlin.de

INTRODUCTION

Laser pulses having intensities of the order of or exceeding the atomic unit of intensity $I_a = 3.5 \times 10^{16}$ W cm^{-2}, corresponding to the atomic unit of electric field strength $\mathcal{E}_a = 5.1 \times 10^9$ V cm^{-1}, can be readily produced in laboratories today. Such fields are strong enough to compete with the Coulomb forces in controlling the electron dynamics in atomic systems. As a result, atoms in intense laser fields exhibit a variety of phenomena, which are collectively referred to as multiphoton processes. In this contribution we shall review the theory of one important process: the multiphoton ionization of atoms. We begin by giving a brief overview of some of the basic features of the multiphoton ionization. Experimental results will be used to illustrate these features. Next, we discuss some general issues concerning the study of the interaction of atoms with intense laser fields. In particular, we introduce the general form of the time-dependent Schrödinger equation to be solved and outline perturbation theory for laser-atom interactions. Two non-perturbative approaches for solving the time-dependent Schrödinger equation, namely the use of Floquet theory and the direct numerical integration of the time-dependent Schrödinger equation, will then be considered. Subsequently, we outline approximation schemes for the low and high frequency regimes. Finally, we make some remarks concerning relativistic effects in intense laser-atom interactions. The reader is referred to the book edited by Gavrila [1], the articles by Burnett et al. [2], Joachain [3], DiMauro and Agostini [4], Protopapas et al. [5] and Joachain et al. [6] as well as the other contributions to this volume for more detailed discussions of multiphoton processes.

Multiphoton ionization (MPI) refers to the process whereby an atom is ionized by absorbing n photons from the laser field:

$$n\hbar\omega + A^q \to A^{q+1} + e^-. \tag{1}$$

Here q is the charge of the target atomic system A, expressed in atomic units (a.u.), $\hbar\omega$ is the photon energy and $n > 1$ is an integer. This process was first observed by Voronov and Delone [8] who used a ruby laser to ionize xenon via seven photon absorption, and by Hall et al. [9] who measured the two-photon electron detachment from the negative ion I^-. In the following years, important results were obtained by several experimental groups, in particular at Saclay where the dependence of the ionization yields on the intensity were studied. A crucial breakthrough was made when experiments detecting the energy-resolved photo-electrons were performed. In this way Agostini et al. [10] discovered that the ejected electron in the reaction (1) could absorb photons in excess of the minimum required for ionization to occur. The study of this excess-photon ionization, known as "above threshold ionization" (ATI), has been one of the central themes of multiphoton physics in recent years.

A typical example of ATI photo-electron energy spectra, obtained by Petite et al. [7], is shown in Fig. 1. The spectra are seen to consist of several peaks, separated by the photon energy $\hbar\omega$. As the intensity I increases [see Fig. 1(b)], peaks at higher energies appear, whose intensity dependence does not follow the lowest order perturbation theory (LOPT) prediction according to which the ionization rate for an n-photon process is proportional to I^n.

Another remarkable feature of the ATI spectra, also apparent in Fig. 1, is that as the intensity increases the low-energy peaks are reduced in magnitude. The reason for this peak suppression is that the energies of the atomic states are Stark-shifted in the presence of a laser field. For low laser frequencies (e.g. a Nd-YAG laser with $\hbar\omega$ = 1.17 eV), the AC Stark shifts of the lowest bound states are small in magnitude. On the other hand, the induced Stark shifts of the Rydberg and continuum states are essentially given by the electron ponderomotive energy U_p, which is the cycle-averaged kinetic energy of a quivering electron in a laser field and is given by

$$U_p = \frac{e^2 \mathcal{E}_0^2}{4m\omega^2}, \tag{2}$$

where m is the mass of the electron, e is the absolute value of its charge and $\mathcal{E}_0$ is the electric field strength. It is worth stressing that the ponderomotive energy U_p is proportional to I/ω^2 and may become quite large. For example, in the case of the Nd-YAG laser the ponderomotive energy is equal to the photon energy, i.e. $U_p = \hbar\omega$ = 1.17 eV at the intensity $I \simeq 10^{13}$ W cm^{-2}. Since the energies of the Rydberg and continuum states are shifted upwards relative to the lower bound states by about U_p, there is a corresponding increase in the intensity-dependent ionization potential $I_p(I)$ of the atom, so that $I_p(I) \simeq I_p + U_p$, where $I_p = -E_i$ denotes the ionization potential of the field-free initial state of energy E_i. If this increase is such that $n\hbar\omega < I_p + U_p$, then ionization by n photons is energetically forbidden (see Fig. 2). However, atoms interacting with smoothly varying pulses experience a range of intensities, so that the

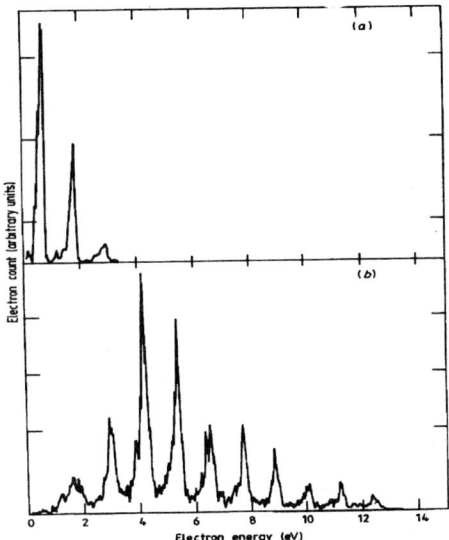

Figure 1: Electron energy spectra showing ATI of xenon at a laser wavelength $\lambda = 1064$ nm. (a) $I = 2 \times 10^{12}$ W cm^{-2}, (b) $I = 10^{13}$ W cm^{-2}. From Petite et al. [7].

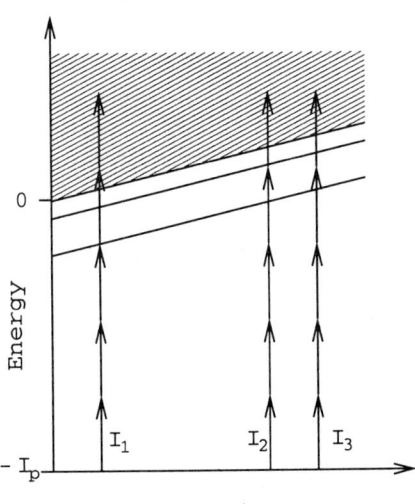

Figure 2: Illustration of the mechanism responsible for the suppression of low-energy peaks in ATI spectra. For low laser frequencies, the intensity-dependent ionization potential of the atom, $I_p(I)$, is such that $I_p(I) \simeq I_p + U_p$, and hence increases linearly with the intensity I. Ionization by 4 photons, which is possible at the intensity I_1 for which $4\hbar\omega \geq I_p + U_p$, is prohibited at the higher intensities I_2 and I_3, where 5 photons are needed to ionize the atom. Also illustrated is the mechanism responsible for the resonantly-induced structures appearing in ATI spectra for short laser pulses. At the intensity I_2, a Rydberg state has shifted into multiphoton resonance with the ground state.

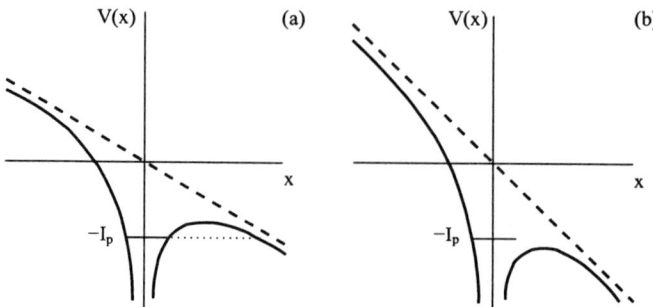

Figure 3: Schematic diagram showing (a) tunneling ionization and (b) over the barrier ionization. The dashed line corresponds to the contribution to the potential energy due to the instantaneous laser electric field. The solid line corresponds to the total effective potential energy.

corresponding peak in the photo-electron spectrum will not completely disappear, as seen in Fig. 1(b).

For short (sub-picosecond) pulses, the ATI peaks exhibit a sub-structure [11] because the intensity-dependent Stark shifts bring different states of the atom into multiphoton resonance during the laser pulse (see Fig. 2). This fine structure is not seen in long pulse experiments because each electron regains its ponderomotive energy deficit from the field as it escapes the laser pulse adiabatically. Highly resolved spectra have been obtained by Hansch et al. [12, 13].

For increasing laser field strengths approaching the Coulomb field binding the electron ($I > 10^{14}$ W cm^{-2}) and for low laser frequencies, the sharp ATI peaks of the photo-electron spectrum gradually blur into a continuous distribution [14, 15]. In this regime, ionization can be interpreted by using a quasi-static model in which the bound electrons experience an effective potential formed by adding to the atomic potential the contribution due to the instantaneous laser electric field (see Fig. 3). This quasi-static approach was used by Keldysh [16] to study tunneling ionization in the low-frequency limit, and pursued by several authors [17–19]. An important quantity in these studies is the Keldysh adiabaticity parameter γ, defined as the ratio of the laser and tunneling frequencies, which is given by

$$\gamma = \sqrt{I_p/2U_p}, \tag{3}$$

where I_p is the field-free atomic ionization potential. For small γ tunneling dynamics dominates, while the transition from multiphoton to tunneling ionization takes place when $\gamma \approx 1$. Above a critical intensity I_c (which is equal to 1.4×10^{14} W cm^{-2} for atomic hydrogen in the ground state), the electron can "flow over the top" of the barrier (over-the-barrier, OTB, ionization), so that field ionization occurs and the atom ionizes in about one orbital period.

The semi-classical, "recollision picture" developed in refs. [20–22] is based on the idea that strong field ionization dynamics at low frequency proceeds via several steps. In the first (bound-free) step, an electron is liberated from its parent atom by tunneling or OTB ionization. In the second (free-free) step, the interaction with the laser field dominates, a fact which was noted earlier by using a simple classical picture of a quivering electron [23, 24]. As the phase of the field reverses, the electron is accelerated

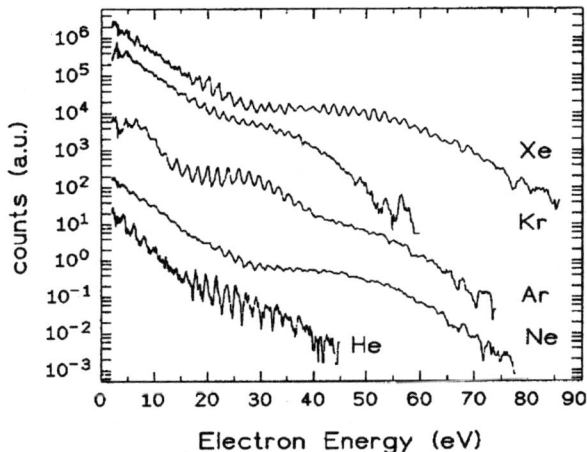

Figure 4: Photo-electron counts as a function of photo-electron energy, for various noble gases, at a laser wavelength of $\lambda = 630$ nm and an intensity $I \simeq 2 \times 10^{14}$ W cm^{-2} (3×10^{14} for He). From Paulus et al. [25].

back towards the atomic core. If the electron returns to the core, a third step takes place in which scattering of the electron by the core then leads to single or multiple ionization, while radiative recombination leads to harmonic generation.

This semi-classical model has been very useful for explaining in terms of classical trajectories and return energies a number of features found in experiments (see the contribution of G. Paulus and H. Walther in this volume). In these experiments, the use of kilohertz-repetition-rate, high-intensity lasers has allowed a precise measurement of photo-electron spectra over many orders of magnitude in yield. These experimental results have revealed the existence of a "plateau" in the ATI photo-electron energy spectra [12, 25]. An example of such spectra is shown in Fig. 4. The prominent groups of ATI peaks that stand out within the plateau have been studied in atomic hydrogen by Paulus et al. [26] and in argon by Hertlein et al. [27] and Muller and Kooiman [28].

THEORY: GENERAL CONSIDERATIONS

The most appropriate theoretical framework for studying the interaction of an atom with a laser field depends on the characteristics of the laser pulse and on the atom or ion being studied. For our purposes, the relevant laser parameters are the frequency, maximum intensity and duration of the laser pulse. With respect to the laser pulse duration, we can, in a non-rigorous way, distinguish between three regimes: the i) long pulse, ii) short pulse and iii) ultra-short pulse regime. In the long pulse regime, which is typically characterized by nanosecond (10^{-9} s; we note that the atomic unit of time is 2.4×10^{-17} s) or longer laser pulses, multiphoton processes in the atom can be characterized by rates which depend only on the angular frequency and maximum intensity of the laser. The exact details concerning the turn-on and turn-off of the pulse are not relevant, except for the requirement that the atom evolves adiabatically in the laser field. In this regime, the rates can be calculated using time-dependent perturbation theory when the field is weak, while for strong or resonant fields the Floquet approach, which will be discussed below, can be used. In the short pulse regime (pulse durations of the order of picoseconds, or 10^{-12} s) the evolution of the atom in the laser field can still be assumed to be adiabatic, apart from isolated resonances between quasi-bound states.

In this regime, the shape of the laser pulse is important and therefore, if a comparison is to be made with experiment, the relevant quantities to be calculated are the transition probabilities which are determined from the instantaneous intensity dependent rates of the atom in the laser field [29]. When comparing with experiment, the spatial profile of the laser in the interaction region must also be considered [29, 30]. Once the rates have been calculated for the range of intensities of the pulse, the final probabilities can be determined for a given laser pulse shape. Finally, in the ultra short pulse regime (of the order of femtoseconds, or 10^{-15}s) the evolution of the atom cannot be considered to be adiabatic so that the only recourse is to obtain information about the multiphoton processes from a direct integration of the time-dependent Schrödinger equation. Obviously, the boundaries which delimit the three regimes cannot be precisely defined.

With respect to the frequency of the laser, we can again distinguish between three regimes. In the low-frequency regime the ionization processes can be viewed in terms of a quasi-static effective potential given by the instantaneous electric field of the laser and the Coulomb potential. In the high-frequency regime the electron orbital frequency is less than the driving laser field frequency and thus an adiabatic approach eliminating the fast laser oscillation proves fruitful. Finally, in the intermediate regime, the photon energy is comparable to relevant atomic transition energies and resonant multiphoton processes will typically play an important role.

Regarding the laser intensity, we can subdivide the low-frequency regime into a multiphoton ionization regime at the low intensity end, an intermediate tunneling ionization regime and the high-intensity over-the-barrier ionization regime. For intermediate and high frequencies, the laser-atom interaction can be described perturbatively when the laser intensity is weak, while semiperturbative methods are useful at moderate intensities and close to resonances, when a relatively small number of atomic states are deemed important in describing the laser-atom interaction. Finally, completely non-perturbative methods are required at high intensities.

The frequency and intensity regimes described above are to a degree inter-dependent. Moreover, while such a classification can usually be made for any atom, the intensities and frequencies which characterize the different regimes will depend strongly on the particular atom and on its initial state, e.g. a frequency that is low for the ground state can be high for an excited state.

THEORY: BASIC EQUATIONS

In order to study the interaction of an atomic system with a laser field, we shall use a semi-classical approach in which the laser field is treated classically, while the atomic system is studied by using quantum mechanics. This approach is entirely justified for the intense fields considered here [3, 31, 32]. We neglect for the moment relativistic effects and treat the laser field in the dipole approximation as a spatially homogeneous electric field $\boldsymbol{\mathcal{E}}(t)$, the corresponding vector potential being $\mathbf{A}(t)$, with $\boldsymbol{\mathcal{E}}(t) = -d\mathbf{A}(t)/dt$. For example, if the field is linearly polarized, we have

$$\boldsymbol{\mathcal{E}}(t) = \hat{\boldsymbol{\epsilon}}\, \mathcal{E}_0 F(t) \cos(\omega t + \phi), \tag{4}$$

where $\hat{\boldsymbol{\epsilon}}$ is the unit polarization vector, $\mathcal{E}_0$ is the electric field strength, $F(t)$ is the pulse shape function and ϕ is a phase. We note that for a chirped pulse, either ω or ϕ are time-dependent.

Our starting point is the time-dependent Schrödinger equation

$$i\hbar \frac{\partial}{\partial t}\Psi(X,t) = H(t)\Psi(X,t) \tag{5}$$

where X denotes the ensemble of the atomic electron coordinates (i.e., their position coordinates $\mathbf{r}_i$ and spin variables). The Hamiltonian $H(t)$ of the system is given by

$$H(t) = H_{at} + H_{int}(t), \tag{6}$$

where $H_{at} = T + V$ is the time-independent field-free atomic Hamiltonian. Here T is the sum of the electron kinetic energy operators and V is the sum of the two-body Coulomb interactions. The laser-atom interaction term is

$$H_{int}(t) = \frac{e}{m}\mathbf{A}(t) \cdot \mathbf{P} + \frac{e^2 N}{2m} A^2(t), \tag{7}$$

where N is the number of electrons and

$$\mathbf{P} = \sum_{i=1}^{N} \mathbf{p}_i, \tag{8}$$

is the total momentum operator. The term in A^2 can be eliminated from the Schrödinger equation (5) by performing the gauge transformation

$$\Psi(X, t) = \exp\left[-\frac{i}{\hbar}\frac{e^2 N}{2m}\int^t A^2(t')dt'\right]\Psi_V(X, t), \tag{9}$$

which gives for $\Psi_V(X,t)$ the Schrödinger equation in the velocity gauge

$$i\hbar\frac{\partial}{\partial t}\Psi_V(X,t) = \left[H_{at} + \frac{e}{m}\mathbf{A}(t) \cdot \mathbf{P}\right]\Psi_V(X,t). \tag{10}$$

On the other hand, if we return to the Schrödinger equation (5) and perform the gauge transformation

$$\Psi(X,t) = \exp\left[-\frac{ie}{\hbar}\mathbf{A}(t) \cdot \mathbf{R}\right]\Psi_L(X,t), \tag{11}$$

where

$$\mathbf{R} = \sum_{i=1}^{N}\mathbf{r}_i, \tag{12}$$

we obtain the Schrödinger equation in the length gauge

$$i\hbar\frac{\partial}{\partial t}\Psi_L(X,t) = [H_{at} + e\,\boldsymbol{\mathcal{E}}(t) \cdot \mathbf{R}]\,\Psi_L(X,t). \tag{13}$$

When performing calculations, the choice of the gauge used will depend on a number of issues. For example, the gauge used has important consequences for the convergence of numerical calculations [33]. It is worth stressing that physical observables do not depend on the gauge used. However, methods for obtaining approximate solutions of the time-dependent Schrödinger equation will give results that are gauge dependent.

As we shall see below, in the high-intensity and high-frequency regime it is useful to study the interaction of an atomic system with a laser field in an accelerated frame, called the Kramers-Henneberger frame [34,35]. Starting from the Schrödinger equation (10) in the velocity gauge, we perform the unitary transformation

$$\Psi_V(X,t) = \exp\left[-\frac{i}{\hbar}\boldsymbol{\alpha}(t).\mathbf{P}\right]\Psi_A(X,t), \tag{14}$$

where
$$\boldsymbol{\alpha}(t) = \frac{e}{m}\int^{t} \mathbf{A}(t')dt' \tag{15}$$

is a vector corresponding to the displacement of a "classical" electron from its oscillation center in the electric field $\boldsymbol{\mathcal{E}}(t)$. The K-H transformation (14) therefore corresponds to a spatial translation, characterized by the vector $\boldsymbol{\alpha}(t)$, to a new frame moving with respect to the laboratory frame in the same way as a "classical" electron in the field $\boldsymbol{\mathcal{E}}(t)$. In this accelerated K-H frame the new Schrödinger equation for a wave function describing a single active electron system, $\Psi_A(\mathbf{r},t)$, is

$$i\hbar\frac{\partial}{\partial t}\Psi_A(\mathbf{r},t) = \left[\frac{p^2}{2m} + V(\mathbf{r}+\boldsymbol{\alpha}(t))\right]\Psi_A(\mathbf{r},t) \tag{16}$$

so that the interaction with the laser field is now incorporated via $\boldsymbol{\alpha}(t)$ into the potential V, which becomes time-dependent. We note that in the case of a linearly polarized monochromatic field

$$\boldsymbol{\mathcal{E}}(t) = \hat{\boldsymbol{\epsilon}}\,\mathcal{E}_0 \cos(\omega t) , \tag{17}$$

we have
$$\boldsymbol{\alpha}(t) = \hat{\boldsymbol{\epsilon}}\,\alpha_0 \cos(\omega t) , \tag{18}$$

where
$$\alpha_0 = \frac{e\mathcal{E}_0}{m\omega^2} \tag{19}$$

is called the "excursion" amplitude of the electron in the field. This parameter is of particular importance in the high-frequency, high intensity regime.

PERTURBATION THEORY

At low intensities (such that the electric field strength $\mathcal{E}_0$ is much smaller than the atomic fields relevant to the process considered) time-dependent perturbation theory (see e.g. Faisal [31]) can in general be used to study multiphoton processes. The simplest form of this approach is called lowest (non-vanishing) order perturbation theory (LOPT). For example, in the case of an n-photon ionization process from an initial (unperturbed) bound state $|\psi_i>$, LOPT predicts that the ionization rate Γ_n is given by

$$\Gamma_n \sim I^n \sum_f |T_{fi}^{(n)}|^2, \tag{20}$$

where $T_{fi}^{(n)}$ is the LOPT transition matrix element for the absorption of n photons and the sum is over allowed final states $|\psi_f>$. Thus, if $H_0 \equiv H_{at}$ is the "unperturbed" (field-free) Hamiltonian and $G_0(E) = (E-H_0)^{-1}$ is the corresponding Green's operator, one has, in the length gauge

$$T_{fi}^{(n)} = <\psi_f|\hat{\boldsymbol{\epsilon}}.\mathbf{R}G_0(E_i+(n-1)\hbar\omega)...\hat{\boldsymbol{\epsilon}}.\mathbf{R}G_0(E_i+\hbar\omega)\hat{\boldsymbol{\epsilon}}.\mathbf{R}|\psi_i>, \tag{21}$$

where E_i is the energy of the unperturbed initial state. Similar LOPT expressions can be written down for other multiphoton processes such as harmonic generation or laser-assisted electron-atom collisions.

The calculation of the LOPT transition matrix element $T_{fi}^{(n)}$ is in general a difficult task, particularly for high order multiphoton process and (or) for complex atoms. The simplest case is that of non-resonant MPI in one-electron atoms for which LOPT has been applied successfully for intensities $I < 10^{13}$ W cm^{-2} and angular frequencies such that $\hbar\omega \gg U_p$ (see e.g. refs. [36, 37]). Discrepancies from the perturbative I^n power law, which are found at higher intensities, signal the breakdown of perturbation theory as do other strong field phenomena such as the "peak suppression" in ATI spectra, the existence of a plateau in high order harmonic generation, or successive FFT peaks of comparable height in laser-assisted electron-atom scattering.

Let us now return to the transition matrix element (21). Using the spectral representation of the Green's operator $G_0(E)$, namely

$$G_0(E) = \sum_k \frac{|\psi_k\rangle\langle\psi_k|}{E - E_k} \qquad (22)$$

with $H_0|\psi_k\rangle = E_k|\psi_k\rangle$, we can write $T_{fi}^{(n)}$ in the more explicit form

$$T_{fi}^{(n)} = \sum_{k_1}\sum_{k_2}\cdots\sum_{k_{n-1}} \frac{\langle\psi_f|\hat{\epsilon}.\mathbf{R}|\psi_{k_{n-1}}\rangle \ldots \langle\psi_{k_2}|\hat{\epsilon}.\mathbf{R}|\psi_{k_1}\rangle\langle\psi_{k_1}|\hat{\epsilon}.\mathbf{R}|\psi_i\rangle}{(E_i + (n-1)\hbar\omega - E_{k_{n-1}})\ldots(E_i + 2\hbar\omega - E_{k_2})(E_i + \hbar\omega - E_{k_1})} \qquad (23)$$

which shows that LOPT always fails for resonant multiphoton processes such that $E_i + r\hbar\omega = E_{k_r}$, for a particular $r \in \{1, 2, ..., n-1\}$. In this case, modifications of the theory are required, in which the resonantly coupled states are treated in a non-perturbative way, while the other states are treated by using perturbation theory.

A FREE ELECTRON IN A LASER FIELD

For a Dirac particle, an exact quantum mechanical solution for a charged particle in a plane electromagnetic wave can be obtained [38]. The solution of the time-dependent Schrödinger equation for an electron and in the dipole approximation,

$$i\hbar\frac{\partial}{\partial t}\Psi(\mathbf{r},t) = \frac{1}{2m}[\mathbf{p} + e\mathbf{A}(t)]^2 \Psi(\mathbf{r},t) \qquad (24)$$

is also readily found, and the reader can verify that the solution is

$$\begin{aligned}\Psi_{\mathbf{k}}(\mathbf{r},t) &= (2\pi)^{-\frac{3}{2}}\exp\left(i\mathbf{k}\cdot\mathbf{r} - \frac{i}{2m\hbar}\int^t dt'\,[\hbar\mathbf{k} + e\mathbf{A}(t')]^2\right) \\ &= (2\pi)^{-\frac{3}{2}}\exp\left(i\mathbf{k}\cdot[\mathbf{r} - \boldsymbol{\alpha}(t)] - \frac{iEt}{\hbar} - \frac{ie^2}{2m\hbar}\int^t dt'\,\mathbf{A}^2(t')\right)\end{aligned} \qquad (25)$$

with $E = (\hbar k)^2/2m$. This so-called Volkov wave function is a momentum eigenstate. Using the gauge transformations discussed above, the Volkov wave function in the length and velocity gauges are easily obtained, while in the K-H frame the Volkov wave function is simply a plane wave. These wave functions have a number of applications. For example, they can be used in low frequency approximations to calculate multiphoton ionization rates when the Coulomb interaction between the electron and the ionic core is, in first approximation, neglected after the laser pulse is applied.

FLOQUET THEORY

We shall now consider a fully non-perturbative approach for studying laser-atom interactions, namely the Floquet theory. Let us restrict our attention to monochromatic laser fields of angular frequency ω and of arbitrary polarization. The Hamiltonian $H(t)$ of the system is then periodic, $H(t+T) = H(t)$, where $T = 2\pi/\omega$. The Floquet method [39–41] can therefore be used to write the wave function $\Psi(X,t)$ in the form

$$\Psi(X,t) = e^{-iEt/\hbar} F(X,t) \tag{26}$$

where the time-independent quantity E is called the quasi-energy, and the function $F(X,t)$ is periodic in time, with period T, so that it can be expressed as the Fourier series

$$F(X,t) = \sum_{n=-\infty}^{+\infty} e^{-in\omega t} F_n(X) . \tag{27}$$

The functions $F_n(X)$ are called the harmonic components of $F(X,t)$. Using eqs. (26) and (27), we obtain for $\Psi(X,t)$ the Floquet-Fourier expansion

$$\Psi(X,t) = e^{-iEt/\hbar} \sum_{n=-\infty}^{+\infty} e^{-in\omega t} F_n(X) . \tag{28}$$

If we also make a Fourier analysis of the interaction Hamiltonian,

$$H_{int} = \sum_{n=-\infty}^{+\infty} e^{-in\omega t} (H_{int})_n \tag{29}$$

and substitute both eqs. (28) and (29) into the Schrödinger equation (5), we obtain for the harmonic components $F_n(X)$ a system of time-independent coupled differential equations:

$$(E + n\hbar\omega - H_{at}) F_n(X) = \sum_{k=-\infty}^{+\infty} (H_{int})_{n-k} F_k(X) . \tag{30}$$

with $n = 0, \pm 1, \pm 2...$. These equations, together with appropriate boundary conditions, form an eigenvalue problem for the quasi-energies, which we can write as

$$(\mathbf{H}_F - E)\, \mathbf{F} = 0 \tag{31}$$

where the Floquet Hamiltonian $\mathbf{H}_F$ is an infinite matrix of operators. In the case of multiphoton ionization, the discrete states of the atom ionize since they are coupled to the continuum states by the laser field. In other words, the discrete states become resonant states in the laser field. The quasi-energies are therefore complex and can be expressed as

$$E = E_i + \Delta - i\frac{\Gamma}{2} \tag{32}$$

where E_i is the energy of the initial unperturbed (field-free) state and Δ is the AC Stark shift of the state. The physical meaning of Γ can be deduced by noting that the integral over a finite volume of the electron density, averaged over one cycle, decreases in time like $\exp(-\Gamma t/\hbar)$. Hence the characteristic lifetime of an atom described by the Floquet state (26) is $\hbar/\Gamma$, which means that $\Gamma/\hbar$ is the total ionization rate of that state. We also note that, in the velocity or the length gauges, the interaction Hamiltonian can be written in the form

$$H_{int}(t) = H_+ e^{-i\omega t} + H_- e^{-i\omega t} \tag{33}$$

where H_+ and H_- are time-independent operators. The coupled equations (30) then take the simpler form

$$(E + n\hbar\omega - H_{at})F_n(X) = H_+ F_{n-1}(X) + H_- F_{n+1}(X) \tag{34}$$

and the Floquet Hamiltonian $\mathbf{H}_F$ is a tridiagonal matrix of operators.

THE STURMIAN-FLOQUET METHOD

The Floquet theory has been used extensively to study multiphoton processes in atomic systems. In particular, detailed calculations have been performed for one-electron atoms. Denoting by $\mathbf{r}$ the electron position vector, and following Maquet et al. [42] the system of coupled equations (34) can be solved by expanding each harmonic component $F_n(\mathbf{r})$ on a discrete basis set:

$$F_n(\mathbf{r}) = \sum_{NLM} c_{NLM}^n \, r^{-1} S_{NL}^\kappa(r) \, Y_{LM}(\hat{\mathbf{r}}), \tag{35}$$

where the Y_{LM} are spherical harmonics and the radial functions S_{NL}^κ are complex Sturmian functions [43]. The parameter κ is chosen to be complex, allowing the computation of quasi-bound (Siegert) states, in complete analogy to the complex-rotation (or dilatation) transformation method [41]. The complex, discrete Sturmian basis is particularly appropriate for obtaining highly accurate solutions to the coupled Floquet equations for one-electron atoms moving in Coulomb or modified Coulomb effective potentials. The Sturmian-Floquet method has been applied extensively to study multiphoton ionization and harmonic generation in atomic hydrogen (see e.g. the review by Potvliege and Shakeshaft [29]) and other systems modeled by effective single-active-electron potentials. As an example, we show in Fig. 5 the calculated photo-electron spectrum for the lowest three ATI peaks [44] for the multiphoton ionization of H(1s) compared with the experimental data of Rottke et al. [45]. The subpeaks are due to Rydberg states moving in and out of resonance when the intensity of the pulse rises and falls [11], as explained above.

R-MATRIX-FLOQUET THEORY

The R-matrix-Floquet (RMF) theory is a non-perturbative approach proposed by Burke, Francken and Joachain [46, 47] to analyze atomic multiphoton processes in intense laser fields. The RMF theory treats multiphoton ionization, harmonic generation and laser-assisted electron-atom collisions in a unified way. It is completely ab-initio and is applicable to an arbitrary atom or ion, allows an accurate description of electron correlation effects, and can be used to describe multiphoton processes involving at most one ejected electron. A more detailed account of the method and applications to two-electron systems is given by M. Dörr in this volume.

Consider an atomic system, composed of a nucleus and N electrons, in a laser field. Neglecting relativistic effects, the atomic system is then described by the time-dependent Schrödinger equation (5). According to the R-matrix method [48,49], configuration space is subdivided into two regions. The internal region is defined by the condition that the radial coordinates r_i of all N electrons are such that $r_i \leq a$ ($i = 1, 2, ...N$),

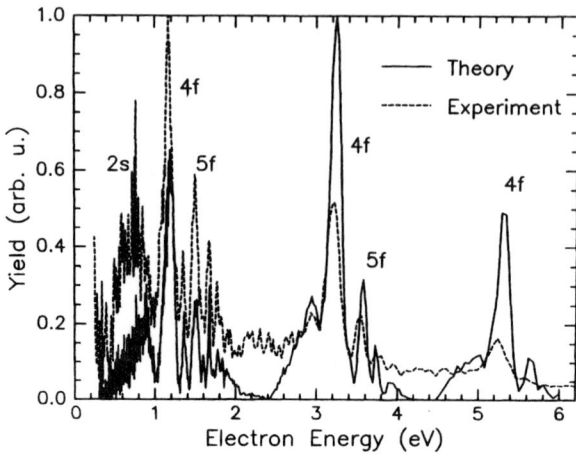

Figure 5: Yield of photo-electrons, into the lowest three ATI channels, versus photo-electron energy, for ionization of H(1s) by a 608 nm pulse whose peak intensity is 6.5×10^{13} W cm^{-2} and whose duration is 0.5 ps. The bold curve is the result of Sturmian-Floquet calculations and the thin curve represents the experimental data (from Rottke et al. [30]). Some of the subpeaks are labelled by the dominant configuration of the resonant Floquet state.

where the sphere of radius a envelops the charge distribution of the target atom states retained in the calculation. In this region, exchange effects involving all N electrons are important. The external region is defined so that one of the electrons (say electron N) has a radial coordinate $r_N \geq a$, while the remaining $N-1$ electrons are confined within the sphere of radius a. Hence in this region exchange effects between the "external" electron and the remaining $N-1$ electrons can be neglected.

Having divided configuration space into an internal and an external region, we must solve the time-dependent Schrödinger equation in these two regions separately. This is done by using the Floquet method, which, as we have seen above, reduces the problem to solving an infinite set of coupled time-independent equations for the harmonic components $F_n(X)$ of the wave function $\Psi(X,t)$. The solutions in the internal and external regions are then matched on the boundary at $r = a$.

The RMF theory has allowed the ab-initio study of a wide variety of resonance effects in multiphoton ionization [50–56]. In particular, a very interesting effect which has been predicted by the RMF theory is the occurrence of laser-induced degenerate states (LIDS) involving autoionizing states in complex atoms [53]. We show in Fig. 6 the results of a RMF calculation [53] in which the influence of a strong laser-induced coupling between the ground state and the $3s3p^64p$ ^1P autoionizing state of Ar has been studied. The trajectories of the complex quasi-energies of the ground state and the autoionizing state are plotted in the complex energy plane as a function of the laser intensity and for fixed values of the angular frequency ω. Two structures are visible, about which the quasi-energy curves associated with the ground state and of the autoionizing state, respectively, exchange their roles. At the center of each of these two structures there is a critical point (to which correspond a critical intensity and angular frequency) such that the two complex quasi-energies are exactly degenerate, i.e., where laser-induced degenerate states (LIDS) occur. The existence of LIDS is a general phenomenon, which has been observed in RMF calculations for multiphoton transitions [53–55,57] and understood by constructing models that retain the essential ingredients of the full RMF calculations. By following an adiabatic path in the frequency and

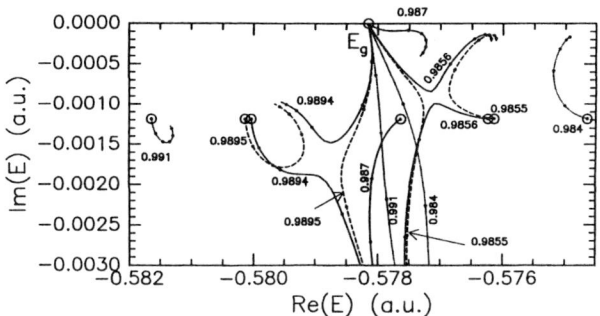

Figure 6: Trajectories of the complex Floquet quasi-energies for the ground state and the 3s 3p^6 4p ^{1}P autoionizing state of argon, for intensities varying from 0 to 5×10^{13} W cm^{-2}. The values of the angular frequency ω are indicated next to the trajectories. The small dots correspond to values of the intensity increasing in steps of 9×10^{12} W cm^{-2}. For each angular frequency ω, there are two trajectory curves: one corresponding to the ground state and the other to the autoionizing state. From Latinne et al. [53].

intensity parameter space one can in principle complete a circuit around the degeneracy, as discussed for degeneracies occuring in atomic hydrogen in a two-colour field [58]. In this sense, LIDS constitute an interesting extension of the work of Berry [59] where the adiabatic passage around degeneracies in a parameter space was described, and which has attracted considerable interest, particularly with respect to the associated geometric phase.

The RMF theory has been extended to describe atoms in two laser fields with incommensurable frequencies, and has been used to analyze light-induced continuum structures (LICS) in helium [60, 61] as well as doubly and triply resonant multiphoton processes involving autoionizing resonances in magnesium [57]. Within the context of these multiply resonant processes, coherent control of the ionization can be exercised in the sense that by changing the laser parameters, the degree of interaction between the resonant processes can be varied.

NUMERICAL SOLUTION OF THE TIME-DEPENDENT SCHRODINGER EQUATION

The non-perturbative Floquet methods discussed above are based on the assumption that the Hamiltonian of the atomic system in the laser field is periodic in time. Although this is not true for a realistic laser pulse, it is still possible to incorporate pulse shape effects into the Floquet or R-matrix Floquet calculations for laser pulses which are very short, even down to a few laser cycles. In particular, if the variation of the laser intensity is slow enough, the atom will remain in the Floquet eigenstate adiabatically connected to the initial state. Numerical studies indicate that this adiabaticity condition is robust for nonresonant, multiphoton ionization by short laser pulses [62–65].

For ultra-short pulses, typically in the femtosecond range, one must in general obtain information about the multiphoton processes by direct numerical integration of the TDSE. Advances in computer technology mean that this is can be done with a moderate amount of effort for atoms or ions with one single active electron (SAE) interacting with laser fields. This approach, pioneered by Kulander [66, 67], has the advantage that the interaction of atoms with laser pulses having a wide range of shapes, frequencies and durations can be studied.

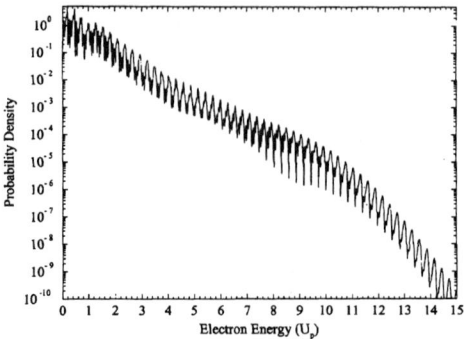

Figure 7: ATI spectrum of atomic hydrogen in a linearly polarized laser pulse of 25 fsec FWHM, $\hbar\omega = 2$ eV and $I = 2 \times 10^{14}$ W cm^{-2}. From Cormier and Lambropoulos [33].

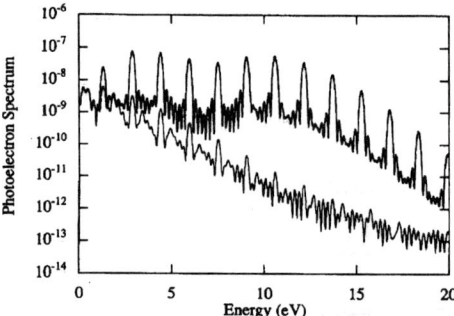

Figure 8: Effect of the presence of the 13th harmonic of a Ti:Sapphire laser on the photo-electron spectrum of atomic hydrogen. The fundamental laser photon energy is $\hbar\omega_L = 1.55$ eV and its intensity is $I_L = 10^{13}$ W cm^{-2}. The thick line corresponds to a 13th harmonic of intensity $I_H = 3 \times 10^8$ W cm^{-2}, the thin line to an intensity $I_H = 0$. From Taïeb, Véniard and Maquet [68].

An example of results obtained for atomic hydrogen is shown in Fig. 7, where a high-order ATI spectrum, obtained by solving the TDSE numerically [33] is displayed. These calculations confirm the existence of a plateau in the spectra, observed in experiments [26, 69]. In Fig. 8 we show the two-colour photo-electron spectrum obtained by Taïeb et al. [68] when hydrogen atoms are submitted to an intense radiation pulse containing the fundamental of a Ti:Sapphire laser operated at $\hbar\omega_L = 1.55$ eV (i.e. $\omega_L = 0.057$ a.u.), and a weaker 13th harmonic having an angular frequency $\omega_H = 0.741$ a.u. (in the UV range) high enough so that the atom can be ionized by a single photon. Interferences arise between multiphoton "above threshold ionization" (ATI) and laser-assisted single photon ionization (LASPI). These interference effects can lead to a partial coherent control of the photo-ionization process [68, 70].

We point out that many features of harmonic generation have been studied by solving the TDSE [71–74]. For more detailed reviews on this subject we refer the reader to the articles of Salières et al. [75], Platonenko and Strelkov [76] and to the contribution of P. Salières in this volume.

A straightforward way of reducing the computational load when numerically integrating the TDSE is to make use of one-dimensional models. Since the one-dimensional models are relatively easy to solve numerically, it is possible to conduct "numerical experiments" by investigating a large range of parameters (see Protopapas et al. [5] and Eberly et al. [77]). In addition, a two-dimensional model has been used to study the dependence of multiphoton ionization and harmonic generation on the ellipticity of the laser field [78, 79].

In order to model multiphoton processes in complex atoms, most TDSE calculations have been performed using the SAE approximation. This model works extremely well, with excellent agreement with experimental ATI spectra being obtained [80]. However, a test for theories of multiphoton processes that include electron correlation effects is to calculate accurately double ionization yields for multi-electron systems in intense laser fields. These quantities have been measured accurately over a wide range of laser intensities in a number of experiments involving different atoms [81, 82]. A striking feature of these experimental results is the existence of two distinct intensity regimes: one in which the double ionization processes proceeds predominantly sequentially (at higher intensities) and the other region in which it is simultaneous (lower intensities).

This phenomenon has been analyzed [83] using a "semi-independent" electron approach, requiring the solution of two single-active-electron problems, the second incorporating the results from the first and thus subject to interelectronic correlation. This calculation reproduces the large enhancement of the double electron ejection at low intensity. A theory of double ionization has been proposed by Faisal and Becker [84, 85] (see also the contribution of F. Faisal in this volume) which relies on a low frequency approximation discussed in the following section. Good agreement has been obtained with several experiments, for different laser frequencies and atomic species.

Finally, we note that using a massively parallel computer Taylor and co-workers [86–88] have studied ab-initio multiphoton processes by integrating the TDSE for helium. A review of the theory of two-electron atoms in intense laser fields has been given by Lambropoulos et al. [89].

LOW FREQUENCY METHODS

When the laser period is much longer than the typical "orbital period" of the bound electron, the laser frequency can be characterized as being in the "low frequency" regime. Thus, most experiments using short, intense pulsed lasers and noble gas atoms fall into this category, since the typical ground state binding energies are of the order of the atomic unit while the corresponding photon energy is typically an order of magnitude smaller.

The continuous passage from multiphoton ionization to static field ("tunnel") ionization has been investigated theoretically [90, 91] and observed experimentally [15]. Thus at sufficiently low frequency and moderately high intensities a tunnel formula given by Keldysh and others [16, 92, 93] describes the ionization rate very well [82]. The Keldysh approach has been modified by Faisal [17] and Reiss [18] and a useful formula for general atoms based on quantum defects has been given by Ammosov et al. [19]. In the limit of low intensities, the ionization process is always a multiphoton one and the intensity scaling agrees with lowest order perturbation theory even for very high orders. At high intensities the tunnel formulae break down, since over-the-barrier ionization occurs.

While total ionization rates at not too high intensities can be adequately described by a tunnel formula, many of the details of the ionization process requires refined models that include processes like the rescattering of the ionizing electron with the residual core [20, 21, 94]. As discussed by G. Paulus and H. Walther in this volume, the interaction of the laser-driven quasi-free electron with the core leads to a plateau in the ATI electron spectrum. For a discussion of low frequency models in the description of harmonic generation, the reader is referred to the contribution of P. Salières in this volume.

THE HIGH FREQUENCY, HIGH INTESITY REGIME

The interaction of atomic hydrogen with with high intensity monochromatic laser fields whose frequency is much larger than the threshold frequency for one-photon ionization was initiated by Gavrila and co-workers using the Floquet theory [1, 95]. Their approach is formulated in the Kramers-Henneberger (K-H) frame introduced above. The Schrödinger equation, for the case of an atom with one active electron, is given by equation (16), with $\boldsymbol{\alpha}(t)$ given by eq. (18) for the case of a linearly polarized, monochromatic field. Since equation (16) now has periodic coefficients, one can seek solutions having the Floquet-Fourier form (28). Making also a Fourier series expansion of the potential

$$V(\mathbf{r} + \boldsymbol{\alpha}(t)) = \sum_{n=-\infty}^{+\infty} e^{-in\omega t} V_n(\alpha_0, \mathbf{r}) , \qquad (36)$$

one obtains for the harmonic components $F_n(\mathbf{r})$ of $\Psi_A(\mathbf{r}, t)$, as defined in eq. (27), the system of coupled equations

$$\left[E + n\hbar\omega - \left(\frac{p^2}{2m} + V_0(\alpha_0, \mathbf{r}) \right) \right] F_n(\mathbf{r}) = \sum_{\substack{k=-\infty \\ (k \neq n)}}^{+\infty} V_{n-k}(\alpha_0, \mathbf{r}) F_k(\mathbf{r}) . \qquad (37)$$

Gavrila and coworkers have shown that in the high-intensity, high-frequency limit, the atomic structure in the laser field is essentially governed by the potential $V_0(\alpha_0, \mathbf{r})$ which is the static, (time-averaged) "dressed" potential associated with the interaction potential V in the K-H frame. (See the contribution of M. Gavrila in this volume.)

Detailed calculations of the energies and eigenstates, performed for the hydrogen atom, show that the atom undergoes "dichotomy", i.e. the electronic cloud in the static part of the potential splits into two disjoint parts. With increasing α_0, the wave function undergoes radiative stretching along the polarization axis. When $\alpha_0=20$ a.u., a saddle and two pronounced maxima around the end points $\pm\alpha_0\hat{\mathbf{e}}$ of the "classical" electron excursion amplitude appear. In addition, the ionization potential of the ground state of $V_0(\alpha_0, \mathbf{r})$ decreases with increasing α_0.

An important prediction of the high-frequency Floquet theory (HFFT) is that at sufficiently high intensity, and when the frequency of the laser field is larger than the threshold frequency for one-photon ionization, the ionization rate decreases when the intensity increases. This phenomenon has been studied in detail using the approximate HFFT [1,95–98], the Sturmian-Floquet theory [99,100] and the R-matrix Floquet theory [101]. As an example, we show in Fig. 9 the total rate Γ for ionization of H(1s), as obtained from the Sturmian-Floquet method [99] and the R-matrix-Floquet theory [101] for an angular frequency $\omega = 0.65$ a.u. which is larger than the threshold value ($\omega = 0.5$ a.u.) for one-photon ionization. The Floquet results are seen to increase linearly at low intensities, as predicted by lowest order (in the present case, first order) perturbation theory (LOPT). They have a peak near 10^{16} W cm^{-2} and then decrease with increasing intensity, thus exhibiting the stabilization behaviour.

In the Floquet analysis discussed above, a monochromatic laser field of constant intensity is assumed to be present at all times. In reality, an atom is subjected to a

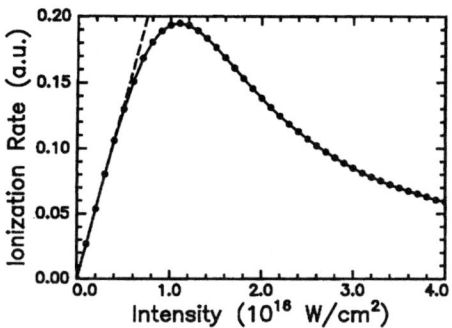

Figure 9: Total ionization rate versus intensity for H(1s) in a linearly polarized laser field of angular frequency $\omega = 0.65$ a.u. The solid line corresponds to the Sturmian-Floquet (from Dörr et al. [99]) and the R-matrix-Floquet (from Dörr et al. [101]) calculations, which are in excellent agreement. The LOPT results are given by the broken line.

laser pulse of finite duration. This implies that during the laser pulse turn on, where the intensity is lower, substantial ionization will occur. In other words, before the atoms experience a super-intense field where they can stabilize, they must pass through a "death valley" where their lifetime is extremely short [102]. Therefore, if adiabatic stabilization is to be observed experimentally, the laser pulse rise-time must not be so slow that saturation (complete ionization) occurs during the turn-on. However it must be slow enough so that the atom adiabatically remains in the ground state in the K-H frame. By comparing Floquet calculations with time-dependent calculations, it has been shown that these criteria can both be fulfilled [64, 65, 103].

Demonstrations of stabilization in short, intense laser pulses were given by Su et al. [104] in one dimension and for atomic hydrogen by Kulander et al. [105], Pont et al. [106] and Latinne et al. [62]. Here stabilization manifests itself as an increase of the survival probability with increasing laser intensity at some fixed, high frequency. For very short pulses, not only the ground state of the static, dressed potential $V_0(\alpha_0, \mathbf{r})$ [see equation (37)] is populated, but also many excited states. For this reason, in the K-H frame the dichotomy of the wave function will not always be present. However, the wave function will remain localized within the region $-\alpha_0$ to α_0 along the polarization axis [105, 107].

Finally, we mention that classical Monte Carlo simulations of stabilization in one and three dimensions have been performed [108–112], and that the issue of stabilization of atoms in laser pulses has been considered for asymptotically large electric field strengths by Fring and coworkers [113–115]. They have remarked that for ultra-short pulses the total momentum transfer to the electron after the pulse is not necessarily zero and thus in the limit of large intensities in general the electron will be ejected by this "kick" in the polarization direction. Numerical evidence exists that, for laser pulses of sufficiently large intensity, no population will survive. This has been seen in classical Monte Carlo simulations [112] as well as in one dimensional calculations [116, 117]. Furthermore at ultra-high intensities stabilization must be modified by relativistic effects including the presence of the magnetic field and the momentum of the photon in the laser propagation direction, as will now be discussed.

RELATIVISTIC EFFECTS

At sufficiently high laser intensities, the atom in the laser field can no longer be described using the Schrödinger equation in the dipole approximation. In particular,

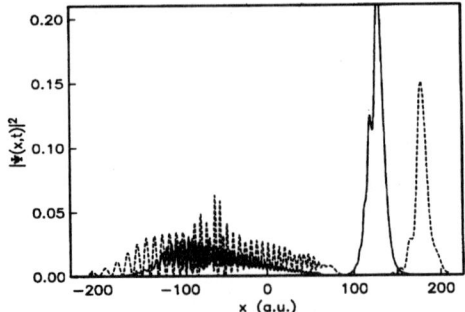

Figure 10: The Dirac (solid line) and Schrödinger (dotted line) probability densities, at the end of the 9th cycle, for a laser pulse with a four cycle sin² turn-on, an angular frequency $\omega = 1$ a.u. and a peak electric field strength $\mathcal{E}_0 = 175$ a.u. From Kylstra, Ermolaev and Joachain [123].

the magnetic field component of the laser will be non-negligible, spin effects become important and the dynamics of the laser-driven electrons become relativistic.

To estimate when the magnetic field component of the laser pulse becomes important, the Lorentz equations for an electron can be solved to lowest order in $1/c$. Non-dipole effects are in general non-negligible when the electron drift per laser cycle in the propagation direction, which is approximately given by $\mathcal{E}_0/(c\omega^3)$, becomes comparable to the width of the electron wave packet in the propagation direction. Taking $\omega = 0.043$ a.u. and the wavepacket width to be one Bohr radius, magnetic field effects can become important when $I \simeq 1.1 \times 10^{16}$ W cm^{-2}, while for $\omega = 1$ a.u. intensities of approximately $I \simeq 1.1 \times 10^{18}$ W cm^{-2} are required.

The dynamics of the system becomes relativistic when the "quiver" velocity of the electron approaches the velocity of light, i.e. when its ponderomotive energy U_p is of the order of its rest mass energy mc^2. Using eq. (2), this means that the quantity

$$q = \frac{U_p}{mc^2} = \frac{e^2 \mathcal{E}_0^2}{4m^2\omega^2 c^2}, \tag{38}$$

must then be of the order of unity. If the electric field strength $\mathcal{E}_0$ and the angular frequency ω are expressed in atomic units (a.u.), we have $q = 1.33 \times 10^{-5} (\mathcal{E}_0/\omega)^2$. Thus, if $\omega = 0.043$ a.u. (corresponding to a Nd-YAG laser), we see that $q = 1$ when $\mathcal{E}_0 = 11.8$ a.u. i.e. when the intensity $I = 4.9 \times 10^{18}$ W cm^{-2}. For a laser of higher angular frequency $\omega = 1$ a.u., we have $q = 1$ when $\mathcal{E}_0 = 274$ a.u., corresponding to the very large intensity $I = 2.6 \times 10^{21}$ W cm^{-2}. Spin effects will not be discussed here. We refer the reader to the contributions of C. Keitel and V. Véniard in this volume who discuss relativistic effects in more detail.

In the low-frequency regime, as discussed above, the laser-atom interaction can in first approximation be viewed in terms of the quasi-static model in which the electron moves in an instantaneous effective potential given by the Coulomb potential and the instantaneous electric field. Therefore, at very high intensities magnetic field and relativistic effects will essentially manifest themselves in the dynamics of the free, laser driven (relativistic) electron wave packets. In this high-intensity, low-frequency regime, classical Monte-Carlo simulations have been carried out [118–120] as well as studies within the framework of the tunneling [121] and KFR theories [122].

The situation is quite different for the case of a high-frequency laser, i.e. in the stabilization regime. As discussed above, studies of the stabilization of atoms in intense, high-frequency laser fields have, with a few exceptions, relied on the non-relativistic quantum theory. For sufficiently high intensities, the magnetic field component of the laser will modify the stabilization dynamics. The validity of the dipole approximation for atomic hydrogen has been tested by Bugacov et al. [124] and Latinne et al. [62]. Using a two-dimensional model, the breakdown of the dipole approximation has been demonstrated by Kylstra et al. [125]. These quantum mechanical calculations confirmed relativistic, classical Monte-Carlo simulations of ionization by Keitel and Knight [112] for atomic hydrogen. The magnetic field component of the laser field induces a significant motion of the electron in the propagation direction of the field. As a result, a breakdown of stabilization can occur.

One long term goal is to study atoms interacting with ultra-intense laser pulses by numerically solving the time-dependent Dirac equation. This is a formidable task since computationally the problem scales approximately as $\mathcal{E}_0^3/\omega^4$. For this reason, nearly all the quantum mechanical calculations of laser-atom interactions in the relativistic domain have been restricted until now to lower dimensional treatments [123, 126–128]. Magnetic field and retardation effects are clearly not included in one-dimensional model calculations. On the other hand, relativistic effects due to the dressing of the electron mass by the laser field [129] can be investigated. We show in Fig. 10 the results of Kylstra et al. [123] for the Dirac and Schrödinger probability densities at the end of the 9th laser cycle, when the electric field is maximum. The peak in the Dirac probability density corresponds to the relativistic "classical" excursion amplitude, $x_0 = 124$ a.u. Likewise, the peak in the Schrödinger probability density occurs at $x_0 = 175$ a.u., the non-relativistic classical excursion amplitude.

OUTLOOK

The study of atoms interacting with intense laser pulses has grown over the past ten years into a major area of research in atomic and optical physics. At present, a range of multiphoton processes are actively being investigated. Examples include the non-sequential double ionization of atoms [130–133], resonances in ATI spectra [134], atto-second pulse [135] and high-harmonic generation using ultra-short pulses [136], magnetic field and relativistic effects [125, 137, 138]. Continuing progress in laser technology, in particular the opening of the DESY free electron laser facility in Hamburg, will lead to the availability of intense laser radiation over an unprecedented frequency and intensity range. This will insure that the investigation of multiphoton processes will continue to be an exciting and rapidly evolving area of reseach in the coming years.

REFERENCES

[1] M. Gavrila, Adv. At. Mol. Opt. Phys. Suppl. **1**, 435 (1992).
[2] K. Burnett, V. C. Reed, and P. L. Knight, J. Phys. B **26**, 561 (1993).
[3] C. J. Joachain, in *Laser-Atom Interactions*, edited by R. M. More, page 39, Plenum Press (New York), 1994.
[4] L. F. DiMauro and P. Agostini, Adv. At. Mol. Phys. **35**, 79 (1995).
[5] M. Protopapas, C. H. Keitel, and P. L. Knight, Rep. Progr. Phys. **60**, 389 (1997).
[6] C. J. Joachain, M. Dörr, and N. J. Kylstra, Adv. At. Mol. Opt. Phys. **42**, 225 (2000).
[7] G. Petite, P. Agostini, and H. G. Muller, J. Phys. B **21**, 4097 (1988).
[8] G. S. Voronov and N. B. Delone, JETP Letters **1**, 66 (1965).
[9] J. L. Hall, E. J. Robinson, and L. M. Branscomb, Phys. Rev. Lett. **14**, 1013 (1965).
[10] P. Agostini, F. Fabre, G. Mainfray, G. Petite, and N. Rahman, Phys. Rev. Lett. **42**, 1127 (1979).
[11] R. R. Freeman, P. H. Bucksbaum, H. Milchberg, S. Darack, D. Schumacher, and M. E. Geusic, Phys. Rev. Lett. **59**, 1092 (1987).

[12] P. Hansch, M. A. Walker, and L. D. van Woerkom, Phys. Rev. A **55**, R2535 (1997).
[13] P. Hansch, M. A. Walker, and L. D. van Woerkom, Phys. Rev. A **57**, R709 (1998).
[14] S. Augst, D. Strickland, D. D. Meyerhofer, S. L. Chin, and J. H. Eberly, Phys. Rev. Lett. **63**, 2212 (1989).
[15] E. Mevel, P. Breger, R. Trainham, G. Petite, P. Agostini, A. Migus, J. P. Chambaret, and A. Antonetti, Phys. Rev. Lett. **70**, 406 (1993).
[16] L. V. Keldysh, Sov. Phys.-JETP **20**, 1307 (1965).
[17] F. H. M. Faisal, J. Phys. B **6**, L89 (1973).
[18] H. R. Reiss, Phys. Rev. A **22**, 1786 (1980).
[19] M. Ammosov, N. Delone, and V. Krainov, Sov. Phys.-JETP **64**, 1191 (1986).
[20] P. B. Corkum, Phys. Rev. Lett. **71**, 1994 (1993).
[21] K. C. Kulander, K. J. Schafer, and J. L. Krause, in *Super-Intense Laser-Atom Physics*, edited by B. Piraux, A. L'Huillier, and K. Rzazewski, page 95, Plenum Press (NewYork), 1993.
[22] M. Lewenstein, P. Balcou, M. Y. Ivanov, A. L'Huillier, and P. Corkum, Phys. Rev. A **49**, 2117 (1994).
[23] M. Y. Kuchiev, JETP Lett. **45**, 404 (1987).
[24] H. B. van Linden van den Heuvell and H. G. Muller, in *Multiphoton Processes*, edited by S. J. Smith and P. L. Knight, Cambridge Univ. Press, 1988.
[25] G. G. Paulus, W. Nicklich, H. Xu, P. Lambropoulos, and H. Walther, Phys. Rev. Lett. **72**, 2851 (1994).
[26] G. G. Paulus, W. Nicklich, F. Zacher, P. Lambropoulos, and H. Walther, J. Phys. B **29**, L249 (1996).
[27] M. P. Hertlein, P. H. Bucksbaum, and H. G. Muller, J. Phys. B **30**, L197 (1997).
[28] H. G. Muller and F. C. Kooiman, Phys. Rev. Lett. **81**, 1207 (1998).
[29] R. M. Potvliege and R. Shakeshaft, Adv. At. Mol. Opt. Phys. Suppl. **1**, 373 (1992).
[30] H. Rottke, B. Wolff-Rottke, D. Feldmann, K. H. Welge, M. Dörr, R. M. Potvliege, and R. Shakeshaft, Phys. Rev. A **49**, 4837 (1994).
[31] F. H. M. Faisal, *Theory of Multiphoton Processes*, Plenum Press (New York), 1986.
[32] M. H. Mittleman, *Introduction to the Theory of Laser-Atom Interactions*, Plenum Press (New York), 2nd edition, 1993.
[33] E. Cormier and P. Lambropoulos, J. Phys. B **30**, 77 (1997).
[34] H. A. Kramers, *Collected Scientific Papers*, page 272, North Holland (Amsterdam), 1956.
[35] W. C. Henneberger, Phys. Rev. Lett. **21**, 838 (1968).
[36] Y. Gontier and M. Trahin, J. Phys. B **13**, 4383 (1980).
[37] M. Crance, Phys. Rep. **114**, 117 (1987).
[38] D. M. Volkov, Z. Phys. **94**, 250 (1935).
[39] G. Floquet, Ann. Ec. Norm. (2) **13**, 47 (1883).
[40] J. H. Shirley, Phys. Rev. B **138**, 979 (1965).
[41] S. I. Chu, Adv. At. Mol. Phys. **21**, 197 (1985).
[42] A. Maquet, S. I. Chu, and W. P. Reinhardt, Phys. Rev. A **27**, 2946 (1983).
[43] M. Rotenberg, Adv. At. Mol. Opt. Phys. **6**, 233 (1970).
[44] M. Dörr, R. M. Potvliege, and R. Shakeshaft, Phys. Rev. A **41**, 558 (1990).
[45] H. Rottke, B. Wolff, M. Brickwedde, D. Feldmann, and K. H. Welge, Phys. Rev. Lett. **64**, 404 (1990).
[46] P. G. Burke, P. Francken, and C. J. Joachain, Europhys. Lett. **13**, 617 (1990).
[47] P. G. Burke, P. Francken, and C. J. Joachain, J. Phys. B **24**, 761 (1991).
[48] E. P. Wigner, Phys. Rev. **70**, 15 (1946).
[49] E. P. Wigner and L. Eisenbud, Phys. Rev. **72**, 29 (1947).
[50] J. Purvis, M. Dörr, M. Terao-Dunseath, C. J. Joachain, P. G. Burke, and C. J. Noble, Phys. Rev. Lett. **71**, 3943 (1993).
[51] M. Dörr, J. Purvis, M. Terao-Dunseath, P. G. Burke, C. J. Joachain, and C. J. Noble, J. Phys. B **28**, 4481 (1995).

[52] N. J. Kylstra, M. Dörr, C. J. Joachain, and P. G. Burke, J. Phys. B **28**, L685 (1995).
[53] O. Latinne, N. J. Kylstra, M. Dörr, J. Purvis, M. Terao-Dunseath, C. J. Joachain, P. G. Burke, and C. J. Noble, Phys. Rev. Lett. **74**, 46 (1995).
[54] A. Cyr, O. Latinne, and P. G. Burke, J. Phys. B **30**, 659 (1997).
[55] N. J. Kylstra, in *Photon and Electron Collisions with Atoms and Molecules*, edited by P. G. Burke and C. J. Joachain, page 205, Plenum Press (New York), 1997.
[56] A. S. Fearnside, J. Phys. B **31**, 275 (1998).
[57] N. J. Kylstra, H. W. van der Hart, P. G. Burke, and C. J. Joachain, J. Phys. B **31**, 3089 (1998).
[58] M. Pont, R. M. Potvliege, R. Shakeshaft, and P. H. G. Smith, Phys. Rev. A **46**, 555 (1992).
[59] M. V. Berry, Proc. Roy. Soc. London A **392**, 45 (1984).
[60] H. W. van der Hart, J. Phys. B **29**, 2217 (1996).
[61] N. J. Kylstra, E. Paspalakis, and P. L. Knight, J. Phys. B **31**, L719 (1998).
[62] O. Latinne, C. J. Joachain, and M. Dörr, Europhys. Lett. **26**, 333 (1994).
[63] M. Dörr, O. Latinne, and C. J. Joachain, Phys. Rev. A **52**, 4289 (1995).
[64] J. Zakrzewski and D. Delande, J. Phys. B **28**, L667 (1995).
[65] B. Piraux and R. M. Potvliege, Phys. Rev. A **57**, 5009 (1998).
[66] K. C. Kulander, Phys. Rev. A **36**, 2726 (1987).
[67] K. C. Kulander, Phys. Rev. A **38**, 778 (1988).
[68] R. Taïeb, V. Véniard, and A. Maquet, J. Opt. Soc. Am. B **13**, 363 (1996).
[69] G. G. Paulus, W. Nicklich, and H. Walther, Europhys. Lett. **27**, 267 (1994).
[70] V. Véniard, R. Taïeb, and A. Maquet, Phys. Rev. Lett. **74**, 4161 (1995).
[71] A. Sanpera, P. Jonsson, J. B. Watson, and K. Burnett, Phys. Rev. A **51**, 3148 (1995).
[72] J. B. Watson, A. Sanpera, and K. Burnett, Phys. Rev. A **51**, 1458 (1995).
[73] P. Antoine, B. Piraux, D. B. Milosevic, and M. Gajda, Phys. Rev. A **54**, R1761 (1996).
[74] S. G. Preston et al., Phys. Rev. A **53**, R31 (1996).
[75] P. Salières, A. L'Huillier, P. Antoine, and M. Lewenstein, Adv. At. Mol. Phys **41**, 83 (1999).
[76] V. T. Platonenko and V. V. Strelkov, Quantum Electronics **2**, 7 (1998).
[77] J. H. Eberly, R. Grobe, C. K. Law, and Q. Su, Adv. At. Mol. Opt. Phys. Suppl. **1**, 301 (1992).
[78] M. Protopapas, D. G. Lappas, and P. L. Knight, Phys. Rev. Lett. **79**, 4550 (1997).
[79] A. Patel, M. Protopapas, D. G. Lappas, and P. L. Knight, Phys. Rev. A **58**, R2652 (1998).
[80] M. J. Nandor, M. A. Walker, L. D. V. Woerkom, and H. G. Muller, Phys. Rev. A **60**, R1771 (1999).
[81] B. Walker, B. Sheehy, L. F. DiMauro, P. Agostini, K. J. Schafer, and K. C. Kulander, Phys. Rev. Lett. **73**, 1227 (1994).
[82] S. F. J. Larochelle, A. Talebpour, and S. L. Chin, J. Phys. B **31**, 1201 (1998).
[83] J. B. Watson, A. Sanpera, K. Burnett, D. G. Lappas, and P. L. Knight, Phys. Rev. Lett. **78**, 1884 (1997).
[84] F. H. M. Faisal and A. Becker, in *Multiphoton Processes 1996, IOP Conf. Ser. No. 154*, edited by P. Lambropoulos and H. Walther, page 118, Institute of Physics (Bristol), 1997.
[85] A. Becker and F. H. M. Faisal, Laser Phys. **7**, 684 (1997).
[86] J. Parker, K. T. Taylor, C. W. Clark, and S. Blodgett-Ford, J. Phys. B **29**, L33 (1996).
[87] K. T. Taylor, J. S. Parker, D. Dundas, E. Smyth, and S. Vivirito, in *Multiphoton Processes 1996*, edited by P. Lambropoulos and H. Walther, page 56, Institute of Physics (Bristol), 1997.
[88] J. S. Parker, E. S. Smyth, and K. T. Taylor, J. Phys. B **31**, L571 (1998).
[89] P. Lambropoulos, P. Maragakis, and J. Zhang, Phys. Reports **305**, 203 (1998).
[90] M. Dörr, R. M. Potvliege, and R. Shakeshaft, Phys. Rev. Lett. **64**, 2003 (1990).
[91] R. Shakeshaft, R. M. Potvliege, M. Dörr, and W. E. Cooke, Phys. Rev. A **42**, 1656 (1990).

[92] A. M. Perelomov, V. S. Popov, and M. V. Terent'ev, Sov. Phys.-JETP **23**, 924 (1966).
[93] A. M. Perelomov and V. S. Popov, Sov. Phys.-JETP **25**, 336 (1967).
[94] P. B. Corkum, N. H. Burnett, and F. Brunel, Adv. At. Mol. Opt. Phys. Suppl. **1**, 109 (1992).
[95] M. Gavrila, in *Photon and Electron Collisions with Atoms and Molecules*, edited by P. G. Burke and C. J. Joachain, page 147, Plenum Press (New York), 1997.
[96] M. Pont, N. Walet, M. Gavrila, and C. W. McCurdy, Phys. Rev. Lett. **61**, 939 (1988).
[97] M. Pont and M. Gavrila, Phys. Rev. Lett. **65**, 2362 (1990).
[98] R. J. Vos and M. Gavrila, Phys. Rev. Lett. **68**, 170 (1992).
[99] M. Dörr, R. M. Potvliege, D. Proulx, and R. Shakeshaft, Phys. Rev. A **43**, 3729 (1991).
[100] R. M. Potvliege and P. H. G. Smith, in *Super-Intense Laser-Atom Physics*, edited by B. Piraux, A. L'Huillier, and K. Rzazewski, page 173, Plenum Press (New York), 1993.
[101] M. Dörr, P. G. Burke, C. J. Joachain, C. J. Noble, J. Purvis, and M. Terao-Dunseath, J. Phys. B **26**, L 275 (1993).
[102] P. Lambropoulos, Phys. Rev. Lett. **55**, 2141 (1985).
[103] M. Pont and R. Shakeshaft, Phys. Rev. A **44**, R4110 (1991).
[104] Q. Su, J. H. Eberly, and J. Javanainen, Phys. Rev. Lett. **64**, 862 (1990).
[105] K. C. Kulander, K. J. Schafer, and J. L. Krause, Phys. Rev. Lett. **66**, 2601 (1991).
[106] M. Pont, D. Proulx, and R. Shakeshaft, Phys. Rev. A **44**, 4486 (1991).
[107] V. C. Reed, P. L. Knight, and K. Burnett, Phys. Rev. Lett. **67**, 1415 (1991).
[108] J. Grochmalicki, M. Lewenstein, and K. Rząźewski, Phys. Rev. Lett. **66**, 1038 (1991).
[109] R. Grobe and C. K. Law, Phys. Rev. A **44**, R4141 (1991).
[110] T. Ménis, R. Taïeb, V. Véniard, and A. Maquet, J. Phys. B **25**, L263 (1992).
[111] M. Gajda, J. Grochmalicki, M. Lewenstein, and K. Rzazewski, Phys. Rev. A **46**, 1638 (1992).
[112] C. H. Keitel and P. L. Knight, Phys. Rev. A **51**, 1420 (1995).
[113] A. Fring, V. Kostrykin, and R. Schrader, J. Phys. B **29**, 5651 (1996).
[114] A. Fring, V. Kostrykin, and R. Schrader, J. Phys. A **30**, 8599 (1997).
[115] C. F. D. Faria, A. Fring, and R. Schrader, J. Phys. B **31**, 449 (1998).
[116] Q. Su, B. P. Irving, C. W. Johnson, and J. H. Eberly, J. Phys. B **29**, 5755 (1996).
[117] N. J. Kylstra, unpublished, 1997.
[118] G. A. Kyrala, J. Opt. Soc. Am. B **4**, 731 (1987).
[119] H. Schmitz, K. Boucke, and H. J. Kull, Phys. Rev. A **57**, 467 (1998).
[120] C. H. Keitel, P. L. Knight, and K. Burnett, Europhys. Lett. **24**, 539 (1993).
[121] V. P. Krainov, J. Phys. B **32**, 1607 (1999).
[122] D. P. Crawford and H. R. Reiss, Phys. Rev. A **50**, 1844 (1994).
[123] N. J. Kylstra, A. M. Ermolaev, and C. J. Joachain, J. Phys. B **30**, L449 (1997).
[124] A. Bugacov, M. Pont, and R. Shakeshaft, Phys. Rev. A **48**, R4027 (1993).
[125] N. J. Kylstra, R. A. Worthington, A. Patel, P. L. Knight, J. R. V. de Aldana, and L. Roso, Phys. Rev. Lett. **85**, 1835 (2000).
[126] M. Protopapas, C. H. Keitel, and P. L. Knight, J. Phys. B **29**, L591 (1996).
[127] U. W. Rathe, C. H. Keitel, M. Protopapas, and P. L. Knight, J. Phys. B **30**, L531 (1997).
[128] A. M. Ermolaev, J. Phys. B **31**, L65 (1998).
[129] L. S. Brown and T. W. B. Kibble, Phys. Rev. **133A**, 705 (1964).
[130] T. Weber et al., Phys. Rev. Lett. **84**, 443 (2000).
[131] R. Moshammer et al., Phys. Rev. Lett. **84**, 447 (2000).
[132] U. Eichmann, M. Dörr, H. Maeda, W. Becker, and W. Sandner, Phys. Rev. Lett. **84**, 3550 (2000).
[133] A. Becker and F. H. M. Faisal, Phys. Rev. Lett. **84**, 3546 (2000).
[134] H. G. Muller, Phys. Rev. Lett. **83**, 3158 (1999).
[135] N. A. Papadogiannis, B. Witzel, C. Kalpouzos, and D. Charalambidis, Phys. Rev. Lett. **83**, 4289 (1999).
[136] G. Tempea, M. Geissler, M. Schnürer, and T. Brabec, Phys. Rev. Lett. **84**, 4329 (2000).
[137] S. X. Hu and C. H. Keitel, Phys. Rev. Lett. **83**, 4709 (1999).
[138] R. E. Wagner, Q. Su, and R. Grobe, Phys. Rev. Lett. **84**, 3282 (2000).

EXPERIMENTS OF MULTIPHOTON DISSOCIATION AND IONIZATION OF MOLECULES

H. Rottke

Max-Born-Insitut
Max-Born-Str. 2a
D-12489 Berlin-Adlershof
Germany

INTRODUCTION

The field of multiphoton dissociation and ionization of molecules is extremely broad. Therefore I will restrict myself to an introduction into experimental investigations on strong field light–molecule interaction, present some of the main experimental results, and their implications on the understanding of the interaction. Strong means that the electric field strength of the light wave becomes comparable to the intramolecular Coulomb interaction, which binds electrons in the valence shell and is responsible for molecular bonding. The electric field strengths of interest will be in the range from $E_0 \approx 10^{10}$ V/m to $E_0 \approx 10^{12}$ V/m. According to the relation

$$I = \frac{1}{2}\sqrt{\frac{\epsilon_0}{\mu_0}} E_0^2 \qquad (1)$$

between the electric field strength and the intensity I of a light wave the corresponding intensity ranges from $I \approx 10^{13}$ W/cm² up to $I \approx 10^{17}$ W/cm². At these light intensities the light–molecule interaction can no longer be analysed theoretically by perturbation methods. Nevertheless, it will be shown that even in this extreme situation quite simple physical mechanisms have been found which enable one to understand excitation and fragmentation of a molecule. Partly, these mechanisms are not molecule specific but apply generally to molecules.

To allow investigations at light intensities beyond 10^{13} W/cm² at least two requirements have to be met, the ionization and dissociation probabilities of the molecule have to be small for the time it takes a light pulse to reach the desired intensity level. This is usually achieved by using ultra–short pulses with a duration between ≈ 10 fsec and a few picoseconds and molecules with excited states which are far off resonance for single photon excitation. A large number of photons necessary to excite the molecule reduces the excitation and ionization cross section.

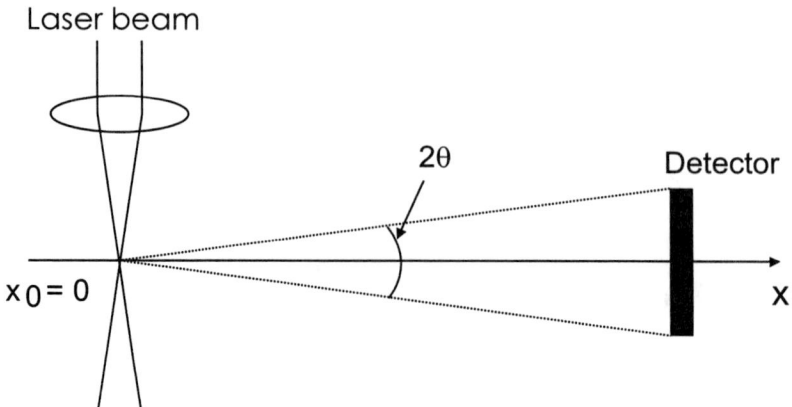

Figure 1. Typical setup of an electron time of flight spectrometer. Photoelectrons are formed in the focal spot of the laser beam at $x_0 = 0$. Those emitted into a cone with acute angle 2θ along the x–axis reach the electron detector.

At the short pulse limit the optical pulse width reaches the time scale of molecular rotation (period: ≈ 200 fsec for H_2, angular momentum $J = 1$) and vibration (period: ≈ 8 fsec for H_2). Assuming a diatomic molecule the light pulse then for example interacts with a molecular ensemble with a frozen (but usually random) orientation of the internuclear axis. If also the vibrational motion is frozen while the pulse is on a lot of energy can first be deposited in electronic motion. This may lead to multiple ionization with a following fragmentation of the ion or, at least partly, first to a redistribution of the energy among electronic and nuclear degrees of freedom of the molecule followed by fragmentation. At intermediate pulse widths an interplay of absorption of energy by the electrons and motion of the nuclei becomes important.

Ultra–short light pulses open the possibility to monitor molecular fragmentation after strong field excitation in real time by the so called pump–probe technique. First, a high intensity (pump) light pulse initiates fragmentation which is then modified by a second (probe) pulse. The probe is separated from the pump pulse by a variable time delay. The modification of molecular motion is detected and gives information on the fragmentation dynamics initiated by the pump pulse. For this method to work the light pulse widths have to be shorter than the characteristic time of nuclear motion of interest.

The following manuscript is divided into three sections where I will present usual experimental setups, results on the topics, ionization, dissociation, and correlation in molecular fragmentation and their interpretation. I will restrict the presentation to diatomic molecules which suffices to present the main fragmentation channels of molecules in a high intensity light pulse. Where possible, I will take H_2, the simplest diatomic molecule, as a representative.

IONIZATION

A typical experimental setup for the investigation of photoionization of molecules is shown in Figure 1. It allows the measurement of the kinetic energy distribution of photoelectrons which is the most important parameter besides the total yield of photoions. The photoelectrons are created in the focal spot of a pulsed laser beam ($x_0 = 0$ in Figure 1). The velocity of these electrons is determined by measuring the time t it takes them to reach a detector which is mounted a certain distance D away from the focal spot. The kinetic energy can be determined from these parameters through the relation:

$$E_{kin} = \frac{m}{2}\left(\frac{D}{t}\right)^2 \qquad (2)$$

if the electrons are not accelerated on their way to the detector. In this setup only electrons emitted into a cone with acute angle 2θ (see Figure 1) reach the detector. Therefore an angle resolved kinetic energy distribution of the electrons can be measured. The scheme described has for example been used in experiments of Zavriyev et al.[1], Yang et al.[2], or Rottke et al.[3].

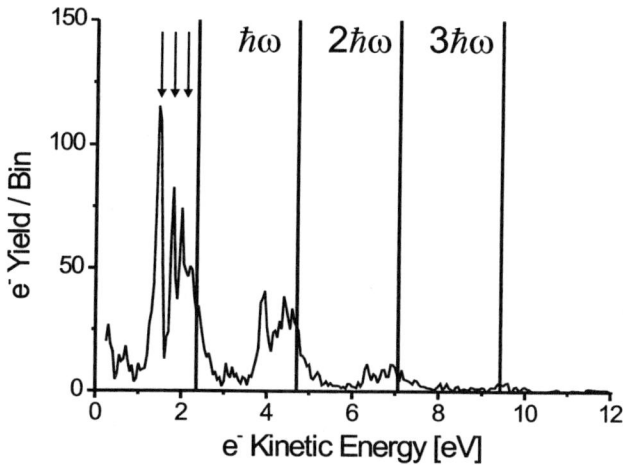

Figure 2. Angle resolved photoelectron kinetic energy distribution of H_2. The molecule was ionized by 527 nm light pulses with a width of 0.7 psec. The light intensity was 6.5×10^{13} W/cm^2.

Strong field photoionization of molecules proceeds in the same way as for atoms. Also two extreme regimes can be identified, one where photoionization can best be understood by assuming that it proceeds by the absorption of many discrete photons and a second one where it is best understood as electric field (tunnel) ionization by the oscillating electric field component of the light wave. The transition between the regimes is governed by the Keldysh γ parameter. High order above–threshold ionization (ATI) was found (see for example Cornaggia et al.[4], Verschuur et al.[5], Rottke et al.[3], Figure 2). Also, similar to atoms, resonance enhancement of multiphoton ionization by excited molecular states was observed. They are shifted into m–photon resonance at certain light intensity levels through their large dynamical Stark–shift[3]. Resonances give rise to a well known line structure in photoelectron spectra, the Freeman resonances[6]. These phenomena have been discussed in the lectures on atomic multiphoton ionization in this book. I will therefore not go into details but only show a photoelectron spectrum in Figure 2 as an example where these phenomena are observable. It was taken with molecular hydrogen exposed to 527 nm, 0.7 psec light pulses with an intensity of 6.5×10^{13} W/cm^2. Different ATI channels are separated by vertical lines and labeled by the number of photons absorbed above the ionization threshold of H$_2$. Up to three photons above the ionization threshold are absorbed in the example spectrum. With the best resolution in the channel where no photon above threshold is absorbed one observes a line structure in the kinetic energy distribution, the Freeman resonances[6] (arrows in Figure 2). They correspond to 7–photon resonant $(7+m)$–photon ionization ($m = 1, 2, ...$) via Rydberg states of the H$_2$ molecule which have principal quantum numbers $n = 4, 5, 6$[3].

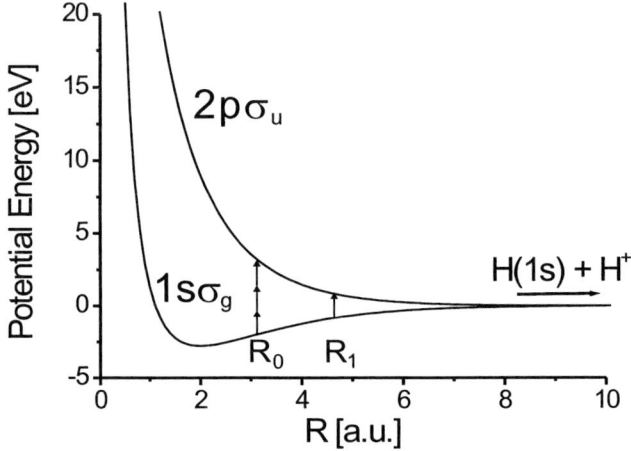

Figure 3. Born–Oppenheimer electronic energy eigenvalues as a function of the internuclear separation R for the H$_2^+$ ion. The eigenvalues, including the internuclear repulsion, are shown for the $1s\sigma_g$ ground state and the $2p\sigma_u$ first excited state (data from Sharp[10]). They form a pair of charge resonance states.

There is one molecule specific ionization mechanism, which in this form is not possible for atoms. It was found in the electric field (tunnel) ionization limit[7-9]. It works for diatomic molecules with a pair of charge–resonance states. I will explain this ionization mechanism taking the H$_2^+$ molecular ion as example. Its lowest lying $1s\sigma_g$ and $2p\sigma_u$ electronic states form such a pair of states. There exist more such pairs but only the two lowest lying ones play an important role in strong field fragmentation of H$_2$. The Born–Oppenheimer electronic energies of these states, including nuclear repulsion, are shown in Figure 3 as a function of the internuclear separation R. The energy eigenvalues become degenerate at infinite R. Using linear combination of atomic orbitals the electronic wavefunctions at large internuclear separation can be approximated by the $1s$ atomic ground state wave functions $\psi_{1s}(\vec{r})$:

$$\psi_{1s\sigma_g}(\vec{r}) = \frac{1}{N}\left[\psi_{1s}\left(\vec{r}+\vec{R}/2\right) + \psi_{1s}\left(\vec{r}-\vec{R}/2\right)\right] \quad (3)$$

$$\psi_{2p\sigma_u}(\vec{r}) = \frac{1}{N}\left[\psi_{1s}\left(\vec{r}+\vec{R}/2\right) - \psi_{1s}\left(\vec{r}-\vec{R}/2\right)\right] \quad (4)$$

with $\vec{R}$ the vector of internuclear separation and $1/N$ a normalizing factor. One wavefunction ($\psi_{1s\sigma_g}$) is symmetric and the other one ($\psi_{2p\sigma_u}$) antisymmetric with respect to inversion of the electronic coordinate $\vec{r}$.

The qualitative dependence on $\vec{r}$ of the square of the electronic wavefunctions along the internuclear axis (z–axis) is shown in Figure 4a (upper part) for an internuclear separation $R = 2$ a.u. (H_2^+ equilibrium internuclear separation). The electronic energy eigenvalues at this internuclear separation are shown in the plot of the potential energy (full line) the electron feels along the internuclear axis in Figure 4a (the two horizontal lines). They are well separated (see also Figure 3). If one now puts H_2^+ in a

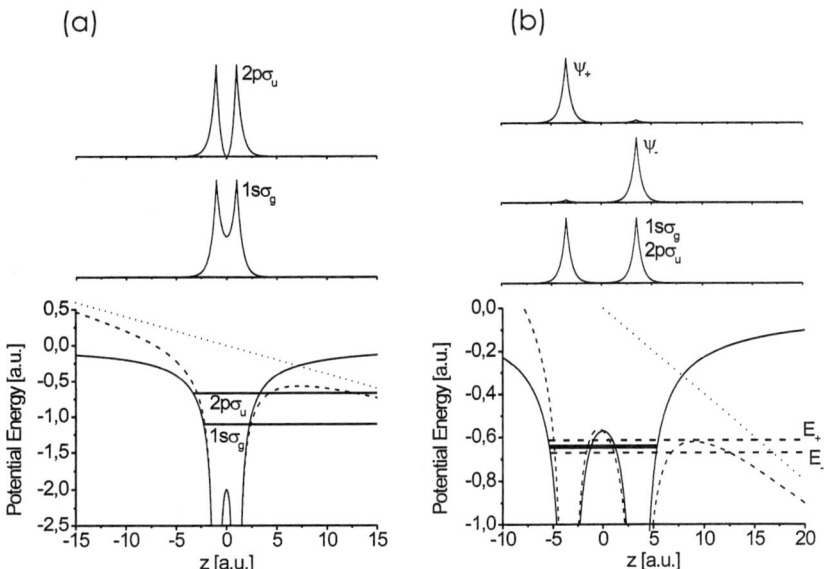

Figure 4. The potential energy of the electron along the H_2^+ internuclear axis (z-axis) at internuclear separations $R = 2$ a.u. (a) and $R = 7$ a.u. (b) are shown in the lower plots. Full lines: unperturbed case, dotted line: potential due to an applied external electric field, dashed lines: the superposition of both potentials. The upper diagrams in (a) show the squared $1s\sigma_g$ and $2p\sigma_u$ electronic wavefunctions along the internuclear axis. In (b) the upper two plots represent the perturbed electronic wavefunctions with the external field applied. They develop from the unperturbed functions shown directly above the potential energy diagram.

light wave with an oscillating electric field component $E_0 \cos \omega t$ along the internuclear axis the potential the electron sees at a certain instant of time (dashed line in Figure 4a) is a superposition of the intramolecular Coulomb potential and the "instantaneous potential" due to the external electric field (dotted line). At the external field strength chosen here the electronic wavefunctions of both states, $1s\sigma_g$ and $2p\sigma_u$, and the corresponding energy eigenvalues are only slightly affected. Nevertheless, a potential barrier appears at large positive values of z. The electron can tunnel through this barrier and thus ionize in both states. The probability to tunnel is small since the barrier, which is similar to the barrier in an atom, is quite broad, especially for the $1s\sigma_g$ ground state of H_2^+.

This situation changes completely in a certain range of internuclear separations. This is shown in Figure 4b at $R = 7$ a.u. for the same external electric field strength as in 4a. Without the external field the electronic energy eigenvalues are nearly degenerate (the full horizontal line in the potential energy plot in Figure 4b). The squared wavefunctions of the states $1s\sigma_g$ and $2p\sigma_u$ are nearly identical (the lowest wavefunction diagram in 4b). The external electric field now affects the nearly degenerate molecular energy eigenvalues more seriously, one is pushed up (E_+) and one depressed (E_-). The electronic eigenfunctions $\psi_\pm$ corresponding to these eigenvalues are sketched in the upper part of Figure 4b. In one state (ψ_+) the electron becomes strongly localized on the left hand nucleus ($R = -3.5$ a.u.) and in the other one (ψ_-) on the right hand nucleus. A central potential barrier at $z = 0$ separates the two localization regimes.

The state localized in the potential well on the right hand side (eigenvalue E_-) can ionize with only a low probability by tunneling through a broad barrier. It is similar to the barrier found in atomic ionization with the hydrogen atom in the $1s$ ground state. A molecule specific ionization mechanism which is impossible for an atom is found for the state where the electron is localized in the left well in Figure 4b (eigenvalue E_+). In this state it suffices that the electron tunnels through the central narrow barrier near $z = 0$ with a small height to become ionized (see the dashed line in the potential energy diagram in Figure 4b). The height of the barrier on the right hand side is smaller than the energy eigenvalue E_+. Therefore the electron is already free after tunneling through the central barrier. This ionization pathway is only open in a certain range of internuclear separations and gives rise to an enhanced ionization probability. If the separation is too small localization of the electron on one or the other nucleus is no longer possible. In both eigenstates it then has to overcome a broad atom like barrier to ionize (Figure 4a). If the internuclear separation becomes too large the central barrier approaches the atomic hydrogen barrier (the barrier on the right hand side of Figure 4b). Accordingly the tunneling probability decreases again.

The enhancement of strong field ionization of molecules with pairs of charge resonance states at intermediate internuclear separations was not directly observed in photoelectron kinetic energy distributions. The mechanism is usually only active in ions and not in the neutral molecule used as starting point for experimental investigations. Multiple ionization of the neutral molecule which is necessary to observe this ionization mechanism usually has a lower probability than single ionization. Therefore there is no easy way to identify the photoelectrons from charge resonance enhanced ionization of the ion in a photoelectron spectrum. The sensitivity of the mechanism to the internuclear separation lead to its identification in an indirect way in ion kinetic energy distributions of charged dissociation fragments. This fragmentation and what can be learned from it about strong field excitation of molecules will be analysed in the following section.

DISSOCIATION

Dissociation of molecules following strong field excitation, single–, or even multiple ionization gives rise to fragments which may have several charge states q. They may also have different masses m, if either a heteronuclear diatomic molecule or a polyatomic molecule breaks into fragments. Additionally, the fragments may have a substantial momentum $\vec{p}$. A kinematical analysis should be able to give all three parameters q, m, and $\vec{p}$. Since an independent measurement of m and q is not easy usually only the ratio q/m is determined in experiments.

The fragment momentum distribution or its equivalent, angle resolved kinetic energy distributions, can be measured with the same setup as used for photoelectron spectroscopy (Figure 1). This field free setup was used in several experiments on molecular hydrogen[11-14]. It can easily reach a high momentum or kinetic energy resolution since no electric fields for acceleration of the ions are applied. Its drawback is the lacking q/m separation. Therefore it is only useful if a very limited number of charge states and fragment masses with momentum distributions known as non-overlapping is found.

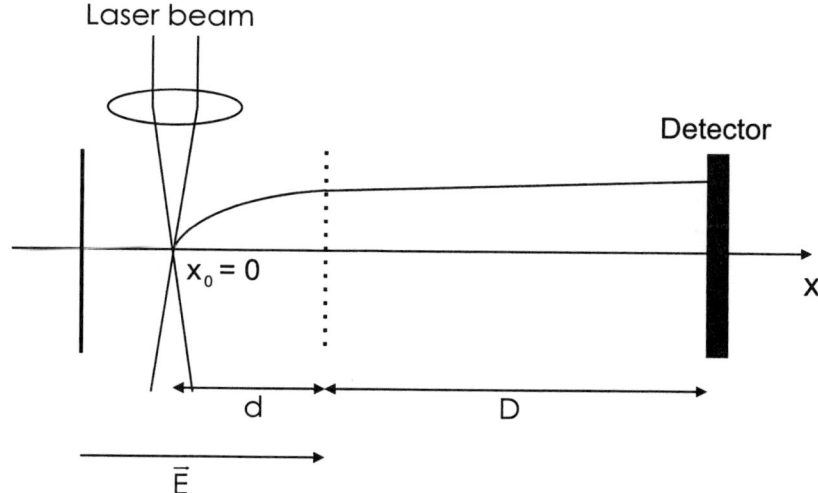

Figure 5. Typical setup of a q/m ("mass") selective ion time of flight spectrometer. An extraction electric field $\vec{E}$ is applied across the light beam focal spot at $x_0 = 0$ to accelerate ions created there over a path length d. A free drift path (length D) to the detector follows. The instrument is capable to measure the initial momentum component p_x the ion is created with and its q/m.

The conventional way to determine the ion momentum p_x along a certain axis x and at the same time discriminate among different q/m values is shown in Figure 5. Similar to Figure 1 a time of flight technique is used. But now the ions are first accelerated in a homogeneous electric field along the x-axis (path length d), then they enter a second field free drift tube (length D) before they reach the ion detector. In the electric field the acceleration the ion experiences is proportional to q/m. Accordingly the total time of flight T to the detector also depends on q/m giving rise to a separation of ions according to their q/m values:

$$T = -\frac{p_x}{qE} + \frac{D}{\sqrt{2\frac{q}{m}Ed}}\left(\frac{1}{\sqrt{1+E/V}} + 2\frac{d}{D}\sqrt{1+E/V}\right) \quad (5)$$

In this relation $V = qEd$ is the energy the ion gains by acceleration through the electric field. $E = p_x^2/2m$ is the "kinetic energy" of the ion due to its momentum component p_x along the electric field vector at the instant of time when it is created by photoionization at $x = 0$. In experiments it is usual to choose the ratio E/V small compared to 1. T may then be approximated by the leading terms in a Taylor series expansion in E/V:

$$T = \frac{D}{\sqrt{2\frac{q}{m}Ed}} - \frac{p_x}{qE} + \frac{D}{\sqrt{2\frac{q}{m}Ed}}\left[\frac{E}{2V}\left(2\frac{d}{D}-1\right) + \frac{E^2}{8V^2}\left(3-2\frac{d}{D}\right) + O\left(\frac{E^3}{V^3}\right)\right] \quad (6)$$

The first term in equ. 6 is responsible for q/m ("mass") separation. It does not depend on the momentum p_x. The second term is proportional to p_x for each discrete q/m. It is the leading term which allows the determination of the momentum component p_x of the ion. If one chooses $D = 2d$ for the lengths of the acceleration and free drift paths only very small corrections to the linear relation between T and p_x of the order $(E/V)^2$ appear in equ. 6. Spectrometers of this type have been used by a lot of research groups.

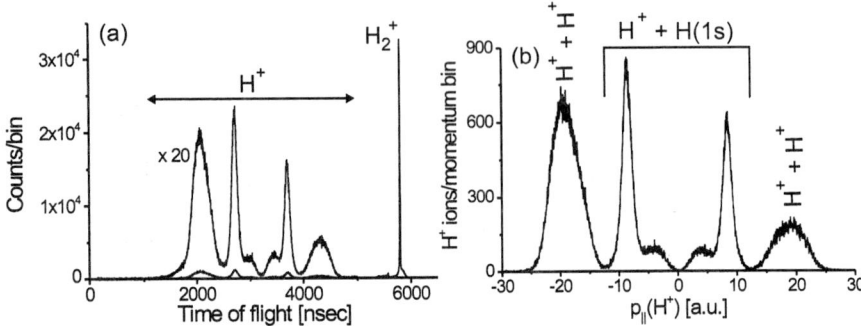

Figure 6. (a) Ion time of flight spectrum measured by strong field excitation of H_2. The light intensity was 6×10^{14} W/cm^2, the light pulse width 25 fsec, and the wavelength 790 nm. The laser beam was linearly polarized with the polarization vector pointing along the spectrometer axis. (b) The H$^+$ ion momentum distribution for the momentum component $P_x (= p_{\|})$ to the time of flight spectrum in (a).

An example of an ion time of flight spectrum and the corresponding momentum distribution is shown in Figure 6. The time of flight spectrum (Figure 6a) was taken for H$^+$ and H$_2^+$ ions formed in strong field ionization and dissociation of H_2. The light intensity used was 6×10^{14} W/cm^2, the wavelength 790 nm and the pulse width 25 fsec. The laser beam was linearly polarized with the polarization vector pointing along the spectrometer axis (x–axis in Figure 5). The narrow line on the right hand side of the spectrum is formed by H$_2^+$ ions. H$^+$ ions generate the group of broad lines in the left part of the spectrum. Their intensity is much lower than that of the H$_2^+$ line therefore I also plotted them scaled by a factor of 20. The width of the H$_2^+$ line is narrow since a well collimated supersonic molecular beam was used as target. The momentum distribution of the H$_2$ molecules in such a beam can be made extremely narrow (see for example Miller[15]). The residual H$_2^+$ line width observed in the time of flight spectrum is completely determined by the photoionization step of the H$_2$ molecule (see for example Moshammer et al.[16,18], Ullrich et al.[17]).

The width of the H$^+$ lines is determined by the photodissociation processes which are responsible for their formation. There is a point of symmetry at the center of the H$^+$ line structure. The meaning of this point becomes obvious in Figure 6b which shows the corresponding H$^+$ momentum distribution for the momentum component p_x along the x-axis. It was calculated from the time of flight spectrum using relation 5 above. The symmetry point corresponds to zero H$^+$ momentum along the spectrometer axis. The momentum distribution has to be symmetric with respect to $p_x = 0$ since the H$^+$ ions may initially be either emitted with momentum p_x directly into the direction where the detector is located or equally well with $-p_x$ into the opposite direction.

The basic strong field molecular dissociation mechanisms are similar for at least the diatomic molecules and molecules consisting of only few atoms. The H$^+$ ion momentum distribution shown in Figure 6b already encompasses them. Charged dissociation products are usually formed in a two step process. In a strong ultra–short light pulse the molecule first looses one electron by multiphoton ionization. The molecular ion is formed near the equilibrium internuclear separation of the neutral molecule with the highest probability usually in the electronic ground state. The ion may be vibrationaly exited, even into the ion's dissociation continuum, already after the photoionization step. Further photoionization at this point is difficult since the ionization potential of the singly charged ion is usually considerably higher than that of the neutral molecule. Its internuclear separation therefore has time to increase. At certain internuclear separations excited electronic states of the ion may become resonant for m–photon excitation (see Figure 3 for examples of internuclear separations where resonant 1– and 3–photon excitation from the H_2^+ $1s\sigma_g$ electronic ground state to the first excited state $2p\sigma_u$ and back is possible for a certain light wavelength). If the example H_2^+ ion was not yet in the dissociation continuum of the electronic ground state excitation to the repulsive $2p\sigma_u$ state at such a point will result in dissociation, for example after 3–photon excitation to this state at R_0 (Figure 3). While dissociating the ion passes R_1 where the electronic states $1s\sigma_g$ and $2p\sigma_u$ are resonantly coupled (1–photon transition). Near this point it may happen that due to this coupling stimulated emission of a photon induces a transition back into the electronic ground state $1s\sigma_g$ and further dissociation proceeds on the corresponding ground state potential energy curve (Figure 3). On its way to dissociation the molecular ion in this case has effectively absorbed two photons.

In case of H_2 the dissociation pathways just described qualitatively are found in the H^+ ion momentum distribution in Figure 6b. The small ion yield enhancements near $p_\| \approx \pm 4$ a.u. can be interpreted to be formed by effective 1–photon dissociation of H_2^+ molecules in vibrational levels $v = 6 \pm 1$ in the electronic ground state $1s\sigma_g$. These vibrational states have before been populated in the H_2 photoionization step. Similarly, the narrow intense lines at $p_\| \approx \pm 9$ a.u. are formed through effective 2–photon dissociation of H_2^+ in states $1s\sigma_g(v = 2 \pm 1)$. The detailed mechanism is the second one described in the previous paragraph. The H_2^+ ion first effectively absorbs three photons at a short internuclear separation, then it starts dissociating on the $2p\sigma_u$ potential, reaches R_1, emits effectively one photon and finally proceeds dissociating on the $1s\sigma_g$ potential. By both of these dissociation mechanisms an H^+ ion and an H atom in the 1s ground state are formed. In case of H_2^+ higher excited electronic states play a minor role in strong field dissociation. This may be different for other diatomic molecules.

For simplicity I first discussed the dissociation mechanisms in an approximate way in well known terms of the language of low order perturbation theory. In doing so I always used the term "effective" above. This is essential, since for example at the light intensity used to measure the momentum distribution in Figure 6b ($I = 6 \times 10^{14}$ W/cm^2) the electronic states $1s\sigma_g$ and $2p\sigma_u$ are strongly coupled. The interaction "potential"

$$V_{gu}(R,t) = d_{gu}(R) E \cos \omega t \qquad (7)$$
$$d_{gu}(R) = \langle \psi_{1s\sigma_g}(\vec{r}, R) |ez| \psi_{2p\sigma_u}(\vec{r}, R) \rangle$$

which is the product of the electronic dipole matrix element $d_{gu}(R)$ and the electric field strength of the light wave E projected on the internuclear axis reaches several eV. This is equal to and at large internuclear separation R even larger than the separation of the diabatic nuclear potentials of these electronic states (see Figure 3). Low order perturbation methods with the light wave as a small perturbation thus are not adequate to analyse photodissociation. Only the overall outcome (product kinetic energy distribution) can usually be interpreted in terms of an effective absorption of a certain number of photons, which may be quite small as the example above shows.

A way to analyse strong field dissociation theoretically was introduced by Pegarkov and Rapoport[19,20] and by Giusti–Suzor et al.[21]. The theoretical aspects are discussed in the lecture by P. Burke in this book. In case of the example H_2^+ molecule the electronic states involved are $1s\sigma_g$ and $2p\sigma_u$. The corresponding potential energy curves are shown in Figure 3 in the Born–Oppenheimer approximation. The 2×2–potential matrix which governs the coupled nuclear motion in these electronic states in the presence of an externally applied light wave then reads:

$$\begin{pmatrix} V_{1s\sigma_g}(R) & V_{gu}(R,t) \\ V_{gu}(R,t) & V_{2p\sigma_u}(R) \end{pmatrix} \qquad (8)$$

Here $V_{gu}(R,t)$ is the interaction potential (equ. 7) due to a linearly polarized light wave. A schematic plot of part of the diabatic (dashed lines) and adiabatic (full lines) dressed potentials governing nuclear motion is shown in Figure 7 (light wavelength $\lambda = 790$ nm). Two light intensities, a low (a) and a high one (b), were chosen. The adiabatic potentials were calculated by diagonalising the R–dependent diabatic Floquet–matrix which is derived from a Floquet–ansatz for a solution of the time periodic Schrödinger equation of nuclear motion (see also the review by Giusti–Suzor et al.[22]). Adiabatic states are coupled by non–adiabatic interaction terms which depend on nuclear momentum. They come in through the R–dependent transformation which diagonalises the Floquet–matrix. In the strong field limit this coupling is treated as a perturbation. It can introduce transitions among adiabatic states. As long as this perturbation is small nuclear motion in the dressed adiabatic electronic states proceeds on the corresponding adiabatic potentials.

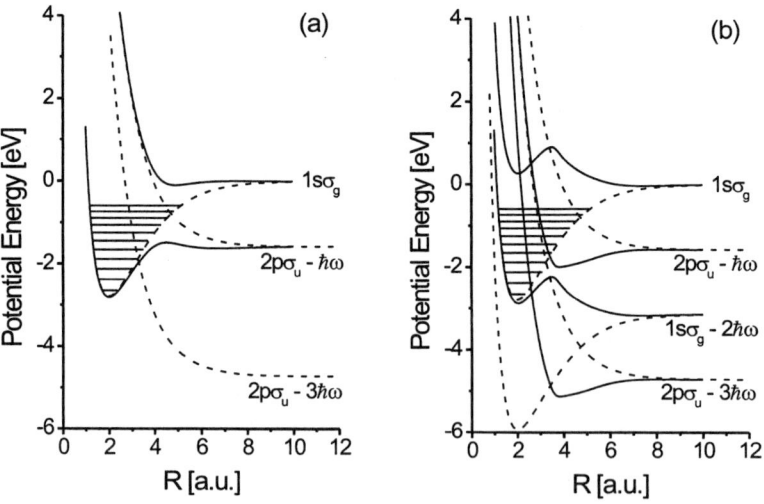

Figure 7. Diabatic (dashed) and adiabatic (full) dressed potential energy curves for the H_2^+ electronic ground state $1s\sigma_g$ and the first excited state $2p\sigma_u$. The light intensity was chosen small (a) to show the avoided crossing of the adiabatic states $1s\sigma_g$ and $2p\sigma_u - \hbar\omega$, and large (b) to show the avoided crossing of states $1s\sigma_g$ and $2p\sigma_u - 3\hbar\omega$ near $R = 3.25$ a.u..

In Figure 7 unperturbed energy eigenvalues of the 11 lowest lying unperturbed H_2^+ vibrational levels ($v = 0, ..., 10$) in the electronic ground state are included. They are are important for the description of the different strong field dissociation mechanisms. Effective 1–photon dissociation which was found in the ion momentum distribution in

Figure 6b proceeds on the dressed adiabatic potential labeled $2p\sigma_u - \hbar\omega$ at large internuclear separations. How it works is best seen in Figure 7a at low light intensity. Ending up in the $2p\sigma_u - \hbar\omega$ state means that finally (after dissociation) the molecule absorbed effectively one photon. At this light intensity one may neglect that the adiabatic potentials $1s\sigma_g$ and $2p\sigma_u - 3\hbar\omega$ do not really cross at $R = 3.25$ a.u.. The three photon coupling of the diabatic dressed states $1s\sigma_g$ and $2p\sigma_u - 3\hbar\omega$ is yet too small (perturbative). The adiabatic potential (full line) is thus assumed to follow the diabatic potential (dashed line) at this point. The bound unperturbed H_2^+ vibrational states indicated in Figure 7a can dissociate in the adiabatic limit if they are lying in the dissociation continuum of the adiabatic dressed potential $2p\sigma_u - \hbar\omega$. As can be seen all unperturbed vibrational states with $v \geq 5$ fulfill this condition at the light intensity chosen. This dissociation mechanism is known as bond–softening[1]. In the course of dissociation on the adiabatic dressed potential a lot of photons are absorbed and re–emitted, especially in the vicinity of the crossing of the corresponding diabatic dressed states $1s\sigma_g$ and $2p\sigma_u - \hbar\omega$ at $R = 4.7$ a.u..

With increasing light intensity the three photon coupling of the dressed diabatic states $1s\sigma_g$ and $2p\sigma_u - 3\hbar\omega$ becomes relevant. Analogous to the 1–photon coupling it gives rise to an avoided crossing of the corresponding adiabatic potentials at $R = 3.25$ a.u.. In the strong field limit a new bond–softening channel opens at this internuclear separation (see Figure 7b). It allows unperturbed vibrational states to dissociate which have lower vibrational quantum numbers than necessary for the bond–softening mechanism discussed in the previous paragraph. At 790 nm excitation wavelength in Figure 7b these states have $v \geq 0$. Dissociation proceeds on the large R adiabatic potential $1s\sigma_g - 2\hbar\omega$ with the effective absorption of two photons after dissociation is complete (see Figure 7b). At the specific light intensity chosen unperturbed states with $v \geq 2$ are lying directly in the continuum of the $1s\sigma_g - 2\hbar\omega$ potential. But also states with $v = 0, 1$ may dissociate by tunneling through the barrier visible in the adiabatic potential near 4.7 a.u.. Dissociation on the large R adiabatic potential $1s\sigma_g - 2\hbar\omega$ means that in the course of dissociation effectively one photon is re–emitted again since dissociation is initiated by effective absorption of three photons near 3.25 a.u.. Dissociation without effective re-emission of a photon is also possible but needs a non–adiabatic transition to the large R $2p\sigma_u - 3\hbar\omega$ adiabatic dressed state. This only happens with significant probability if the relative momentum of the dissociating nuclei and therefore the non–adiabatic coupling of the adiabatic states is large enough (see for example Zavriyev et al.[1] and the review by Giusti-Suzor et al.[22]). In the H^+ fragment momentum distribution in Figure 6b (excitation wavelength $\lambda = 790$ nm) no hint is found on the presence of effective 3–photon dissociation. The non–adiabatic transition probability is too small in this case to be detectable experimentally.

The effective 2–photon dissociation channel on the large R adiabatic dressed potential $1s\sigma_g - 2\hbar\omega$ may also give rise to so called above threshold dissociation. This happens for unperturbed vibrational states if already the absorption of one photon suffices for dissociation of the molecule. In Figure 7b the unperturbed vibrational states must have $v \geq 5$ for this to happen. The excess energy of the dissociation fragments is then larger than the energy of one photon. Experimentally above threshold dissociation was found for example in H^+ ion kinetic energy distributions measured by Zavriyev et al.[1].

Besides dissociating molecules the strong light field may also give rise to suppression of dissociation by the formation of so called light–induced bound vibrational states. They were predicted by Zavriyev et al.[23] (see also the review by Giusti–Suzor et al.[22]). In a light pulse these states may exist in transient light–induced wells of the adiabatic potential curves. An example of such a well can be found in Figure 7b. The adiabatic potential $2p\sigma_u - \hbar\omega$ (full line) develops such a well near $R = 4$ a.u. which may support bound vibrational levels.

One feature in the H^+ ion momentum distribution in Figure 6b cannot be explained by assuming simply strong field dissociation of H_2 or H_2^+ as described above. It consists of the two broad lines with maxima near $p_\| = \pm 19$ a.u.. Experiments showed that Coulomb explosion of the nuclei after double ionization of H_2 is their origin. The main contribution to the finally measured kinetic energy of the two H^+ fragment ions may be assumed to come from the Coulomb repulsion of the protons at the instant of time when both electrons are removed from the molecule. The final kinetic energy of the protons and correspondingly their momentum then is determined by the internuclear separation R at that time. In atomic units:

$$E_{kin} = 1/2R \tag{9}$$

$$|\vec{p}| = \sqrt{2mE_{kin}} \tag{10}$$

With this relation it is possible to determine R from the measured ion momentum. For Coulomb explosion of H_2 with excitation at $\lambda = 790$ nm (Figure 6b) one calculates that both electrons are lost at internuclear separations between $R \approx 3$ a.u. and $R \approx 11$ a.u.. The equilibrium internuclear separation of the H_2 molecule in the electronic ground state where double photoionization starts from is $R = 1.4$ a.u.[10]. It is considerably smaller than the starting interval for Coulomb explosion. Before removal of both electrons is finished the internuclear separation therefore has to increase. The mechanism behind Coulomb explosion therefore is expected to be a sequential double photoionization at different internuclear separations. First H_2 looses one electron at the H_2 equilibrium internuclear separation $R = 1.4$ a.u.. Then R increases before the second electron is lost and Coulomb explosion is initiated. The experimental data (Figure 6b) show that the second electron is most easily lost between $R \approx 3$ a.u. and $R \approx 11$ a.u.. The reason for this behaviour is charge resonance enhanced ionization of H_2^+ which was explained in the chapter on molecular ionization. For H_2^+ this was first shown in a paper by Zuo and Bandrauk[9]. As expected charge resonance enhanced photoionization has a considerably higher probability than photoionization of atomic hydrogen in the electronic ground state even though the H_2^+ ionization potential in the relevant range of internuclear separations is higher than that of the atom.

The photoionization mechanism behind Coulomb explosion of H_2 is a quite general one. It applies also to other diatomic and even three-atomic molecules as experimental investigations showed. This was first pointed out by Seideman et al.[7] and Posthumus et al.[8] as I already said above. It is not restricted to double photoionization (see for example the refs. cited in Posthumus et al.[8]). In strong field laser-molecule interaction the formation of the highly charged atomic fragments with a characteristic fragment kinetic energy distribution was always observed experimentally at surprisingly low light intensities. Starting from the neutral atomic fragment a considerably higher intensity would be necessary to reach the same charge state. Posthumus et al.[8] pointed out that the atomic fragment ion kinetic energies for all atomic charge states found in the experiments may be explained by assuming multiple ionization of the molecules at a certain critical internuclear separation R_c. R_c is approximately the same for all charge states. Similar to H_2 it is considerably larger than the equilibrium internuclear separation of the corresponding neutral molecule. As for H_2 fragmentation into ions in high charge states is at least a two step process. First again the molecule is singly ionized at the equilibrium internuclear separation then the ionic internuclear separation increases to R_c where further ionization starts. At R_c tunneling of electrons through the central potential barrier as shown in Figure 4b or above barrier ionization suffices for further ionization of the molecules. It leads to a high probability for the formation of high ionic charge states.

In strong field multiple ionization and dissociation of molecules a broad spectrum of atomic fragment charge states is observed (for examples see Normand and Schmidt[24], Gibson et al.[25]). In the case of a diatomic molecule AB for example the Coulomb explosion into ions A^{p+} and B^{q+} just described gives rise to fragment kinetic energies which depend on the specific explosion channel named (p,q) according to the charge states formed. An unambiguous identification of the explosion channels from simply measuring q/m selective ion kinetic energies or momenta is usually difficult or even impossible. It became possible with the advent of ion correlation measurements. They will be the subject of the following chapter.

CORRELATION IN MOLECULAR FRAGMENTATION

The investigation of correlation in strong field molecular fragmentation was initiated by Frasinski et al.[26]. The experimental method they used is called covariance mapping. It was since then applied in a lot of experiments to investigate correlation between ionic fragments in different charge states. Also one experiment is reported where this technique was applied to look for electron–ion and electron–electron correlation in strong field double ionization of molecular hydrogen[27].

I will describe covariance mapping as it is used to investigate ion–ion correlation. The experimental setup is identical to that for simple q/m selective ion time of flight spectroscopy (see Figure 5). The extraction electric field is chosen in such a way that nearly all ions formed in the focal spot of the laser beam reach the detector, irrespective of the momentum they may get in the process of dissociation. For each single laser pulse the complete ion time of flight spectrum $S_\alpha(t)$ is measured and stored ($\alpha = 1, ..., n$ enumerates the laser shots). One then looks for correlation in the statistical fluctuations of the measured signal $S_\alpha(t)$ at different instants of time, for example at t_1 and t_2. This is done by calculating the covariance $C(t_1, t_2)$ of the time of flight spectra at these times. It is given by the relation:

$$C(t_1, t_2) = \frac{1}{n} \sum_{\alpha=1}^{n} [S_\alpha(t_1) - \langle S(t_1)\rangle][S_\alpha(t_2) - \langle S(t_2)\rangle] \qquad (11)$$

$$= \frac{1}{n} \sum_{\alpha=1}^{n} [S_\alpha(t_1) S_\alpha(t_2) - \langle S(t_1)\rangle \langle S(t_2)\rangle]$$

with

$$\langle S(t)\rangle = \frac{1}{n} \sum_{\alpha=1}^{n} S_\alpha(t) \qquad (12)$$

the mean time of flight spectrum which is the average over all n single shot spectra.

The covariance $C(t_1, t_2)$ is equal to zero if the statistical fluctuations in the ion time of flight spectra at the two instants of time are independent. On the other hand correlated fluctuations give rise to a non–zero value of $C(t_1, t_2)$. Suppose that at times t_1 and t_2 the measured signal is due to ions A^{p+} and B^{q+}, respectively, from a Coulomb explosion channel (p, q):

$$AB \longrightarrow AB^{(p+q)+} \longrightarrow A^{p+} + B^{q+} \qquad (13)$$

Then for every laser shot where A^{p+} ions are formed and detected at t_1 one also expects to detect ions B^{q+} at t_2 if the detection efficiency of the system is sufficiently high.

This means the signal statistics at times t_1 and t_2 in the ion time of flight spectra are correlated (the statistical fluctuations are not independent). Therefore the covariance $C(t_1, t_2)$ is different from zero. If the ions detected at times t_1 and t_2 originate from two different Coulomb explosion channels of the molecule AB the time of flight signals fluctuate independent of each other and the covariance $C(t_1, t_2)$ is zero. Examples of ion–ion covariance maps can be found in Frasinski et al.[26], Frasinski et al.[27], Codling et al.[28], Frasinski et al.[29], Cornaggia et al.[30], and Cornaggia[31].

Covariance mapping is a powerful tool to detect correlation in ion time of flight spectra. It is straightforward to adapt the method to uncover also electron–ion and electron–electron correlation in strong field ionization[27]. Ion–ion covariance mapping was used by several research groups to successfully uncover the Coulomb explosion channels (p, q) in strong field ionization and dissociation of many diatomic[27,28] and even polyatomic molecules[29–31]. These systematic studies allowed the identification of the general mechanism behind molecular multiple ionization and subsequent Coulomb explosion described in the previous chapter.

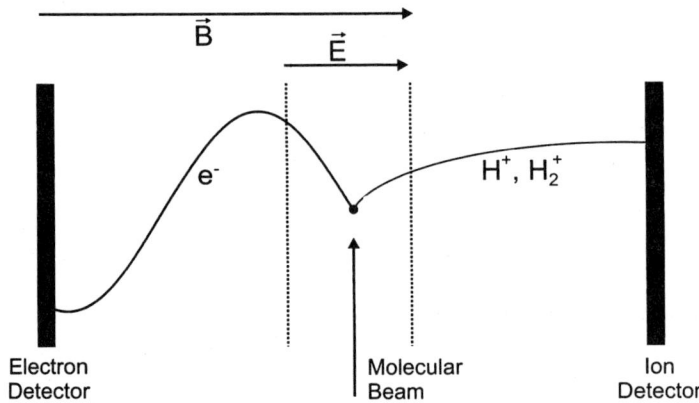

Figure 8. Schematic view of the COLTRIM spectrometer setup. The laser beam intersects a molecular beam at the central point. Electrons and ions are extracted by an electric field $\vec{E}$ to opposite position sensitive detectors. A magnetic field $\vec{B}$ directs photoelectrons to the detector which are initially emitted with a large momentum component perpendicular to the spectrometer axis.

Recent experiments on strong field multiple photoionization of atoms showed that a complete kinematical analysis of this process is feasible[18,32,33]. The technique used to accomplish this is named "COLd Target Recoil Ion Momentum Spectroscopy" (COLT RIMS)[16,17]. We started to apply this technique to the analysis of strong field multiple ionization and dissociation of molecules. Besides fully revealing the correlation present in these processes COLTIMS even allows the determination of the full vector–momenta $\vec{p}$ of all charged particles, electrons and ions, after molecular fragmentation with high momentum resolution. This is equivalent to a full kinematical analysis of the final state. In the final part of this chapter I will present this promising experimental method and show some results for strong field ionization and fragmentation using H_2 as an example.

The basis for a COLTRIM spectrometer is a time of flight spectrometer (see Figure 5). Details are shown schematically in Figure 8. Starting point is a "cold target" atomic/molecular beam. This means, the momentum distribution for all three translational degrees of freedom is significantly smaller than that for a gas target at room temperature. This is achieved by using a beam formed in an adiabatic expansion of high pressure gas through a small nozzle into vacuum[15]. The momentum distribution along the beam axis is reduced significantly in the adiabatic expansion. The distribution along the two directions perpendicular to the axis is adjusted by beam collimation using suitable orifices. Widths of the residual momentum distribution which can be achieved for not too heavy atoms and molecules are ≤ 0.1 a.u.. The molecular beam is intersected at right angles by a focused laser beam. In the focal spot the molecules are photoionized and dissociated.

A homogeneous constant electric field is applied along the spectrometer axis to extract photoelectrons and ions to opposite directions. After acceleration the charged particles drift freely over a certain distance before they reach the respective detectors. For the ions the acceleration path d and the free drift path D are usually chosen to fulfill the condition $D = 2d$. In this way one gets a nearly linear relation between the initial particle momenta p_x along the extraction electric field and the time of flight if the particles have an initial kinetic energy which is not too high (see the previous chapter). Besides the time of flight also the position is determined where each particle hits the respective detector. The information on the position (y, z) allows one to determine the two momentum components p_y and p_z perpendicular to the extraction electric field via:

$$p_y = (y - y_0)/T \tag{14}$$
$$p_z = (z - z_0)/T$$

with (y_0, z_0) the starting point of the particles motion in the laser focal spot and T the time of flight to the detector (see equ. 5). This completes the determination of the full momentum vector of each individual particle.

Determination of the position is done by using a special anode design for the microchannel plate detectors. It consists of two independent delay lines, one for the y– and one for the z–coordinate. A particle impinging on the front of the plate releases a localized electron charge cloud at its back. The position where this charge cloud then impinges on the anode delay lines is determined from the difference in the times it takes the induced electrical signal to reach opposite ends of each delay line (see Ali et al.[34] and refs. cited there).

To be able to collect almost all photoelectrons, including those with a large momentum component perpendicular to the spectrometer axis, a homogeneous magnetic field B is applied parallel to the electric field E over the complete flight path of the electrons. It forces them on cyclotron orbits in the plane perpendicular to the field vector. In the presence of the magnetic field the relation 14 between the momentum components (p_y, p_z) and the position (y, z) where the electron hits the detector is modified. The equation of motion of the electrons in the combined fields is given by (in atomic units):

$$\ddot{\vec{r}} = -E(z)\hat{e}_x - \frac{\dot{\vec{r}} \times B\hat{e}_x}{c} \tag{15}$$

with $\hat{e}_x$ the unit vector along the x–axis and c the speed of light. The relation between the initial momentum vector components in the yz–plane and the position where the electron hits the detector can be derived from equ. 15. In matrix form it reads:

$$\begin{pmatrix} y \\ z \end{pmatrix} = \frac{1}{\omega_c} \begin{pmatrix} \sin \omega_c T & -(1 - \cos \omega_c T) \\ 1 - \cos \omega_c T & \sin \omega_c T \end{pmatrix} \begin{pmatrix} p_y \\ p_z \end{pmatrix} \tag{16}$$

In this relation $\omega_c = B/c$ is the electron cyclotron frequency in atomic units. It was assumed in the calculation that the motion starts at $t = 0$ at the origin $\vec{r} = 0$. With the exception of discrete points of time the 2 × 2–matrix in equ. 16 is invertible. One can thus determine the initial momentum components from the measured yz–coordinates on the position sensitive detector and the time of flight T:

$$\begin{pmatrix} p_y \\ p_z \end{pmatrix} = \frac{\omega_c}{2} \begin{pmatrix} \frac{\sin \omega_c T}{1-\cos \omega_c T} & 1 \\ -1 & \frac{\sin \omega_c T}{1-\cos \omega_c T} \end{pmatrix} \begin{pmatrix} y \\ z \end{pmatrix} \qquad (17)$$

The relation between the point where the electron hits the detector and the initial momentum components p_y and p_z in the plane perpendicular to the magnetic field is not unique if the denominators of the two fractions appearing in the matrix in equ. 17 become zero. As can be seen in equ. 16 this is always the case after each complete revolution on the cyclotron orbit. Irrespective of its initial momentum the electron then comes back to the origin where its motion started.

The magnetic field influences only the electron motion. The ions are nearly unaffected. The reason for this is the only small acceleration ions experience by the magnetic force. It is by a factor m_e/m_I smaller than that for the electron. The relation between the position where the ions hit the ion detector and their initial momenta is yet given with a high accuracy by relation 14.

For the COLTRIMS experiments described below a Ti:Sapphire laser (center wavelength: 795 nm) equipped with an amplifier was used. The repetition rate of the amplified pulses was 1 kHz and their pulse energy adjustable up to ≈ 0.6 mJ at a pulse duration of 25 fsec. The laser beam was focused within the COLTRIM spectrometer with a focal spot diameter of ≈ 8 μm. In this way light intensities in the focal spot of up to $\approx 3 \times 10^{16}$ W/cm^2 were possible. In the experiment we only used 6×10^{14} W/cm^2.

The H$_2$ molecular beam density in the laser beam focal spot was adjusted to a level where the probability to detect an electron together with an ion was less than 0.02 per laser shot. This setting ensures that most of the charged particles we detect are from ionization– and/or dissociation events of single H$_2$ molecules in the focal spot. The momenta of the particles then are correlated and allow the complete kinematical analysis of each single ionization/dissociation event.

A simple example to demonstrate the correlation between particle momenta is found in the strong field photoionization channel of the hydrogen molecule:

$$H_2 \longrightarrow H_2^+ + e^- \qquad (18)$$

The momenta of the photoelectron and the H_2^+ ion are correlated. Assume for simplicity strong field ionization of H_2 can be viewed as induced by multiphoton absorption of n discrete photons. In this case the momentum balance reads:

$$\vec{p}(H_2) + n\vec{p}_\gamma = \vec{p}(e^-) + \vec{p}(H_2^+) \qquad (19)$$

with $\vec{p}_\gamma$ the momentum of a single photon. One thus has exact momentum conservation of the mechanical system (H$_2$, H$_2^+$, e$^-$) transverse to the light beam propagation direction which is the direction of the vector $\vec{p}_\gamma$. Along this direction the momentum of the mechanical system alone is not conserved since the momentum of the absorbed photons is transferred to the molecule and therefore the fragmentation products. In our experimental situation where photoelectrons reach a kinetic energies which are less than ≈ 200 eV the transfer of momentum to the molecule remains small (less than ≈ 0.05 a.u.). This is well below the resolution limit of the COLTRIMS setup. With only a negligible error complete conservation of the momentum of the mechanical system may therefore be assumed and this should be found in a measurement of the momentum of the H$_2^+$ ion and the corresponding photoelectron.

The above analysis of the relation between the momenta of the mechanical system can also be done by treating the intense light beam as a classical electromagnetic wave. The momentum transfer to the mechanical system in this case can be shown to be caused by the magnetic field component of the electromagnetic wave via the Lorentz force term in the equation of motion. It is directed along the light beam propagation direction. The effect on the mechanical system remains small as long as the velocity of the particles involved (especially of the electron) remains small compared to the velocity of light.

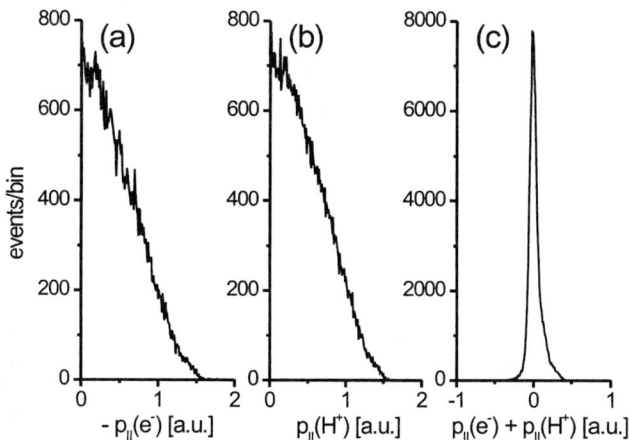

Figure 9. (a) Electron momentum distribution parallel to the light polarization direction for the photoionization channel: $H_2 \longrightarrow H_2^+ + e^-$. (b) Momentum distribution of the H_2^+ photoions for this channel. (c) The corresponding distribution of the sum–momentum of the ion and electron.

Conservation of momentum of the mechanical system is demonstrated experimentally in Figure 9 for the e^- and H_2^+ momentum components $p_\parallel$ along the polarization vector of the linearly polarized light wave used for multiphoton ionization. The measurement was done with the COLTRIM spectrometer described above. Figure 9a shows the momentum distribution of the photoelectron ($p_\parallel(e^-)$), 9b of the H_2^+ ion ($p_\parallel(H_2^+)$), and 9c the distribution of the sum–momentum ($p_\parallel(e^-) + p_\parallel(H_2^+)$) of the ion and the corresponding photoelectron. All particles with initial momentum pointing to the respective detector are included in Figure 9 irrespective of their momentum components perpendicular to the light polarization vector. The ion and electron momentum distributions are practically identical. The only difference is a slightly higher momentum resolution in the e^- distribution (Figure 9a).

The sum–momentum distribution shows a narrow peak at zero sum–momentum. Its full width at half maximum is 0.095 a.u.. It is formed by those events where the electron and the ion originate from the same H_2 molecule. In this case the sum–momentum is equal to the momentum component $p_\parallel(H_2)$ of the hydrogen molecule (see equ. 19) which is photoionized. $p_\parallel(H_2)$ is equal to zero in the experiment since the light beam polarization vector is perpendicular to the molecular beam axis. The width of the peak at zero sum–momentum is a measure of the resolution achieved with the COLTRIM spectrometer. Events where no correlation between the detected ion and electron momenta exists would contribute a broad background in Figure 7c. Such a background is practically missing. This means, events where more than one H_2 molecule in the focal spot or an H_2 molecule and a background gas molecule are photoionized do practically not contribute to the measured momentum distributions.

In strong field ionization the final momentum component of the photoelectron along the polarization vector of the light wave shown in Figure 9a is not determined by the initial ionization step of the H_2 molecule. Most of the momentum is gained by the already free photoelectron through acceleration by the electric field component of the light wave. Therefore the final momentum of the H_2^+ ion which compensates the electron momentum exactly (Figure 9b) is not recoil momentum of the ion gained in the photoionization step. Similar to the electron also the final ion momentum is determined by the acceleration it experiences in the light wave. In terms of classical mechanics it

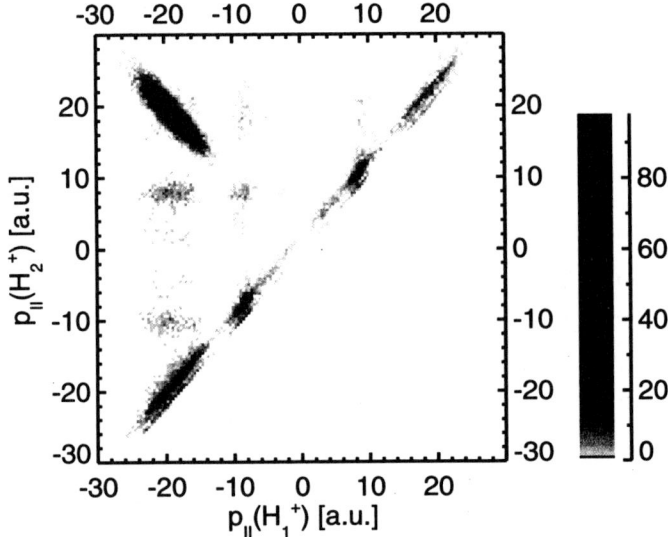

Figure 10. Correlation plot for the momentum components $p_{||}$ parallel to the light polarization vector of two hydrogen ions (H_1^+, H_2^+) formed by strong field ionization and dissociation of H_2. The excitation light wavelength was 790 nm, the pulse width 25 fsec, and the intensity 6×10^{14} W/cm^2. Correlated ion momenta from Coulomb explosion are found near $p_{||}\left(H_1^+\right) \approx -19$ a.u. and $p_{||}\left(H_2^+\right) \approx 19$ a.u..

is easily seen that the free electron and the ion experience the same force along the light polarization vector which only points into opposite directions. It is exerted by the electric field of the wave alone ($E_0\left(t - z/c\right)\cos\omega\left(t - z/c\right)$ for a plane wave propagating in z–direction). This means that in strong field ionization ion momenta can be used to measure photoelectron momenta. This becomes relevant for multiple ionization for example of atoms where the momentum distribution of the multiply charged ion then is a measure of the sum-momentum of all the the emitted photoelectrons[18,32].

The ion–electron momentum balance is destroyed if the light pulse width becomes too long. In this situation the charged particles experience the spatial gradient of the electric field of the wave in the focal spot (especially in the direction perpendicular to the light beam axis). Due to their different velocities electrons and ions travel different distances in the gradient which additionally varies in time. They experience a different net acceleration and therefore different final momenta. One expects that the width of the sum–momentum distribution in Figure 9c increases with an increasing light pulse width due to this effect.

For the representative molecule H_2 used here it can be shown that COLTRIM spectroscopy is able to verify that really Coulomb explosion is responsible for the formation of the fast broad peaks in the ion momentum distribution in Figure 6b at $p_{||} \approx \pm 19$ a.u.. The momenta of the two H^+ ions should be correlated. If the momenta of the two emitted photoelectrons are neglected in comparison to the large ion momenta (see Figure 6b) the momenta of the two ions should be nearly equal in magnitude but point into opposite directions. For all events where we detected two H^+ ions per laser shot I plotted in Figure 10 the 2–dimensional distribution function of the momentum components of both ions parallel to the light polarization vector. The main accumulation point of events is found in the form of an elliptical island at $p_{||}\left(H_1^+\right) \approx -19$ a.u. and $p_{||}\left(H_2^+\right) \approx 19$ a.u.. It consists of true coincidences formed by ions from single Coulomb explosion events. All other accumulation points with significantly smaller numbers of events are accidental coincidences caused by the fragmentation of two H_2 atoms or, along the main diagonal of the plot $(p_{||}\left(H_1^+\right) = p_{||}\left(H_2^+\right))$, due to double triggering of the detection electronics by the same ion. The correlation of the ion momenta thus identifies the fast ions in Figure 6b as originating from Coulomb explosion. In the case of H_2 this is the only fragmentation channel which can be responsible for correlation of the momenta of two ions.

Figure 10 not only allows the identification of the Coulomb explosion channel it also contains information on the momentum distribution of the two photoelectrons created prior to Coulomb explosion by double ionization of H_2. Cutting through the island of events corresponding to Coulomb explosion along lines parallel to the main diagonal of the plot $(p_{||}\left(H_1^+\right) = p_{||}\left(H_2^+\right))$ gives the distribution function of the sum-momentum of the two H^+ ions along the light polarization vector. Momentum conservation of the mechanical system along this direction

$$0 = p_{||}\left(H_2\right) = p_{||}\left(e_1^-\right) + p_{||}\left(e_2^-\right) + p_{||}\left(H_1^+\right) + p_{||}\left(H_2^+\right) \qquad (20)$$

shows that this is equal to the sum-momentum of the two photoelectrons emitted after double ionization of H_2. As one can see in Figure 10 the electron sum–momentum distributions along the cuts specified above all have their maximum at zero sum–momentum and a width which is less than ≈ 2.5 a.u.. The electron sum–momentum distribution calculated in this way can for example be used as a starting point for the analysis of the double ionization mechanism of the H_2 molecule.

CONCLUDING REMARKS

In this lecture it was only possible to present some of the main experimental methods used to investigate strong field molecular photoionization and dissociation and the main physical ionization and dissociation mechanisms which can be extracted from experimental results. I did not discuss molecular alignment through a torque exerted by the strong electric field of the light wave on a dipole moment which is induced by the same wave (see for example Posthumus et al.[35,36]). Also pump–probe experiments with ultrashort light pulses can shed light on strong field dissociation as experiments for example on molecular hydrogen showed[12,37]. The light–induced adiabatic molecular states I discussed above to interpret ion kinetic energy distributions formed by strong field dissociation have also been investigated using different techniques as those mentioned. Pairs of strongly dipole coupled electronic states with a spacing of approximately the photon energy of visible or near infrared radiation are usually found in molecular ions. They are best suited to investigate strong field dissociation phenomena. A controlled population of these states is possible in a molecular ion beam. Exposing this beam to intense light pulses allows the investigation of the strong field phenomena under defined initial conditions. Experiments along this direction have been done by Wunderlich et al.[38–40]. In ultrashort light pulses recently non–sequential double photoionization of molecules has been identified experimentally by Cornaggia and Hering[41,42]. They investigated the dependence of the integral yield of the doubly charged molecular ions on the light intensity. Similar to atomic double ionization (see the lecture of Agostini in this book) the question concerning the mechanism behind molecular double ionization arises. Is it the same as for atoms or are there differences? The COLTRIM spectroscopy described above is certainly well suited to shed light also on molecular non–sequential multiple ionization as it already did for atoms.

REFERENCES

1. Zavriyev, A., Bucksbaum, P.H., Muller, H.G., and Schumacher, D.W., Ionization and dissociation of H_2 in intense laser fields at 1.064 μm, 532 nm, and 355 nm, *Phys. Rev. A* 42:5500 (1990).
2. Yang, B., Saeed, M., and DiMauro, L.F., High-resolution multiphoton ionization and dissociation of H_2 and D_2 molecules in intense laser fields *Phys. Rev. A* 44:R1458 (1991).
3. Rottke, H., Ludwig, J., and Sandner, W., H_2 and D_2 in intense sub-picosecond laser pulses: Photoelectron spectroscopy at 1053 and 527 nm, *Phys. Rev. A* 54:2224 (1996).
4. Cornaggia, C., Normand, D., Morellec, J., Mainfray, G., and Manus, C., Resonant multiphoton ionization of H_2 via the E,F $^1\Sigma_g^+$ state: Absorption of photons in the ionization continuum, *Phys. Rev. A* 34:207 (186).
5. Verschuur, J.W.J., Noordam, L.D., and van Linden van den Heuvell H.B., Anomalies in above-threshold ionization observed in H_2 and its excited fragments, *Phys. Rev. A* 40:4383 (1989).
6. Seideman, T., Ivanov, M. Yu., and Corkum, P.B., Role of electron localization in intense–field molecular ionization, *Phys. Rev. Lett.* 75:2819 (1995).
7. Freeman, R.R., Bucksbaum, P.H., Milchberg, H., Darack, S., Schumacher, D., and Geusic, M.E., Above-threshold ionization with subpicosecond laser pulses, *Phys. Rev. Lett.* 59:1092 (1987).
8. Posthumus, J.H., Frasinski, L.J., Giles, A.J., and Codling K., Dissociative ionization of molecules in intense laser fields: a method of predicting ion kinetic energies and appearance intensities, *J. Phys. B: At. Mol. Opt. Phys.* 28:L349 (1995).
9. Zuo, T. and Bandrauk, A.D., Charge–resonance–enhanced ionization of diatomic molecular ions by intense lasers, *Phys. Rev. A*, 52:R2511 (1995).
10. Sharp, T.E., Potential-energy curves for molecular hydrogen and its ions, *At. Data* 2:119 (1971).
11. Ludwig, J., Rottke, H., and Sandner, W., Dissociation of H_2^+ and D_2^+ in an intense laser field, *Phys. Rev. A* 56:2168 (1997).
12. Trump, C., Rottke, H., and Sandner, W., Multiphoton ionization of dissociating D_2^+ molecules, *Phys. Rev. A* 59:2858 (1999).
13. Trump, C., Rottke, H., and Sandner, W., Strong–field photoionization of vibrational ground–state H_2^+ and D_2^+ molecules, *Phys. Rev. A* 60:3924 (1999).

14. Trump, C., Rottke, H., Wittmann, M., Korn, G., H., Sandner, W., Lein, M., and Engel V., Pulse–width and isotope effects in femtosecond–pulse strong–field dissociation of H_2^+ and D_2^+, *Phys. Rev. A* 62:063402 (2000).
15. Miller, D.R., Free jet sources, in: *Atomic and Bolecular Beam Methods, Volume 1*, G. Scoles, ed., Oxford University Press, New York (1988).
16. Moshammer, R., Unverzagt, M., Schmitt, W., Ullrich, J., and Schmidt–Böcking, H., A 4π recoil–ion electron momentum analyser: a high–resolution "microscope" for the investigation of the dynamics of atomic, molecular and nuclear reactions, *Nucl. Instr. Meth. Phys. Res. B* 108:425 (1996).
17. Ullrich, J., Moshammer, R., Dörner, R., Jagutzki, O., Mergel, V., Schmitt–Böcking, H., and Spielberger, L., Recoil–ion momentum spectroscopy, *J. Phys. B: At. Mol. Opt. Phys.* 30:2917 (1997).
18. Moshammer, R., Feuerstein, B., Schmitt, W., Dorn, A., Schröter, C.D., Ullrich, J., Rottke, H., Trump, C., Wittmann, M., Korn, G., Hoffmann, K., and Sandner, W., Momentum distribution of Ne^{n+} ions created by an intense ultrashort laser pulse, *Phys. Rev. Lett.* 84:447 (2000).
19. Pegarkov, A.I. and Rapoport, L.P., Absorption of laser light by a diatomic mole-cule in the case of double resonance, *Opt. Spectrosk.* 63:501 (1987).
20. Pegarkov, A.I. and Rapoport, L.P., Two–photon dissociation of a diatomic mole-cule in an intense laser field, *Opt. Spectrosk.* 63:751 (1987).
21. Giusti–Suzor, A., He, X., Atabek, O., and Mies, F.H., Above–threshold dissociation of H_2^+ in intense laser fields, *Phys. Rev. Lett.* 64:515 (1990).
22. Giusti–Suzor, A., Mies, F.H., DiMauro, L.F., Charron, E., and Yang, B., Dynamics of H_2^+ in intense laser fields, *J. Phys. B: At. Mol. Opt. Phys.* 28:309 (1995).
23. Zavriyev, A., Bucksbaum, P.H., Squier, J., and Saline, F., Light–induced vibrational structure in H_2^+ and D_2^+ in intense laser fields, *Phys. Rev. Lett.* 70:1077 (1993).
24. Normand, D. and Schmidt, M., Multiple ionization of atomic and molecular iodine in strong laser fields, *Phys. Rev. A* 53:R1958 (1996).
25. Gibson, G.N., Li M., Guo, C., and Nibarger, J.P., Direct evidence of the generality of charge–asymmetric dissociation of molecular iodine ionized by strong laser fields, *Phys. Rev. A* 58:4723 (1998).
26. Frasinski, L.J., Codling, K., and Hatherly, P.A., Covariance mapping: a correlation method applied to multiphoton multiple ionization, *Science* 246:1029 (1989).
27. Frasinski, L.J., Stankiewicz, M., Hatherly, P.A., Cross, G.M., Codling, K., Langley, A.J., and Shaik, W., Molecular H_2 in intense laser fields probed by electron–electron, electron–ion, and ion–ion covariance techniques, *Phys. Rev. A* 46:R6789 (1992).
28. Codling, K., Cornaggia, C., Frasinski, L.J., Hatherly, P.A., Morellec, J., and Normand, D., Charge–symmetric fragmentation of diatomic molecules in intense picosecond laser fields, *J. Phys. B: At. Mol. Opt. Phys.* 24:L593 (1991).
29. Frasinski, L.J., Hatherly, P.A., Codling, K., Larsson, M., Persson, A., and Wahl-ström, C.-G., Multielectron dissociative ionization of CO_2 in intense laser fields, *J. Phys. B: At. Mol. Opt. Phys.* 27:L109 (1994).
30. Cornaggia, C., Schmidt, M., and Normand, D., Coulomb explosion of CO_2 in an intense femtosecond laser field, *J. Phys. B: At. Mol. Opt. Phys.* 27:L123 (1994).
31. Cornaggia, C., Carbon geometry of $C_3H_3^+$ and $C_3H_4^+$ molecular ions probed by laser–induced Coulomb explosion, *Phys. Rev. A* 52:R4328 (1995).
32. Weber, Th., Weckenbrock, M., Staudte, A., Spielberger, M., Jagutzki, O., Mergel, V., Afaneh, F., Urbasch, G., Vollmer, M., Giessen, H., and Dörner, R., Recoil–ion momentum distributions for single and double ionization of helium in strong laser fields, *Phys. Rev. Lett.* 84:443 (2000).
33. Weber, Th., Giessen, H., Weckenbrock, M., Urbasch, G., Staudte, A., Spielberger, L., Jagutzki, O., Mergel, V., Vollmer, M., and Dörner, R., Correlated electron emission in multiphoton double ionization, *Nature* 405:658 (2000).
34. Ali, I., Dörner, R., Jagutzki, O., Nüttgens, S., Mergel, V., Spielberger, L., Khayyat,

Kh., Vogt, T., Bräuning, H., Ullmann, K., Moshammer, R., Ullrich, J., Hagmann, S., Groeneveld, K.-O., Cocke, C.L., Schmidt-Böcking, H., Multi-hit detector system for complete momentum balance in spectroscopy in molecular fragmentation processes, *Nucl. Instrum. Meth. Phys. Res. B* 149:490 (1999).
35. Posthumus, J.H., Plumridge, J., Thomas, M.K., Codling, K., Frasinski, L.J., Langley, A.J., and Taday, P.F., Dynamic and geometric laser-induced alignment of molecules in intense laser fields, *J. Phys. B: At. Mol. Opt. Phys.* 31:L553 (1198).
36. Posthumus, J.H., Plumridge, J., Frasinski, L.J., Codling, K., Langley, A.J., and Taday, P.F., Double pulse measurements of laser-induced alignment of molecules, *J. Phys. B: At. Mol. Opt. Phys.* 31:L985 (1998).
37. Posthumus, J.,H., Plumridge, J., Taday, P.,F., Sanderson, J.,H., Langley, A.,J., Codling, K., and Bryan, W.A., Sub-pulselength time resolution of bond softening and Coulomb explosion using polarization control of laser-induced alignment, *J. Phys. B: At. Mol. Opt. Phys.* 32:L93 (1999).
38. Wunderlich, C., Figger, H., and Hänsch, T.,W., Ar_2^+ molecules in intense light fields, *Chem. Phys. Lett.* 256:43 (1996).
39. Wunderlich, C., Kobler, E., Figger, H., and Hänsch, T.,W., Light-induced molecular potentials, *Phys. Rev. Lett.* 78:2333 (1997).
40. Wunderlich, C., Figger, H., and Hänsch, T.,W., Tunneling through light-induced molecular potentials in Ar_2^+, *Phys. Rev. A* 62:023401 (2000).
41. Cornaggia, C. and Hering, Ph., Laser-induced non-sequential double ionization of small molecules, *J. Phys. B: At. Mol. Opt. Phys.* 31:L503 (1998).
42. Cornaggia, C. and Hering, Ph., Non-sequential double ionization of small mole-cules induced by a femtosecond laser field, *Phys. Rev. A* 62:023403 (2000).

TWO-COLOR AND SINGLE-COLOR ABOVE THRESHOLD IONISATION

Pierre Agostini

Service des Photons Atomes et Molécules, CEA,
Centre d'Etudes de Saclay 91191 Gif Sur Yvette

PART ONE: TWO-COLOR ABOVE THRESHOLD IONISATION AND APPLICATIONS

INTRODUCTION

My aim for this first lecture is to give an account of some of current interest aspects of Above-Threshold Ionisation (ATI) and, more specifically, of two-color ATI of atoms which is in some sense much simpler than single-color ATI. In two-color ATI (Fig.1), one field (in the XUV –X-ray domain) is too weak to induce multiphoton processes but its photons are energetic enough to photoionize the atom. The other field, on the contrary (in the near IR) is relatively more intense and the probability for exchanging several photons with the atom is significant although it is too weak to ionize the atom by itself.

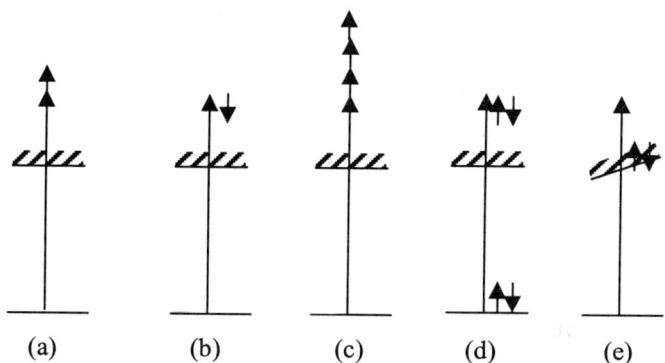

Fig. 1 Types of two-color ATI processes: two-photon ATI or laser-assisted photoeffect (a),(b); multiphoton case (c); photo ionisation in presence of ac-Stark shifts and intensity-dependent ionisation energy and angular distribution(d), (e).

The interest for this kind of transitions arose a few years ago with the rapid development of XUV sources (X-ray sources from laser plasma and high-harmonic generation) naturally synchronized with an intense infrared laser which create them. Two-color transitions were not only easy to study in the laboratory due to this natural synchronization but also proved very useful to characterize ultrashort the XUV pulses.

Fig.1 displays a few energy diagrams for such processes. The first one (a) is the simplest two-photon, two-color ATI in which one photon from each field is absorbed by the atom while the optically-active electron is transferred into the continuum. The IR photon may be emitted instead thus decreasing the kinetic energy of the electron (b). Thus in the photoelectron energy spectrum, besides the main photo ionisation peak, a satellite, or sideband separated from it by the IR photon energy appears. If the intensity of the IR field is higher then several photons may be absorbed (or emitted) together with the XUV photon, the (positive or negative) excess energy being converted into kinetic energy for the electron (c) giving rise to several sidebands in the energy spectrum on both sides of the main peak. In general the case (c) involves some ac-Stark shifts of the atomic energy levels (d) and an increase of the ionisation potential (e) function of the IR intensity. The main peak in the spectrum, under certain conditions, may acquire a width which is correlated to the time duration of the XUV pulse and the shape of the IR pulse. All these transitions may also apply to inner-shell electrons provided the XUV photon energy is high enough.

As multiphoton processes, these types of transitions are particularly simple since most of the strong field effects are concentrated in the continuum part of the spectrum, that is to say on a quasi-free electron. In particular, the influence of the intense IR field on the excited states, (responsible for Freeman transient resonances and the accompanying complications in the case of single-color ATI) may be ignored. There is always a clear separation between the bound-free and the free-free steps in this type of transitions while there are often entangled in a complex way in single-color strong-field ATI. From the standpoint of applications transitions of type (a) are well suited to cross-correlation measurements of XUV short (ps or below) pulses. Methods making use of the change in ionisation potential in a strong IR field, and removing some of the limitations of the cross-correlation have been recently developed. In this Chapter, after recalling the properties of Photo ionisation or single-photon ionisation, I will proceed to two-photon in the case where the IR field is also weak perturbation before investigating the strong field limit and its classical approximation, the so-called Simpleman's theory and its quantum counterpart based on the description of the final state by a Volkov state. We will be in position to explore the various applications for two-color ATI to short pulse XUV characterization.

For a general introduction to ATI and to the general theory of this process the reader is referred to the Chapter by N. Kylstra in the present Book and the references therein. I will limit the theoretical considerations hereunder to the two-color case.

THEORETICAL APPROACHES TO TWO-COLOR ABOVE-THRESHOLD IONISATION

Photoeffect

As it is well-known in perturbation theory, an atom submitted to a time-dependent perturbation $W(t)$ can make a transition from an initial state $|i\rangle$ of energy E_i to a final state $|f\rangle$ of energy E_f. If the energy of the final state belongs to a continuum, like in the case of photo ionisation by an electromagnetic field, the transition probability is proportional to time and the probability per unit time is given by the Fermi Golden rule:

$$w(t) = \frac{2\pi}{\hbar} \left| \langle f, E_f = E_i | W(t) | i \rangle \right|^2 \rho(E_f = E_i) \tag{1}$$

where $\rho(E)$ is the continuum density of states. In the case of a monochromatic perturbation of pulsation ω, energy conservation implies $E_f = E_i + \hbar\omega$, or equivalently the Einstein's law:

$$E_k = \hbar\omega - I_0 \qquad (2)$$

where I_0 is the atom ionisation energy and E_k the photoelectron kinetic energy. The matrix element in Eq. (1) in the case of a dipole transition induced by an electromagnetic wave is proportional to the square of the electric field E or its intensity I

$$I = \frac{\varepsilon_0 c E^2}{2} \qquad (3)$$

photo ionisation rate w and the cross-section σ are related by :

$$w = \sigma F \qquad (4)$$

where F is the photon flux $I/\hbar\omega$. These concepts are easily extended to the two-photon, two-color case.

Two-color ATI in the perturbative limit

The case of Fig. 1(a) the photoelectron energy spectrum is determined by the first-order and second-order ground state transition rates $w^{(1)}$ and $w^{(2)}$. While the first is determined from Eq.(1), the second is given by[1]:

$$w^{(2)} = \frac{2\pi}{\hbar}\left|M^{(2)}_{E,1}\right|^2 \qquad (5)$$

where ω_X and ω_{IR} are the photon pulsation from the XUV and IR fields respectively and $M^{(2)\pm}$ is the second-order matrix element given by:

$$M^{(2)}_{E,1} = \sum_k \frac{V_{Ek}V_{k1i}}{E_k - E_1 \pm \hbar\omega_j} + \int dE' \frac{V_{EE'}V_{E'1}}{E' - E_1 \pm \hbar\omega_j + i\delta} \qquad (6)$$

where V_{fk} and V_{ki} $V_{EE'}$ and $V_{E'1}$ are dipole matrix elements and ω_j may be either ω_X or ω_{IR} and the upper and lower signs corresponding to the emission or absorption of photons respectively. Here δ is a small imaginary part which allows to handle the pole occurring since $\hbar\omega_X > |E_1| = I_0$. The singularity appearing in the integral of (6) is evaluated by means of a principal value (P):

$$\frac{1}{x+i\delta} = P\frac{1}{x} - i\pi\delta(x) \qquad (7)$$

It turns out that two-photon ATI cross-section are of the same order of magnitude that ordinary two-photon ionisation where the photon energies are both below threshold. Moreover, from expressions (5) and (6) it is clear that the ionisation rate $w^{(2)}$ is proportional to the product of the XUV and IR intensities. This property will be used later on in the cross-correlation methods.

Multiphoton two-color ATI

Expressions (5) and (6) are easily generalized to transitions where the number of absorbed or emitted IR photons is larger than one. The rate for the absorption of one XUV photon and n IR photons is found proportional to the product $I_{XUV}I_{IR}^n$. Although the validity of the perturbative calculation above breaks down rapidly this situation is actually met in cross-correlation measurements described below.

The Keldysh-Faisal-Reiss approximation

This approximation was first proposed by Keldysh and later on reformulated more rigorously by Faisal and Reiss (Keldysh 1965, Faisal 1973, Reiss, 1981) and now widely known as the KFR strong field theory. This approach is aimed at deriving the ionisation amplitude in the strong-field, low-frequency limit where the Keldysh parameter:

$$\gamma = \sqrt{\frac{I_0}{2U_P}} \qquad (8)$$

is much smaller than one. In this limit of the KFR approximation, ionisation is a quasi-static tunneling and the effect of the strong electromagnetic field on the free electron is taken into account by describing the final state of the ionisation process as solution of the Schrödinger equation for a free electron in the electromagnetic field (known as a Volkov state) rather than an eigen state of the atomic potential. In the two-color situation, it is mainly this last aspect that is important. This case has been treated by several authors and it has been shown (Leone et al., 1988, Cionga et al., 1993, Schins et al., 1995) that the amplitude of the nth sideband is proportional to the square of the generalized Bessel function defined as:

$$J_n(a,b) = \sum J_{n+2l}(a) J_l(b) \qquad (9)$$

where $\quad a = \dfrac{\sqrt{8U_0 U_P \cos\theta}}{\omega_{IR}} \quad$ and $\quad b = \dfrac{U_P}{\omega_{IR}} \qquad (10)$

where θ is the angle between the direction of the laser field and the that of the initial electron velocity. The physical meaning of the arguments (11) will appear clearly in the classical hereunder. In particular a is related to the average kinetic energy of the electron in the field while b is the ponderomotive energy in units of photons energy.

Simpleman's theory of two-color ATI

The essence of this approximation[2] is to treat classically the interaction of the free electron with the intense IR field an to neglect both its interaction with the Coulomb potential and with the low-intensity field. Simpleman's theory may be used both in single and two-color ATI but is particularly well suited to the two-color case where all the complexity due to the discrete spectrum is removed from the problem. Quantum mechanically ATI is characterized by peaks separated by the IR photon energy in the electron energy spectrum on both sides of the main peak due to photo ionisation. The average kinetic energy of the free electron acquired in the IR field is the classical counterpart of ATI. It is interesting to determine the classical limits of the spectrum or alternatively, the mimimum IR intensity required to produce at least one sideband.

Classically, the problem is determined by the spatio-temporal characterization of the IR field and by the initial conditions for the electron motion. Let us specify low-frequency field by $E_0\cos\omega_{IR}t$, the kinetic energy contains of course an oscillatory part (the so-called ponderomotive energy) but also a secular term which arises from the conservation of the canonical momentum. The ponderomotive energy, or average kinetic oscillatory energy is given by (in atomic units, $\hbar=m=e=1$) $U_P=E_0^2/4\omega^2$. Noting t' the time at which the electron is freed in the low-frequency field and $\varphi=\omega t'$ the corresponding phase, the kinetic energy averaged over one optical cycle writes:

$$\langle E_k \rangle = U_0 + U_P(1+\sin^2\varphi) - \sqrt{8U_P U_0}\sin\varphi \qquad (11)$$

The first term in Eq. (11) is the initial kinetic energy resulting from the primary process releasing the electron: in case of the photoeffect it is given by (2) but for instance in the Auger decay it depends only on atomic states energies. The second term, proportional to the field intensity, is the oscillation energy plus a drift energy function of the initial phase, the third term finally is a cross term which depends both on the initial phase and the drift and initial velocities. The relative magnitude of these terms varies with the initial velocity and the field strength. The cross-term in Eq.(11) is found in the argument a of Eq. (10) (the factor $\cos\theta$ has been omitted here) and the classical limit for the kinetic energy change is seen to correspond to the quantum expression (10) since Bessel functions decrease exponentially for orders larger than the argument. The next two sections are about two-color ATI situations met in Auger decay or photoeffect in a strong IR field and applications to the metrology of ultrashort XUV pulses.

EXPERIMENTS IN TWO-COLOR ATI

Laser-assisted Auger decay and Photoeffect

Situations relevant to Eq.(11) with a large initial energy are met in Laser-Assisted Auger Decay [3] or Laser-assisted photoeffect [4]. In the first case, a hard photon removes one inner-shell electron leaving a core-hole ion which very rapidly decays to the ground state of the doubly charged ion through the cooperation of two outer-shell electrons. Although the physics of these two processes is quite different, they are quite equivalent from the view point of interest here: they both promote a bound electron into a continuum dressed by a strong electromagnetic field. The case of argon (electronic configuration $1s^2 2s^2 2p^6 3s^2 3p^6$) has been studied experimentally [5]. One X-ray photon from a broad band laser plasma source extracts one L-shell p electron. Two electrons from the M-shell collide and one of them fills the L-shell hole while the second one is emitted in the continuum as an Auger electron of well defined energy (independent of the absorbed photon energy). In presence of IR radiation (800 nm) with an intensity of the order of 10^{12} W.cm^{-2} several photons from this field are emitted and absorbed. The resulting energy spectrum therefore consists of a central line at the Auger electron energy and several "sidebands" on both sides separated from each other by the IR photon energy. In reality, the Auger "line" has a number of components due to the fine structure of both the ion excited state and the doubly charged ion ground state. It is necessary to deconvolute from this fine to extract the spectrum reduced to a central line surrounded by n sidebands separated by the IR photon energy $\hbar\omega_{IR}$ (Fig.1). The number n is well predicted by Eq.(11), which reduces to

$$\langle E_k \rangle = U_0 \pm \sqrt{8 U_p U_0} \sin\varphi \qquad (12)$$

from which one defines n as $\sqrt{8 U_p U_0} \sin\varphi / \hbar\omega_{IR}$ in which $\sin\varphi$ is set to 1. In this process the appearance of the Auger electron in the IR field is completely random with respect to the optical cycle and all values of φ between 0 and 2π are equally probable.

A similar situation is encountered when irradiating atoms simultaneously with the fundamental and high-harmonics of a Ti:Saphire laser in the 20-30 eV range of energy. Only outer-shell electrons are touched this time but since the incident UV radiation is narrow band, sidebands may be detected on the energy spectrum of the primary electrons. Expression (4) provides a good prediction for the intensity required to detect the two-color process. For example, with an initial kinetic energy of 1 atomic unit (27.21 eV), at the Ti:Saphire wavelength (800nm), to observe at least one sideband (n=1) an IR peak intensity of $2.7 \, 10^{10}$ W.cm^{-2} is needed. If the photoionized atom is a rare gas, such an intensity is much too low for direct ionisation (multiphoton transition ground state-continuum) but sufficient to induce continuum-continuum transitions.

A natural application of such two-color ATI processes is the cross-correlation measurement of X-ray ultrashort (subpicosecond) pulses for which there is currently no other available method. Cross-correlation is an extension of well-know autocorrelation While the classical theory above provides an easy way to estimate the number of sidebands at a given IR intensity and wavelength, it says nothing of their amplitudes. In the low intensity limit (n≈1), one may evaluate the probability for the two-photon, two-color process through perturbation theory. From perturbation theory the probability is proportional to the product of the X-ray and IR intensities. Hence the amplitude of a given sideband is proportional to the convolution:

$$\int_{-\infty}^{\infty} I_{IR}(t) I_X(t-\tau) d\tau \qquad (13)$$

which can be recorded by varying the delay τ and the unknown function $I_X(t)$ extracted from it by deconvolution from the knowledge of $I_{IR}(t)$. Note that the two-photon ATI process has no intrinsic time lag since the intermediate state in the continuum has an extremely short lifetime. The method has been applied to X pulses in the 250-400 eV range as well as to XUV pulses from harmonic generation at longer wavelength (30 eV).

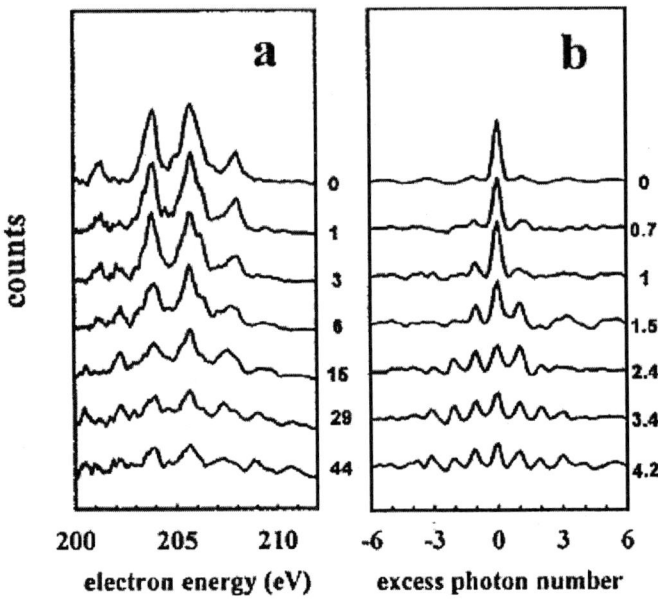

Fig.2 Electron Energy spectra from Laser-Assisted Auger decay in Argon. The left box shows the raw spectra for increasing IR intensities from top to bottom (numbers on the right of each spectra in relative values). The right box shows the same spectra after deconvolution from the zero-intensity one. The numbers on the right of each spectra is n calculated from Eq. (11).

The classical model (a classical electron dropped in a classical oscillating field at a given phase and with a given initial velocity) is strongly connected with the KFR theory in which the ground state is coupled to a Volkov state (exact solution of the Schrödinger equation for a free electron in an electromagnetic field).

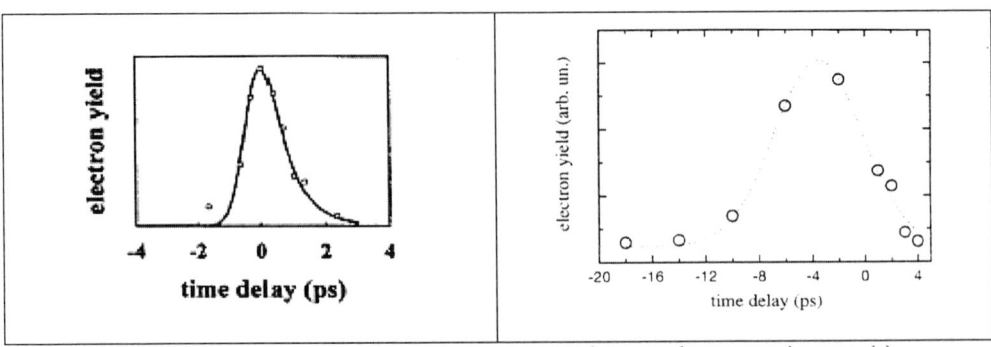

Fig. 3 Cross-correlation signal recorded Laser-assisted Auger decay experiments.: (a) $L_{23}M_{23}M_{23}$ in argon. The solid line is a fit assuming a duration of 700 fs for the X-ray pulse. (Schins et al. 1994).; (b) N00 in xenon (R. Constantinescu Ph D thesis, unpublished).

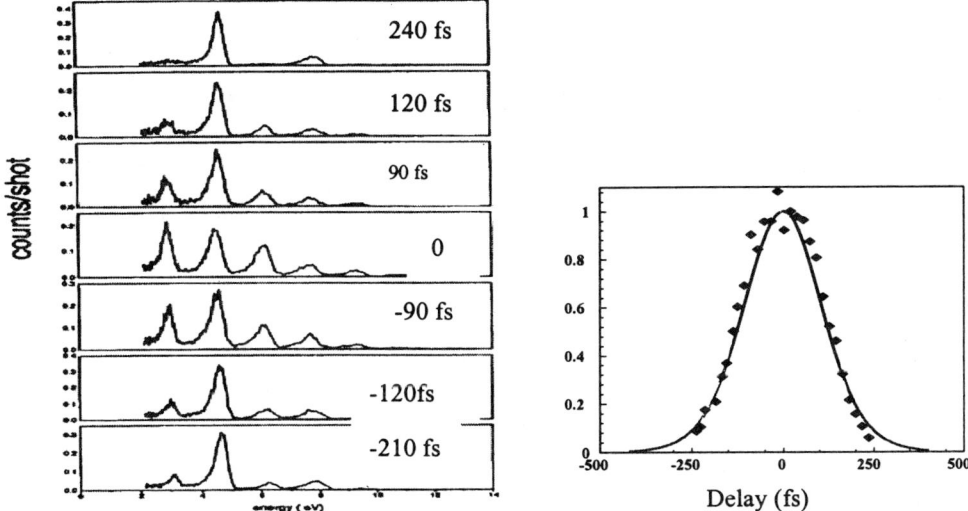

Fig.4 Electron energy spectra from photoionisation of helium by the harmonics 19 and 21 of a Ti:S laser for various time delays between the harmonic and IR pulses The first spectrum on the top right is for a large delay and hence no temporal overlap. As the delay is scanned sideband peaks appear between and around the initial two-peak spectrum. The right box contains the cross-correlation signal vs. delay (Bouhal et al., 1998).

The amplitude of the first sideband is no longer proportional to the IR intensity and extraction of the unknown duration from Eq.(13) is no longer straightforward. This question has been discussed by Bouhal et al.[4].

Similar experiments have been conducted with the photoeffect from high-harmonics in the VUV range [5] with similar conclusions about the amplitude and the number of sidebands.

Ponderomotive streaking of the ionisation potential[6]

Even though the intensity necessary to induce continuum-continuum transitions and create sidebands in two-color ATI is normally insufficient to ionize the atom, it reaches values for which shifts of the spectra become visible. As the ionisation potential is increased by U_P, the peaks of two-color ATI are shifted by this amount towards low energies. The increase of the ionisation potential may be seen either as the result of AC-Stark shifts of the Rydberg states which converge to U_P or alternatively as a consequence of the fact that the electron is freed in an electromagnetic field in which it must have a quiver energy U_P therefore replacing the Einstein's law (2) by:

$$E_k + U_P = \hbar\omega - I_0 \qquad (14)$$

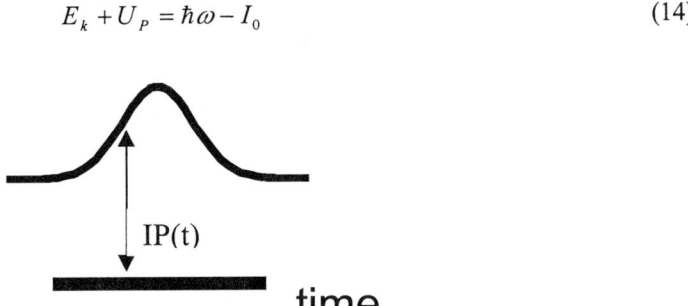

Fig. 5 Schematic representation of an atom in a strong laser pulse: the ionisation potential is increased proportionally to intensity during the pulse.

In fact this latter perspective is better suited to the two-color case in which the hard photon allows to jump over all the bound states of the atom and, therefore, to ignore their energies. In both cases, the energy levels of an atom in a laser pulse is looking like Fig. (5).

If photo ionisation is produced during a short period of time during the rise time (or the fall time) of such a pulse (Fig. (6)), the photoelectron energy spectrum will be directly affected. The corresponding peak will acquire an extra broadening due to the change of ionisation energy.

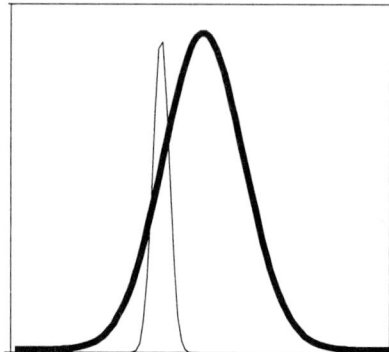

Fig. 6 The broad pulse (thick line) is an IR pulse inducing a streaking of the ionisation potential during the short pulse (thin line) which induces photo ionisation.

Roughly speaking, if the rate of intensity change during the short pulse is $x=dI/dt$ and the corresponding broadening of the photoelectron energy peak y, the duration of the short pulse may be extracted as y/x. The following is an application of this simple idea to the measurements of high harmonics femtosecond pulses.

High-harmonic generation (see P. Salières Chapter in this Book), one of the most utilized in applications among XUV secondary sources, produces extremely short pulses. For a time characterization of such pulses few methods are available. Cross-correlation techniques described above suffer from a number of limitations and drawbacks.

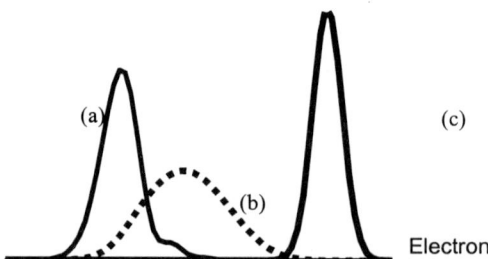

Fig. 7 Theoretical plots of one electron peak for three values of the delay between the two pulses. If the two pulses come peak-to-peak the shift is maximum and the extra broadening is very small because the IR intensity does not change too much (a). If the XUV pulse comes on the slope of the IR pulse the broadening is very large (b). If the delay is large then there neither shift nor broadening (c).

The disadvantages come from the fact that the probe pulses that can be used are longer than the XUV pulse and this imposes a serious limitation to the temporal duration that can be measured.

The new method is based on the change of the photoelectron energy spectrum resulting from ionisation with the high-frequency pulse due to the ponderomotive shift of the atomic ionisation potential induced by a more intense pulse of longer wavelength. We have demonstrated this method by measuring the duration of high-harmonics generated by an infrared laser beam in an Ar gas jet. The harmonics are sent on a second Ar jet to ionize atoms that are exposed at the same time to part of the fundamental beam.

The XUV photons are energetic enough to ionize the atoms in a one-photon process. The method requires temporal overlap between the two pulses and also a spatially homogeneous distribution of infrared-beam intensity over the size of the harmonic beam. The energy of the photoelectrons is dependent on the instantaneous value of the ionisation potential, which is changing rapidly in time according to the variation of the infrared-pulse intensity.

The infrared modifies in three ways in the electron energy spectrum. First, a down-shift of the mean energies due to the increased value of the ionisation potential as discussed above. Second, a broadening of the peaks due to the fact that the intensity of the dressing pulse is not constant during the presence of the XUV pulse. Finally, apparition of the "sidebands" as a consequence of multiphoton free-free transitions induced by the IR radiation. The characteristics of the shift, broadening and sideband amplitude depend on the delay of one pulse with respect to the other (Fig. 7). The case where the XUV arrives during the rising edge of the infrared pulse, where the intensity changes very fast, the broadening is large, while the shift is less (Fig. 7b). This region of quasi linear variation of the infrared intensity in time is the most suited for time measurements of short XUV pulses, because it implies a linear dependence on time for the U_P mapping of the pulse duration into broadening of electron peaks. Theoretically the method can have a very high resolution. An infrared Gaussian pulse of 50 fs duration and 10 TW/cm^2 peak intensity changes the ionisation potential by 20 meV per fs around the inflection point. An electron spectrometer having a resolution of 20 meV can measure broadening differences of this magnitude, implying a resolution of 1 fs for the time measurement. The natural width of the peaks and the presence of sidebands in the spectrum set a lower limit on the pulse duration for which this method can be applied to about ≈3 fs.

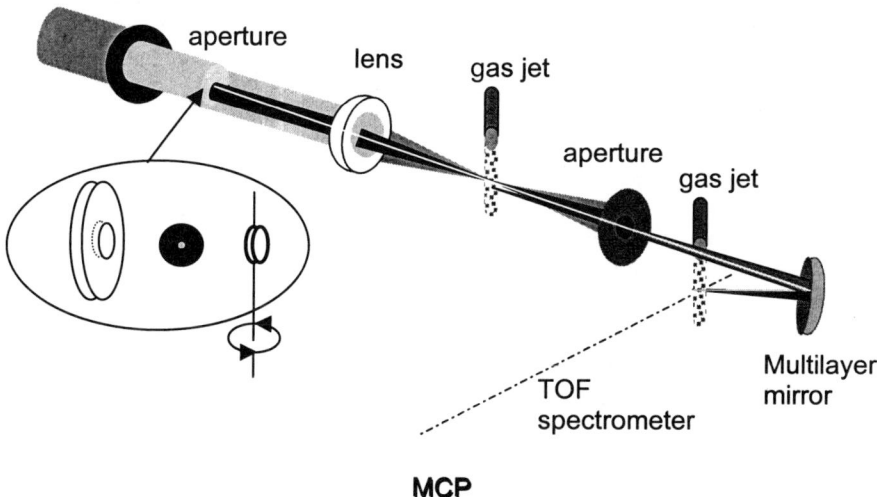

Fig. 8 Schematic of the experimental setup. The insert shows the ensemble of two plates of the same thickness and a sphere in which an octagonal tunnel has been drilled as described in the text. This ensemble realizes the spatial and temporal separation of the IR pulses. The time delay is adjustable by rotation f one plate.

For the demonstration, a Titanium:sapphire laser system has been used, producing pulses of 40 fs, at a central wavelength of 800 nm and a repetition rate of 1 kHz, with energies up to 8 mJ. The beam was divided by a mask and an iris in an annular part used for the generation of the harmonics and a small central part used as a dressing beam[4]. The mask was implemented as metal ball with a diameter of 8 mm, having a central octagonal hole of up to 2 mm diameter (Fig. 8). This hole was designed such that on rotation of the ball about a vertical axis all the sides of the hole move symmetrically towards the center thus keeping the shape the same while reducing the size of the hole. The energy in the annular beam can be changed by modifying the iris diameter, and it was usually kept at 3 mJ for the best harmonic signal. The masked beam was focused by a 1 m focal-distance lens on a continuous-flow Ar jet, and the position of the focus with respect to the gas jet was adjusted for a maximum harmonic yield. The harmonics propagate in the direction of the fundamental. Somewhere after the focus an image of the masking device forms, and the infrared recovers a sharply delimited annular shape. The outer ring is blocked there by a 2.5 mm pinhole, and only the generated harmonics and the central infrared beam are transmitted into a magnetic-bottle spectrometer chamber. At this position of the pinhole, the mask image is demagnified by a factor of 2. The two beams are refocused on a second Ar jet by a spherical off-axis tungsten-coated mirror with a focal length of 35 mm, and the photoelectrons are detected at the end of the time-of-flight (TOF) tube by microchannel plates. This set-up guaranties optimal alignment on the same axis and makes possible the determination of the absolute zero of the delay. Data acquisition has been done using an transient-digitizer board at a sampling rate of 500 MHz, which allowed the acquisition of a full time-of-flight trace at 980 shots per second. The diameters of the two beams on the spherical mirror are 2.5 mm for the XUV and 0.8 mm for the infrared. Consequently, the two FWHM in the focus are 0.88 µm (for the 15th harmonic) and respectively 40 µm for the infrared. This large ratio assures an homogeneous spatial distribution of the dressing-beam intensity over the focus of the XUV beam. The confocal parameter of the infrared beam is very long, and the diameter hardly changes on propagation in the volume seen by the electron spectrometer. The XUV beam, on the other hand, diverges much faster due to its small focal size and numerical aperture. At 65 µm from the XUV focus, the ratio of the two beam diameters is still large, ≈ 20. The two pulses could be delayed with respect to each other using two anti-reflection coated glass plates of 6 mm thickness, a small one for the central infrared beam and a large one with a hole in the center for the annular part of the infrared beam. The plates were cut from the same window, to ensure equal thickness, and thus knowledge about the absolute timing of the beams with respect to each other. The path of a beam through its plate depends on the incidence angle, so by tilting one plate or the other any of the pulses can be delayed. Close to normal incidence, the time delay has second-order dependence on the incidence angle, and very small delays can be obtained at easily measurable angles. For the dimensions of our plates, a delay of 1 fs could be obtained by a tilt angle of 1 degree. Another advantage of this way of delaying is that it does not affect the spatial overlap of the beams. The W-coated mirror is not selective and, hence, ionisation of argon by all odd harmonics above order 11 is possible. Fig. 9a shows photoelectron energy spectra taken at different intensities of the infrared dressing beam and at zero delay between the two pulses. Harmonics 13, 15 and 17 as well as sidebands one IR photon apart from those are observed at non-zero intensity. In Fig.9b is shown spectra taken at constant intensity of the infrared beam but at different time-delays between the two pulses. The maximum shift (0.7 eV) corresponds to a peak intensity of ≈ 13 TW/cm^2 in the infrared beam focus. The peak asymmetry present at small delays comes from the non-linearity of the temporal profile of I_{IR} around the top.

The spectra can be modeled by a single bound level (the initial state) with a binding energy $-I_0 + \alpha I_{IR}(t)$ dependent on the instantaneous infrared intensity coupled by a XUV pulse of electric field $E_X(t)$ to a flat continuum. In this case the final population of the continuum level with energy E is proportional to :

$$\left| \int_{-\infty}^{\infty} e^{[-i(\lambda\omega_X - I_0 - E)t + i\alpha \int_0^t I_{IR}(t')dt']} E_X(t) dt \right|^2 \qquad (15)$$

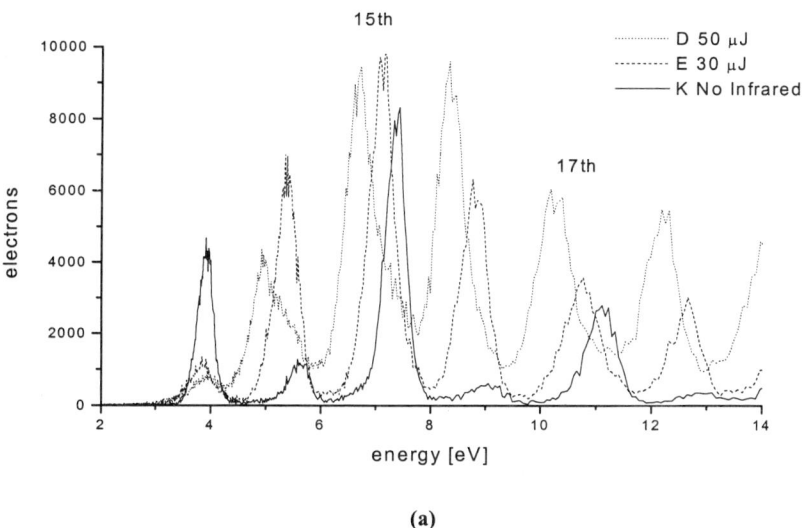

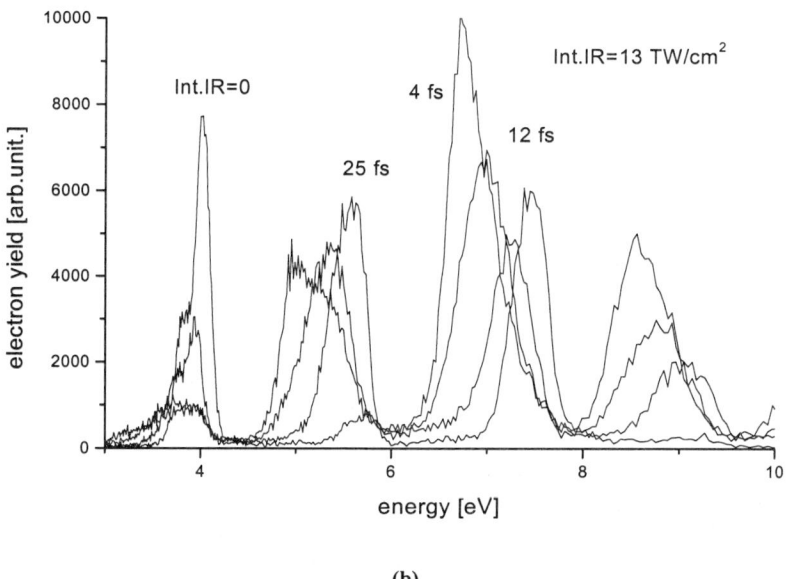

Fig. 9 (a) Electron energy spectra from different intensities of the IR pulse. (b) Electron energy spectra for different time delays between the two pulses.

By fitting the experimental shifts and the broadenings of the 15th-harmonic peak versus delay, a duration of 10±2 fs (FWHM) has been determined for this harmonic.

In conclusion, the ponderomotive streaking of the ionisation potential during a photo ionisation process by a short XUV pulse was proved to be an efficient method for measuring duration down to 10 fs with a lower limit for the current conditions (λ_1=800 nm, λ_2=53 nm) of about 3 fs. One of the limits is of course the IR radiation period (2.5 fs) since it takes at least that time to define the quiver motion and the ponderomotive energy. Another limit is imposed by the requirement that the broadening must be less than the separation of the main peak from one of the sidebands, which is the IR photon energy, i.e. again the IR period.

Methods based on two-color ATI have been proposed for measuring attosecond pulses, much shorter than the IR period. The principle is this time to measure the change in the electron momentum due to the initial phase at which it is born in the field.

Other processes related to two-color ATI

The extra degrees of freedom provided by the second color (polarization, intensity etc) open the possibility of exploring many new processes intimately related to two-color ATI. The following are currently under studies, essentially for their academic interest or potential applications.

- **Angular distributions of photo ionisation in presence of a strong IR field**

Just as the photoelectron energy spectra are modified in a time-dependent ionisation potential (see previous section) the angular distributions of the photoelectrons ejected by a XUV radiation is changed by the presence of relatively strong electromagnetic field (Fig. 1d). Angular distributions from single-photon ionisation are described by a Legendre polynomial and a single parameter β. The differential cross-section for a linearly polarized field and an electron ejected in the direction q with respect to the direction of polarization writes:

$$\frac{d\sigma}{d\Omega} = \frac{\sigma_0}{4\pi}(1+\beta\frac{3Cos^2\theta-1}{2}) \qquad (16)$$

In the presence of the IR field transitions to the same final state imply a minimum of three photons (see Fig. 1d) and therefore a Legendre polynomial of order 4.

It should be relatively easy to measure such distributions from photo ionisation of argon for instance by a combination of high harmonics and fundamental with the setup described above, replacing the magnetic bottle by a straight time-of-flight and rotating the linear polarization or, better, using a toroidal spectrometer and a position-sensitive detector.

- **Circular dichroism in two-color ATI**

Optical dichroism may be observed in molecules with no center of symmetry. It is at first glance surprising that a centro-symmetric system displays dichroism. It is known on the other hand that single-photon double ionisation shows circular dichroism or single-color two-photon ionisation shows elliptical dichroism which, however vanishes for circular polarization. Perturbation theory however does predict that circular dichroism should be observable in two-color, two-photon ionisation of atoms[7]. The effect is very small though as soon as the electron initial energy is not very close to zero.

- **Two-photon double ionisation**

If the XUV photon energy is above the second ionisation threshold single-photon double ionisation takes place. The resulting photoelectron energy spectrum is a continuum, the two ejected electron sharing the excess energy above the threshold:

$$E_{k_1} + E_{k_2} = \hbar\omega_X - I_0' \qquad (17)$$

In presence of an intense IR field two-photon double ionisation (a generalization of two-color ATI) will occur. By detecting the two electron in coincidence it may be possible to observe for the first time such transitions which can be further extended to multiple IR photon absorption and choosing an XUV photon energy below the double ionisation threshold. Such experiments are difficult though due to the relatively low repetition rate of the current lasers (in general 1 kHz) compared to the 10 or 100 MHz of synchrotron sources.

CONCLUSION

Two-color, XUV-IR ATI has been studied in a number of situations both theoretically and experimentally for its academic interest and applications to the metrology of ultrashort XUV pulses. Laser-assisted Auger decay or photoeffect have been investigated and used for cross-correlation measurements of fs XUV pulses. Beyond the limitations of cross-correlation methods, a new method based on the streaking of the ionisation potential could be single-shot with potentially 3fs resolution. Methods based on the spectrometry of photoelectron produced in a strong IR pulse will be very important in the near future for the characterization of attosecond pulse trains or single pulses for example by extension of methods like SPIDER to the XUV[8].

PART TWO: SINGLE-COLOR ABOVE THRESHOLD IONISATION AND DOUBLE IONISATION

INTRODUCTION

High-Order Above-Threshold Ionization, which is the subject of the present Section, is a process in which Simpleman's theory provides much insight (Fig. 10). Since it occurs at high intensity, the Keldysh parameter (Eq. (8)) is close or less than one which means that the ionisation itself occurs in the tunneling regime. It is therefore reasonable to separate the production of the free electron with a very small initial velocity from its subsequent acceleration in the electromagnetic field and to treat the second part classically. This yields immediately the classical limit of $2U_P$ for the maximum average kinetic energy of the electron.

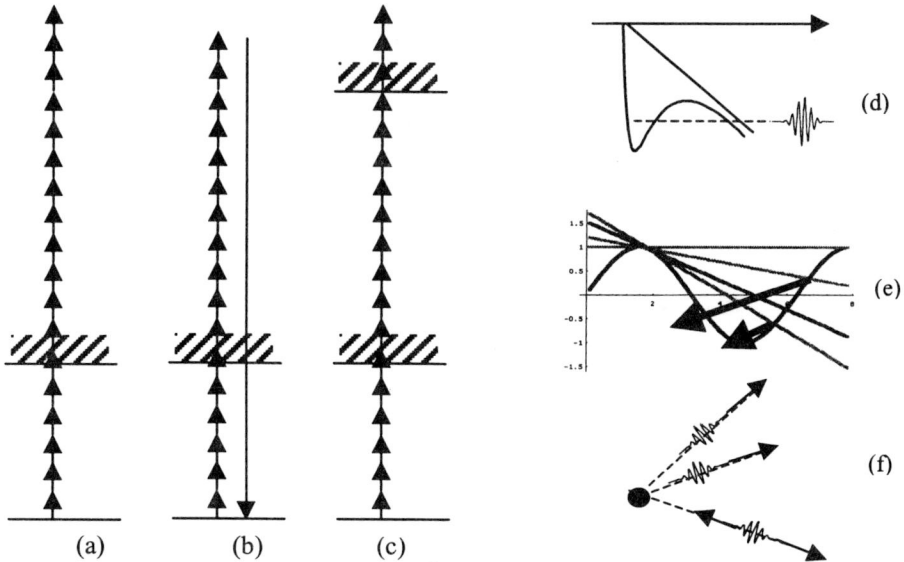

Fig. 10 High-Order ATI (a), High-Harmonic Generation (b), Double ionisation (c) are respectively interpreted in terms of a classical electron tunneling into the continuum at zero velocity (d); accelerates in the field and backscatters(e), recombines with its parent ion (e) or ejects a second electron in e-2e collision (f).

Experiments using (relatively) high-repetition rate lasers by considerably increasing the dynamic range of ATI measurements have revealed that ATI spectra were in fact extending up to 10 U_P, much beyond this limit (the ATI "plateau") (Fig. 10a). At the same time it was found in numerical simulations and experimentally that the High-Harmonic generation cutoff was determined by the intensity and given by approximately 3.2 U_P (Fig.10b). Non sequential double ionisation by long wavelength, intense laser pulses was also carefully investigated during the last few years (Fig. 10c). Corkum[2] proposed a unified approach to the physics of these three phenomena by considering that a bound electron tunnels into the continuum (Fig. 10d), is accelerated in the field and re-encounters its parent ion during a collision where it either elastically scatters or recombines emitting the harmonic photon or extracts a second electron (Fig.10e,f). Several other chapters in this book (see P. Salières, G. Paulus, H. Horst and F. Faisal contributions among others) are dealing with different aspects of high-order ATI, harmonic generation and double ionisation. To avoid repetitions the next Section will contain only a very brief account of the Simpleman's theory of the ATI plateau while in the following Section (the most important one) I summarize a recent work on channel closure. I give in the last Section a brief summary of the history of double ionisation which may be taken as an introduction to the most recent developments to be found in H. Horst Chapter in this Book.

HIGH-ORDER SINGLE COLOR ATI

As in the two-color case, the perturbation theory breaks down as soon as the number of ATI peaks becomes larger than one or, alternatively, for intensities above 10^{13} W.cm^{-2}. Other approaches (Keldysh-Faisal-Reiss (KFR), essential states, singular matrix elements) are reviewed elsewhere (see G. Paulus Chapter and M. Fedorov in the Bibliography). Here I only want to apply the classical approach outlined in Part One to the case where the initial phase is close to peak of the field and to include the rescattering of the electron on its parent ion.

If the frequency is low enough or the field strength high enough the ionization rate may be approximated by the static limit tunneling rate which, in the case of hydrogen atom writes (Landau and Lifshitz) in atomic units (E electric field):

$$w = \left(\frac{4}{E}\right) \exp\left[-\frac{2}{3E}\right] \qquad (18)$$

easily generalized to another potential E_i (in atomic units) (Corkum et al):

$$w = 4(2E_i)^{5/2} \frac{1}{E} \exp\left[-\frac{2}{3}(2E_i)^{3/2} \frac{1}{E}\right] \qquad (19)$$

Because of the exponential factor it is clear that tunneling will take place close to the peaks of the field during the half cycle when it lowers the potential barrier:

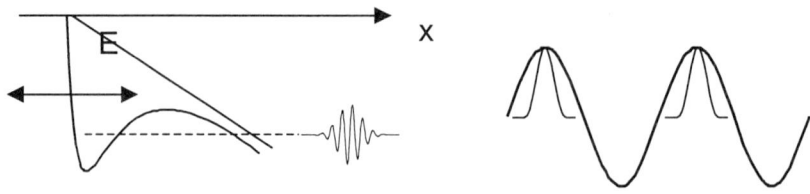

Fig. 11 Tunneling in a 1D potential and time dependence of the rate in the optical cycle.

The electron wavepacket are emitted in periodic bursts whose Fourier transform are periodic in frequency with a period ω (the ATI spectrum). The KFR theory allows to calculate the amplitude of the successive peaks through the generalized Bessel functions (Eq. 9). The classical calculation of the average kinetic energy in the field:

$$\left\langle \frac{1}{2}\dot{x}^2 \right\rangle = 2U_P(\frac{1}{2} + \cos^2 \omega t_0) \tag{20}$$

predicts a maximum energy of the ATI spectrum at $3U_P$ including one U_P of oscillating energy (first term on the right hand side of Eq. (20)) which normally lost in short pulses when the electric field turns off before the electron leaves the focal spot. (In long pulses all or part of this U_P may be conserved through the work of the gradient force as the electron traverses the field gradient). Coupled to the rate from Eq.(19) which gives the distribution of initial phases (quite different from the distribution from Auger decay discussed in Part One), Eq.(20) predicts a spectrum envelope with an exponential decay and a maximum at 2 U_P under the usual conditions.

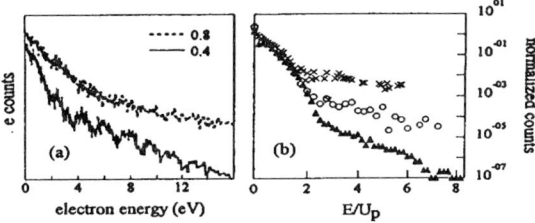

Fig. 12 ATI spectra from helium at 800 nm. The right box shows the spectra for different intensities on an energy scale in U_P units. Most of the electrons have an energy below 2 U_P. The tail beyond the classical limit is the plateau. (from Walker et al., 1994).

Fig. (10e) represents classical electron trajectories in the frame oscillating at the optical frequency (some time called the Kramers-Henneberger frame) in which the electron is either at rest or travels in straight lines. The nucleus trajectory in this frame is the sine

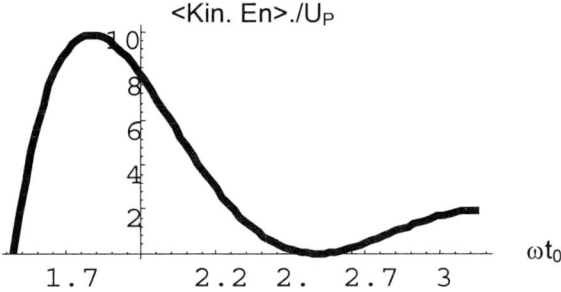

Fig. 13 Classical average kinetic energy after a backscattering on the nucleus as a function of the initial phase

curve. The electron trajectory is tangent to the sine curve at the instant when it appears in the field through the tunneling process.

The possible trajectories do or do not encounter the sine curve depending on the initial slope or the initial phase. With the phase chosen in Fig. (10e), the electron "born" before $\pi/2$ do not come back to the nucleus. If the electron re-encounters its parent nucleus during its oscillating motion it may backscatter and its subsequent average kinetic energy may reach $10U_P$ as shown in Fig. 13. This is the origin of the limit in the extension of the ATI plateau. One example of ATI spectrum in helium as well as some angular distributions are shown in Fig.14. Several other characteristics of the ATI spectra and angular distributions are closely related to this rescattering event (like the intensity-dependent rings) will not be commented here. The reader is referred to the literature for more details (See G. Paulus Chapter in this Book).

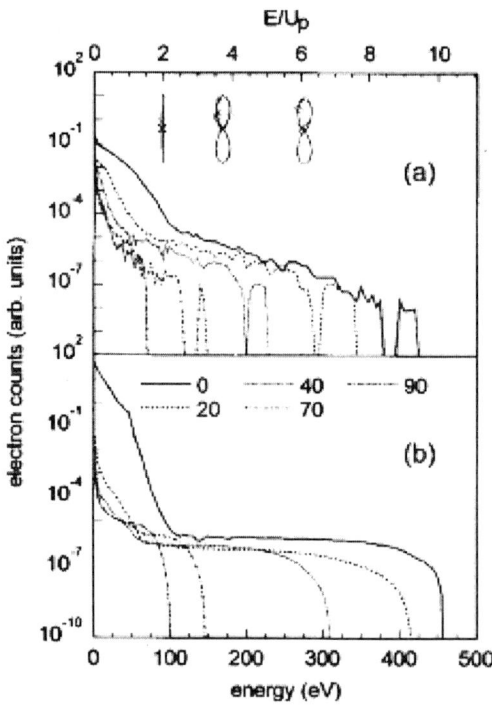

Fig. 14 ATI spectrum from helium at 800 nm (a) and theories (b). For details the reader is referred to ref. [9].

Channel closure and ATI contrast[10]

Laser pulses in the visible or infra-red frequency range intense enough to induce multiphoton ionization in rare gases, also strongly distorts the atomic structure. As already discussed in Part One, the Rydberg states are up-shifted in energy due to AC-Stark effect. The amount of energy by which the states are displaced converges to the so-called ponderomotive energy U_P in the limit of high lying Rydberg states. Typical field parameters correspond to shifts more than *three times* the photon energy. It follows that when the intensity is varied over a large enough range, the minimum number of photons required to ionize the atom may increase by several units. This effect is known as Channel Closure (CC).

Besides, the AC-Stark shifts necessarily induce *transient resonances* during the light pulse when the energy of shifted states matches the energy of an integer number of photons absorbed from the ground state. In the dressed atom picture, such resonances (known as Freeman resonances Freeman et al., 1987) correspond to avoided crossings of the ground state dressed by an integer number of photons and a Rydberg state. The resonant states are populated through Landau-Zener transitions during the time-dependent level crossing and are subsequently ionized during the rest of the pulse. Such resonances appear as substructures of the ATI peaks obtained with short pulses (1 ps or below)

As discussed in the previous section the ATI orders up to energies of 10 U_P are explained by the backscattering of the electron on its parent ion. However, recently that, it was established in the case of argon that the plateau was arising from a resonant (i.e; quantum) process, albeit related to the electron rescattering[10]. The nature of these resonances though is still under debate.

Alternatively, calculations using a delta-potential, for which no Freeman resonance exist, can reproduce the envelop enhancement of the plateau (see the Chapter by Paulus in this Book). The work reported here contributes to this debate by investigating an aspect which has not been explicitely discussed so far in this context.

We have recorded [9] the angle-integrated ATI spectra in the ionization of argon by 800 nm, 40 fs pulses over a large energy range including the so-called plateaus. The spectra show a conspicuous drop in the peak contrast around an intensity of 6×10^{13} W.cm^{-2}. Numerical simulations using Muller's model potential for argon and a B-spline package for solving TDSE provides insight into the correlation between the contrast changes and channel closure, the simultaneous occurrence of two series of resonances with opposite parity as well as the nature of the plateau resonances.

The experiment reduces to measuring the electron energy spectra for different intensities. Details may be found elsewhere.

Extensive simulations of the experimental ATI spectra have been carried out. The calculation includes the atomic structure of argon, the temporal dependence of the pulsed laser beam and, most important, the spatial dependence which is accounted for by averaging over the intensity distribution inside the interaction volume. Practically, we solve the Time-Dependent Schrödinger Equation (TDSE) for the interaction of an argon atom with a time-dependent laser field having a given peak intensity, yielding what is referred to as single-intensity spectra. We then build the electron yield by summing each single-intensity spectra, each one weighted by a factor equal to the interaction volume corresponding to a particular intensity. One should, in principle, consider the multi-electron character of the system when investigating the interaction with external fields. It is well established however that high-order multiphoton processes are well described in the approximation of a single active electron.

Figures 15 (a)-(c) display both the experimental and simulated spectra for three intensities. The spectra show the typical features of ATI: a succession of peaks (separated by a photon energy) whose amplitudes first decrease for low energy electrons and then level off around E_k=15 eV thus giving rise to the plateau. An interesting and remarkable feature is clearly observable on Fig.11a-c. As the intensity is increaseds above 5×10^{13} W/cm^2, the overall peaked structure of the spectrum is blurred around 6×10^{13} W/cm^2 before reappearing with even higher dynamics at 7×10^{13} W/cm^2. This is more quantitatively analyzed if one defines the contrast of the spectra on the energy scale of a photon as:

$$m(E) = \frac{f_{\max}(E) - f_{\min}(E)}{f_{\max}(E) + f_{\min}(E)} \tag{21}$$

where $f_{\max}(E) = \max(f(E'))$ for $E' \in [E - \hbar\omega/2; E + \hbar\omega/2]$ and $f(E)$ is the number of electrons collected with a kinetic energy E ($f_{\min}(E)$ is defined similarly). In Fig.12 we have plotted the contrast as a function of the field peak intensity for 4 sampled electron energies located near the ionization threshold (E_k=5.2 eV and E_k=7.6 eV), at the beginning of the first plateau (E_k =14.4 eV) and in the middle of the plateau (E_k =21.2 eV). The graph clearly confirms for all electron energy cited above a large reduction in the contrast at around $I = 6 \times 10^{13}$ W/cm² followed by a revival.

The sharp minimum observed in the contrast is the signature of a channel closure. Because of the ponderomotive shift of the threshold, above an intensity of $I = 2 \times 10^{13}$ W/cm² a minimum of 12 photons are required to ionize argon. The 12-photon ionization channel closes at $I = 4.6 \times 10^{13}$ W/cm² and the 13-photon channel at $I = 7.2 \times 10^{13}$ W/cm². These intensities agree only within a factor 2 with the estimated absolute values of experimental ones based on the pulse energy; the relative values though are in excellent agreement. The minimum in the contrast however is not observed (see Fig.12 at the CC intensity (neither 4.6 nor $I = 7.2 \times 10^{13}$ W/cm²) but midway in between. As shown below this is due to the resonant character of the ATI spectra. The spectra are build up from electrons collected from an interaction volume i.e. averaged over a large intensity distribution: therefore they should be broad and shift ponderomotively when the peak intensity changes. As they do not shift and display narrow substructures they clearly signal resonant multiphoton processes occurring always at the same intensity. This can be checked in details in Fig.13 (upper graphs) were we have plotted the computed spectra in the direction of the polarization (θ= 0°) for low electron energy (left graph) and for electrons in the first plateau (right graph). Because most of the ionization occurs along the laser polarization, angle integrated measurements are close to spectra computed at $\theta = 0$ °. One can see that as the intensity is increased from 5×10^{13} W/cm² to 7×10^{13} W/cm², narrow resonances are developing on the side of the main peaks. Eventually, for an intensity around 6×10^{13} W/cm² (thick solid curve in Fig. 13) the new comb of resonances are competing with the main peaks thus reducing the contrast. Further increase of the intensity leads to the domination of only one of the substructures. The channel closure scenario is validated by the examination of the ATI peaks angular distribution. As it is well known, the dipole selection rules involved in multiphoton processes sets the final state parity. (To our knowledge this simple property of ATI has never been checked experimentally). Thus the probability of observing an ATI peak electron at 90 degrees of the laser polarization is a signature of the parity of the resonance. We have indeed verified that the ATI peaks change their parity when the intensity is raised from 5×10^{13} W/cm² to 7×10^{13} W/cm².

Although the resonant behavior is seen in the entire spectrum, the nature of the resonances is probably different. Freeman Rydberg resonances, associated to the substructures observed in the spectra for low electron energy below 12 eV, cannot explain the triplet structure of the plateau. Let us first discuss the low energy part of the spectrum (left upper graph in Fig. 13). There the effect can be traced back to the "usual" Freeman resonances combined to the integration over the intensity distribution inside the interaction volume. For example, consider electrons from the S=4 ATI peak (6.2-7.75 eV). For a spatio-temporal pulse with a peak intensity 5×10^{13} W/cm², we have checked by examining single-intensity spectra that the main contribution to the intensity averaged spectrum comes from electrons emitted from the region in the interaction volume where the intensity at the maximum of the pulse is around 4×10^{13} W/cm² (i.e. below the channel closure intensity). At this intensity, the spectra are dominated by resonances which are necessarily of even-parity at the 11-photon level where they are the most likely to occur at this intensity. Now, for peak intensities of about 5.5×10^{13} W/cm² a new set of resonances clearly emerges. The simplest explanation is that because the 12th channel is now closed, states of odd parity may become resonant through 12-photon absorption. Since the kinetic energies of the corresponding photoelectrons are apparently different, a new comb of ATI peaks, shifted with respect to the previous one, arises, blurring the contrast of the spectra. As the intensity is further increased, eventually the odd-parity resonances dominate and the contrast increases again, albeit with a global shift of the ATI comb of almost the photon energy.

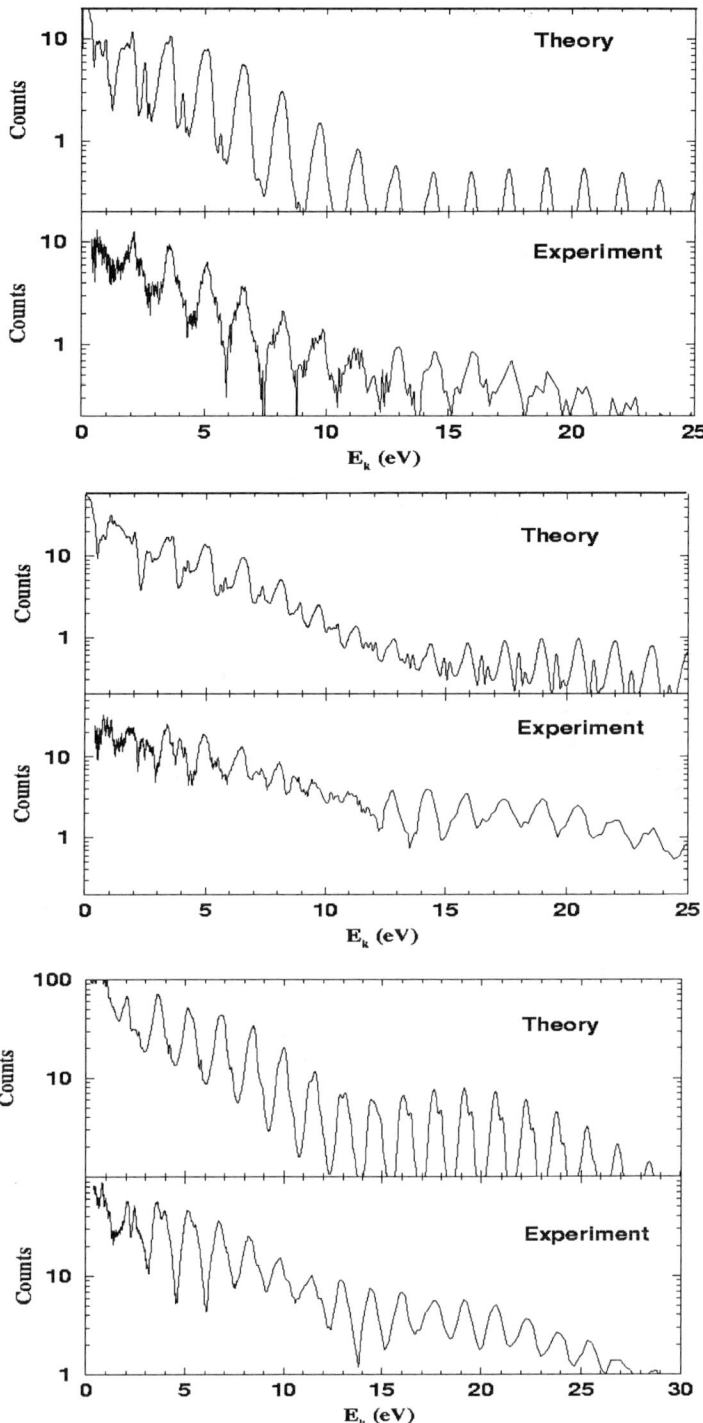

Fig. 15 ATI spectra for three intensities 5×10^{13}, 6×10^{13}, 7×10^{13} W.cm^{-2} from top to bottom. Each box, theory and experiment as indicated.

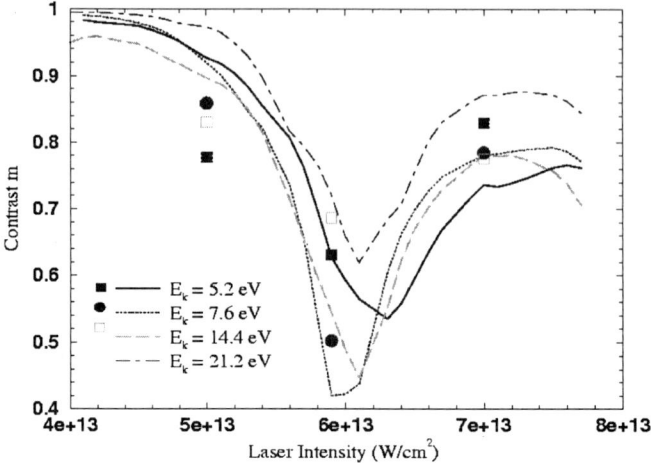

Fig. 16 Contrast (as defined by Eq. (20)) vs intensity for different electron kinetic energies. Solid lines: simulation, Marks: experiment.

At a first glance, a very similar behavior is observed for the substructures in the plateau (right upper graph in Fig.13 although the number of resonances per peak is less. The minimum of contrast is also observed at 6×10^{13} W/cm² and is also clearly due to a new series of substructures growing after the channel closure. However, significant differences exist between the two regions of the ATI spectra. The main one appears in the spectra at $\theta = 90$ (right lower graph in Fig.13): the plateau peaks are free of substructures

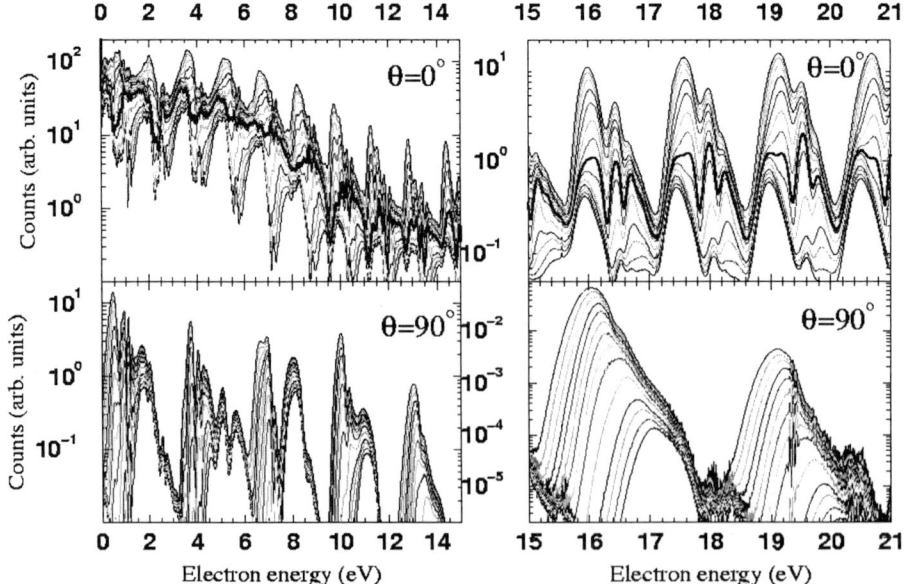

Fig. 17 Simulated electron energy spectra for electron ejected in the direction of the laser polarization ($\theta=0$) and perpendicular to it ($\theta=90$) and in two kinetic energy ranges (left and right). See text for discussion.

and very weak with respect to the θ= 0 ones. Actually the ratio between the two is of the order of 10^{-3} while it is about 10^{-1} in the low energy part of the spectra. Note that because the ATI peaks are successively odd and even as the order S is incremented, only one ATI peak out of two appears in a spectrum of electrons ejected at 90°. If the plateau structures were to be interpreted as the same Rydberg resonances, they would also appear at 90° as it is the case for low electron energy (left lower graph in Fig.16). Instead, the ATI peaks shift ponderomotively towards lower energy as the intensity is increased indicating that the process ejecting electrons at this energy and angle is typically non-resonant. On the other hand, electrons emitted at the same energy but in the direction of the polarization do reveal a resonant behavior. It is now well adopted that plateau electrons (and actually electrons emitted at energies as high as $10U_P$) are backscatterred electrons that have been driven back and forth around the nucleus by the field. The multiplet structure thus observed in the direction of the polarization could be interpreted as interferences in the backscatterring process as proposed by Kopold and Becker [11] or as interferences between wave-packets as proposed by Muller[12].

In conclusion the loss of contrast in argon ATI spectra in a small intensity range around 6×10^{13} W/cm^2 close to the intensity at which the 12th ionisation channel closes is observed in the whole spectrum including the plateaus. The channel closure and the related possibility of simultaneous even- and odd-parity Freeman resonances with comparable weight due to the averaging over the intensity distribution inside the interaction volume explain the effect in the low energy part of the spectrum. The sub-peaks ATI structure observed in the plateau region [13] (between 13 and 25 eV) undergoes qualitatively similar changes. However, according to the present calculation, the spectra computed for θ = 90° are free of such structures which therefore must have a different origin although clearly related to the channel closure too. This view seems to be in agreement with the conclusions of a very recent work of another group (see G. Paulus Chapter in this book).

NON-SEQUENTIAL DOUBLE IONISATION: BRIEF HISTORY AND RECENT DEVELOPMENTS

The observation of multiphoton double ionisation of alkaline earth atoms goes back to the mid-seventies and was first reported by an Ukrainian group[13]. The rate was obviously too large when compared to a typical single ionization rate with the same number of photons. This discrepancy was solved by photoelectron spectroscopy which revealed that alkaline earth atom, due to the structure of their energy level spectrum, have a strong tendency to absorb a few photons above the first threshold and ionize leaving the ion in an excited state. The ion subsequently ionizes in a generally resonant multiphoton process. The existence of a multiphoton transition directly coupling the ground state to the double continuum, suspected to be the explanation of the high rate was therefore ruled out. However the problem was posed again a few years later by the observation of such an enhanced rate in double ionization of rare gas atoms by a Nd glass laser[14]. However very little progress was achieved beyond a general belief that a process involving a correlation between the two outgoing electrons was taking place. Helium was the benchmark in this problem and an experimental breakthrough occurred with the advent of high repetition rate high intensity lasers. With the dynamic range of the experiment jumping to twelve orders of magnitude, the total rate for singly and doubly charged ions productions from helium irradiated by 800nl femtosecond pulses became accurate enough for a solid comparison with theories[15]. Many others and qualitatively similar data have been since then obtained in multiple ionization of rare gases and other atoms. One of the most simple interpretation of the double ionization process came again from the quasi-classical picture illustrated in Fig.(10f) [2]. If the instantaneous (NOT the average) kinetic energy is larger that the ionization energy of the second electron, the latter may be extracted in an e-2e collision. Quantitatively, however the rate and in particular the ratio He^{++}/H^+ as a fuction of intensity was net well reproduced by this simple approach. It appears though that taking into account *multiple collisions* and the refocusing of the wavepacket by the Coulomb potential yields an excellent agreement with the theory[16].

These precise data on total ionisation rates was recently completed by measurements of the ion recoil momenta. Both the latter and an account of one of the most successful theories are reported in this Book:(see the Chapter by H. Horst for the experiments and the Chapter by F. H. Faisal).

Coincidence measurements of the double ionization electron energy spectrum currently in progress at Brookhaven [17] will complete the already available data.

CONLUSION

ATI spectra in the tunneling limit extend in first approximation to 2 U_P. A plateau extending to 10 U_P is related to the rescattering of the wavepacket on the parent ion. This rescattering also gives rise to intensity-dependent rings in the angular distributions. Classical ATI, i.e. the classical average kinetic energy of a free electron the electromagnetic field accounts for onset and extension of ATI plateau. Heights are well predicted by quantum calculations [12]. The Quantum nature of ATI is also reflected in Channel Closure and the opening of a new set of Freeman resonances with different parity as seen through ATI spectra contrast. The contrast drop in the plateau seems to be correlated with another type of resonance (see also G. Paulus Chapter in this Book).

The Non-sequential process with a much higher rate in a certain intensity range is a general feature of double or multiple ionization. Most data including the ion momenta distribution (see H. Horst Chapter in this Book) are well reproduced by the Many-Body S-Matrix calculations of Becker and Faisal (with the exception of neon) and short wavelength data (see Faisal Chapter in this Book). Rescattering e-2e model does account easily for many aspects of NSDI but fails quantitatively for the prediction of the onset on DI. New developments including multiple rescattering and Coulomb refocusing of the wavepacket improve the comparison. Recent experiments (COLTRIMS and electron-ion coincidence measurement) should rapidly put a final touch to this 10-year old puzzle. The coupling of lasers with XUV sources like HHG will provide tools to investigate the problem of (moderately) multiphoton double ionization.

ACKNOWLEDGMENTS
to the people who have done the work: the Brookhaven National Laboratory, *Upton NY*, L F DiMauro, Baorui Yang, B Walker, B Sheehy, R Lafon and J Chalupka ; to my coworkers from the SPAM, *Saclay,* E. Mével (now professor at the Bordeaux University), P Breger, P M Paul, M Cheret, O. Gobert, P. Meynadier and M. Perdrix; to the LOA team LOA *Palaiseau,* A. Bouhal (now working at the Ecole Centrale Paris*),* J P Chambaret, C Leblanc, G Grillon; the FOM team in *Amsterdam* R. Constantinescu (now working at Philips Co), J. Schins (now professor at the Twente university), H. G. Muller also professor at the Free University in Amsterdam and the famous Prof "HG" on TV, E. Toma; the group of Prof. Maquet at Pierre and Marie Curie, *Paris,* V Veniard, R Taieb and N Manakov from the Voronezh University in Russia. We benefited from fruitful discussions with G. Paulus and W. Becker whom we thank for communication of unpublished results and a submitted preprint.
The EEC TMR program provided partial support (under Contract # FMRX-CT96-0080). Louis DiMauro and myself acknowledge support for a NATO collaborative Research Grant # CRG.910678
Support from the ESF FEMTO program is also gratefully acknowledged.

REFERENCES

[1] M. V. Fedorov *Atomic and Free Electrons in a strong Laser Field* World Scientific, Singapore 1997.
[2] H. B. van Linden van den Heuvell and H. G. Muller in Multiphoton Processes, Studies in Modern Optics, No. 8, 25 Cambridge (1988); P. Corkum Phys. Rev. Lett. **71**, 1994 (1993); T. Gallagher Phys. Rev. Lett. **61**, 2304.(1988).
[3] J. Schins et al. Phys. Rev. Lett.**73**, 2180 (1994)
[4] A. Bouhal et al. J. Opt. Soc. Am. B **4**, 950 (1997).
[5] J. Schins et al. Phys. Rev. A **52**, 1272 (1995).
[6] E. Toma, H. G. Muller, PM. Paul, P. Breger, M. Cheret, P. Agostini, G. Chériaux and G. Mulot Phys. Rev. A **62**, 061801-1,. 2000.
[7] R. Taieb et al. Phys. Rev. A **62**, 13402 (2000).
[8] I. Walmsley Gordron Conference on Multiphoton Processes June 2000, Tilton, NH unpublished.
[9] B. Sheehy et al. Phys. Rev. A **83**, 5270 (1999).
[10] E. Cormier, D. Garzella, P. Breger, P. Agostini, G. Chériaux, C. Leblanc to be published in J. Phys. B At. Mol. Opt. Phys. (2000).
[11] H. G. Muller Phys. Rev. A **60**, 1341 (1999).
[12] R. Kopold and W. Becker J. Phys. B. At. Mol. Opt. Phys. **32**, L419 (1999).
[13] H. G. Muller and F. C. Kooiman Phys. Rev. Lett. **81**, 1207 (1998).
[14] V. V. Suran and I. P. Zapesnoshnii, JETP Lett (Engl. Transl.) **1**, 973 (1975).
[15] A. L'Huillier et al. Phys. Rev. A **27**, 2503 (1983)
[16] B. Walker et al. Phys. Rev. Lett. **73**, 1227 (1994).
[17] M. Yu. Ivanov et al. LPhys2000 July 2000 Bordeaux
[18] R. Lafon et al submitted to Phys. Rev. Lett. Nov. 2000.

HIGH-ORDER HARMONIC GENERATION

Pascal Salières

CEA/DSM/DRECAM, Service de Photons, Atomes et Molécules
Centre d'Etudes de Saclay
91191 Gif-sur-Yvette, France
psalieres@cea.fr

INTRODUCTION

Thanks to the invention of the laser in 1960, intensities never reached before became available allowing the observation of a number of new phenomena. Among them, the discovery in 1961 of optical harmonic generation by Franken et al. [1] marked the birth of nonlinear optics, one of the richest field in optics that has found applications in nearly all areas of science. By generating the second harmonic of a Ruby laser in a quartz crystal, they demonstrated that atoms could absorb simultaneously two optical photons and emit a photon of doubled energy. This opened wide perspectives for the extension of *coherent* light sources to shorter wavelengths. Indeed, harmonic orders from 3 to 11 could be generated in the following years, but the quickly decreasing conversion efficiency with the order prevented the observation of higher orders. The intuitive explanation is that in this perturbative (weak-field) regime, the probability of absorbing simultaneously n photons decreases quickly when n increases, so that the extension of the process to shorter wavelengths seemed hopeless.

In 1987 was observed for the first time the phenomenon of *high-order harmonic generation* in two laboratories, Chicago [2] and Saclay [3]. The context was the study of the behavior of atoms submitted to strong laser fields, brought about by the progress in short-pulse intense laser technology, with focused intensities of 10^{13}-10^{15} W/cm^2 and pulse durations in the picosecond range. When focusing such a laser pulse into a rare gas jet, the spectrum of the radiation emitted on axis presents a characteristic behavior: after the expected rapid decrease for the first orders, the harmonic conversion efficiency remains constant up to relatively high orders (region of the spectrum called "plateau") and then drops in the "cutoff" region (Figure 1).

The generation of the 25th harmonic, for example, becomes as probable as fifth-order harmonic generation, which means that the lowest-order perturbation theory, so successful in describing the (nonresonant) nonlinear optical processes till then, does not apply any more. Indeed, the laser field is now so strong that it is no longer a weak perturbation of the atomic system. Note that only odd harmonic orders are generated in atomic gases due to the symmetry of the interaction.

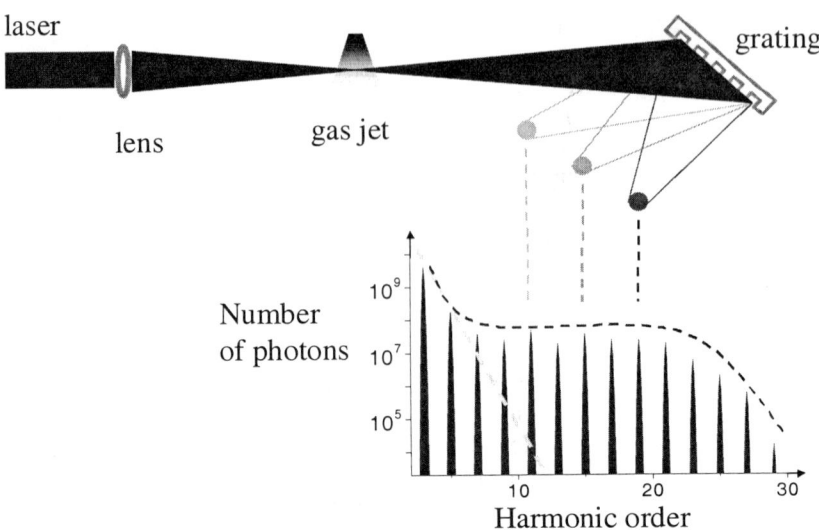

Figure 1. Experimental setup for the generation and spectral analysis of harmonics. Typical spectrum obtained in the strong-field regime. The heavy dashed line indicates the shape of the spectrum in the perturbative (weak-field) regime.

From that discovery, numerous studies have been carried out in order to understand the physics of the process and to extend the plateau to very high orders. Thanks to the progress in laser technology, and in particular to the pulse shortening in the femtosecond range, the plateau extension has been continuously increased, without apparently reaching the limit of the process. In 1992, the 135th harmonic of a 1 ps Neodymium:Glass laser was generated in neon by Anne L'Huillier et al. [4] in Saclay, and the 109th harmonic of a 125 fs Titanium:Sapphire (Ti:S) laser was observed by Macklin et al. [5] in Stanford. In 1997, by focusing a 27 fs Ti:S laser pulse in a helium jet, the group in Michigan [6] could observe discrete harmonic peaks up to the 221st order, corresponding to a wavelength of 3.6 nm, and unresolved radiation down to 2.7 nm. Coherent radiation in the water window (below the carbon K-edge at 4.4 nm) was reported simultaneously by the group in Vienna [7] using a 5 fs Ti:S laser. In 1998, the same group [8] observed 500 eV photons (equivalent to 323rd harmonic at a wavelength of 2.5 nm). In parallel to these studies, many experiments have investigated the spatial and temporal characteristics of the harmonic emission, revealing the unique properties of this XUV source: good coherence, ultrashort (femtosecond) pulse duration, high brightness, high repetition rate, tunability, compactness... During the last years, these properties have been used in a growing number of applications, ranging from atomic and molecular spectroscopy to solid-state and plasma physics.

High-order harmonic generation is thus an interesting phenomenon for two reasons. From the fundamental point-of-view, this spectacular process is a probe of the behavior of a strongly perturbed atom. From an applied point-of-view, this XUV source presents unique properties of coherence and ultrashort duration, giving rise to a number of new and interesting applications.

We will first describe the current understanding of the physical process leading to the harmonic emission. Then, we will review the properties of this XUV radiation, and finally give some examples of applications.

PHYSICAL PROCESS OF THE HARMONIC EMISSION

There are two aspects, microscopic and macroscopic, in harmonic generation. First, the source of the (microscopic) emission is the nonlinear response of the atoms submitted to the intense laser field. Second, the macroscopic harmonic field measured in the experiments is the *coherent* superposition of the fields emitted by all the atoms of the generating medium. Thus the theoretical description of harmonic generation involves both the calculation of the single-atom response and the propagation of the fields in the nonlinear medium.

Single-Atom Response

The first observations of high harmonic orders raised many theoretical questions such as: What is the origin of the plateau in the harmonic spectrum? How far can it extend and in what conditions? How can the atom absorb so many photons above its photo-ionization threshold without being ionized, which should kill any harmonic emission? In order to answer these questions, one has to solve the Time-Dependent Schrödinger Equation (TDSE) for the atom submitted to the short intense laser pulse, and calculate the induced dipole moment. Numerical methods for the solution of TDSE have been developed in the Single Active Electron (SAE) approximation and for a linearly polarized laser field, initiated by Kulander et al.[9].

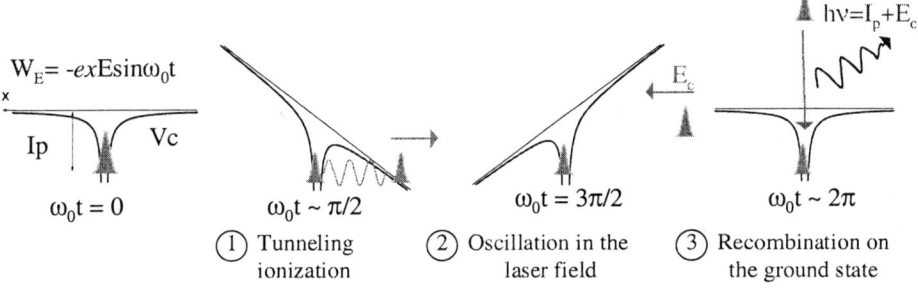

Figure 2. Semi-classical model for harmonic generation: the electron is submitted to the Coulomb potential Vc and to the interaction W_E with the strong laser field.

A breakthrough in the theoretical understanding of the emission process was initiated by the discovery of a "cutoff law"[10] for the extension of the harmonic plateau as a function of the parameters of the interaction. The cutoff position in the harmonic plateau was found to follow the universal law Ip+3Up, where Ip is the ionization potential and $U_p = e^2 E^2 / 4m\omega^2$ is the ponderomotive potential, that is the mean kinetic energy acquired by a free electron oscillating in the laser field. Here e and m are the electron charge and mass, E and ω are the laser electric field and frequency, respectively. The explanation of this universal law was found soon afterwards in the frame of a semi-classical model proposed by Corkum[11] and Kulander et al.[12]. It describes the harmonic emission as a three-step process (Figure 2). In a strong field low frequency regime, the electron can tunnel out from the core through the Coulomb barrier lowered by the (relatively slowly varying) electric field of the laser. It then undergoes oscillations in the field, during which the influence of the Coulomb force from the nucleus is negligible. Some of the trajectories come back in the vicinity of the nucleus and may lead to a radiative recombination back to the ground state, thus producing a photon of energy Ip plus the kinetic energy Ec acquired during the oscillatory motion. According to classical mechanics, the maximal kinetic energy that the electron can gain is indeed about 3Up. As a result, the extension of the plateau increases linearly with the laser peak intensity, but the ionization sets an upper limit on this extension. Indeed, when the saturation intensity Isat is reached, the atom is completely ionized, and the harmonic emission is stopped.

The predictions of the cutoff law have been verified by a number of experiments. The most energetic photons are indeed produced in light rare gases, that present a large ionization potential and are thus difficult to ionize, with infrared laser pulses of ultrashort duration, so that the atoms "survive" to higher intensities before ionizing (high Isat). This is the reason why the recent progress in laser technology towards pulse shortening has resulted in the observation of still higher harmonic orders. This model thus provides a nice physical picture of the underlying physics, even though it mixes quantum and classical elements. A fully quantum mechanical theory, based on the Strong Field Approximation (SFA) to the TDSE, was developed soon after by Lewenstein et al.[13]. It recovers and justifies the semi-classical model, while taking rigorously into account tunnel ionization, quantum diffusion and interferences.

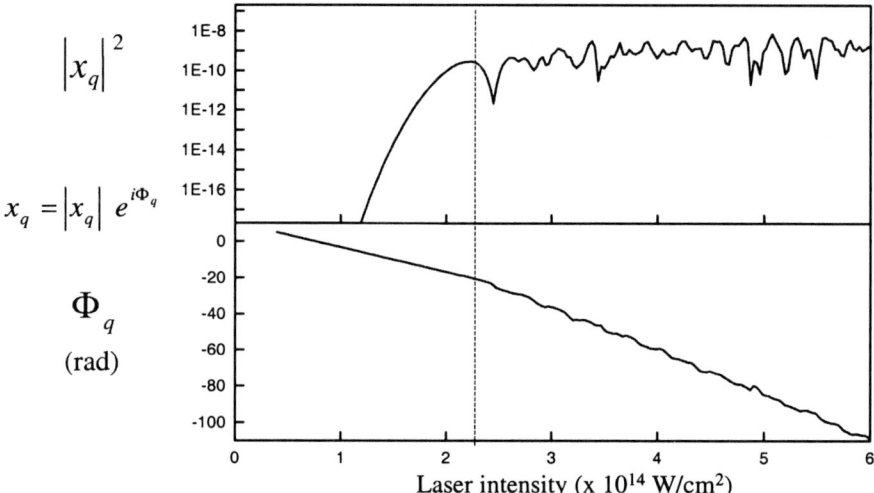

Figure 3. Strength and phase of the 45th harmonic dipole moment generated in neon as a function of the laser intensity.

The harmonic dipole moments calculated in this approach exhibit a generic behavior for sufficiently high orders. It is shown in Figure 3 for the 45th harmonic generated in neon by a 800 nm laser pulse. At low intensity, when the considered harmonic is in the cutoff region (I<2.3 10^{14} W/cm^2), the dipole strength increases steeply with intensity while the dipole phase decreases linearly. At higher intensity (I>2.3 10^{14} W/cm^2), when the harmonic enters the plateau region, the strength saturates and exhibits many interferences, while the phase decreases twice as fast as in the cutoff with superimposed oscillations. The reason for such a behavior is that in the cutoff region, there is a single trajectory that contributes to the harmonic emission, hence a regular evolution of the dipole, whereas in the plateau region at least two trajectories have significant contributions, whose phases behave differently with intensity (see below) leading to interferences and distortions of the total dipole moment.

The rapidly decreasing phase with intensity is an extremely important physical parameter for the macroscopic emission as we shall see later. It is very specific to the strong field regime (in the perturbative regime, there is no "intrinsic" phase). The origin of this phase is the action acquired by the electron along the trajectory in the laser field. In a first approximation, it is the product of the mean kinetic energy Up by the return time τ (the time spent in the continuum). The action thus varies linearly with intensity, with a slope all the more important as the trajectory is long. In the plateau, the two main trajectories have return times shorter than one laser period T such that: $0<\tau_1<\tau_2<T$ and are called respectively 'short' and 'long' trajectory.

Macroscopic Response

The harmonic macroscopic emission is optimized when phase-matching is achieved, that is, when the phase velocity of the generated harmonic field is equal to that of the driving polarization, allowing an efficient energy transfer. (The nonlinear polarization P_q is simply the product of the harmonic dipole moment x_q by the atomic density N_a). In the perturbative regime (low-order harmonics), this results in the well-known phase matching condition on the wave vectors k_1 and k_q of the laser and harmonic fields respectively: $k_q=qk_1$. In the strong-field regime, this condition is modified by the presence of the dipole phase that exhibits a spatial distribution in the focal region through its dependence on the laser intensity [14]. The phase matching condition [15] is now: $k_q=qk_1+grad\,\Phi q$.

The focusing of the laser also introduces a geometrical dephasing (Gouy phase) that modifies the total wave vector k_1 compared to the plane wave case. These two phenomena (geometric phase and dipole phase) may compensate each other for some positions in the focal region, and result in a good phase matching either on axis (emission of a centered harmonic beam) or off axis (annular beam). Note that phase matching depends on the electron trajectory leading to the harmonic emission. The "long" trajectory, associated to a rapidly-varying dipole phase with intensity, is favored for a focusing after the jet while the "short" one will be selected for a focusing before the jet. Thus, the position of the generating medium (rare gas jet) relative to the laser focus has a strong influence on the way phase matching is achieved and on the coherence properties of the harmonic emission.

Another cause of dephasing is the dispersion of the refractive index. Due to the low pressure in the rare gas jet (a few tens of Torr), the atomic dispersion is in general negligible (except in the case of very defocused geometries [16] or generation in hollow-core fibers [17]). On the other hand, the dispersion due to the free electrons generated by the ionization of the medium becomes important as soon as the intensity approaches the saturation intensity. Note that the free electron density gradient may cause a defocusing of the laser beam, which is detrimental to the harmonic emission since the effective intensity experienced by the nonlinear medium is lowered. When perfect phase matching is not achieved, one can define a coherence length as the length over which the harmonic field is constructed efficiently: $L_{coh} = \pi / |\Delta k|$. Passed this distance, the locally generated field interferes destructively with the propagating harmonic field, resulting in a decrease of the harmonic intensity.

Finally the re-absorption of the harmonics by the generating medium itself is also a limiting factor for the conversion efficiency, in particular for harmonics close to the photoionization threshold. The absorption length is $L_{abs} \sim 1/\sigma N_a$, where σ is the photoionization cross section. Since the maximum efficiency is proportional to the squared product of L_{abs} and P_q (strength of the polarization constructed efficiently over the absorption length), it does not depend on the atomic density N_a. Therefore, as soon as L_{abs} becomes the limiting length (compared to the other characteristic lengths L_{coh} and L_{med} the medium length), an optimal macroscopic efficiency is obtained. Constant et al.[16] have worked out a "rule of thumb" for this; one must fulfil: $L_{med} > 3\ L_{abs}$ and $L_{coh} > 5\ L_{abs}$.

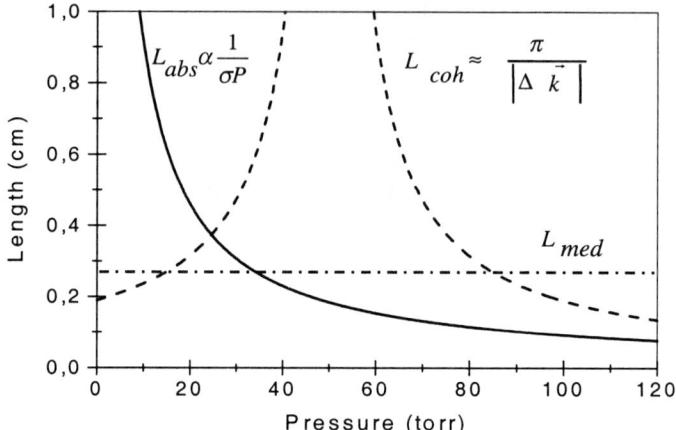

Figure 4. Example of pressure dependence of L_{abs} (solid line), L_{coh} (dashed line) and L_{med} (dot-dashed line) in the case of the 47th harmonic generated in neon.

The interplay between these characteristic lengths is illustrated in Figure 4 for the 47th harmonic generated in neon. In the chosen focusing conditions, the geometric dispersion is balanced around 50 Torr by the atomic dispersion plus the dipole phase variation. Perfect phase matching is then realized and the coherence length becomes infinite. However, at this pressure, the absorption length is still too large compared to the considered medium length (2.7 mm) to fulfil the first condition. A higher pressure, around 75 Torr, is needed for both conditions to be fulfilled. Above this pressure, the coherence length becomes the limiting factor. This analysis thus indicates that, in these generating conditions, there should be an optimal pressure where an absorption-limited emission is achieved. Note however that this simple analysis considers only on-axis phase-matching, while off-axis phase-matching may also lead to an efficient harmonic emission.

A propagation code taking into account all these phenomena was developed by L'Huillier et al.[18] in Saclay. The propagation equations for the fundamental and harmonic fields are solved using the slowly varying envelope and paraxial approximations. Coupled to the harmonic dipole data from the Lewenstein's model, this code has been very successful in reproducing the harmonic experimental results. More recently, a number of numerical codes have been developed in order to describe the harmonic emission by ultrashort laser pulses. Indeed, for durations shorter than 30 fs and high enough intensities, nonadiabatic phenomena are expected to come into play due to the fact that the laser field amplitude varies significantly over a period. This modifies the electron trajectories leading to the harmonic emission, so that the dipole moment is distorted. These phenomena are currently under investigation and will not be discussed here.

PROPERTIES OF THE HARMONIC SOURCE

Generation Efficiency/ Spectral Range

As stated by the cutoff law, the most energetic photons are produced in light rare gases. However the conversion efficiency in heavier rare gases is higher in their allowed spectral range, so that the choice of the optimal generating gas will depend on the desired photon energy. We give in Figure 5 typical numbers of photons as a function of the photon energy (generation by a Ti:S laser with a 250 fs pulse duration loosely focused at 10^{15} W/cm^2). In xenon (15-25 eV), 10^{10} photons per pulse are typically generated. In argon, the efficiency decreases from 10^9 photons in the plateau to 10^7 photons at 45 eV. In neon, between 10^7 and 10^5 photons are generated in the plateau that extends to 160 eV. Note that higher photon energies have been obtained with shorter duration pulses (less than 30 fs). In helium, the spectral range has been extended to 500 eV but with a lower generation efficiency. For these short pulse durations and high photon energies, the spectra are not any more resolved and resemble a blank spectrum: the harmonics are so broad that they overlap.

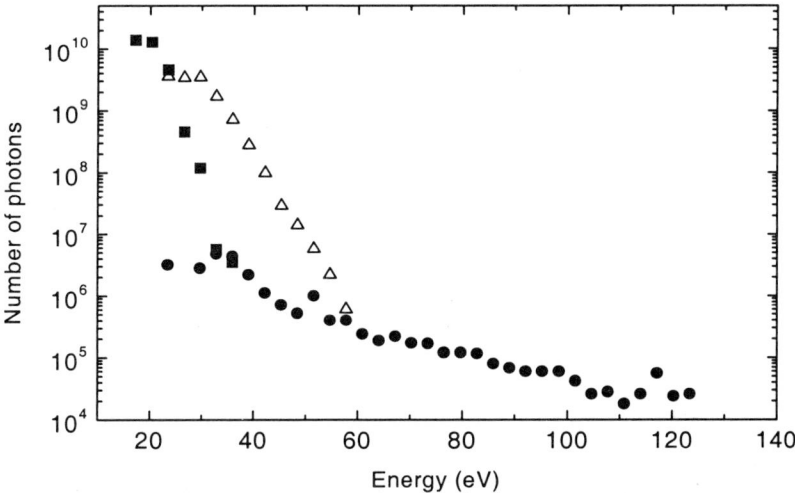

Figure 5. Typical number of harmonic photons generated in different rare gases: xenon (squares), argon (triangles) and neon (circles).

During the last years, the optimization of the conversion efficiency has become a hot topic since many applications need a high photon flux. Focusing as weakly as possible has been used in order to optimize phase matching and to increase the transverse size of the generating medium. Different geometries for the latter have been tried besides the gas jet: hollow-core fibers [16,17] or cells filled with gas [19,20]. The density of atoms in the interaction region has also been varied [21]. Harmonic generation being a coherent process is expected to vary as the square of the atomic density. However, the dispersion of the medium (due to atoms, ions and free electrons) as well as the absorption increase with density and may alter this simple scaling in particular by changing the phase matching conditions.

In conclusion, the conversion efficiency is very sensitive to the experimental conditions. In particular, it strongly depends on the spectral range of interest. Optimized conditions yield an efficiency of 10^{-6} around 40 eV and 10^{-8} around 100 eV [17,20].

Spatial Properties

The harmonic spatial profiles strongly depend on the way phase matching is achieved in the generating medium [14,22,23]. This is shown in Figure 6 in the case of the 39th harmonic generated in neon for different positions of the laser focus relative to the gas jet.

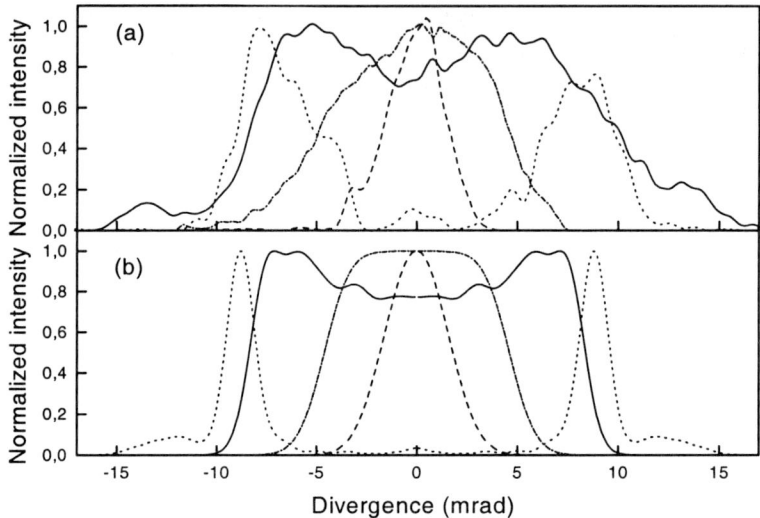

Figure 6. Evolution of the spatial profile of the 39th harmonic as a function of the position of the laser focus relative to the neon gas jet: (a) experiment, (b) theory (From Salières et al. [14]).

When the laser is focused sufficiently *before* the gas jet, the profile is narrow (dashed curve). When the focus is drawn closer to the jet, the profile first broadens (dot-dashed curve), then presents a dip at the center (solid curve), and finally becomes annular (dotted curve) when the laser is focused *after* the jet. The good agreement between the experimental results (Figure 6a) and the simulations (Figure 6b) allowed us to interpret this evolution: for a focus before the jet, phase matching is achieved on axis, whereas for a focus after the jet, phase matching can only be realized off axis, resulting in an annular profile. This behavior is a result of the interplay between the geometrical phase due to focusing and the dipole phase of the harmonic emission (see preceding chapter). It is thus possible to generate very regular, close to Gaussian profiles presenting a divergence smaller than that of the laser (55 mrad FWHM).

The regularity of the phase front has been investigated by measuring the 13th harmonic spot size in the focal region of a multilayer spherical mirror [24]. A beam quality $M^2 \sim 2$ corresponding to a two times diffraction-limited beam was obtained at low pressure and small jet thickness. This quality is degraded when increasing the pressure, because the free electron dispersion resulting from the ionization of the medium induces a laser defocusing, a change in phase matching, etc.

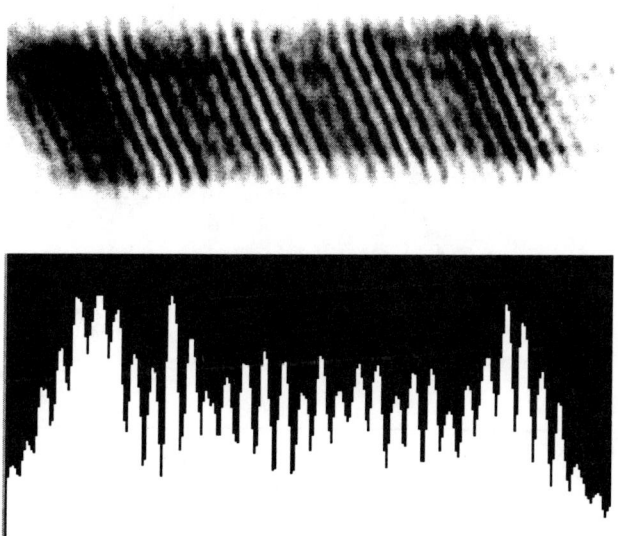

Figure 7. Interference pattern of the 13th harmonic beam obtained with the Fresnel's mirrors interferometer. The horizontal profile shows the good fringe contrast (From Le Déroff et al. [26]).

The spatial coherence of the harmonic beam has been studied with the Young's slits setup [25] and with the Fresnel's mirrors interferometer [26]. In the latter, two slightly tilted mirrors reflect the two halves of the harmonic beam, making them cross under a small angle and overlap, resulting in an interference pattern. This allowed us to perform a systematic study of the coherence inside the 13th harmonic beam produced in xenon. Figure 7 shows the interference pattern corresponding to the interference between all the rays distant of d=2 mm in the incident beam (of diameter ~3mm). A good fringe contrast is observed for this distance, but also for distances d=1 and 3 mm, indicating that a high coherence degree (γ>0.5) can be obtained over the entire harmonic beam. This coherence can however be distorted in the focal region both by the ionization-induced free electron dispersion and by the intensity-dependent dipole phase. They are time- and space-dependent factors that degrade the correlation between the harmonic fields particularly in the focal region where the laser intensity is high.

Temporal And Spectral Properties

The duration of the harmonic emission is expected to be shorter than that of the generating laser pulse. This raises the technical problem of the measurement of XUV pulses in the femtosecond range. Agostini *et al.* have developed a cross correlation method where the harmonic beam produces the photoionization of atoms in a continuum dressed by an intense laser field. The intensity of the sidebands that appear in the photoelectron energy spectrum on either side of the main energy peak scales as the cross-correlation of the harmonic and laser pulses. Figure 8 shows such a cross-correlation signal for the 21st harmonic generated in argon. The deconvolution gives a pulse duration of ~150 fs, for a 190 fs generating laser pulse [27]. Pulses as short as a few 10 fs have been measured with this technique. With an autocorrelation method using the two-photon ionization of helium, a duration of 27 fs has been obtained for the 9th harmonic of a 34 fs Ti:S laser [28].

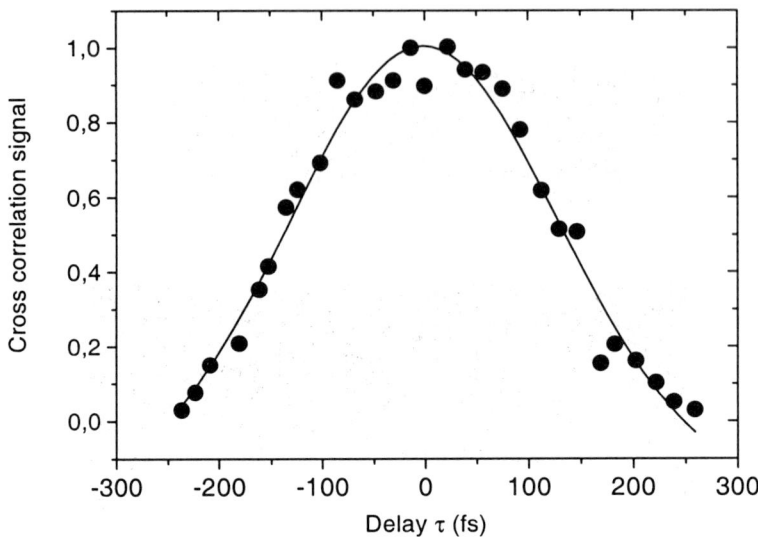

Figure 8. Cross correlation signal of the 21st harmonic with the laser beam as a function of the delay between the two beams (From Bouhal et al. [27]).

To characterize the temporal coherence, we have to compare the pulse duration with the coherence time that is inversely proportional to the spectral width. We have measured quasi-simultaneously the pulse duration and the spectral width of the 21st harmonic emission in argon. This is done by coupling an electron spectrometer to an XUV spectrometer whose grating is set either in the 0th order or in the 1st order of diffraction. In the best conditions, we find a product $\Delta t \Delta \nu \sim 2$, which is reasonably close to the Fourier transform limit, indicating a good temporal coherence.

This coherence is degraded in some generating conditions: when the laser is focused after the gas jet, the harmonic spectrum broadens considerably and symmetrically, while the pulse duration increases slightly. This indicates a chirp of the emission. Indeed, we have seen in the preceding chapter that the dipole phase varies rapidly with intensity, so that the harmonic emission by the short laser pulse presents a temporally varying phase. This phase modulation is all the more important as the electron trajectory leading to the emission is long. Since the "long" trajectory is favored when the laser is focused after the medium, this position leads to a larger chirp and thus a larger spectral width. This result corroborates the measurements of the coherence times inside the harmonic beam, reported by Bellini et al. [29].

Other Properties

The tunability of the radiation is an important parameter for many applications. Besides the coarse tunability obtained by changing the harmonic order, a fine adjustment of the wavelength can be obtained by generating the harmonics with a chirped laser [30] or with a laser mixed with an OPA [31]. Moreover, using ultrashort laser pulses (5 fs) and high intensities results in a blank spectrum [7] where the harmonics are so broad that they overlap, leading to a full tunability.

The polarization of the harmonic emission may be useful. When the laser is linearly polarized, the harmonics are also linearly polarized in the same direction [32]. Note that for symmetry reasons, there is no harmonic emission from a circularly polarized laser. In fact, the harmonic efficiency drops as soon as a small ellipticity is introduced in the laser polarization.

Finally, harmonic generation is a table-top XUV source, naturally synchronized with the generating laser and at the same repetition rate (10 Hz to 1 kHz).

APPLICATIONS

During the last few years, a number of applications of the harmonic emission have been reported in various areas of physics, making use of the unique properties of this new XUV source. We give below some examples of current applications.

Atomic And Molecular Spectroscopy

The ultrashort pulse duration and the natural synchronization with the generating laser are well suited to pump-probe experiments for studying ultrafast processes in atomic [33,34], molecular and solid-state physics. An example of application is the spectroscopy of excited states (2p, 3p) in helium, where lifetimes and ionization cross sections have been measured with a good accuracy [35,36].

Solid-State Physics

One of the first applications of the harmonics was in solid-state physics for the spectroscopy and the study of the dynamics of surface states on a Ge(111):As surface [37]. Recently, a time-resolved Ultraviolet Photoelectron Spectroscopy (UPS) experiment has been performed with the harmonics in Saclay. The aim was to study the relaxation of electrons in the conduction band of an insulator. The 25th harmonic was used to excite electrons, whose energy was probed at different delays with the laser pulse. Unexpectedly long relaxation times were measured for high energy electrons, compared to that of electrons in the bottom of the conduction band [38].

Nonlinear Physics

The good beam quality should allow to focus the harmonic beam to tight spots (less than 1 μm) and thus to reach the high XUV intensities necessary to produce nonlinear processes in this spectral range. A non-resonant two-photon ionization has already been reported with the 9th harmonic in helium [28]. The observation of nonlinear phenomena with higher harmonic orders in the near future relies on our ability to increase the number of generated photons and to focus this XUV beam to very small focal spots. The latter problem is currently investigated with new XUV optics such as Bragg-Fresnel lenses and zone plates.

XUV Interferometry

The unprecedented coherence properties find applications in XUV interferometry. Particularly interesting is the diagnostic of dense plasmas, since they refract (by the steep density gradients), absorb and reflect (above the critical density) the long wavelengths. A first study using the difference of transmission of different harmonics through an exploding foil has been reported [39]. However the deconvolution of these results to extract the electronic density is quite sensitive to the model chosen for the absorption mechanisms. In contrast, the phase shift observed in an interferometry experiment is directly connected to the electronic density.

We have performed recently two interferometry experiments using two different schemes. In the first one, performed in collaboration with the Lund group, we have generated two spatially-separated but phase-locked harmonic sources by focusing the two halves of a laser beam at different locations in a gas jet. The plasma was created on the path of one harmonic beam, the second beam being the reference. The interference fringes observed in the far-field when the beams overlap are locally shifted [40] allowing the determination of electronic densities as high as 2×10^{20} electrons/cm^3. In the second experiment, we have generated two phase-locked harmonic sources delayed in time, that interfere in the spectral domain after dispersion on a grating. This XUV spectral interferometry was used to measure the dynamics of the ionization of a high density helium jet at the femtosecond timescale [41]. These feasibility experiments show that it should be possible in the near future to probe densities as high as 10^{23} electrons/cm^3 with a temporal resolution of a few 10 fs.

Attosecond Pulse Generation

Finally, the possibility of generating sub-femtosecond pulses, so called attosecond pulses, by harmonic generation has attracted considerable interest during the last years. Indeed, the shortest pulses achieved today (4.5 fs in the infrared) are limited by the long period of the radiation (2.7 fs), so that shorter wavelengths are required for further pulse shortening. High-order harmonics have become the most promising way to generate attosecond pulses. Different schemes, based on simulations, have been proposed to reach these ultrashort pulse durations [42-46].

The first one deals with the relative phase of the harmonics generated in the plateau region. If they were emitted in phase, the corresponding temporal profile would consist of a train of ultrashort pulses separated by half the laser period, the duration of each pulse being in the attosecond range. There is a clear analogy here with mode-locked lasers. However, early calculations of the single-atom response showed that the high harmonics were, in general, not in phase, due to the interference of various energetically allowed electronic trajectories leading to the harmonic emission. Recently, Antoine et al. [43] revived this proposal by showing that the propagation in the atomic medium could select one of these trajectories, resulting in the macroscopic emission of a train of ~200 attosecond pulses. Furthermore, Corkum et al. [42] proposed a way of generating a *single* pulse by using the high sensitivity of harmonic generation in gases to the laser polarization [32]. By creating a laser pulse whose polarization is linear only during a short time, close to a laser period, the emission could be limited to this interval. Recently, this idea has been used in an experiment aiming at manipulating in time the harmonic emission. Using chirped pulses, Altucci et al. [44] have reduced the emission to a few fs.

Another scheme of ultrashort pulse generation uses the phase properties of a single harmonic pulse. In the preceding chapter, it was shown that the dipole phase induces a frequency chirp on the harmonic emission. This chirp can be used to compress the pulse duration to ~1 fs [45,46] by means of a pair of gratings, as is done in the Chirped Pulse Amplification technique.

Finally, another promising scheme for the production of *isolated* attosecond pulses is the harmonic generation by a few-cycle (5 fs) laser pulse [47]. Thanks to the rapid ionization of the medium (less than one cycle), the emission can be confined to a few 100 attoseconds. However, the reproducibility of the emission is highly dependent on the absolute phase of light in the laser pulse which is not controlled so far.

The technical realization of these schemes is currently pursued in many laboratories. In addition to the generation, the detection of these attosecond pulses is a challenging task. So far, the literature on this subject hardly exists, except for some theoretical studies. Recently, it has been proposed to perform a sort of autocorrelation of the harmonic pulses by generating with two delayed laser pulses focused at the same location in a gaseous medium. By this way, attosecond temporal structure in the harmonic emission could be observed [48].

CONCLUSION

We have first presented the current understanding of the harmonic generation process. This is a rich phenomenon involving two fields: atomic physics in strong laser fields for the single-atom response, and nonlinear optics for the propagation of the generated harmonic fields. In this strong field regime, the intensity dependence of the dipole phase has dramatic consequences on phase matching and on the coherence properties of the harmonic beam. Another important phenomenon is the free electron dispersion due to the ionization of the nonlinear medium. It can induce both a phase front distortion and a loss of coherence. However, we show that there are conditions where the generated harmonic beam is well collimated and highly coherent both spatially and temporally. This makes harmonics a very useful source for XUV interferometry. Moreover, the ultrashort pulse duration, the natural synchronization with the driving laser and the high repetition rate are particularly suited to pump-probe investigations of ultrafast dynamics.

Finally, we have given some examples of current applications of the harmonic radiation in atomic and molecular spectroscopy, solid-state physics, diagnostics of dense plasmas... Future applications will undoubtedly make use of the possibility of generating attosecond pulses, that open the way to probing matter on an unprecedented time scale.

We thus believe that the unique properties of this XUV source will be useful in a growing number of applications.

REFERENCES

1. P.A. Franken et al., Generation of optical harmonics, *Phys. Rev. Lett.* 7:118 (1961).
2. A. McPherson et al., Studies of multiphoton production of vacuum-ultraviolet radiation in the rare gases, *J. Opt. Soc. Am. B* 4:595 (1987).
3. M. Ferray et al., Multiple-harmonic conversion of 1064 nm radiation in rare gases, *J. Phys. B* 21:L31 (1988).
4. A. L'Huillier et al., High-order harmonic generation in rare gases with a 1-ps 1053-nm laser, *Phys. Rev. Lett.* 70:774 (1993).
5. J.J. Macklin et al., High-order harmonic generation using intense femtosecond pulses, *Phys. Rev. Lett.* 70:766 (1993).
6. Z. Chang et al., Generation of coherent soft x-rays at 2.7 nm using high harmonics, *Phys. Rev. Lett.* 79:2967 (1997).
7. Ch. Spielmann et al., Generation of coherent X-rays in the water window using 5-femtosecond laser pulses, *Science* 278:661 (1997).
8. M. Schnürer et al., Coherent 0.5-keV x-ray emission from helium driven by a sub-10-fs laser, *Phys. Rev. Lett.* 80:3236 (1998).
9. K.C. Kulander et al., Calculations of multiple-harmonic conversion of 1064-nm radiation in Xe, *Phys. Rev. Lett.* 62:524 (1989).
10. J.L. Krause et al., High-order harmonic generation from atoms and ions in the high intensity regime, *Phys. Rev. Lett.* 68:3535 (1992).
11. P.B. Corkum, Plasma perspective on strong-field multiphoton processes, *Phys. Rev. Lett.* 71:1994 (1993).

12. K.C. Kulander et al., Dynamics of short-pulse excitation, ionization and harmonic conversion, in: Super-Intense Laser-Atom Physics, B. Piraux, Anne L'Huillier, and K. Rzcazewski, eds., Plenum Press, New York, NATO ASI Series B, 316:95 (1993).
13. M. Lewenstein et al., Theory of high-harmonic generation by low-frequency laser fields, *Phys. Rev. A* 49:2117 (1994).
14. P. Salières et al., Coherence control of high-order harmonics, *Phys. Rev. Lett.* 74:3776 (1995).
15. Ph. Balcou et al., Generalized phase-matching conditions for high harmonics: the role of field gradient forces, *Phys. Rev. A* 55:3204 (1997).
16. E. Constant et al., Optimizing high harmonic generation in absorbing gases: model and experiment, *Phys. Rev. Lett.* 82:1668 (1999).
17. A. Rundquist et al., Phase-matched generation of coherent x-rays, *Science* 280:1412 (1998).
18. A. L'Huillier et al., Calculations of high-order harmonic-generation processes in xenon at 1064 nm, *Phys. Rev. A* 46:2778 (1992).
19. Y. Tamaki et al., Highly efficient phase-matched high harmonic generation by a self-guided laser beam, *Phys. Rev. Lett.* 82:1422 (1999).
20. M. Schnürer et al., Absorption-limited generation of coherent ultrashort soft-X-ray pulses, *Phys. Rev. Lett.* 83:722 (1999).
21. C. Altucci et al., Influence of atomic density in high-order harmonic generation, *J. Opt. Soc. Am. B* 13:148 (1996).
22. J.W.G. Tisch et al., Angularly resolved high-order harmonic generation in helium, *Phys. Rev. A.* 49:R28 (1994).
23. J. Peatross et al., Angular distribution of high-order harmonics emitted from rare gases at low density, *Phys. Rev. A.* 51:R906 (1995).
24. L. Le Déroff et al., Beam-quality measurement of a focused high-order harmonic beam, *Opt. Lett.* 23:1544 (1998).
25. T. Ditmire et al., Spatial coherence measurement of soft x-ray radiation produced by high-order harmonic generation, *Phys. Rev. Lett.* 77:4756 (1996).
26. L. Le Déroff et al., Measurement of the degree of spatial coherence of high-order harmonics using a Fresnel's mirrors interferometer, *Phys. Rev. A* 61:043802 (2000).
27. A. Bouhal et al., Temporal dependence of high-order harmonics in the presence of strong ionization, *Phys. Rev. A* 58:389 (1998).
28. Y. Kobayashi et al., 27-fs extreme ultraviolet pulse generation by high-order harmonics, *Opt. Lett.* 23:64 (1998).
29. M. Bellini et al., Temporal coherence of ultrashort high-order harmonic pulses, *Phys. Rev. Lett.* 81:297 (1998).
30. J. Zhou et al., Enhanced high-harmonic generation using 25 fs laser pulses, *Phys. Rev. Lett.* 76:752 (1996).
31. M.B. Gaarde et al., High-order tunable sum- and difference frequency mixing in the XUV region, *J. Phys. B* 29:L163 (1996).
32. Ph. Antoine et al., Polarization of high-order harmonics, *Phys. Rev. A* 55:1314 (1997).
33. Ph. Balcou et al., High-order harmonic generation in rare gases : a new source in photoionization spectroscopy, *Z. Phys. D* 34:107 (1995).
34. A. L'Huillier et al., High-order harmonics : a coherent source in the XUV range, *J. Nonlinear Opt. Phys. and Mat.* 4:647 (1995).
35. J. Larsson et al., Two-colour time-resolved spectroscopy of helium using high-order harmonics, *J. Phys. B* 28:L53 (1995).
36. M. Gisselbrecht et al., Absolute photoionization cross sections of excited He states in the near-threshold region, *Phys. Rev. Lett.* 82:4607 (1999).
37. R. Haight et al., Antibonding state on the Ge(111):As surface: spectroscopy and dynamics, *Phys. Rev. Lett.* 70:3979 (1993).
38. F. Quéré et al., Hot electron relaxation in quartz using high-order harmonics, *Phys. Rev. A* 61:9883 (2000).
39. W. Theobald et al., Temporally resolved measurement of electron densities ($>10^{23}$ cm^{-3}) with high harmonics, *Phys. Rev. Lett.* 77:298 (1996).
40. D. Descamps et al., Extreme ultraviolet interferometry measurements with high-order harmonics, *Opt. Lett.* 25:135 (2000).
41. P. Salières et al., Frequency-domain interferometry in the XUV with high-order harmonics, *Phys. Rev. Lett.* 83:5483 (1999).
42. P.B. Corkum et al., Subfemtosecond pulses, *Opt. Lett.* 19:1870 (1994).
43. Ph. Antoine et al., Attosecond pulse trains using high-order harmonics, *Phys. Rev. Lett* 77:1234 (1996).
44. C. Altucci et al., Frequency-resolved time-gated high-order harmonics, *Phys. Rev. A* 58:3934 (1998).
45. K.J. Schafer et al., High harmonic generation from ultrafast pump lasers, *Phys. Rev. Lett.* 78:638 (1997).

46. P. Salières et al., Temporal and spectral tailoring of high-order harmonics, *Phys. Rev. Lett.* 81:5544 (1998).
47. N.A. Papadogiannis et al., Observation of attosecond light localization in higher order harmonic generation, *Phys. Rev. Lett.* 83:4289 (1999).
48. F. Krausz et al., Extreme nonlinear optics: exposing matter to a few periods of light, *Opt. Photon. News* 9:46 (1998).

CLUSTERS IN INTENSE LASER FIELDS

John W.G. Tisch

Blackett Laboratory
Imperial College
London SW7 2BW, UK

INTRODUCTION

Atomic clusters have provoked great interest since their first observation in the mid 1950s [1]. Physicists' and chemists' fascination with them derives from the unique position they hold as an intermediate state between molecules and solids. We are interested here in clusters comprising about 500-10^5 atoms, having diameters of a few nanometres, making them considerably smaller than the micron size droplets [2] and micro-structured targets that have been studied for some time as laser targets [3].

In the last five or so years, the study of laser-cluster interactions has been extended to laser intensities in excess of 10^{15} Wcm^{-2} [4-22] (laser pulse-widths in the range 0.1-10 ps). This is the strong-field interaction regime, where the electric field of the laser is no longer small compared to the atomic field, and the interaction becomes highly nonperturbative [23]. This regime, made widely accessible by the development of chirped pulse amplification (CPA) lasers [24], had been studied already in atoms [25,26,27], small molecules [28,29] and bulk solids [30] since the late 1980s. In stark contrast to earlier studies of laser-cluster interactions that had revealed dynamics similar to those seen in molecules - with relatively inefficient coupling of laser energy to electrons and ions - studies of X-ray generation from gases of clusters (>1000 atoms) [4,5,6,8] at high laser intensities ~10^{16} Wcm^{-2} began to show startling evidence of a laser-cluster interaction that was very much more energetic. In fact, it was more reminiscent of the laser interaction with a bulk solid (*i.e.*, displaying plasma-like behaviour).

In the face of this new evidence, a plasma model of the laser-cluster interaction was developed [9], in which the cluster is treated as a spherical "nanoplasma", subject to the standard processes of a laser-heated plasma, such as collisional heating and collisional and tunnel ionisation, but also taking into account the spherical geometry and sub-wavelength size of the cluster. The model showed that the collisional heating in a single cluster was greatly enhanced, resulting in plasma electron temperatures that exceeded even those obtained in laser-solid interactions. It also predicted that the hot electrons then drove an extremely rapid and energetic explosion of the nanoplasma, creating a shrapnel of fast electrons and highly charged, fast ions. These predictions were subsequently confirmed by a variety of experiments [10-17,20,21]. It has thus became very clear that clusters can not simply be viewed as a bridge between molecules and solids, but as a target medium with unique properties which allow the coupling of laser energy into matter with an unprecedented high efficiency.

In this chapter, we will outline experimental and theoretical aspects of laser-cluster interactions and review the experimental results and emerging applications. The chapter is organised as follows. First, we will look at cluster formation in gas-jets, and the measurements of cluster sizes. Then we will describe the experimental apparatus and techniques used to study laser-cluster interactions. We will then move onto a discussion of the physics of the interaction, in terms of the successful nanoplasma model, and look at the predictions of the model. The major experimental results are reviewed in the following sections, concentrating on the electron and ion energies and ion charge states measured from the explosion of isolated clusters and how they scale with the experimental parameters. Two exciting applications of the laser-cluster interaction, X-ray generation and pulsed neutron production from thermonuclear fusion are outlined in the final section.

CLUSTER FORMATION IN GAS-JETS

Cluster formation is a broad and well-studied topic, so here we shall only touch on the important points that are relevant to engineering and characterising a laser target medium. We consider van der Waal bonded clusters that can be readily formed in high pressure gas-jets (*i.e.*, free-jet expansions). Clustering occurs due to the cooling associated with the adiabatic expansion of the gas into vacuum ($T \propto V^{-2/3}$ for a monatomic gas). The cooling leads to a phase-transition and under the right conditions a large fraction (known as the condensation fraction) of the gas atoms form into clusters. The properties of the resulting medium (which we shall refer to as the cluster medium) are remarkable. It has the average density of a gas, but the individual, nanometre scale clusters are at near solid density.

The standard pulsed valves (piezo-electric or solenoid driven) that have been used for many years to deliver gas targets for laser-interaction experiments are used to produce cluster media. It is the properties of the nozzle, the gas species and the pre-expansion temperature and pressure of the gas that determine the extent of clustering.

The Hagena Parameter

There is no rigorous theory to predict cluster formation in a free-jet expansion. However, the onset of clustering and size of clusters produced can be described by an empirical scaling parameter, Γ^*, known as the Hagena parameter [31]:

$$\Gamma^* = k \frac{(d/\tan\alpha)^{0.85} p_0}{T_0^{2.29}} \qquad (1)$$

where d is the nozzle diameter (in mm), α the expansion half angle ($\alpha = 45°$ for sonic nozzles, $\alpha < 45°$ for supersonic), p_0 the backing pressure (in mbar), T_0 the pre-expansion temperature (in Kelvin) and k, the condensation parameter, is a constant related to bond formation (see Table 1 for a range of k values for common atomic and molecular gases).

Table 1. Condensation parameters for a variety of gases

Gas	H_2	D_2	N_2	O_2	He	Ne	Ar	Kr	Xe
k	184	181	528	1400	3.85	185	1650	2890	5500

Gas-jets with the same Γ^* tend to form clusters of the same average size. Cluster formation is a statistical process and therefore there will be a distribution of cluster sizes. This distribution is not very well studied, but is expected to be quite broad [32]. The *average* number of atoms per cluster $<N_c>$ scales approximately as

$$<N_c> \sim \Gamma^{*2.0}. \tag{2}$$

Experiments have also shown that clusters of more than about 100 atoms begin forming for values of $\Gamma^* \sim 1800$. This point is known as the onset of massive condensation. By varying T_0 and p_0, it is possible to control the value of Γ^* and engineer a cluster medium of arbitrary average density and cluster size.

Given the distribution of sizes, measuring cluster sizes turns out to be very difficult. Optical Rayleigh scattering provides a convenient technique to infer average cluster sizes, in situ. A low power visible laser is used to irradiate the region of interest in the gas-jet. The Rayleigh scattered light at 90° to the incident beam is collected with a lens and detected with a filtered photo-multiplier tube. The dependence of the scattered signal on the temperature and pressure of the backing pressure and observation of the onset of massive condensation can be used to infer the average cluster size (within a factor of two) for arbitrary conditions. Details are given in Ref.[17].

Rayleigh scattering measurements carried out at Imperial College give the following empirical formulae relating $<N_c>$ to the gas-jet backing pressure:

$$<N_c> (\text{Xe 300 K}) \approx 100(p_0[\text{bar}])^2$$
$$<N_c> (\text{Kr 300 K}) \approx 30(p_0[\text{bar}])^2. \tag{3}$$

The cluster radii, R_c, proportional to $(N_c)^{1/3}$, are calculated assuming that the clusters have the same density as the bulk solid [33]. This gives

$$R_c(\text{Xe}) \approx 2.39(N_c)^{1/3} \text{ Å}$$
$$R_c(\text{Kr}) \approx 2.26(N_c)^{1/3} \text{ Å}. \tag{4}$$

EXPERIMENTAL DETAILS

In this section we look at some important experimental issues related to laser-cluster interactions. The set-up used at Imperial College is described to provide a concrete example. A variety of laboratories around the world have similar set-ups.

Dense versus Diffuse Targets

Laser-cluster interactions fall into two broad categories. One involves the interaction with a very low density cluster medium and the other with dense cluster medium. The former provides an approximation to the interaction of intense laser pulses with single clusters which is important for understanding the physics without the complications introduced by macroscopic effects. The latter is required for applications, where a high density of clusters is needed to produce a high energy-density plasma, or where macroscopic effects are essential to the process. One should bear in mind that the applications in dense media are underpinned by the understanding of the single cluster response that has been gained through detailed experimental investigations using low density cluster beams.

The study of these two density regimes demands quite different experimental approaches. From Equation 1, it becomes clear that it is not possible to produce a low density of clusters from a gas-jet simply by reducing the backing pressure. Instead, one takes advantage of the fact that the density of clusters drops rapidly as a function of distance below the nozzle of the pulsed valve, and the laser is focused many hundreds of nozzle diameters beneath the nozzle. In practice, a skimmer and a differential pumping scheme is required to produce a low density (*i.e.*, ~10^{10} cm^{-3}) cluster beam. Also, because studies of single cluster interactions rely on time-of-flight measurements of ions and electrons, ultra-high vacuum is required in the interaction chamber to reduce background signal and minimise space-charge effects.

Interactions with dense cluster media are more straightforward. The laser is focused a few nozzle diameters beneath the nozzle of the gas jet, where the density of clusters can be as high as 5×10^{16} cm^{-3}. Ultra-high vacuum is not required.

Laser System

We will be describing a number of experimental results obtained at Imperial College, so it is useful to briefly describe the femtosecond CPA laser used for these studies. The laser is based on titanium doped sapphire (Ti:S), and produces femtosecond pulses of energy up to ~60 mJ centred at a wavelength of ~780 nm. A Kerr lens mode-locked oscillator generates near transform-limited pulses of duration ~90 fs which are amplified in the rest of the system by a factor of ~10^8. The oscillator pulses are stretched in a diffraction-grating stretcher to a duration of ~250 ps, thus reducing the pulse power by a factor ~3000. This permits safe amplification first to the millijoule level in a Ti:S regenerative amplifier operating at 10 Hz and then to ~120 mJ in a multi-pass Ti:S power amplifier. The amplified pulses are then recompressed in a grating compressor to yield ~60-mJ/150 fs pulses (~0.5 TW) at a repetition rate of 10 Hz. More details on the system can be found in Ref.[34].

In conjunction with standard (single-shot) autocorrelation pulse duration measurements, and equivalent-plane focal spot characterisations, the focused laser intensity is inferred from ion appearance data using over the barrier ionisation thresholds [35]. This is carried out in the same interaction chamber (described below) used for the low-density cluster studies, providing *in situ* intensity measurement.

Interaction Chamber for Low-Density Experiments

This chamber, illustrated in Figure 1, is essentially a time-of-flight particle spectrometer coupled to a skimmed, gas-jet cluster source. It is used to measure the energies of electrons and ions as well as the charge states of ions produced in the laser induced explosion of isolated clusters.

As outlined above, to achieve a relatively low density of clusters in the interaction region, a cluster jet produced from a solenoid pulsed-valve at the top of the chamber is skimmed to produce a much lower density cluster beam. The cluster beam intersects the laser focus at the centre of the high vacuum section of the chamber, some 50 cm below the nozzle. Differential pumping is employed to maintain the high vacuum in the main chamber. Typically, the density of clusters is ~10^{10} cm^{-3} corresponding to ~10^3 1000-atom clusters in the laser focal volume.

The linearly-polarised laser is focused at ~$f/10$ with plano-convex lens which gives an approximately Gaussian focal spot of ~20 μm (e^{-2} radius). By rotating a half-wave plate before the lens, the polarisation vector can be rotated through 360°, enabling the angular distribution of particles in the plane perpendicular to the laser axis to be studied.

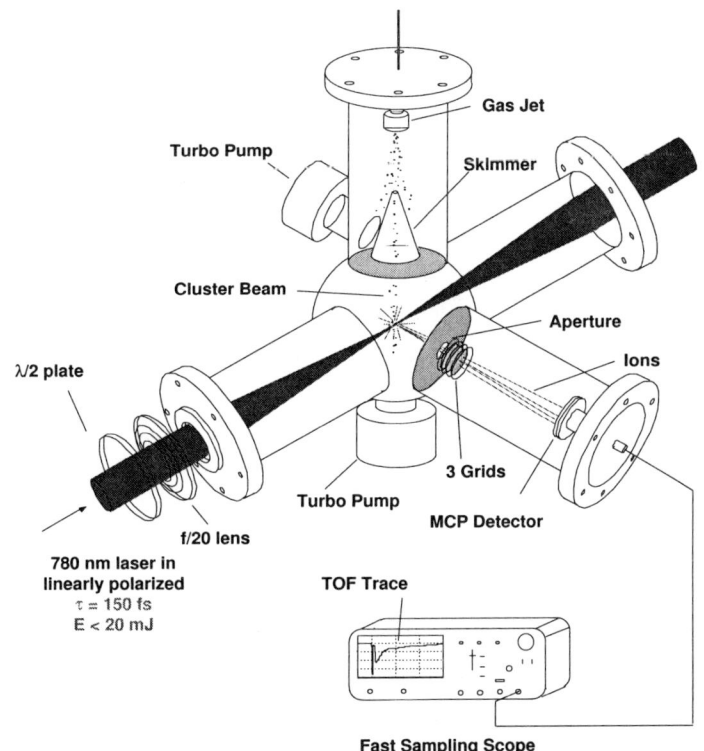

Figure 1. Interaction chamber used for laser-cluster experiments at Imperial College.

Measuring Ion and Electron Energies

The energies of ions produced in the interaction are determined from their time-of-flight (TOF) from the laser focus to a micro-channel plate (MCP) detector, without the use of an extraction field. Consider the arrangement shown in Figure 2. For an ion of mass m_i, travelling a field-free distance L (typically, ~0.5m) between the focus and the detector, the ion kinetic energy, E, is given by

$$E = \tfrac{1}{2} m_i L^2 T^{-2}, \tag{5}$$

where T is the TOF. T can be measured with a fast oscilloscope as the time interval between a trigger coincident with the laser pulse reaching the interaction region and the ion hit on the detector. Upon averaging ion hits over many laser shots (typically thousands), a TOF spectrum is built up. The ion energy spectrum, $g(E)$, is related to the measured TOF spectrum, $f(T)$, by

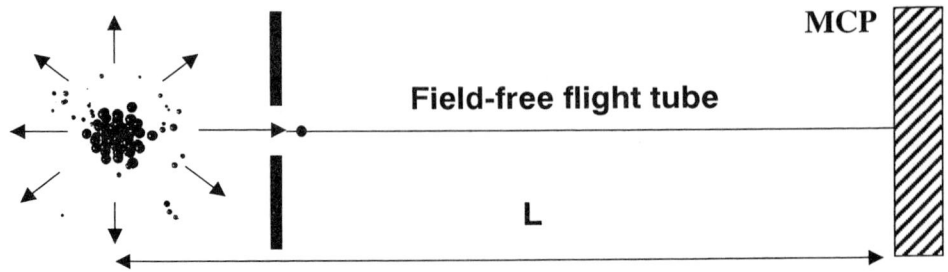

Figure 2. TOF geometry.

$$g(E) = \frac{f(T)}{m_i L^2 T^{-3}}, \qquad (6)$$

which comes from the requirement that the total number of ions from integration over T or E must be the equal. The mean ion energy is defined as

$$\langle E_{ion} \rangle = \frac{\int Eg(E)dE}{\int g(E)dE}. \qquad (7)$$

The electrons produced in cluster explosions are too fast for TOF techniques to be used. Instead, a retardation method is employed. Two closely spaced grids are place in the flight path. To the first, a voltage Φ is applied, the second is grounded. This introduces a potential barrier to electrons with energy less than $e\Phi$, where e is the electron charge. The MCP signal is recorded as a function of Φ. The electron energy spectrum is then found by differentiating this distribution with respect to Φ.

Measuring Ion Charges States

With a small modification to the standard TOF apparatus, information can also be gained on the ion charge state, Z. Three closely-spaced metal grids are placed in the flight tube. Applying a potential Φ to the middle grid, while keeping the front and back grids at earth introduces a barrier to ions with energy less than $Ze\Phi$, without significantly altering the flight time of higher energy ions. By measuring the number of ions reaching the detector as a function of Φ, and then differentiating with respect to Φ, the charge state distribution of the ions as a function of their kinetic energy can be determined. However, this technique can not measure the energy distribution of a given charge state. The complete charge state information can be gained by incorporating magnetic deflection into the ion flight path, as demonstrated in Ref [16].

THE NANOPLASMA MODEL

In this section we look at the physics of laser-cluster interaction. We shall see that is it essentially the physics of a laser-solid interaction with some important differences owing to unique cluster properties. The nanometre scale cluster is converted into a tiny, high-density, spherical plasma – a nanoplasma - into which laser energy is very efficiently coupled. The nanoplasma model has been successful in reproducing many of the features observed experimentally. We shall look at this model in some detail and the predictions it makes, but first we need to look at the laser heating of an optically ionised monatomic gas, following Ref. [9].

Laser-Plasma heating of Monatomic Gases

Consider a monatomic gas at a density of 10^{18}-10^{19} cm^{-3} irradiated by a ~100 fs pulse of wavelength ~800 nm, in the 10^{15}-10^{17} Wcm^{-2} range. We note that this is the average density of our cluster medium, so this example serves as a useful point of comparison. Clearly, for these intensities a plasma will be formed. By what mechanisms are the plasma electrons heated, and how hot will they get? For this parameter regime, there are three main plasma heating mechanisms; above threshold ionisation (ATI), stimulated Raman scattering (SRS) and collisional heating, known also as Inverse Bremsstrahlung (IB). We look at each of these in turn.

In ATI, an electron absorbs more photons than necessary for ionisation, leaving the electron with an increased kinetic energy. A quasi-classical analysis of this process shows that the energy distribution of the electrons has the following form:

$$E_{ATI} \approx 2U_p \sin^2 \Delta\phi, \qquad (8)$$

where U_p is the ponderomotive energy and $\Delta\phi$ is the phase of the laser at ionisation relative to the peak field. At 10^{16} Wcm^{-2}, $U_p \approx 1$ keV for $\lambda = 800$ nm, but the tunnel ionisation rate depends very strongly on the laser field strength, so most electrons are ionised around $\Delta\phi = 0$. Hence the distribution is peaked around $E_{ATI} = 0$, with an average energy, $<E_{ATI}> \approx$ 50 eV, which is confirmed in experiments.

In SRS, the laser drives a plasma wave which can store energy after the laser pulse has gone, leading to heating of the plasma electrons. We do not need to go into more detail here, because this heating mechanism is negligible for intensities below 10^{17} Wcm^{-2} unless the electron density exceeds 10^{20} cm^{-3}.

The third heating mechanism is IB. Consider an electron initially at rest oscillating in the laser field. In the absence of a perturbation, the electron kinetic energy returns to zero when the pulse has passed. However, if the electron suffers a dephasing collision with an ion during its oscillation, it can keep some of its wiggle energy after the laser pulse has gone. This is shown in Figure 3.

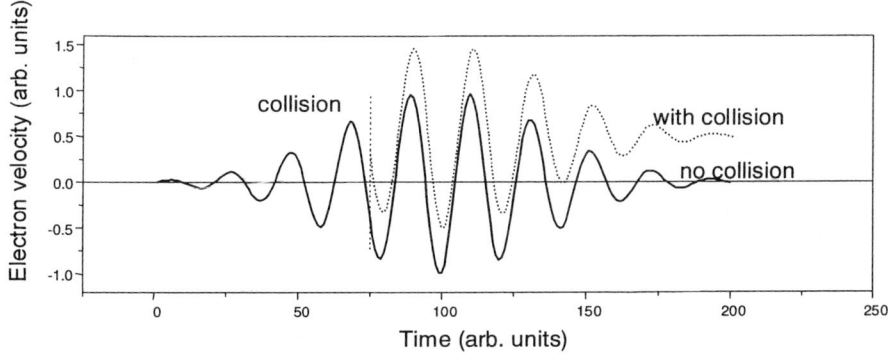

Figure 3. Collisional absorption: In the absence of a collision (solid line) the electron velocity returns to zero. Quiver energy is retained if a dephasing collision occurs during the laser pulse (dotted line).

In the strong-field limit, the cycle-averaged heating rate by IB scales as $n_e n_i Z^2/\sqrt{I}$, where n_e, n_i are the electron and ion densities, Z is the ion charge state and I the laser intensity. For the gas densities considered here and for $I = 10^{16}$ Wcm^{-2}, it can be shown that for a fully stripped helium plasma, the plasma electrons are only heated to a few eV, owing to the low density.

In summary, at gas densities, heating of the plasma electrons by the laser is very inefficient, dominated by ATI heating which leads to effective electron temperatures of <50 eV. In contrast, we shall see that if the same average gas density is rearranged into near solid density clusters, then IB heating becomes the dominant heating mechanism, leading to multi-keV plasma temperatures.

Requirements for Validity of Nanoplasma Model

It is not immediately obvious that a nanometre scale ionised cluster can be treated as a plasma, as is assumed in the nanoplasma model. Here we look at the requirements for validity of such a model.

First, a plasma treatment only makes sense if the vast majority of the ionised electrons in the cluster are confined to the cluster during the laser pulse, to ensure quasi-neutrality. While for molecules and small clusters (less than a few hundred atoms) the electrons can escape, for the large clusters we are interested in here (thousands of atoms), the electrons are confined by the space charge of the ions, as is the case for a solid. In fact, modelling (detailed later) shows that a significant number of electrons do eventually escape from the cluster, but only after the laser heating has reached a maximum near the peak of the laser pulse and they have sufficient energy to do so.

The second technical requirement for a plasma treatment is that the cluster size, R_c, has to be much larger than the Debye length

$$\lambda_D = \left(\frac{kT_e}{4\pi n_e e^2}\right)^{1/2}, \qquad (9)$$

where n_e is the electron density, T_e the electron temperature, e the electronic charge and k is Boltzmann's constant. For a solid density xenon plasma ionised to 5$^+$ with an electron temperature of 1 keV (we shall see that these are realistic values), $\lambda_D \approx 0.5$ nm. For the 5-10 nm clusters we are interested in, the $R_c \gg \lambda_D$ condition is clearly met.

The final requirement is that the nanoplasma holds together during the laser pulse. Let us define a disassembly time as the time taken for the nanoplasma to expand to such an extent that its density has dropped to the background gas density ($\sim 10^{17}$ cm^{-3}). Assuming that the nanoplasma expands at the plasma sound speed,

$$c_s = \sqrt{ZkT_e/m_i}, \qquad (10)$$

where Z is the ion charge state, kT_e is the electron thermal energy and m_i is the ion mass, then using the previous values (Xe cluster, $Z = 5$, $kT_e = 1$ keV) the disassembly time is of order 1 ps. This shows that femtosecond laser pulses are required for the peak of the laser pulse to interact with the nanoplasma.

Physics of the Nanoplasma Model

Following these introductory considerations, we are now in a position to look in more detail at the physics of the nanoplasma model which was developed by Ditmire and co-workers at Lawrence Livermore National Laboratory [9]. A number of simplifying assumptions are made in this model, and we look at these now.

Simplifying Assumptions. The model assumes that the electric field is uniform across the cluster and that all the atoms in the cluster simultaneously experience the same laser electric field strength. This approximation is valid provided the clusters considered are much smaller than both the laser wavelength, λ, and the plasma skin depth given by

$$\delta = (c/\omega_p), \tag{11}$$

where c is the speed of light and ω_p is the plasma frequency in the nanoplasma, given by

$$\omega_p = \left(\frac{4\pi e^2 n_e}{m_e}\right)^{1/2}, \quad \omega_p[s^{-1}] = 5.64 \times 10^4 \sqrt{n_e[cm^{-3}]}, \tag{12}$$

where m_e is the electron mass. To see that this is indeed the case, consider a very large cluster of ~5×10^6 Xe atoms, with $R_c \sim 40$ nm. This is still only $\lambda/20$ for $\lambda = 800$ nm, and much smaller that the skin depth for realistic value of n_e. In fact, $R_c \leq \delta$ only for an unrealistically high electron density of $\geq 400 n_{crit}$, where n_{crit} is the critical electron density given by

$$n_{crit} = \frac{\pi c^2 m_e}{e^2 \lambda^2}, \quad n_{crit}[cm^{-3}] = \frac{1.1 \times 10^{21}}{(\lambda[\mu m])^2}. \tag{13}$$

It is also assumed that the temperature distribution across the plasma is isotropic. This is a good assumption, because the nanoplasma is small enough and collisional enough to ensure that the temperature gradients are negligible. Further, the ion density distribution is assumed to be uniform across the cluster and the expansion is assumed to be self-similar, so the density remains uniform across the cluster throughout its expansion. The final important assumption is that the electron energy distribution is Maxwellian.

Ionisation. Because of the high density in the nanoplasma, ionisation arising from electron ion collisions (collisional ionisation) must be accounted for. The model uses the empirical formula of Lotz [36] to calculate the collisional ionisation rate, taking into account not only the thermal energy of the electrons (thermal collisional ionisation), but also their quiver energy in the laser field (laser-driven collisional ionisation).

To see the importance of collisional ionisation, consider an argon cluster with $n_e = 10^{23}$ cm^{-3} ionised to Ar^{8+} which has an ionisation potential $I_p \approx 150$ eV. If the thermal and quiver energy of the electrons are $\approx I_p$, then the Lotz collisional ionisation rate is about 0.1 fs^{-1}, implying 100% ionisation to Ar^{9+} in 10 fs.

Heating. We saw earlier that collisional heating (or IB) will become important in the high density nanoplasma. The collisional heating of the cluster is treated as the heating of a uniform dielectric sphere in a time-varying electric field representing the laser field. This field is assumed to be spatially uniform, since the clusters are much smaller than the laser wavelength. By using the Drude model for the plasma dielectric constant, the plasma nature of the sphere is introduced. Inside the dielectric sphere, the electric field is [37]

$$E = \frac{3E_0}{|\varepsilon + 2|}, \qquad (14)$$

where E_0 is the external field and the Drude dielectric constant is

$$\varepsilon = 1 - \frac{n_e/n_{crit}}{1 + i\nu/\omega}, \qquad (15)$$

where ω is the laser frequency and ν the electron-ion collision frequency calculated from the Silin formulae [38]. When $n_e/n_{crit} = 3$, $|\varepsilon+2|$ goes through a minimum and the field inside the cluster is greater than the external field. At this resonance, the heating rate is also increased. While for a bulk plasma, an increase in the laser field actually results in a decrease in the collisional heating rate at high intensity, for the nanoplasma near this resonance, the field enhancement results in a significant increase in the heating rate. Another consequence of the geometry of the cluster is that there is no heat conduction possible to surrounding "cold" material, as is the case in a bulk plasma, suggesting that very high plasma temperatures may be attained in the nanoplasma.

Expansion. The expansion of the nanoplasma is driven by two pressures. The Coulomb pressure arising from repulsion between ions following a charge build-up Qe on the cluster of radius r:

$$P_{Coul} = \frac{Q^2 e^2}{8\pi r^4}, \qquad (16)$$

and the hydrodynamic pressure from the hot electrons (the hot electrons expand outwards, dragging the ions with them):

$$P_H = n_e k T_e. \qquad (17)$$

The Q/r^4 scaling of P_{Coul} shows that it will be important for small clusters or for low Z clusters (such as deuterium – see later) where the electrons are not confined and Q becomes large. The hydrodynamic pressure scales as r^{-3} (since $n_e \sim volume^{-1}$), so is therefore more important for larger clusters. In fact, it appears to dominate over P_{Coul} for most cases we have studied in Ar, Kr and Xe clusters ($r > 2$ nm). For realistic nanoplasma conditions ($n_e = 10^{23}$ cm^{-3}, $kT_e = 1$ keV), $P_H \approx 100$ Mbar. The nanoplasma expansion during the laser pulse is very important since it results in the rapid decrease in n_e required to reach the $n_e = 3n_{crit}$ heating resonance near the peak of the pulse.

Charge Buildup. The charge build up on the cluster is due to electrons free-streaming from the surface of the sphere. To free-stream, the electrons must be less than one mean free path (in the direction of travel) from the surface and have an energy greater than the escape energy of the cluster. The escape energy is calculated from the potential energy on the surface of a sphere with charge Q, assuming that the charge is distributed isotropically over the sphere. Obviously, free-streaming depletes the hot tail of the Maxwellian electron distribution. The assumption of a Maxwellian distribution at all times is based on a rapid rethermalisation of the distribution.

Ion Energies. Electron-ion collisions heat up the electrons efficiently, but not the ions. This can be seen from the electron-ion equilibration time. For typical nanoplasma conditions and for initially cold ions, it is in excess of 30 ps, much longer than the nanoplasma disassembly time. In other words, there is insufficient time for the electron energy to be transferred to the ions through collisions.

Instead, the ions gain energy in the hydrodynamic expansion (where the thermal energy of the electrons is converted to direction ion kinetic energy) and/or through Coulomb explosion. For a hydrodynamic expansion at the plasma sound speed, the mean ion energy will be of order

$$<E_{ion}> \approx <Z>kTe, \qquad (18)$$

where $<Z>$ is the mean ion charge state. This can be substantial for keV electron temperatures and highly charged ions ($<Z> \approx 20$ – see later). Coulomb explosion [29] is the dominant process by which ions gain kinetic energy for molecules and small clusters in intense laser fields. For large clusters, it seems only to becomes important for low Z clusters where the charge build up is substantial.

A Numerical Simulation

We now look at a simulation of a laser-cluster interaction carried out using a numerical nanoplasma code that embodies the physics described above. The calculated time-history of an exploding cluster of 5000 Xe atoms irradiated by a 200-fs, 780-nm laser pulse with a peak intensity of 1×10^{16} Wcm^{-2} is illustrated in Figure 4.

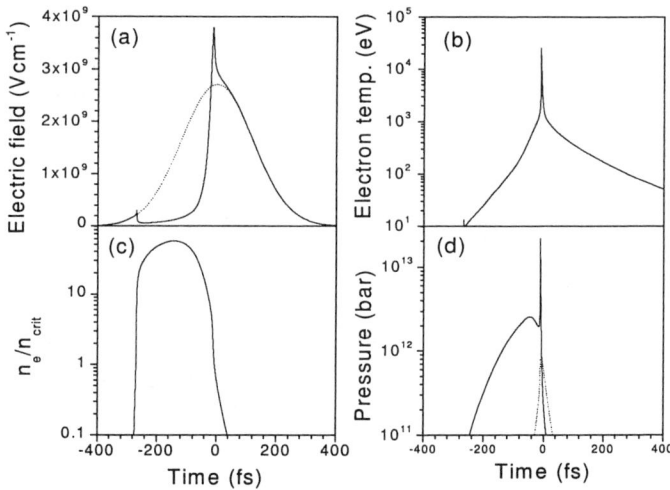

Figure 4. Calculated time history of an exploding Xe cluster. (a) Envelope of the laser pulse (dotted line) and field inside the cluster (solid line). (b) Electron temperature inside the cluster. (c) Electron density/critical density. (d) Hydrodynamic (solid line) and Coulomb (dotted line) pressure.

The peak of the laser pulse is at $t = 0$ fs. Early on in the laser pulse, at around $t = -280$ fs when the intensity is $\sim 4 \times 10^{13}$ Wcm^{-2}, a few free electrons are created through tunnel ionisation. The electron density rises to reach $3n_{crit}$ at $t = -270$ fs (Figure 4c). At this point the field in the cluster is enhanced (Figure 4a) and more electrons are liberated through tunnel, laser-driven and thermal ionisation. The electron density is now higher than $3n_{crit}$ and the field inside the cluster is shielded from the external laser field. The tunnel and laser-driven ionisation rates fall off, but electrons are still created through thermal collisions.

From $t = -50$ fs onwards, some electrons are able to leave the cluster, as the mean electron temperature is in the region of 100 – 1000 eV and the escape energy is ~ 200 – 2000 eV. The combined effect of the free-streaming of electrons out of the cluster and the hydrodynamic expansion of the cluster is that the electron density starts to fall, after peaking at over $50 n_{crit}$. The field in the cluster again starts to rise as the electron density drops, so the tunnel and laser-driven ionisation rates increase while the thermal ionisation rate falls. Near the peak of the laser pulse, at $t = -12$ fs, the electron density in the cluster drops to $3 n_{crit}$. The resonantly increased heating rate causes the electron temperature in the cluster to soar to 25 keV (Figure 4b). The field in the cluster is also strongly enhanced and the peak intensity in the cluster reaches 2×10^{16} Wcm^{-2}, twice the intensity outside. The electron free-streaming rate increases sharply as a significant number of electrons have energies above the then 4-keV escape energy.

The total charge on the cluster increases to 5.5×10^4 e, resulting in the Coulomb pressure increasing to 10 Mbar (Figure 4d). However, this is small compared to the hydrodynamic pressure due to the hot electrons of 200 Mbar. This pressure causes a sharp increase in the cluster expansion velocity. This is the explosion of the cluster. Once the nanoplasma density has dropped to $\approx 10^{17}$ cm^{-3} – the typical background density in an extended cluster gas medium - the final expansion velocity of electrons and ions is 3.3×10^7 cms^{-1}, which corresponds to a maximum ion energy of ~80 keV. The final electron energy is much lower, only 30 eV, a consequence of their much lighter mass. However, electrons that free streamed away from the nanoplasma have energies in the 0.2-2 keV range.

This simulation shows an extremely energetic laser-cluster interaction, with ion energies close to 100 keV and electron energies of several keV. These energies are much higher than those observed in the Coulomb explosion of molecules and small clusters, and are more reminiscent of a laser-solid interaction – which of course is the case at the microscopic level. In the following review of the experimental data, we will see that these predictions are borne out in experiments.

REVIEW OF EXPERIMENTAL DATA

In this section we examine a range of experimental data from the Imperial College Laboratories which will bring out the important features of the interaction and allow the predictions of the nanoplasma model to be tested. As well as Imperial College, a number of other groups around the world are also active in this field.

Electron Data

The existence of extremely energetic electrons (>1 keV) from the laser-cluster interaction had been inferred since the first X-ray spectra were recorded from cluster plasmas [4] and is also predicted by the nanoplasma model.

The spectrum shown in Figure 5 is the result of the first direct measurement of the electron energies from the explosion of isolated clusters [10]. using the apparatus and techniques described earlier. This spectrum was recorded for ~2100-atom Xe clusters irradiated with ~150 fs Ti:S pulses at an intensity of ~2×10^{16} Wcm^{-2}.

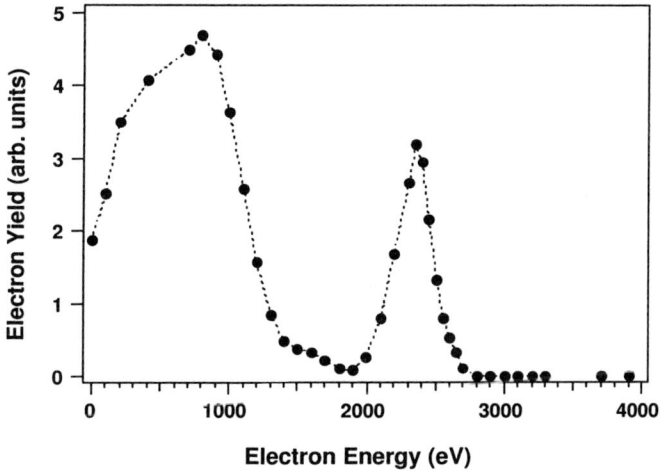

Figure 5. Electron energy spectrum from the explosion of 2100-atom Xe clusters at an intensity of 2×10^{16} Wcm^{-2}.

The spectrum is striking for two reasons. First, the production of electrons with energy out to nearly 3 keV is radically different to the spectra one would see from laser interactions with single Xe atoms, where energies of no more than 100 eV would be expected [39]. In fact, the electrons are even hotter than for a laser-solid interaction at this intensity [40]. This highlights the remarkably efficient coupling of the laser energy into the cluster.

Second, the spectrum shows two peaks. The first, referred to as the "warm" electrons comprises electrons with energies in the range 0.1 to 1 keV. A sharper "hot" electron peak is also seen at an energy of ~2.5 keV. Warm and hot electrons are produced under different conditions at different times in the cluster expansion. This conclusion was reached by examining the angular dependence of the electron emission with respect to the laser polarisation (Figure 6). The measured angular distribution of warm electrons is shown in part (a), and that of the hot electrons in part (b) of this figure.

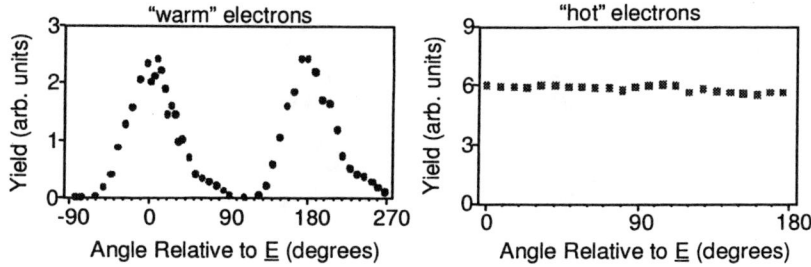

Figure 6. Angular distribution of (a) warm and (b) hot electrons.

The angular distributions of the two groups of electrons are markedly different. The hot electron emission is isotropic, while the warm electron emission is peaked along the laser polarisation, with a full-width at half-maximum of about 60°, significantly broader than expected for ATI electrons from single atoms [41].

Simulations from the nanoplasma code reproduce the prominent features. Figure 7 shows the calculated electron energy spectrum using parameter values close to the experimental ones. A broad warm peak and a sharper hot peak near 2.5 keV is seen, in reasonable agreement with the experimental data.

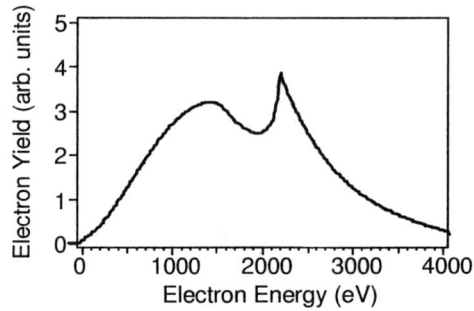

Figure 7. Calculated electron spectrum.

Simulations show that the warm electron peak is the result of collisional heating of electrons near the surface of the cluster on the rising edge of the laser pulse. The hot electrons result from rapid heating of the remaining electrons in the bulk of the cluster later in the pulse when the electron density drops to a point to bring the heating into resonance. This is consistent with the observed angular distribution data. As illustrated in Figure 8, the warm electrons have undergone a limited number of collisions, broadening the angular distribution from that of purely tunnel ionised electrons. The hot electrons result from extensive collisional heating of the electrons in the bulk of the cluster. Consequently, their velocity distribution has been completely randomised, accounting for the isotropic distribution observed. It is clear that electron spectroscopy provides an extremely useful channel of information for understanding the laser-cluster dynamics.

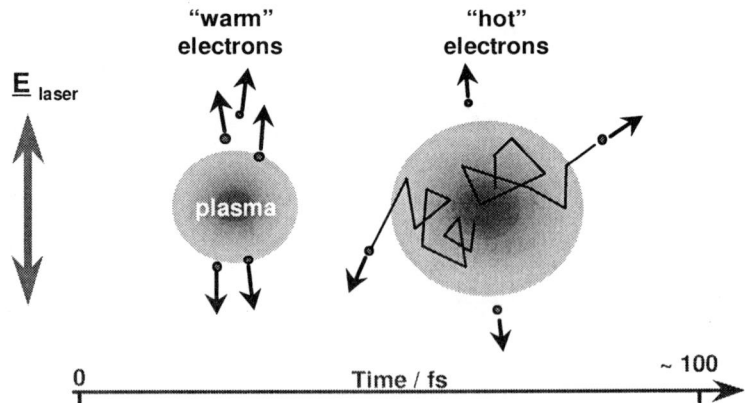

Figure 8. Schematic showing "warm" and "hot" electron production in the nanoplasma.

Ion Data

Attention is now focused on the ions from the same interaction. TOF spectra (obtained without an extraction field – see earlier) averaged over several thousand shots are shown in Figure 9a, and the ion energy spectrum in part (b) of this figure. The fast unresolved feature in the TOF spectrum is due to the electrons, the broader peak to the ions. By varying the half-wave plate in the beam path before the focusing lens, it is seen in that the ion energy distribution is almost isotropic with respect to the direction of the laser polarisation; a clear signature of a spherically symmetric cluster explosion.

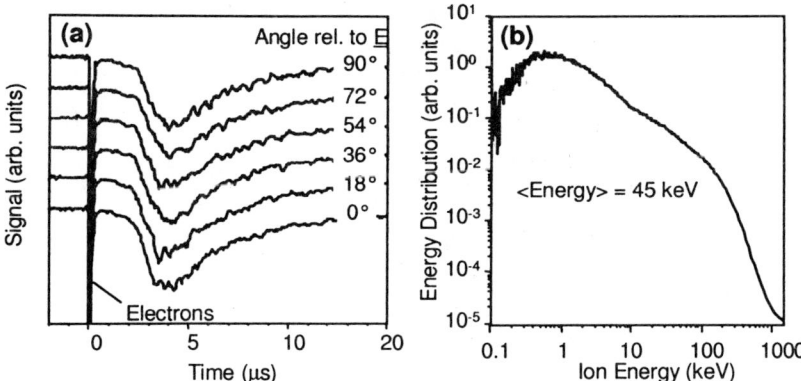

Figure 9. (a) Ion TOF spectra for a range of laser polarisation angles. (b) Corresponding energy spectrum.

The mean ion energy is 45±5 keV, showing that the average laser energy deposited per ion is substantial. This is consistent with Equation 10 if we assume that the electron temperature is given by the hot electron feature i.e., $kT_e \sim 2.5$ keV, and the average charge state is $Z \sim 20^+$ (see next section).

A remarkable aspect of the ion spectrum is the presence of ions with energies up to 1 MeV. This energy is four orders of magnitude higher than has previously been observed in the Coulomb explosion of molecules [42] and about 1000 times higher than the average energy of the highest charge state Ar ions ejected in the disintegration of small (<10 atoms) clusters [7]. In fact 1 MeV is about a factor four higher than the predictions of the nanoplasma model for these parameter values. In Ref.[16] it is proposed that the MeV ions do not result from the hydrodynamic expansion, but are the result of an electrostatic shock wave that Coulomb ejects the most highly charged ion that are close to the cluster surface.

The production of highly stripped ions is another characteristic feature of the laser-cluster interaction, already observed for lower intensity laser-cluster interactions [5,8,14]. The charge state distributions of the Xe ions above is shown as a function of their kinetic energy in Figure 10. For high energy Xe ions (>100 keV), the peak charge state is at $Z = 18^+$ to 25^+, with some ions, remarkably, having charge states as high as 40^+. These are much higher than the $\sim 12^+$ expected from field ionisation of single atoms at these intensities [43]. In fact, over the barrier ionisation to Xe^{40+} would require a Xe atom to be exposed to an intensity of nearly 10^{20} Wcm^{-2}.

These very high charge states can be explained in terms of collisional ionisation in the nanoplasma by high temperature electrons. The experimental data show that the charge states depend only weakly on ion kinetic energy, contrary to what would be expected from a pure Coulomb explosion. An alternate model for the enhanced ionisation (and efficient heating of the cluster) is the ionisation ignition model [44]. In this model, the ionisation rate is increased due to the close packing of the ions in the cluster, which lowers their ionisation potentials. However, this model does not predict an optimum cluster size for cluster heating. The existence of an optimum size (for given laser pulse parameters) emerges naturally from the nanoplasma model, and has been confirmed experimentally, as we shall see in the next section.

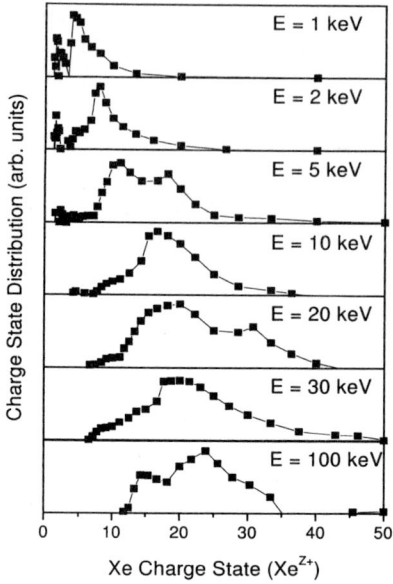

Figure 10. Measured ion charge state distributions.

Parameter Scalings

In this section we look at how the ion energies from cluster explosions scale with the two important experimental parameters, the cluster size and the laser intensity. These scaling provide tests of the models and help identify the conditions which maximise the explosion energy which is important for applications. The data presented was obtained with the 200 fs, 780 nm Ti:S laser.

Cluster Size. Figure 11 shows how the maximum ion energy varies with the number of atoms in a Xe cluster in the range 200 to 74000 Xe atoms [22].

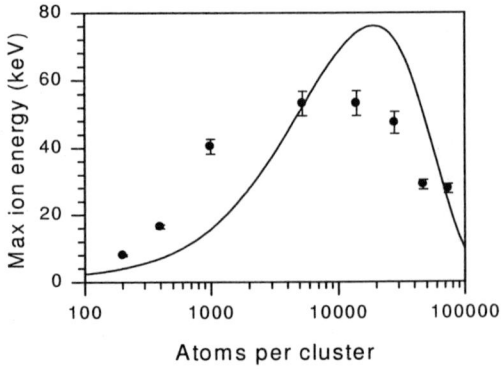

Figure 11. Measured (points) and calculated (line) maximum ion energies as a function of cluster size.

A laser intensity of 3×10^{15} Wcm^{-2} was used. The threshold size for energetic ion production (>1 keV) appears to be ~ 200-400 atoms. The maximum ion energy rises from 8 keV at 200 atoms/cluster to a peak of ~53 keV for clusters of ~ 8,000 atoms before falling to ~28 keV as the cluster size is increased beyond 50,000 atoms/cluster. Kr and Ar clusters show a qualitatively similar behaviour.

The curve in this figure is the result of the nanoplasma model, which is in reasonable agreement with the experimental data, though the calculated optimum size is somewhat larger. Recall that there is a factor of two uncertainty in the cluster size measurements and further, the calculation does not take into account the distribution of cluster sizes present in the laser focus.

The existence of an optimum cluster size is easy to understand. The maximum ion energy is determined largely by the laser intensity when the cluster experiences the resonant heating. The highest ion temperatures are obtained when the cluster passes through the $n_e = 3n_{crit}$ point near the peak of the laser pulse. Small clusters expand too quickly and reach this point before the peak. Larger clusters pass through $3n_{crit}$ well after the peak intensity is reached. This also explains the existence of an optimum laser pulse-width for a given cluster size seen in [9,18].

Laser Intensity. Figure 12 shows how the maximum ion energies vary with the peak laser intensity (230 fs, 780 nm pulses) for 5300-atom Xe clusters and 6200-atom Kr clusters [22].

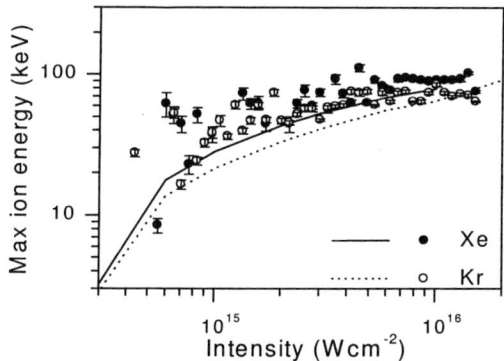

Figure 12. Maximum ion energies measured as a function of laser intensity for Xe clusters (solid points) and Kr clusters (open points) together with simulated scalings for Xe clusters (solid line) and Kr clusters (dotted line).

In Xe, a sharp onset of fast ion production is seen at 6×10^{14} Wcm^{-2}. The ion energies rise steeply up to ~ 10^{15} Wcm^{-2}, to an energy of ~ 50 keV. Above 10^{15} Wcm^{-2}, the ion energies saturate, increasing as $\sim I^{0.2}$ (the integrated ion yield scale as $\sim I^{1.4}$, consistent with the increase in focal volume). At 10^{16} Wcm^{-2}, the maximum ion energy is 90 keV. The Kr ion energies follow a similar trend to the Xe, with a sharp increase up to a mean energy of 9 keV at 10^{15} Wcm^{-2} followed by a slow increase up to energies of 75 keV at 10^{16} Wcm^{-2}. For Kr the ion energies are about 20% lower. The solid and dotted curves are the nanoplasma calculations for Xe and Kr, respectively. Again, the agreement is reasonable.

The modelling provides explanations for the well-defined "appearance intensity" for energetic ion production, as well as the eventual saturation in the ion energies with increasing intensity. Referring back to Figure 4c, we see that the energetic cluster explosion is triggered by the second occurrence of the $n_e = 3n_{crit}$ resonance, since the first one occurs very early in the laser pulse when the intensity is low. Modelling shows that the appearance intensity is just the minimum peak intensity required to ensure two occurrences of the resonance during the laser pulse. Below this intensity, the resonance is achieved only once, early in the laser pulse, because the expansion velocity is insufficient compared to the ionisation rate to drop n_e back down to $3n_{crit}$. The saturation in the ion energies with increasing intensity is a product of the increased ionisation that occurs on the rising edge of the pulse. This results in the cluster expanding faster and reaching the second resonance earlier in the pulse. It therefore does not experience a substantially higher laser intensity at the point of resonant heating.

APPLICATIONS

So far, we have concentrated on the explosions of individual clusters that have been studied in experiments with low density cluster beams and described theoretically in terms of the nanoplasma model. Now we turn to applications which are carried out in extended, high density cluster media (see earlier discussion for details). Here, a hot, underdense ($n_e < n_{crit}$) plasma is formed following the explosions of the individual clusters in the focal volume. This plasma cools and decays on a much longer time scale than the femtosecond laser pulse.

X-ray Generation

These plasmas containing highly charged ions, emit copious quantities of X-rays in the 0.1-10 keV range. In fact, it was the observation of "anomalous" bright X-rays from gas-jets that triggered the interest in clusters in strong fields. Conversion of up to 10% of the incident laser energy into X-rays in the 17-30 nm range in Ar clusters was reported in Ref.[9] using 30 mJ, 130 fs laser pulses. The X-ray spectra show resonant line emission from excited plasma ions (e.g. 4p-3d in Kr^{10+}). Time-resolved measurements show that the X-ray emission consists of a prompt sub-picosecond spike followed by a lower intensity emission that lasts for several nanoseconds. The spike is attributed to intense X-ray emission from the individual nanoplasmas immediately after heating by the laser, while the nanosecond time-scale radiation is emitted from the resulting underdense plasma. The main attractions of cluster X-ray sources, particularly for lithographic applications, are that they produces very little debris (which is a major problem with solid targets) and that they can be operated at a very high repetition rate, or even CW with a continuous gas flow.

Nucluear Fusion

The very large ion energies observed in cluster interaction experiments and the high efficiency of energy coupling to bulk cluster media (almost 100% absorption of laser energy has been observed [45]) suggest a new method of driving nuclear fusion that can be achieved with small scale (table-top) laser systems operating at high repetition rates. The $D + T \rightarrow n + He^4$ has by far the largest cross-section for the range of ion energies accessible in laser-cluster experiments, but tritium poses a serious radiological hazard and so the D-D reaction has been investigated instead. Half the D-D fusion events follow the $D + D \rightarrow n(2.45 \text{ MeV}) + {}^3He$ reaction, which is convenient for study since the neutron can escape the interaction chamber and being of well defined energy can be identified unambiguously through TOF measurements.

The first definitive observation of a cluster based fusion reaction was reported in Ref.[46], in a dense D_2 cluster target irradiated with a 35 fs, 200 mJ laser pules from a Ti:S CPA system. The low value of the condensation parameter, k, for D_2 requires it to be cooled below 80 K to allow large cluster formation. Around 50,000 neutrons were observed per shot (at a repetition rate of 10 Hz), which is a large number when normalised to the input laser energy. In contrast to high Z clusters (Ar, Kr, Xe), D_2 clusters can be stripped of almost all their electrons by a sufficiently short laser pulse (<50 fs is required). Hence a Coulomb explosion picture, rather than a nanoplasma model is appropriate. For a fully stripped D_2 cluster of radius r, the maximum ion energy from the Coulomb explosion is given by

$$E_{max} = \frac{2n_D e^2 r^2}{3\varepsilon_0}, \qquad (19)$$

where n_D is the initial ion density (~3×10^{22} cm^{-3} for D_2). From this it can be see that a D_2 cluster of radius >2.5 nm is required to produce multi-keV deuterons necessary for fusion.

Fusion in clusters has attracted considerable interest as a pulsed, point-like source of neutrons. The neutron pulse duration is set by the time taken for an energetic deuteron to leave the laser focus, and is estimated to be of order 100 ps, which is considerably shorter than spallation sources. With improvements to the neutron yield and operation at high repetition rate (*e.g.*, 1 kHz), such table-top neutron source may well have application in new short pulse neutron imaging techniques along with material probing and long term testing of materials under high fluxes of neutrons.

Acknowledgements

Many people have contributed to the work described in this chapter. In particular, the author acknowledges the contributions of T. Ditmire, N. Hay, H. Hutchinson, K. Mendham, J. Marangos, M. Mason, P. Ruthven, Y. Shao, R. Smith and E. Springate. The Imperial College programme is supported by the UK Engineering and Physical Sciences Research Council (EPSRC) and the author is supported by an EPSRC Advanced Fellowship.

References

[1] E.W. Becker, K. Bier and W. Henkes, *Z. Phys.* **146** 333 (1956).
[2] L. Rymell and H.M. Hertz, *Rev. Sci. Inst.* **66** 4916 (1995).
[3] S.P. Gordon, T. Donnelly, A. Sullivan, H. Hamster and R.W. Falcone, *Opt. Lett.* **19** 484 (1994).
[4] A. McPherson, T.S. Luk, B.D. Thompson, K. Boyer, and C.K. Rhodes, *Appl. Phys. B* **57** 337 (1993).
[5] A. McPherson, B.D. Thompson, A.B. Borisov, K. Boyer, and C.K. Rhodes, *Nature* (London) **370** 631 (1994).
[6] A McPherson, TS Luk, BD Thompson, AB Borisov, OB Shiryaev, X Chen, K Boyer, and CK Rhodes, *Phys. Rev. Lett.* **72** 1810 (1994).
[7] J. Purnell, E.M. Snyder, S. Wei, and A.W. CastlemaN Jr, *Chem. Phys. Lett.* **229** 333 (1994).
[8] T. Ditmire, T. Donnelly, R.W. Falcone, and M.D. Perry, *Phys. Rev. Lett.* **75** 3122 (1995).
[9] T. Ditmire, T. Donnelly, A.M. Rubenchik, R.W. Falcone, and M.D. Perry, *Phys. Rev. A* **53** 3379 (1996).
[10] Y.L. Shao, T. Ditmire, J.W.G. Tisch, E. Springate, J.P. Marangos, and M.H.R. Hutchinson, *Phys. Rev. Lett.* **77** 3343 (1996).
[11] T. Ditmire, J.W.G. Tisch, E. Springate, M.B. Mason, N. Hay, R.A. Smith, J. Marangos, and M.H.R. Hutchinson, *Nature* (London) **386** 54 (1997).
[12] T. Ditmire, J.W.G. Tisch, E. Springate, M.B. Mason, H. Hay, J.P. Marangos, and M.H.R. Hutchinson, *Phys. Rev. Lett.* **78** 2732 (1997).
[13] E.M. Snyder, S. Wei, J. Purnell, S.A. Buzza, and A.W.C. Jr., *Chem. Phys. Lett.* **248** 1 (1996).
[14] E.M. Snyder, S.A. Buzza, and A.W. Castleman, *Phys. Rev. Lett.* **77** 3347 (1996).
[15] M. Lezius, S. Dobosz, D. Normand and M. Schmidt, *J. Phys. B: At, Mol. Opt. Phys.* **30** L251 (1997).
[16] M. Lezius, S. Dobosz, D. Normand and M. Schmidt, *Phys. Rev. Lett.* **80** 261 (1998).
[17] T. Ditmire, E. Springate, J.W.G. Tisch, Y.L. Shao, M.B. Mason, N. Hay, J.P. Marangos and M.H.R. Hutchinson, *Phys. Rev. A* **57** 369 (1998).

[18] J. Zweiback, T. Ditmire and M.D. Perry, *Phys. Rev. A* **59** R3166 (1999).
[19] S. Dobosz, M. Schmidt, M. Perdrix, P. Meynadier, O. Gobert, D. Normand, K. Ellert, A. Ya. Faenov, A.I. Magunov, T.A. Pikuz, I. Yu. Skobelev, N.E. Andreev, *J. Expt. And Theor. Phys.* **88** 1122 (1999).
[20] J.W.G. Tisch, N. Hay, E. Springate, E.T. Gumbrell, M.H.R. Hutchinson and J.P. Marangos, *Phys. Rev. A* **60** 3076 (1999).
[21] E. Springate, N. Hay, J.W.G. Tisch, M.B. Mason, T. Ditmire, J.P. Marangos and M.H.R. Hutchinson, *Phys. Rev A* **6104** (4) 4101 (2000).
[22] E. Springate, N. Hay, J.W.G. Tisch, M.B. Mason, T. Ditmire, M.H.R. Hutchinson and J.P. Marangos, *Phys. Rev A* **6106** (6) 3201 (2000).
[23] see, *e.g.*, M.D. Perry and G. Morou, *Science* **264** 917 (1994).
[24] D. Strickland and G. Morou, *Opt. Comm.* **56** 219 (1985).
[25] P. Agostini and G. Petite, *Contemp. Phys.* **29** 57 (1988).
[26] K. Burnett, V.C. Reed and P.L. Knight, *J. Phys. B*, **26** 561 (1993).
[27] M. Protopapas, C.H. Keitel and P.L. Knight, *Rep. Prog. Phys.* **60** 389 (1997).
[28] B. Sheehy and L.F. DiMaurio, *Annual Rev. Phys. Chem.* **47** 463 (1996).
[29] K. Codling and L.J. Frasinski, *Contemp. Phys.* **35** 243 (1994).
[30] see, *e.g.*, P. Gibbon and S. Forster, *Plas. Phys. Control Fusion* **38** 769 (1996).
[31] O.F. Hagena and W. Obert, *J. Chem. Phys.* **56** 1793 (1972).
[32] M. Lewerenz, B. Schilling and J.P. Toennies, *Chem. Phys. Lett.* **206** 381 (1993).
[33] C.L. Briant and J.J Burton *J. Chem. Phys.* **63** 2045 (1975).
[34] D.J. Fraser and M.H.R. Hutchinson, *J. Mod. Opt.* **43** 1055 (1996).
[35] S. Augst, D.D. Meyerhofer, D. Strickland and S.L. Chin, *Phys. Rev. Lett.* **63** 2212 (1989).
[36] W. Lotz, *Z. Phys.* **216** 241 (1968).
[37] J.D. Jackson, "Classical Electrodynamics", (New York: John Wiley & Sons 1975).
[38] V.P. Silin, *Sov. Phys. JETP* **20** 1510 (1965).
[39] T.E. Glover, T.D. Donnelly, E.A. Lipman, A. Sullivan and R.W. Falcone, *Phys. Rev. Lett.* **73** 78 (1994).
[40] G. Güthlein, M.E. Foord and D. Price, *Phys. Rev. Lett.* **77** 1055 (1996).
[41] U. Mohideen, M.H. Sher, H.W.K. Tom, G.D. Aumiller, O.R.W. II, R.R. Freeman, J. Bokor, and P.H. Bucksbaum, *Phys. Rev. Lett.* **71** 509 (1993).
[42] C. Cornaggia, M. Schmidt, and D. Normand, *J. Phys. B: At. Mol. Opt. Phys.* **27** L123 (1994).
[43] S. Augst, D.D. Meyerhofer, D. Strickland, and S.L. Chin, *J. Opt. Soc. B* **8** 858 (1991).
[44] C. Rose-Petruck, K.J. Schafer, K.R. Wilson and C.P.J. Barty, *Phys. Rev. A* **55** 1182 (1997).
[45] T. Ditmire, R.A. Smith, J.W.G. Tisch and M.H.R. Hutchinson, *Phys. Rev. Lett.* **78** 3121 (1997).
[46] J. Zweiback, R.A. Smith, T.E. Cowan, G. Hays, K.B. Wharton, V.P. Yanovsky and T.Ditmire, *Phys. Rev. Lett.* **84** 2634 (2000).

INTRODUCTION TO LASER-PLASMA INTERACTION AND ITS APPLICATIONS

Stefano Atzeni

Dipartimento di Energetica
Università degli Studi di Roma "La Sapienza" and INFM
Via A. Scarpa, 14 - 00161 Roma, Italy
E-mail: atzeni@uniroma1.it

INTRODUCTION

Pulsed lasers delivering intensities $I > 10^{11}$ W/cm^2 are unique tools for concentrating energy into matter, which is soon brought to the plasma state. Since the early 1960s the interaction of laser light with a plasma proved an extremely rich topic, with many applications. Wide presentations of the status of the field around 1990 are given by a textbook[1] and a handbook.[2] Recently, new frontiers in laser-plasma investigation[3,4,5] have been opened by the introduction of the chirped-pulse-amplification (CPA) technique,[6] which has made feasible subpicosecond pulses with intensity 10^{17}–10^{21} W/cm^2 and power up to a Petawatt.[7] Parameters of lasers used in laser-plasma research are listed in table 1.

Both the more conventional lasers delivering energy $E > 100$ J in nanosecond pulses and CPA lasers delivering $E > 0.1$ J in subpicosecond pulses allow to reach extreme conditions in the heated matter. Laboratory record values of pressure, matter density, electric and magnetic field in the matter have been achieved in laser-produced plasmas. We shall see later that nanosecond pulsed lasers easily generate multi-megabar collisional shock waves, which can be used to compress matter to very high density. Subpicosecond pulses of a fraction of a joule, focused at $I \approx 10^{17}$ W/cm^2 in a few micron spot onto a solid target, can create solid state plasmas with temperatures of a

Table 1. Main parameters of pulsed lasers for plasma research [a]

	nanosecond pulse	ultrashort pulse	
		table-top	large-size
pulse energy E	0.1–60 kJ	< 1 J	≤ 800 J
pulse duration t_p	1–5 ns	10–200 fs	≤ 1 ps
peak power P	0.05–30 TW	< 10 TW	≤ 1300 TW
focal spot radius r_f	0.1–0.5 mm	≈ 10 μm	≈ 10 μm
peak intensity I	10^{11}–10^{15} W/cm^2	10^{17}–10^{19} W/cm^2	≤ 10^{21} W/cm^2

[a] Laser wavelengths $\lambda = 0.25$–1.33 μm

Atoms, Solids, and Plasmas in Super-Intense Laser Fields
Edited by D. Batani *et al.*, Kluwer Academic/Plenum Publishers, 2001

few hundred electronvolts.[8] Pulses with $I > 10^{18}$ W/cm² and wavelength $\lambda = 1$ μm, focused onto gas targets or preformed plasmas, instead, create relativistic plasmas. This is immediately seen by considering the momentum of an electron oscillating in the laser field

$$p_{osc} = m_e v_{osc} \left(1 - \frac{v_{osc}^2}{c^2}\right)^{-1/2} = \frac{q_e \mathcal{E}}{\omega}, \tag{1}$$

where $\mathcal{E}$ is the amplitude of the electric field, ω is the laser (angular) frequency, q_e and m_e are the electron charge and rest mass, respectively, v_{osc} is the electron quiver velocity and c is the speed of light. Since (in vacuum)

$$\mathcal{E} = \left(\frac{8\pi I}{c}\right)^{1/2} = 2.73 \times 10^{10} \ I_{18}^{1/2} \ \text{V/cm}, \tag{2}$$

we can write

$$a_L^2 = \left(\frac{p_{osc}}{m_e c}\right)^2 = 0.73 \ I_{18} \lambda_{\mu m}^2, \tag{3}$$

which shows that electrons behave relativistically for $I_{18}\lambda_{\mu m}^2 > 1$. In the above equations I_{18} is the intensity in units of 10^{18} W/cm² and $\lambda_{\mu m}$ is the laser wavelength in μm. In this regime also the magnetic field B and the radiation pressure p_L associated to the laser field take very large values, given by

$$B = \left(\frac{8\pi I}{c}\right)^{1/2} = 91 \ I_{18}^{1/2} \ \text{MG} = 107 \ \frac{a_L}{\lambda_{\mu m}} \ \text{MG} \tag{4}$$

and

$$p_L = \frac{I}{c} = 333 \ I_{18} \ \text{Mbar} = 456 \ \frac{a_L^2}{\lambda_{\mu m}^2} \ \text{Mbar}. \tag{5}$$

Laser-produced plasmas have a number of applications.[9] In the longer pulse regime, these include studies of thermodynamic and optical properties of matter at very high density and pressure,[10, 11] inertial confinement fusion (ICF),[12, 13] the simulation of astrophysical processes in the laboratory,[14, 15] coherent and incoherent X-ray sources.[16] Plasmas produced by ultrashort and ultraintense laser pulses can be used as intense sources of particles and photons (relativistic electrons, MeV ions, positrons, X- and gamma-rays, neutrons etc.) and for particle acceleration.[4, 17, 18, 19] One can now perform nuclear physics experiments driven by a table-top photon source.[20] A fascinating, but still speculative use of ultrapowerful lasers is the *fast ignitor*[21] ICF concept.

In this tutorial paper, we first briefly outline basic concepts on the interaction of laser light with a plasma, and characterize different interaction regimes. Since many other chapters of this book deal with relativistic plasmas created by ultraintense, sub-picosecond lasers, here we focus on the generation of ablative pressure in plasmas produced by moderate intensity, nanosecond pulses. In the subsequent sections we survey physics issues and applications concerning plasmas created by both nanosecond and subpicosecond pulses. The last part of the paper deals with laser-driven inertial confinement fusion.

LASER-PLASMA INTERACTION

Here, we present elementary notions about the interaction of a pulsed laser beam with a plasma. These are also relevant to irradiation of solid targets, when the pulse duration is longer than the time required for the formation of a plasma. Also notice that in many laser-solid experiments ultrashort pulses actually interact with a plasma formed in front of the solid surface by the relatively low intensity foot which often preceeds the main laser pulse. The readers interested in the interaction of an ultrashort and ultraintense pulse with a solid, which is not dealt with here, can refer to recent papers.[22, 23, 24]

Collisional absorption in underdense plasma

Laser light can propagate through a plasma if the laser frequency ω exceeds the plasma frequency $\omega_p = (4\pi n_e q_e^2/m_e)^{1/2}$, where n_e is the electron density. In terms of the mass density $\rho = m_i n_i = A m_p n_e/Z$, the condition for propagation is $\rho < \rho_c$, where $\rho_c = 1.865 \times 10^{-3} A/Z\lambda_{\mu m}^2$ g/cm^3 is the so-called critical density. (Here A is the mass number, Z is the average ion charge, m_p is the proton mass, and m_i and n_i are the ion mass and number density, respectively.) This propagation condition follows from the refraction index of an unmagnetized plasma for electromagnetic waves, $\mu \simeq (1 - \omega_p^2/\omega^2)^{1/2}$.

At relatively low intensity the main coupling mechanism of laser light to a plasma is inverse bremsstrahlung.[1, 25] The electrons performing quiver oscillations in the laser electric field collide with the ions, thus converting the coherent motion into thermal motion. As a result, the plasma is heated. The rate of absorption of the laser energy is then computed by equating the power lost by the radiation field to that lost collisionally by the oscillating electrons:[1]

$$\frac{\mathcal{E}^2}{8\pi}\nu_{\text{abs}} \approx n_e \frac{m_e v_{\text{osc}}^2}{2}\nu_{\text{ei}}, \tag{6}$$

where ν_{ei} is an appropriately averaged electron-ion collision frequency[1, 26] and ν_{abs} is the damping frequency of the electromagnetic field energy. Inserting the non-relativistic limit of $v_{\text{osc}} = q_e\mathcal{E}/m_e\omega$ into Eq. (6) we find

$$\nu_{\text{abs}} = \frac{\omega_p^2}{\omega^2}\nu_{\text{ei}}. \tag{7}$$

In a damping period $1/\nu_{\text{abs}}$ the light travels a distance $L = c_{\text{gr}}/\nu_{\text{abs}}$, where $c_{\text{gr}} = c\mu$ is its group velocity in the plasma. Therefore, as light propagates by a distance dx its intensity decreases by an amount $dI = -I dx/L = -K_{\text{ib}} I dx$, with inverse bremsstrahlung absorption coefficient

$$K_{\text{ib}} \approx \frac{\nu_{\text{abs}}}{c_{\text{gr}}} = \frac{\nu_{\text{ei}}}{c}\frac{\omega_p^2}{\omega^2}\left(1 - \frac{\omega_p^2}{\omega^2}\right)^{-1/2}. \tag{8}$$

Equation (8) can also be obtained from the linear theory of wave propagation. We refer to plane Fourier modes, so that the transverse electric field is $\mathcal{E} \sim e^{-i(kx-\omega t)}$ and the intensity is $I \sim e^{-2i(kx-\omega t)}$. The absorption coefficient is then obtained from the dispersion relation $k = k(\omega)$ by

$$K = -2\Im[k(\omega)] = -\Im[k(\omega)^2]/\Re[k(\omega)], \tag{9}$$

Here the symbols $\Im$ an $\Re$ indicate imaginary and real parts, respectively. The dispersion relation for electromagnetic waves propagating in a homogeneous, unmagnetized, cold, collisional plasma is[1, 25]

$$\frac{k^2 c^2}{\omega^2} = 1 - \frac{\omega_p^2}{\omega^2 + \nu_{ei}^2}\left(1 + i\frac{\nu_{ei}}{\omega}\right). \tag{10}$$

Inserting Eq. (10) into Eq. (9) we get

$$K = \frac{\nu_{ei}}{c}\frac{\omega_p^2}{\omega^2}\left(1 - \frac{\omega_p^2}{\omega^2 + \nu_{ei}^2}\right)^{-1/2}, \tag{11}$$

from which we recover Eq. (8), since in all interesting cases $\nu_{ei} \ll \omega$. By writing $\omega = 2\pi c/\lambda$ and introducing the expressions of ν_{ei}, ω_p and ρ_c as functions of the laser wavelength and of the plasma density, electron temperature and average ion charge, Eq. (8) becomes

$$K_{ib} = \frac{64\pi^3 e^6 Z n_e^2 \ln\Lambda}{3\omega^2 c(2\pi m_e k_B T_e)^{3/2}}\left(1 - \frac{\rho}{\rho_c}\right)^{-1/2} \propto \left(\frac{\rho}{\rho_c}\right)^2 \lambda^{-2} T_e^{-3/2} Z \left(1 - \frac{\rho}{\rho_c}\right)^{-1/2}, \tag{12}$$

where $\ln\Lambda$ is the Coulomb logarithm for electron-ion collisions.[26]

Regimes of interaction

In writing Eq. (8) we have implicitly assumed that the laser field does not distort appreciably the Maxwellian distribution function of the electrons. This requires that electron-electron collisions equilibrate the electron distribution at faster pace than electron-ion collision cause the distribution to heat, i.e.[1]

$$\nu_{ee} v_{th} \gg \nu_{ei} v_{osc}, \tag{13}$$

where $\nu_{ee} = \nu_{ei}/Z$ is the electron-electron collision frequency,[26] $v_{th} = (2 k_B T_e/m_e)^{1/2}$ is the electron thermal velocity, and k_B is the Boltzmann constant. Making use of the dimensionless ratio $a_L = v_{osc}/c$ (in the non relativistic limit), we can write Eq. (13) as $a_L \ll 6.3 \times 10^{-2} T_{keV}^{1/2}/Z$ or $I_{14}\lambda_{\mu m}^2 \ll 53\, T_{keV}/Z^2$, where T_{keV} is the electron temperature in keV, and I_{14} is the intensity in units of 10^{14} W/cm^2; this justifies a widely used rule of thumb for collisional interaction:

$$I_{14}\lambda_{\mu m}^2 \leq 1, \tag{14}$$

corresponding to $a_L \leq 8.5 \times 10^{-3}$. When $I_{14}\lambda_{\mu m}^2 > 1$ absorption becomes weaker, due to distortions of the electron distribution. In addition, as a substantial fraction of the incoming light approaches the critical density, plasma waves can be excited by several parametric instabilities. In particular, we refer to stimulated Raman and Brillouin scattering, and two two-plasmon decay. These processes reduce absorption and produce undesired hot electrons. At the critical density, also resonance absorption and ion-acoustic instability occur. Detailed, pedagogical discussions of these interaction mechanisms can be found elsewhere.[1, 27, 28] Finally, as anticipated in the introduction, when $a_L > 1$ or $I_{14}\lambda_{\mu m}^2 > 10^4$, laser-plasma interaction is dominated by relativistic effects.

The interaction regimes identified in this section are represented in Fig. 1 in the wavelength-intensity plane. In the rest of this paper we shall consider the two extreme cases of collisional absorption of nanosecond pulses and of relativistic interaction of subpicosecond pulses.

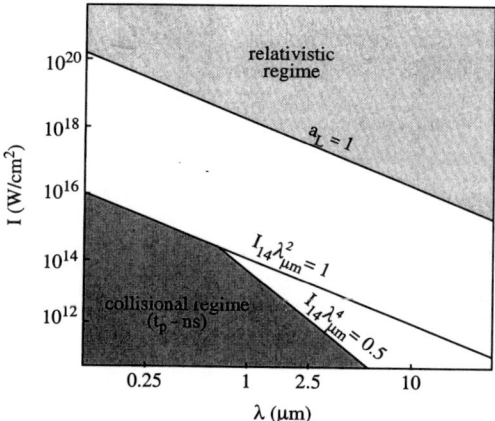

Figure 1. Laser-plasma interaction regimes in the wavelength-intensity plane.

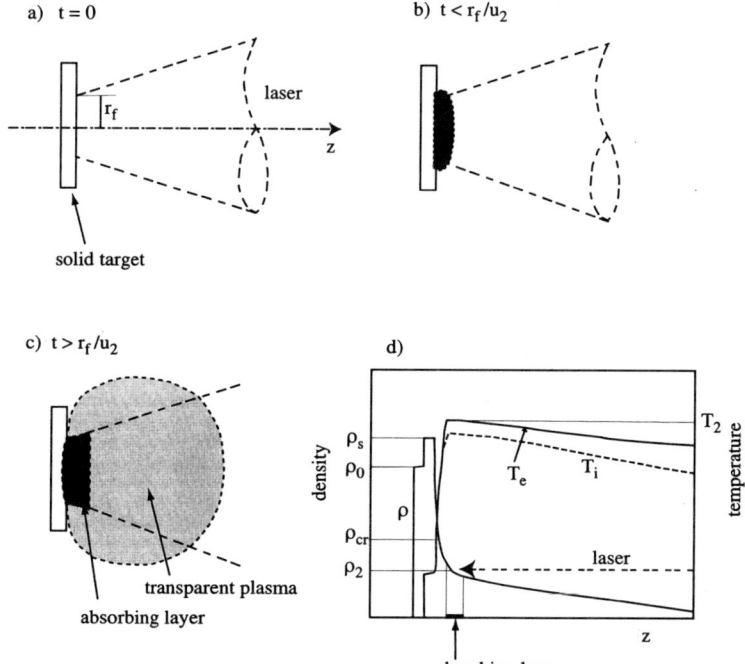

Figure 2. Irradiation of a solid target by a moderate intensity, nanosecond laser pulse. a)-c) axial cross-sections at different times; d) axial profiles of density and temperature (in logarithmic scale) during the quasi-stationary ablation stage.

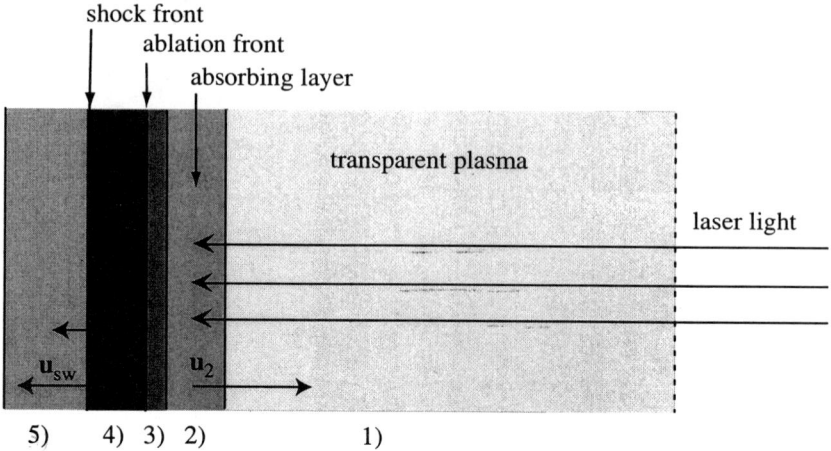

Figure 3. Model for the stationary stage of evolution of a laser irradiated solid target.
1) transparent plasma; 2) absorbing layer; 3) thermal conduction-dominated layer;
4) shock-compressed solid material; 5) unperturbed solid. The absorbing layer expands toward the right hand side with velocity u_2; the shock propagates to the left hand side with velocity u_{sw}; the shocked matter moves with velocity u_s.

Ablation physics and ablation pressure

Here, we refer to the intensity regime $I_{14}\lambda_{\mu m}^2 < 1$, and consider the irradiation of a solid target with density ρ_0 by a laser pulse with wavelength λ, intensity I and duration t_p, focused onto a spot of radius r_f (see Figure 2). After a transient stage, in which a surface layer of the solid is heated, vaporized and ionized, thus forming a plasma cloud, a self-regulating nearly stationary ablative process occurs.[29] As the hot tenuous plasma expands, it becomes nearly transparent [because $K_{ib} \propto \rho^2 T^{-3/2}$; see Eq. (12)], so that laser light penetrates deeper into the plasma and additional matter is heated. Laser light is absorbed in a sub-critical density plasma layer with thickness $\Delta z \simeq K_{ib}^{-1}$, and energy is transported from this region toward the solid by thermal conduction, which causes further ablation of material. As a reaction to such an ejection of matter on the laser side, a high pressure shock wave is driven towards the dense phase

This process is described by a simple and elegant model by Caruso and Gratton.[29] With reference to Fig. 3, we distinguish five regions of the target: 1) the very tenuous expanded plasma, transparent to laser light; 2) the absorbing plasma layer, with characteristic density $\rho_2 < \rho_c \ll \rho_0$, electron temperature T_2, mean expansion velocity u_2, and thickness $\Delta z \simeq K_{ib}^{-1}$; 3) a thin layer of ablated material, where the density rises steeply from $\rho = \rho_2$ to $\rho = \rho_s$; 4) the shock-compressed layer, with density $\rho_s > \rho_0$, and 5) the unperturbed solid.

The ablation pressure, i.e. the pressure generated as a result of the ablation of material from the solid surface, is then given by

$$p_{abl} = p_2 = \rho_2 u_2^2 \qquad (15)$$

For mass conservation, the ablation rate, i.e. the rate of variation of the areal mass σ of the target, is

$$|\dot{\sigma}| = \rho_2 u_2. \qquad (16)$$

In order to find expressions for ρ_2 and u_2 in terms of the laser and target parameters, we analyze the physics of the absorbing layer. We assume that the laser absorbing layer (2) expands sonically, so that $u_2^2 \simeq c_s^2 \simeq (Z/Am_p)k_B T_2$, where we have neglected the ionization energy and the ion thermal energy with respect to the electron thermal energy. We then rewrite the absorption coefficient [Eq. (12)] as

$$K_{\rm ib} = a\frac{\rho_2^2}{u_2^3}\left(1 - \frac{\rho_2}{\rho_c}\right)^{-1/2}, \qquad (17)$$

where we have used $u_2 \propto T_2^{1/2}$, and have introduced the quantity

$$a = 2 \times 10^{29} Z^{9/2} A^{-7/2} \lambda_\mu^2 \quad \text{cm}^8\text{g}^{-2}\text{s}^{-3}. \qquad (18)$$

Here we have taken a constant value of the Coulomb logarithm $\ln \Lambda = 7$. From Eq. (17) we see that when laser light is absorbed at $\rho_2 \ll \rho_c$, the critical density does not characterize the interaction.

We also asssume that the thermal conductivity, although essential to transport energy from the absorption layer to the solid surface and then to cause ablation, does not affect the properties of the absorption region.

Next, we now consider the geometry of the expanding plasma. When the axial size of the hot plasma bubble $L_z \approx u_2 t$ is smaller than the spot size r_f, the expansion can be taken as plane one-dimensional. As L_z becomes comparable to the spot size, instead, the expansion becomes three-dimensional and the expanding plasma becomes transparent very soon. Therefore, for $t > r_f/u_2$ plasma production can be described by a stationary model, according to which the plasma interacts with laser light only in a region of typical dimension r_f. In the following, we consider the case of pulses of duration $t_p > r_f/u_2$, i.e. spherical geometry and steady ablation. The treatment of the complementary case $t_p < r_f/u_2$ can be found in the literature.[29, 30]

From the preceding discussion we conclude that the whole problem under consideration is characterized by the unknowns ρ_2 and u_2, and by the known dimensional quantities a, I, and r_f. The plasma thermal conductivity is not relevant for the assumption made above, while the pulse time is not relevant since for $t_p > r_f/u_2$ the properties of the absorbing layer do not depend on time. We can then write

$$u_2 \propto r_f^{b_1} a^{b_2} I^{b_3}, \qquad (19)$$
$$\rho_2 \propto r_f^{b_4} a^{b_5} I^{b_6}, \qquad (20)$$

where the constant exponents b_i, with $i = 1,...6$, are determined uniquely by requiring that Eqs. (19) and (20) be dimensionally homogeneous. We thus find

$$u_2 \propto r_f^{1/9} a^{1/9} I^{2/9} \propto r_f^{1/9} \lambda^{2/9} I^{2/9}, \qquad (21)$$
$$\rho_2 \propto r_f^{-1/3} a^{-1/3} I^{1/3} \propto r_f^{-1/3} \lambda^{-2/3} I^{1/3}, \qquad (22)$$

from which we get

$$p_{\rm abl} = C_p f_1(Z, A) r_f^{-1/9} \lambda^{-2/9} I^{7/9}, \qquad (23)$$
$$|\dot{\sigma}| = C_\sigma f_2(Z, A) r_f^{-2/9} \lambda^{-4/9} I^{5/9}, \qquad (24)$$

where f_1 and f_2 are weak functions of the material, and C_p and C_σ are constants, which have been determined by a more detailed analysis[30] and by numerical simulations.[31] The resulting expressions for the ablation pressure and the ablation rate are

$$p_{\text{abl}} \simeq 11\, r_{\text{mm}}^{-1/9} \lambda_{\mu m}^{-2/9} I_{14}^{7/9} \tilde{f}_1(Z, A) \quad \text{Mbar}, \tag{25}$$

$$|\dot{\sigma}| \simeq 1.6 \times 10^5\, r_{\text{mm}}^{-2/9} \lambda_{\mu m}^{-4/9} I_{14}^{5/9} \tilde{f}_2(Z, A) \quad \text{g cm}^{-2}\text{s}^{-1}, \tag{26}$$

where r_{mm} is the focal spot radius in mm, and $\tilde{f}_1$ and $\tilde{f}_2$ are the functions f_1 and f_2 normalized in such a way as to take values $\tilde{f}_1 = \tilde{f}_2 = 1$ for fully ionized CH plastic. The characteristic temperature of the plasma corona scales as $T_2 \propto u_2^2 \propto r_f^{2/9}(I\lambda)^{4/9}$. Low-Z coronal plasmas, produced by laser pulses with $I \approx 10^{14}$ W/cm^2 and $\lambda = 0.5$–1 μm have electron temperature of 1–2 keV.

We can now study the limits of validity of the model. The condition for the formation of a stationary corona is

$$t_p \geq \frac{r_f}{u_2} \simeq 1.5\, \frac{\tilde{f}_2}{\tilde{f}_1}\, r_{\text{mm}}^{8/9}(I_{14}\lambda_{\mu m})^{-2/9} \quad \text{ns}, \tag{27}$$

which is well satisfied in experiments with nanosecond pulses and focal spot radius of a few hundred microns. The condition $\rho_2 \ll \rho_c$, stating that light must be absorbed before reaching the critical density, can be written as

$$I_{14} \ll I_{14}^* = 4\lambda_{\mu m}^{-4} r_{\text{mm}} \left(\frac{A\,\tilde{f}_1}{2Z\,\tilde{f}_2^2}\right)^3, \tag{28}$$

from which we see that the allowed intensity is a strong function of the wavelength. Actually, it can be shown[30] that the fraction of absorbed light η_{abs} is a function of the ratio I/I^*:

$$-\eta_{\text{abs}} \ln(1 - \eta_{\text{abs}}) = (I/I^*)^{-1}, \tag{29}$$

which has asymptotic behavior $\eta_{\text{abs}} \simeq 1 - \exp(-I^*/I)$ for $I \ll I^*$ and $\eta_{\text{abs}} \sim (I/I^*)^{-1/2}$ for $I \gg 1$. For $I = I^*$, $\eta_{\text{abs}} = 0.74$. To account conservatively for this last limitation to the intensity, in Fig. 1 we have set an additional limit to the classical absorption regime by drawing the line $I_{14}\lambda_{\mu m}^4 = 0.5$. This condition is more stringent than Eq. (14) for $\lambda > 0.7$ μm.

We observe that according to Eq. (25) the ablation pressure in the collisional regime scales weakly with the laser wavelength. However, since the allowed intensity depends strongly on the wavelength, the maximum achievable pressure p_{max} is also a strong function of the wavelength. For instance, at $\lambda \leq 0.7$ μm, where the intensity is limited by $I_{14} \leq \lambda_{\mu m}^{-2}$, we have

$$p_{\text{abl}} \leq p_{\text{max}} \approx 130 \left(\frac{\lambda}{0.35\,\mu m}\right)^{-16/9} r_{\text{mm}}^{-1/9} \quad \text{Mbar}. \tag{30}$$

The above model is supported by a large body of experimenta data. Measured values[32, 33] of the absorption efficiency for a wide range of intensities and at several wavelengths are shown in Fig. 4, and found qualitatively in agreement with the model above. It is confirmed that when Eqs. (27) and (28) are satisfied, light is absorbed efficiently. In the regime of good collisional absorption the ablation pressure is well approximated by Eq. (25), as shown by Fig. 5.[34, 35] These results motivate the choice of

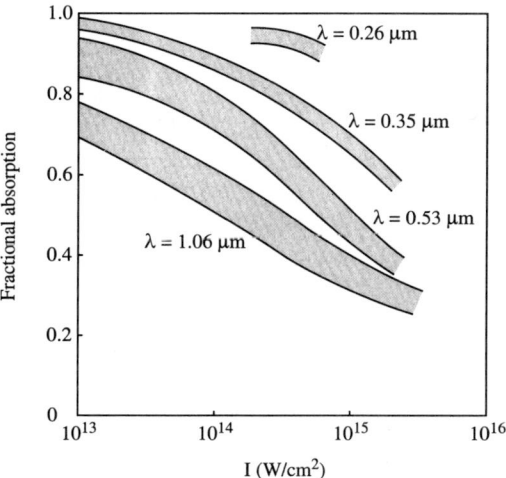

Figure 4. Fractional absorption of low-Z materials against intensity at different laser wavelengths. The figure summarizes data from Refs. 32 and 33.

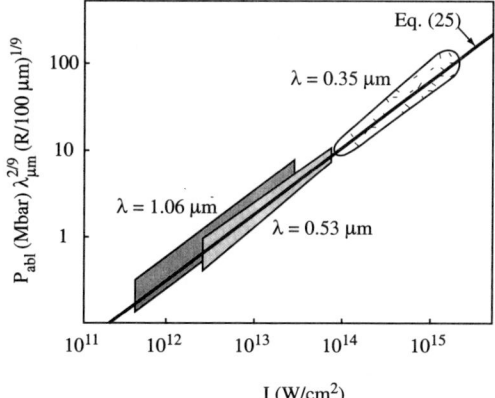

Figure 5. Ablation pressure in low-Z materials against laser intensity at different laser wavelengths, for laser pulses satisfying Eqs. (28) and (27). The figure summarizes data from experiments performed at LLE-Rochester, RAL-Oxford, NRL-Washington DC, UBC-Vancouver, CEL-V-Limeil, MPQ-Garching (see Ref. 34 for references to original works) and LLNL-Livermore.[35]

$\lambda = 0.35$ μm for the next generation of laser fusion experiments[12], in which pressures larger than 100 Mbar are required. Wavelength $\lambda = 0.25$ μm is assumed in most laser fusion reactor studies.[13] We stress that pressures obtained by short wavelength laser irradiation are more than one order of magnitude higher than the pressures achieved by any other means in the laboratory.[10]

At intensities close to or exceeding I^* a non-negligible fraction of the laser light reaches the critical density, so that one of the assumptions of the model is not fulfilled, and other theories should be used. In the limit in which all laser light were absorbed at the critical density one would have $p_{abl} \propto I^{2/3}\lambda^{-2/3}$, which follows from $I \approx \rho_c u_c^3$, $p_{abl} \approx \rho_c u_c^2$, and $\rho_c \propto \lambda^{-2}$. (Here u_c is the expansion velocity of the heated matter at the critical density.) We recall that when laser light reaches the critical density absorption is incomplete and considerable amounts of hot electrons are produced.[27, 28]

Pulsed lasers can also be used to generate intense fields of thermal X-rays (see below), which in turn irradiate a target. The physics of this process, called *indirect-drive* by laser is discussed, e.g., by Lindl.[12] Here we only mention that the ablation induced by a Planckian photon distribution with temperature T_r generates a pressure

$$p_{abl}^{rad} \approx 140 \left(\frac{T_r}{300 \text{ eV}}\right)^{7/2} \text{ Mbar}. \tag{31}$$

(again for a plastic material). Interestingly, in presently conceived indirect-drive inertial confinement fusion experiments T_r is limited to 300–350 eV, so that maximum radiation driven ablative pressures are comparable to those achieved by direct laser irradiation. At fixed p_{abl}, mass ablation rates are instead larger for radiation driven ablation, because $|\dot{\sigma}| \approx \rho_{abs} u_{abs} \approx (\rho_{abs} p_{abl})^{1/2}$, and the characteristic absorption density ρ_{abs} is higher for X-rays than for laser light. (Here u_{abs} is a characteristic expansion velocity of the radiation absorbing layer). It is worth stressing that both in direct-drive and in indirect-drive the ablation pressure is not the radiation pressure $p_r = I/c$ of the laser or thermal photon field, but the kinetic pressure of the plasma heated by such radiation. Indeed, in the present regime

$$\frac{p_r}{p_{abl}} \approx \frac{u_{abs}}{c} \ll 1. \tag{32}$$

APPLICATIONS OF PLASMAS CREATED BY NS LASER PULSES

Two distinctive features of plasmas produced by irradiating a solid target are the intense pressure generated by ablation and the high temperature of the laser heated corona. Concerning applications of laser-produced plasmas, we can just distinguish between those exploiting the hydrodynamics induced by the strong shock waves associated to the ablative pressure, and those exploiting the radiation emitted by the plasma.

High pressure physics

Laser generated multi-megabar ablative pressures are employed for studying the properties of dense, compressed matter, for driving inertial confinement fusion implosions, radiation hydrodynamics experiments and scaled-down simulations of astrophysical phenomena. Parameters achieved in laser-plasma experiments and in laser-driven implosions[36, 37] and parameters required for inertial fusion are represented in Fig. 6 in

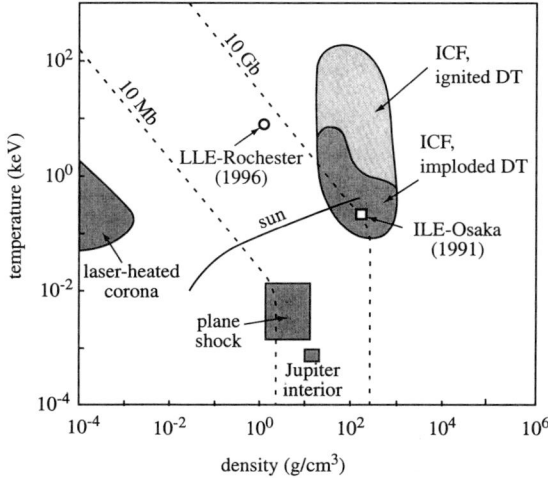

Figure 6. Parameters achieved in the density-temperature plane by laser-heated matter, matter compressed by laser-driven shocks, and by laser-driven spherical implosions performed at ILE-Osaka[36] and LLE-Rochester.[37] Also shown are the regions aimed at by ICF, a curve referring to the interior of the Sun and a point in the interior of Jupiter.

the density-temperature-pressure space. For comparison, also conditions in the interior of the sun and of Jupiter are reported.

Laser-driven Equation-of-State (EOS) measurements[9, 10, 38] are performed by sending a well characterized plane shock wave through a sample of the material to be analyzed. Both direct- and indirect-drive can be used. The shock velocity and/or the velocity of the shocked material are obtained by optical or X-ray diagnostics. The Hugoniot relations[39] for strong shock waves are then used to compute EOS points. In this way EOS points at $P = 10\text{--}50$ Mbar have been measured with good accuracy.[9, 38] Recently, an insulator-metallic phase transitions has been evidenced in deuterium at about 0.5 Mbar.[40] Experiments are currently planned for materials under conditions of relevance to planetary interiors. In some experiments, amplification of the ablative pressure has been obtained by the collision of a plane target accelerated by a laser pulse with another target at rest. In this way a record pressure for plane geometry of 750 Mbar has been achieved.[41] Further pressure amplification is obtained in converging geometry, where pressures higher than 5 Gbar have already been attained;[36, 37] this topic will be discussed later in the section devoted to inertial fusion.

Laser-driven plasmas are an essential component of the rapidly growing laboratory astrophysics.[14, 15] Some experiments concern shock propagation and intensification through regions of decreasing density, and the collision of a strong shock propagating through a tenuous material against a denser material. Their purpose is the simulations of the emergence of shock waves from supernovas and their impact with other objects. Other experiments address the non linear evolution of Richtmyer-Meshkov and Rayleigh-Taylor instabilities in targets designed to simulate environments of relevance to supernovas. Also, supersonic radiative jets of astrophysical interest have been produced using the NOVA laser.[42] Laser-produced dense plasmas also provide unique test beds for theoretical models of optical properties of dense and hot matter.[9]

Laser-plasmas as X-ray sources

Laser-plasmas can be used as sources of X-rays, with application to lithography, microscopy, biology, etc. A comprehensive presentation of this topic can be found in a recent book.[16]

Incoherent sources Plasmas produced by the irradiation of high-Z solid materials emit X-rays efficiently.[16, 43, 44] Primary X-rays are emitted by the region of hot, tenuous and optically thin plasma where laser light is absorbed. Additional thermal X-rays are emitted by a colder, denser and optically thick *re-emission layer* closer to the solid surface. The relative importance of the contributions from these two layers depends on pulse and target parameters.

A large body of experimental data, collected at a number of laboratories, shows that the overall X-ray conversion efficiency η_x, i.e. the ratio of the energy emitted as X-rays to the absorbed laser energy, generally improves with decreasing laser wavelength, and with increasing Z. At $\lambda = 0.35$ μm, η_x is maximum for $I \approx 10^{14}$ W/cm^2. For high-Z materials, $\eta_x = 80\%$ can be achieved.

Thermal radiation sources can be obtained by enclosing the laser-produced plasma within a cavity with high-Z walls.[44] Here, multiple absorption-emission processes thermalize the primary radiation. Such sources find important applications in radiation hydrodynamics, in the study of the properties of dense plasmas, and in indirect-drive inertial confinement fusion.

Coherent sources: X-ray lasers Several X-ray laser concepts employ laser-produced plasmas as active media.[16, 43, 45] Actually, the first demonstration of soft X-ray amplification (at $\lambda = 20.9$ nm) was obtained in 1984 by a plasma generated by the NOVA laser.[46] In such X-ray lasers a laser beam is cylindrically focused onto a thin solid foil to generate a sort of cylindrical rod of lasing plasma. The hydrodynamic evolution then brings the plasma into some conditions where population inversion occurs, and lasing ions are created. Normal incidence multilayer mirrors may be used to create a cavity. Several pumping schemes, active media and cavity geometries have been proposed and tested. Lasing action has been observed at a number of wavelengths, also falling into the water window ($2.3 < \lambda < 4.4$ nm), of particular interest for biological applications. X-ray laser uses include X-ray microscopy and holography, and diagnostics of high density plasmas.

RELATIVISTIC LASER-PLASMAS

The advent of CPA lasers has opened the rapidly growing field of relativistic plasma physics. Here, we briefly survey a few physical effects and list possible applications. Deeper treatments can be found in other chapters of this book; several interesting papers can be found in the proceedings of two recent conferences.[17, 47]

Interaction physics

The complex phenomena occurring in a relativistic plasma essentially follow from the ultraintense laser field, in which the kinetic energy of the quivering electrons becomes comparable to or larger than the rest energy. A first important effect is the relativistic self-focusing[48] of a laser beam propagating through an underdense plasma.

Since the electron mass $m = m_e(1+a_L^2)^{1/2}$, m_e being the rest mass, is now a function of the laser intensity, the mass of the electrons varies in the plane normal to the symmetry axis, being maximum on the axis, where the laser intensity is higher. As a consequence, the refraction index exhibits a radial gradient, so that the beam is refracted towards the axis. Such a self-focusing occurs when the beam power exceeds $17.3(\omega/\omega_p)^2$ GW. The ponderomotive force, proportional to the gradient of the light pressure, which causes density depletion near the axis, further enhances this effect.

Accelerated electrons propagate inside the channel and a large static magnetic field is created;[49] at the same time a return current is generated to neutralize the channelled current. The Weibel instability then causes the electron current to break into filaments.[50] However, according to theory and to 2-D[49] and 3-D simulations,[51] at $I \geq 10^{19}$ W/cm² the self-generated 100 MG magnetic field pinches the electrons again into a single very narrow channel. Thin channels, such as observed in the simulations have been evidenced experimentally.[52] This effect might turn useful for increasing the acceleration length in laser-plasma accelerators. It is also promising for the fast ignitor ICF scheme.

Different mechanisms have been proposed to explain the generation of forward-directed multi-MeV electrons observed both in experiments[53, 54, 55] and in simulations. (Notice that the laser electric field would accelerate the electrons in a direction orthogonal to the beam.) These include the laser-plasma-wakefield acceleration,[56] i.e. acceleration due to the plasma wave caused by the displacement of the quivering electrons, and the inverse-free-electron-laser effect.[57] The same mechanisms could be exploited to build particle accelerators based on laser-plasmas.[56, 58, 59]

The relativistic electrons emit an intense flux of X- and γ-rays.[60] Photoreactions (as photo-fission[61, 62]) and electron-positron pairs generated by such γ-rays have been detected in recent experiments. Such findings open the way to table-top nuclear physics.[20]

The ponderomotive force ejects the bulk electrons from the above light propagation channel. The resulting charge separation generates a radial electric field, which in turn accelerates radially the ions remaining in the channel. This mechanism is invoked to explain the production of neutrons by D-D fusion reactions in a laser irradiated deuterated target.[63] In an experiment employing pulses of 200 mJ in 160 fs, 160 neutrons/shot were detected. More efficient neutron production (10^5 neutrons/J) was obtained from laser-irradiated deuterium clusters.[18] Here, laser heating causes the explosion of the cluster, and the production of energetic ions which can react with ions from nearby clusters.

Applications

In addition to the already mentioned applications to particle acceleration and to the generation of X-rays, neutrons and positrons, other uses of ultraintense lasers concern X-ray laser schemes, spectroscopy, generation of higher harmonics.[4] An application pushing the development of multi-kJ CPA lasers is the fast-ignitor ICF concept, which will be discussed later. We also find it worth mentioning that CPA lasers have already proved successful in material processing (microwelding and micromachining) and in microsurgery. For these applications a distinctive advantage of ultrashort pulses over longer ones consists in the elimination of the deleterious side effects due to thermal conduction, which may spread laser energy over distances $L \approx (Dt)^{1/2}$ considerably larger than the focal spot. (Here D is the thermal diffusion coefficient.)

LASER-DRIVEN INERTIAL FUSION

The main driving force for the development of large powerful lasers has been inertial confinement fusion (ICF). In this section, we first briefly recall the basic principles of ICF, then discuss laser fusion schemes, experimental achievements and prospects. It is worth noticing that many of the issues for the feasibility of ICF are common to any high energy density physics experiment driven by intense ablative pressure.

Inertial Confinement Fusion

ICF aims at achieving fusion burn in highly compressed and appropriately heated samples of fuels.[12, 13, 65, 66] Since no external means can constrain the motion of the high-pressure reacting fuel, the compressed configuration only survives for a time τ set by its own inertia. For a compressed sphere of radius R_c, $\tau \simeq R_c/4c_s$, where c_s is the relevant sound speed. The fuel of most immediate interest is the equimolar deuterium-tritium mixture (DT), since the reaction $D + T \to n + \alpha + 17.6$ MeV is the easiest to achieve. It has *ideal ignition temperature* of about 4.3 keV and specific yield $Q_{DT} = 3.4 \times 10^{18}$ erg/g. Since the burn of 1 mg of DT fuel releases 340 MJ, equivalent to 85 kg of TNT, the fuel mass m of a *target* is limited to a few mg. An inertial fusion reactor would then act as a sort of engine, burning targets at the frequency of 1–10 Hz.

Energy production by ICF (sometimes referred to as inertial fusion energy, IFE) requires that the energy gain of a target, i.e. the ratio $G = E_{fus}/E_d$ of the fusion energy released by the target to the *driver* energy delivered to compress and heat it, satisfies $G\eta_d \geq 10$, where η_d is the efficiency of the driver. For simple evaluations, we introduce the fuel energy gain

$$G_F = \frac{E_{fus}}{E_F} = \frac{E_{fus}}{\eta E_d} = \frac{G}{\eta}, \tag{33}$$

where E_F is the energy delivered to the fuel, which is a fraction η (overall coupling efficiency) of the driver energy. Since for foreseen IFE schemes one can assume $\eta\eta_d \approx 1\%$,[12, 13, 66] it follows that G_F must be of the order of 1000 or larger.

Necessary conditions for high gain[12, 64] are nearly isentropic compression of most of the fuel to density $\rho = 200$–500 g/cm^3, and heating to 5–10 keV of a small portion of it (a *hot spot*, with mass $m_c \ll m$). The hot spot should self-heat due the power released by the 3.5 MeV fusion α-particles, and then trigger a propagating burn-wave reaching the whole fuel. According to simulations, the hot spot temperature T_h, density ρ_h and radius R_h must satisfy $T_h > 5$–10 keV, and $\rho_h R_h > 0.2$–0.5 g/cm^2, with the actual numbers depending on details of the igniting configuration.[67]

The above conditions concerning compression and selective heating are recovered by the following simple model. We write

$$G_F = \frac{E_{fus}}{E_F} = \frac{m\Phi Q_{DT}}{E_c + E_h} \tag{34}$$

where E_c is the internal energy of the compressed fuel, E_h is the additional energy contained in the hot spot and Φ is the fraction of burned fuel, which is well approximated by

$$\Phi \simeq \rho R_c/(\rho R_c + 7 \text{ g/cm}^2). \tag{35}$$

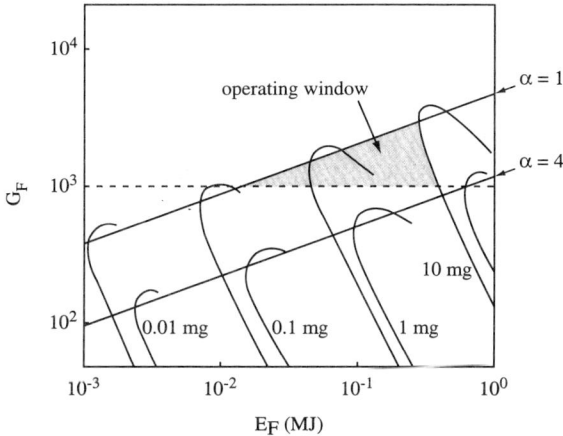

Figure 7. Fuel energy gain vs fuel energy for two values of the isentrope parameter α and several fuel masses, computed from the isobaric fuel gain model [69, 70]. IFE conditions are represented by the grey area indicated as *operating window*.

Computations indicate that for IFE one must achieve $\Phi \geq 0.2$ and hence $\rho R_c > 2$ g/cm². The need for strong compression follows immediately from the relationship $m = (4\pi/3)(\rho R_c)^3/\rho^2$, with m limited to a few mg. For instance, for $\rho R_c = 3$ g/cm² and $m = 3$ mg, one has $\rho = 194$ g/cm³. Notice that the product ρR_c in ICF plays a role analogous to the Lawson confinement parameter[68] $n\tau$ in magnetically confined plasmas. Indeed $\rho R_c \approx m_i c_s n_i \tau$, where n_i and m_i are the number density and the average mass of the plasma ions, respectively.

Compression has an energetic cost, with a lower bound the specific energy being set by zero-temperature compression, $e_{\text{deg}} = (3Z/5Am_p)(\hbar/2m_e)(3\pi^2 n_e)^{2/3} = C_d \rho^{2/3}$, with $C_d = 3.2 \times 10^{12}$ erg $/[\text{g}(\text{g}/\text{cm}^3)^{2/3}]$ for an equimolar DT mixture. The actual internal energy of the compressed fuel can then be parametrized as $E_c = me_c = m\alpha C_d \rho^{2/3}$, where the *isentrope parameter* $\alpha > 1$ measures deviations from perfect degeneracy. One of the main issues in ICF is just that of keeping the isentrope of the bulk fuel as low as possible to reduce the energy investment and increase the target gain. Since the pressure of the compressed fuel can be written as $p = (2/3)\rho e_c = (2/3)\alpha C_d \rho^{5/3}$, the smaller α, the smaller is also the pressure at a given density.

The contribution due to the internal energy of the hot spot can be evaluated by using the ideal gas equation-of-state, giving $E_h = m_h e_h = m_h C_v T_h$, where $C_v = 1.15 \times 10^{15}$ erg/(g keV) is the relevant specific heat. Hot spot ignition, rather than ignition from a homogeneously heated target, is just aimed at for making E_h as small as possible.

The simple model above, with additional information concerning the spatial profiles of density and temperature, and an appropriate ignition condition provides useful insight on gain trends. In Fig. 7 we show the result of a computation based on the assumption (supported by theory and simulations) that at ignition the fuel is isobaric, with a central hot spot surrounded by denser and colder fuel.[69] Gain curves $G_F = G_F(E_F)$ for different fuel masses and two values of the isentrope α are plotted. Along each individual curve the fuel density is varied. The dashed curve $G_F \simeq 5000[E_F(\text{MJ})/\alpha^3]^{0.3}$ joins *optimal operating points* for different masses. We find

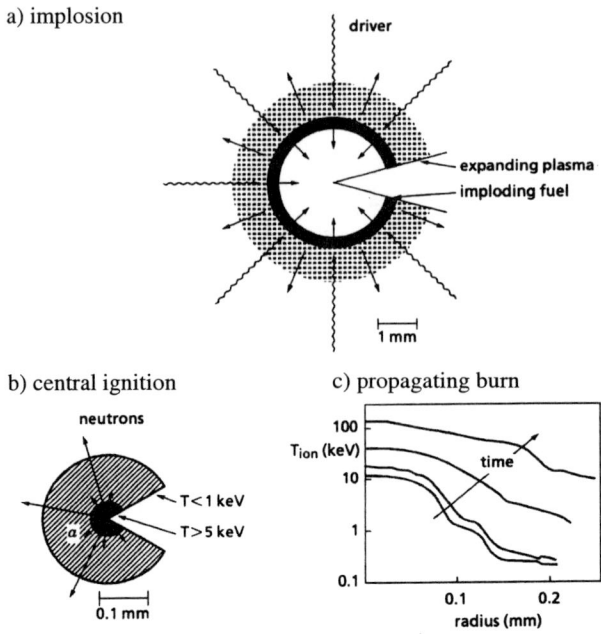

Figure 8. Essential features of the standard approach to ICF: a) ablation-driven implosion of the fuel containing shell; b) central ignition of the compressed fuel; c) propagating burn, illustrated by a sequence of radial profiles of the ion temperature during a time interval of about 100 ps.

that IFE relevant gains require fuel energy in excess of about 50 kJ and mass larger than about 0.5 mg. The pressure on the optimal gain curve is[70]

$$p = 240 \, \alpha^{1/5} \, m_{mg}^{-8/15} \text{ Gbar,} \tag{36}$$

where m_{mg} is the fuel mass in mg, which shows that the smaller the fuel mass (and hence the driver energy), the higher is the fuel pressure. Analogous considerations apply to the specific energy and the density. From Eq. (36) we see that the pressures required for IFE are at least three orders of magnitude larger than the laser- or radiation-driven ablative pressure, and even larger than those occurring at the center of the sun; see Fig. 6.

The standard laser fusion scheme

The scheme used in ICF to achieve the needed compression is illustrated in Figs. 8 and 9. Figure 9 shows a simulation performed by the author of a target originally designed by the LLE-Rochester group.[71] A certain number of laser beams irradiate uniformly the outer surface of a spherical target, delivering a total energy of 1.6 MJ laser light with $\lambda = 0.25$ μm in a time-tailored pulse reaching the peak power of 600 TW. The target consists of a spherical hollow shell of radius $R_0 \simeq 2$ mm, containing

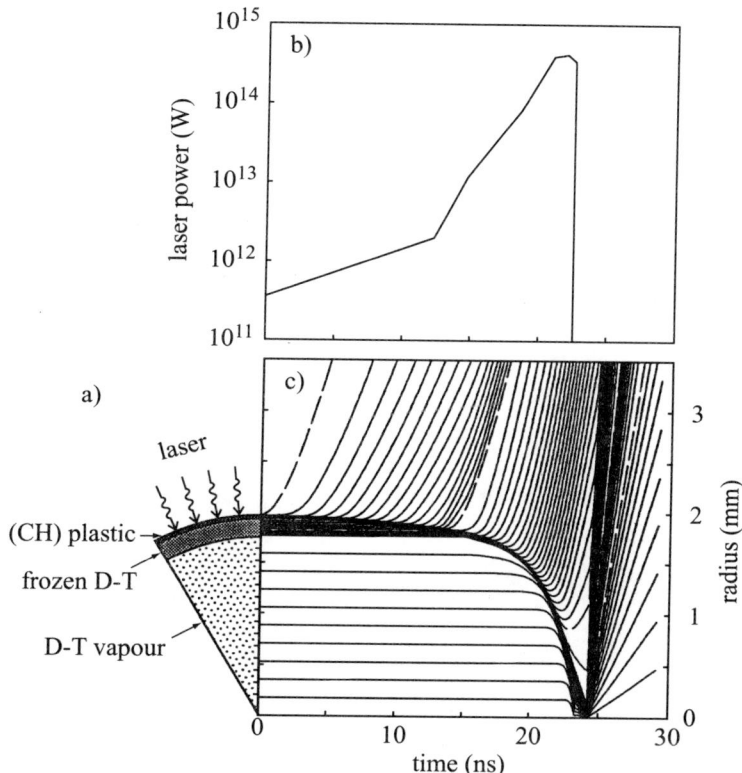

Figure 9. High gain direct-drive laser fusion target, irradiated by 1.6 MJ of laser light, at $\lambda = 0.26\ \mu m$: a) shell sector, b) laser power vs time, c) radius vs time evolution of selected lagrangian markers.

an inner layer of $m = 1.7$ mg of frozen DT fuel (with thickness $\Delta R_0 = R_0/10$). As a reaction to ablation, first a shock is sent through the target, and later the whole non-ablated material is accelerated inward to implosion velocity u_{imp}. This process is analogous to a rocket, where ejection of material causes acceleration of the payload in the opposite direction. In this way, a fraction η_h, called hydrodynamic efficiency, of the absorbed laser energy is converted into kinetic energy of the imploding fuel $E_k = (1/2)mu_{\mathrm{imp}}^2 = \eta_{\mathrm{abs}}\eta_h E_d$. In the present target $\eta_{\mathrm{abs}} \simeq 80\%$ and $\eta_h \simeq 10\%$. Notice that the use of a thin shell is essential to deliver the needed specific energy $e = u_{\mathrm{imp}}^2/2 \geq 50$ kJ/g to the fuel, i.e. to achieve implosion velocity $u_{\mathrm{imp}} = 3 \times 10^7$ cm/s. Indeed, this is simply seen by equating the fuel energy to pdV work and neglecting mass variation due to ablation; we have $e \simeq u_{\mathrm{imp}}^2/2 \propto p_{\mathrm{abl}} R_0^2 (R_0/\Delta R_0)$. The time profile of the pulse is designed so as to gradually reach the needed value of the pressure,[12, 64] thus minimizing the entropy generated by unwanted shock waves.

As the shell collapses at the centre, at $t = 23$ ns, the kinetic energy of the imploding matter is converted into internal energy. The fuel is compressed and heated, and the central hot spot is also created. At this stages the fuel pressure reaches the value $p \simeq 3.6 M^3 p_{\mathrm{abl}}$, where $M = u_{\mathrm{imp}}/c_s$ is the Mach number of the collapsing shell.[72] In the case of the simulation of Fig. 9, $M \simeq 9$, so that collapse amplifies the ablation pressure by a factor about 2000. Notice that since $c_s \propto p_{\mathrm{abl}}^{1/5} a^{3/10}$, we get another argument in favor of low entropy compression. Once ignition occurs, a violent burn

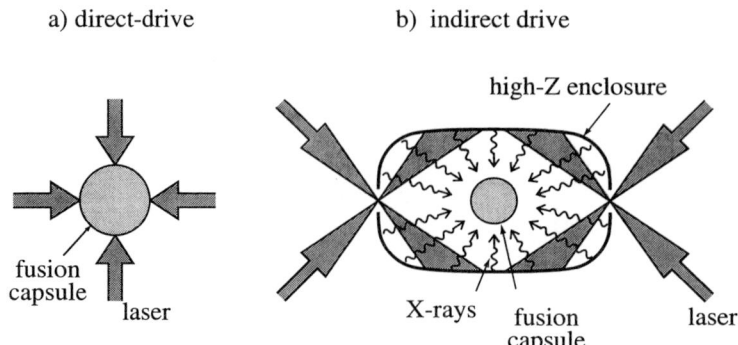

wave propagates to the whole fuel, which burns in a time interval of the order of 100 ps, much shorter than the capsule irradiation and implosion time, releasing about 160 MJ of fusion energy.

The scheme illustrated in Fig. 9 is called direct-drive, because laser beams directly hit the fusion capsule. In alternative, one can use indirect-drive,[12, 73] where the implosion is induced by the thermal X-rays generated by laser beams[12, 44] or by heavy-ion beams[12, 74] and confined within a suitable cavity (see Fig. 10). Implosion physics is analogous to that of a direct-drive target; a notable difference concerns the scaling of the implosion velocity with the aspect ratio $R/\Delta R$. In indirect-drive, where ablation is substantial, one has[12] $v \propto T_r^{9/10} R/\Delta R$, where $R/\Delta R$ is the *in-flight-aspect-ratio* or IFAR of the target. The reader interested in a detailed discussion of indirect-drive ICF should refer to a recent review by Lindl.[12]

Indirect-drive is pursued for symmetry and stability reasons. Conversion of laser light into X-rays in principle makes the symmetrical illumination of the capsule partially independent on the geometrical disposition of the driver beams. In addition, in indirect-drive one overcomes the problems due to the unavoidable intensity fluctuations at the beginning of a laser pulse, inducing a hydrodynamic *imprint* onto the fusion capsule. Finally, the Rayleigh-Taylor instability is believed to be weaker for X-ray driven ablation than for laser-driven ablation. On the other hand, these advantages have an energetic cost, because laser light has first to be converted (with efficiency η_x) into X-rays, only a fraction η_{tr} of which is absorbed by the capsule. Even if these inefficiencies are partially compensated by higher hydrodynamic efficiency, which can reach 18%, overall coupling efficiencies $\eta = \eta_{abs}\eta_x\eta_{tr}\eta_h$ foreseen for indirect-drive range from $\eta \simeq 0.02$ in the next generation of experiments to $\eta \simeq 0.04$ in advanced targets. For direct-drive, instead, $\eta = 0.05$–0.10 seems feasible.

Issues in ablative acceleration and compression

A general issue in laser-driven ablation concerns the spatial uniformity of the pressure front in the direction normal to the laser. This is essential in ICF, to allow for high-convergence symmetrical implosion of the fuel with the generation of a central hot spot.[12, 76] Numerical simulations [71, 75] indicate that r.m.s. relative amplitude perturbations of the ablation pressure should be limited to about 1%. Shorter wavelength perturbations must also be kept small because they cause mass displacements, which

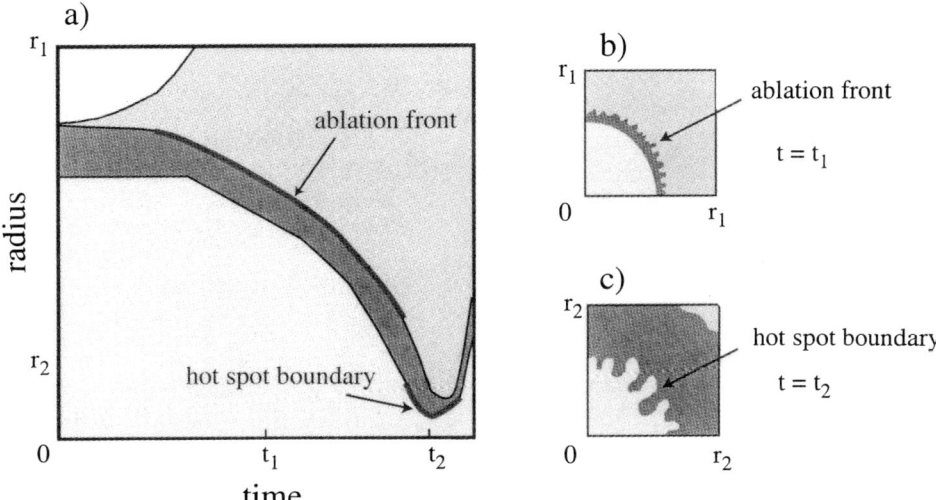

Figure 11. Sketch showing Rayleigh-Taylor unstable regions in the implosion of an ICF shell. a) radius versus time evolution for a spherically symmetric implosion; b) and c) 2-D cuts of a shell showing Rayleigh-Taylor deformed surfaces at different stages of the implosion.

may seed hydrodynamic instabilities. Analogous requirements are set by basic physics experiments, where one-dimensional, well characterized shock fronts are needed. A number of beam homogeneization techniques have been developed in the past fifteen years to produce uniform laser beams. Basically they consists in splitting a beam into a number of beamlets, breaking the coherence between beamlets, and recombining them into a single homogeneized beam (see Ref. 76 and the refs. cited therein). Use of smoothing techniques has resulted in drastically increased performance of ICF targets (smaller growth of Rayleigh-Taylor instabilities, larger neutron yield, higher density of the compressed fuel, and also reduced level of plasma instabilities, both in direct-drive[77] and in indirect-drive[78]) and has allowed for direct-drive EOS experiments with accuracy of a few percent.[38] Solutions[79, 80] have also been proposed to mitigate the effects of the *imprint*[81] generated at the early stages of interaction, when the homogeneization techniques are not yet effective.

A related, even more crucial issue, is hydrodynamic stability. Indeed, a solid target, accelerated by the pressure exerted by a hot, tenuous gas is subjected to the Rayleigh-Taylor instability (RTI),[82] analogous to that occurring when a fluid of density ρ_2 is supported by a lighter fluid, with density $\rho_1 < \rho_2$, in a gravitational field.[83] In spherical ICF targets the instability also occurs at the final stage of the implosion, when the dense shell material is slowed down by the inner gas forming the ignition hot spot (see Fig. 11). Similar instabilities are believed to occur during some stages of the violent evolution of supernovas.[14]

Linear theory of the classical RTI[82] shows that perturbations of wavelength λ_w and amplitudes $\xi \ll \lambda_w$ grow exponentially, with growth rate $\gamma = (A_t k_w g)^{1/2}$, where $A_t = (\rho_2 - \rho_1)/(\rho_2 + \rho_1)$, $k_w = 2\pi/\lambda_w$ and g is the gravity or the acceleration. For surfaces characterized by a large number of modes, when the amplitudes of the perturbations exceed some linear saturation threshold, the different modes couple non linearly, and after some time give rise to a turbulent mixing layer whose size grows in time as[84] $h \approx 0.05 A_t g t^2$.

Classical RTI growth during inward acceleration would be incompatible with ICF, because it would limit the IFAR ($= R/\Delta R$) to less than 10. However, effects associated with ablation reduce the RTI linear growth at the outer surface of ICF shells. Theoretical and computational studies agree in showing that, at any k_w, the linear growth of ablative RTI modes is smaller than classical, and that the shortest wavelengths are stable. Extensive set of results are fitted by the so-called modified-Takabe[85] linear dispersion relation (see e.g. Refs. 12 and 86 and refs. therein)

$$\gamma = \left(\frac{gk_w}{1 + k_w L}\right)^{1/2} - \beta k_w v_a, \tag{37}$$

where L is the minimum scale-length of the density gradient at the ablation front, $v_a = |\dot{\sigma}|/\hat{\rho}$ is the ablation velocity, $|\dot{\sigma}|$ is the areal mass ablation rate, $\hat{\rho}$ is the peak density of the non-ablated fuel, and β is a numerical coefficient. Depending on the laser and target parameters, $\beta = 1.7$–4 for ablation driven by laser and $\beta = 1$–2 for ablation driven by radiation. Experimental growth rates roughly agree with theory in the region around the maximum growth rate, where growth rates are about 50% the classical value.[87, 88] Direct experimental evidence for full stability at very short wavelengths is instead yet to be provided. Notice that since in indirect-drive the ablation velocity is about five time larger than in direct-drive,[12] indirect-drive appears less unstable than direct-drive, despite the smaller value of β.

In ICF the initial seed for the instability is provided by target imperfections as well as by laser non-uniformities. Perturbations growing at the outer surface during the implosion stage are fed to the inner surface, with amplitude reduced by a factor $\exp-k_w\Delta R = \exp-l\Delta R/R$ (l being the spherical mode number). Here they add to imperfections of the fuel target layer and seed the instability occurring at shell stagnation. At this stage growth is reduced by the presence of finite density gradients, while convective effects are negligible. It is found that spherical modes $l = 100$–1000 are the most dangerous during shell acceleration, and modes with $l = 10$–30 dominate the RTI at collapse.

In conclusion, although less violent than the classical one, instability is unavoidable, and one has to live with it. This is done by choosing operating conditions where stabilizing effects are enhanced, by limiting as far as possible initial seeds, and by designing targets with a not too large IFAR. This latter requirement limits the achievable implosion velocity, thus setting an upper bound to the specific energy deliverable to the fuel, and hence to the achievable density and pressure of the compressed fuel. The consequence is the non-accessibility of the left hand side portion of Fig. 7, and the existence of a threshold energy for ignition (see below).

Achievements and prospects

All individual hydrodynamics issues thought to be crucial for the feasibility of ICF have been addressed experimentally with lasers with energy up to 60 kJ. Among the main achievements we mention the nearly total absorption of laser light at the third and fourth harmonic of the Nd laser and intensity of 10^{15} W/cm^2, with minimal generation of undesired non-thermal particles, the achievement of drive pressures in excess of 100 Mbar (both in direct- and in indirect-drive) and of implosion velocity of 500 km/s. A plastic target has been compressed to density 600 times its initial density,[36] while temperatures exceeding 10 keV have instead been reached in fuels at relatively low density which have released 1.4×10^{14} neutrons.[37] Pressures of 5–10 Gbar have been achieved in the compressed material.[36, 37] Surveys of experimental results, also concerning plasma instabilities in laser interaction, and all aspects of indirect-drive physics can be found in the review paper by Lindl[12] and in more recent papers presented at the 1999 IFSA Conference.[47]

Great progress has resulted from improved beam balancing and beam smoothing. E.g. in the late 1980s neutron yields of targets converging by a radial factor about 20–30 were systematically smaller by 3–4 orders of magnitude than the values predicted by 1-D simulations,[36] while presently they are within a factor of a few units and are accurately reproduced by the 3-D codes which are used to design ignition experiments.[89] The record values of the whole target confinement parameter and of hot spot confinement product, are respectively, $\rho R_c = 0.1$ g/cm^2,[36] and $\rho_h R_h T_h = 0.02$ g/cm^2 keV,[90] limited by the very small mass of the targets which can be driven by present lasers.

According to present understanding,[12, 91] ignition of an indirect-drive target requires delivering about 150–250 kJ to a fusion capsule with initial aspect ratio $R_0/\Delta R_0 = 10$, outer surface r.m.s. defects below 1000 Å, and r.m.s. defects of the cryogenic layer below 1 μm. Such a capsule has to be driven to an implosion velocity of 400 km/s, with in-flight-aspect ratio kept below 40. The required driver energy and power are 0.8–1.5MJ, and 300–500 TW, respectively. The same laser parameters appear adequate also for the ignition of direct-drive targets.[12, 92] Two facilities for achieving ignition and gain $G = 10$ are now under construction, the National Ignition Facility (NIF) in the US[12, 93] and Laser Megajoule (LMJ) in France.[94] Both are frequency-tripled Nd:glass lasers ($\lambda = 0.35$ μm) with about 200 beams delivering on target 1.5–1.8 MJ with peak power of 500 TW. Special care is put in the control of beam quality, pulse shaping and positioning. Safety margins are included in target design to account for physics uncertainties concerning RTI, the control of the implosion symmetry, and the level of damaging parametric instabilities in the laser-plasma interaction. The expected gain from the baseline indirect-drive target is $G = 10$, but recent studies indicate that substantially higher gains may be obtained.[80, 95]

The NIF and LMJ lasers will have efficiency well below that required for IFE, operate on single shot basis, and with targets designed to achieve $G = 10$ only. This points to the development path towards IFE, with the need for improved target concepts, more efficient, reliable and repetitive drivers, and reactor chambers operating at frequency in the Hertz range. Concerning the drivers, actively studied IFE candidates include diode-pumped-solid-state lasers, excimer lasers and heavy-ion accelerators.[13]

The fast ignitor

In the ICF scheme described so far a central hot spot is generated by hydrodynamic cumulation at shell collapse. This process concentrates energy in space, but demands a high degree of spherical symmetry and is particularly sensitive to instabilities. An alternative scheme, known as *fast ignitor*,[21] consists in creating the hot spot by external beam interaction applied after implosion. The concept is illustrated by the simulation shown in Fig. 12. Here a beam of fast particles creates a hot spot close to the surface of a compressed sphere. The hot spot self heats due to the deposition of the energy of the fusion alpha-particles, and drives a burn wave. Such an approach would considerably relax symmetry and stability requirements, and would also allow for higher energy gain at given driver energy.[21] However, since the energy concentration cannot rely on hydrodynamic convergence, ultraintense drivers are necessary. The pulse parameters (energy E_p, power P, and intensity I) for fast ignition can be evaluated as follows. The pulse energy should be of the order of the energy required to create a hot spot in the compressed fuel, $E_p = m_h C_v T_h$; the hot spot mass m_h and temperature T_h are then computed by using the ignition condition for an isochoric DT assembly[67]

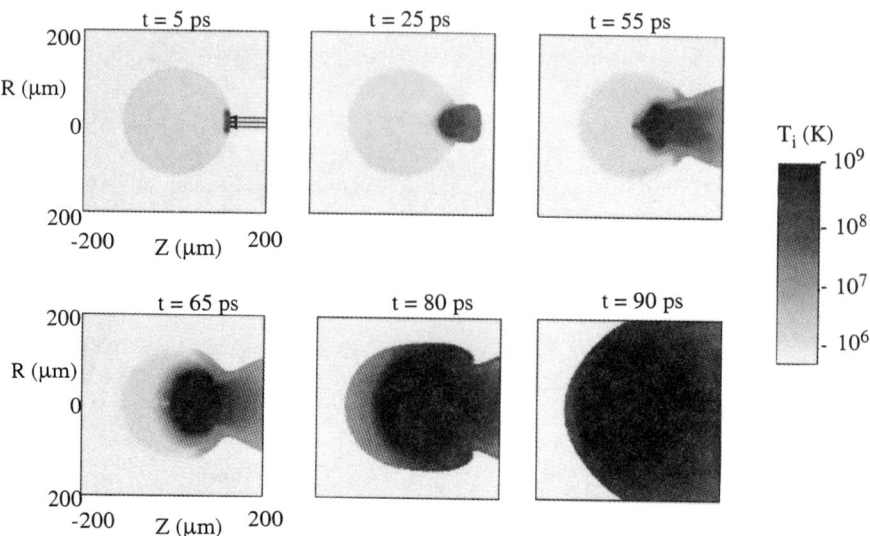

Figure 12. The fast ignition concept illustrated by a 2-D simulation. An intense beam of fast particles creates a hot spot close to the surface of a DT sphere, precompressed to density $\rho = 300$ g/cm^3. The beam delivers 20 kJ in 20 ps onto a spot of radius of 20 μm. The figure shows a sequence of temperature maps at different times.

($\rho r_h = 0.5$ g/cm^2; $T_h = 12$ keV), and taking $m_h \approx (4\pi/3)\rho r_h^3$. We thus get $E_p \geq 72/\hat{\rho}^2$ kJ, where $\hat{\rho} = \rho/(100$ g/cm$^3)$. The pulse duration should be of the order of the confinement time of the hot spot $t_p \approx \tau_h \approx r_h/c_s \approx 40/\hat{\rho}$ ps, and the beam radius of the order of the hot spot radius. Pulse power and intensity follow immediately from $P = E_p/t_p$ and $I = P/\pi r_h^2$. A more accurate evaluation has been performed by 2-D numerical simulations (analogous to the one shown in Fig. 12) in which fast particles with assigned range impinge onto a DT sphere precompressed at high density ρ. It is found that for $50 \leq \rho \leq 3000$ g/cm^3 the minimum beam parameters for ignition are accurately fitted by[96]

$$E_p = 140\, \hat{\rho}^{-1.85} \text{ kJ}, \tag{38}$$
$$P = 2.6 \times 10^{15}\, \hat{\rho}^{-1} \text{ W}, \tag{39}$$
$$I = 2.4 \times 10^{19}\, \hat{\rho}^{0.95} \text{ W/cm}^2. \tag{40}$$

Equations (38)–(40) above refer to the energy, power and intensity actually delivered to the fuel, and to heating particles with range comparable to the hot spot diameter. For the reference density value of 300 g/cm^3 the fast-ignition beam should deliver about 20 kJ of fast particles in 20 ps onto a 20 μm spot. The corresponding power and intensity are $P = 10^{15}$ W, and $I = 7 \times 10^{19}$ W/cm^2, respectively. Even assuming a highly efficient mechanisms of conversion of the driver energy into kinetic energy of suitable forward-directed fast particles, extremely high intensities are required. An option to achieve them is now provided by CPA lasers. Actually, the original fast ignition concept[21] relies on (1) superintense laser beam hole-boring into the strongly overdense plasma surrounding the imploded core and (2) efficient production of a collimated beam of relativistic (1–10 MeV) electrons to heat the ignition spot. Both points involve relativistic laser-plasma interaction and are currently actively investigated both theoretically and experimentally. For overviews of recent work on fast ignitors see, e.g.,

Refs. 47, 53 and 97. Laser beam self-channeling in an underdense plasma [52] and the generation of forward peaked hot electron populations have been demonstrated.[53, 54, 55] Evidence has also been presented for self-focusing in overdendense plasmas.[98] However, parameters are still far from those required for fast ignition, and many issues are still to be solved. These include, in particular, channeling into 10^5 overcritical plasma, transport of a current exceeding the Alfvèn limit, and the efficiency of hot electron generation.

Recently, a variant[99] to the previous scheme has been proposed, in which the ignition hot spot is created by the fast protons produced by the interaction of an ultraintense beam with a suitable target. This suggestion follows from experimental evidence[100] for production of collimated MeV proton beams, with energy efficiency about 12%.

It is worth mentioning that simulations indicate that fast ignitors may also offer an option to burn nearly pure deuterium targets.[101] Here, a small quantity of tritium is required for ignition, but a comparable amount is bred inside the target itself. Therefore no external tritium breeding blanket would be needed, with positive consequences on reactor safety and environmental impact.

OUTLOOK

Forty years after the demonstration of the first lasers, the field of laser-plasma interaction still provides opportunities for fundamental research and is giving rise to a growing number of applications. As we have briefly surveyed in this paper, the basic plasma physics of the interaction of nanosecond pulses at intensity $I_{14}\lambda_{\mu m}^2 \leq 1$ is well established. Still, parametric instabilities raise issues and deserve fundamental research.[12, 28] In this field, however, the major developments follow from the beam requirements set by a number of unique applications (ICF, ultrahigh pressure material studies, laboratory astrophysics, X-ray lasers). These require accurate control of beam properties and highly sophisticate diagnostics. As an example, we cite the ultrasmooth and balanced beams needed for the success of direct-drive inertial fusion, and the diagnostics required to analyze with sufficient spatial and temporal resolution the hydrodynamic processes they induce. Challenging enterprises are the construction and operation of the megajoule-class lasers designed to achieve ICF ignition.

Regarding ultraintense lasers, the situation is quite different. This field is just blooming, due to the opportunities offered by the CPA technique, and by the feasibility of facilities of moderate size and cost, affordable by many institutions throughout the world. In this case, most of the activity concerns the understanding of fundamental processes. The virtually new brand of relativistic plasma physics is now amenable to experimental investigation. Despite incomplete knowledge of even basic facts, a number of applications have been proposed, and the principles of some of them have been demonstrated. Nuclear physics processes induced by table-top lasers raise great interest. Finally, the fast ignitor ICF concept may offer an alternative to the standard central ignition scheme.

Acknowledgment

I thank Mr. Bottomei (ENEA, Frascati) for preparing some of the figures appearing in this paper.

REFERENCES

1. W. L. Kruer, *Physics of Laser Plasma Interactions*, Addison Wesley, Redwood City, CA (1988).
2. A. Rubenchik and S. Witkowski, eds., *Physics of Laser Plasma* (vol. 3 of *Handbook of Plasma Physics*, M. N. Rosembluth and R. Z. Sagdeev, general eds.), North-Holland, Amsterdam (1991).
3. G. A. Morou and D. Umstadter, *Phys. Fluids B* 4:2315 (1992).
4. G. A. Morou, C. P. J. Barty, and M. D. Perry, *Phys. Today* 51(1):22 (January 1998).
5. M. D. Perry and G. Morou, *Science* 264:917 (1994).
6. D. Strickland and G. Morou, *Opt. Commun.* 56:219 (1985).
7. M. D. Perry et al., *Opt. Lett.* 24:160 (1999).
8. A. Saemann et al., *Phys. Rev. Lett.* 82:4843 (1999).
9. R. W. Lee et al., *Science on High Energy Lasers: from Today to the NIF*, Lawrence Livermore National Laboratory, report UCRL-ID-119710, Livermore, CA (1996); see also the web-site http://www.llnl.gov/science_on_lasers
10. S. I. Anisimov et al., *Sov. Phys. Usp.* 142:395 (1984).
11. E. M. Campbell and W. J. Hogan, *Plasma Phys. Controll. Fusion* 41:B39 (1999).
12. J. D. Lindl, *Phys. Plasmas* 2:3933 (1995); also: J. D. Lindl, *Inertial Confinement Fusion: the Quest for Ignition and Energy Using Indirect Drive*, Springer, New York, USA (1998).
13. W. Hogan et al., *Energy from Inertial Fusion*, IAEA, Vienna (1994).
14. B. A. Remington, D. Arnett, R. P. Drake and H. Takabe, *Science* 284:1488 (1998).
15. B. A. Remington, R. P. Drake, H. Takabe and D. Arnett, *Phys. Plasmas* 7:1641 (2000).
16. I. C. E. Turcu and J.B. Dance, *X-Rays from Laser Plasmas: Generation and Applications*, Wiley, London (1998).
17. M. Lontano, G. Morou, F. Pegoraro and E. Sindoni, eds., *Superstrong Fields in Plasmas*, AIP Conf. Proc. 426, AIP, Woodsbury, NY (1998).
18. T. Ditmire et al., *Nature (London)* 398:489 (1999).
19. N. Bloembergen, *Rev. Mod. Phys.* 71:S283 (1999).
20. K. W. D. Ledingham and P. A. Norreys, *Contemp. Phys.* 40:367 (1999).
21. M. Tabak et al., *Phys. Plasmas* 1:1626 (1994).
22. P. Gibbon and E. Foerster, *Plasma Phys. Controll. Fusion* 38:769 (1996).
23. P. Mulser, D. Bauer, S. Hain and F. Cornolti, Ultraintense laser-solid interaction phenomena, in: Ref. 17, p. 201; see also the whole part 3 of Ref. 17.
24. K. Eidmann, J. Meyer-ter-Vehn, T. Schlegel and S. Hüller, *Phys. Rev. E* 62:1202 (2000).
25. H. Motz, *The Physics of Laser Fusion*, Academic Press, New York, NY (1979).
26. L. Spitzer, Jr., *Physics of Fully Ionized Plasmas*, 2nd ed., Interscience, New York (1967).
27. H. A. Baldis, E. M. Campbell and W. L. Kruer, Laser-plasma interactions, in Ref. 2, p. 361.
28. W. L. Kruer, *Phys. Plasmas* 7:2270 (2000).
29. A. Caruso and R. Gratton, *Plasma Phys.* 10:687 (1968).
30. P. Mora, *Phys. Fluids* 25:1051 (1982).
31. J. H. Gardner and S. E. Bodner, *Phys. Rev. Lett.* 47:1137 (1981).
32. C. Garban-Labaune et al., *Phys. Rev. Lett.* 48:1018 (1982).
33. F. Amiranoff et al., in: *Plasma Physics and Controlled Fusion Research 1986, proc. 11th Int. Conf.*, vol. 3, IAEA, Vienna, (1987), p. 79.
34. S. Atzeni, *Plasma Phys. Cont. Fusion* 29:1535 (1987).
35. J.D. Kilkenny, Ablation pressure measurement at 0.35 μm, in: *Laser Program Annual Report*, Lawrence Livermore National Laboratory, report UCRL 50021-86, p. 3-6, Livermore, CA (1987).
36. H. Azechi et al., *Laser Part. Beams* 9:193 (1991).
37. J. M. Soures et al., *Phys. Plasmas* 3:2108 (1996).
38. M. Koenig et al., *Europhys. News* 27:210 (1996).
39. Ya. Zeldovich and Yu. P. Raizer, *Physics of Shock Waves and High Temperature Hydrodynamic Phenomena*, Academic Press, New York (1967).
40. P. Celliers et al., *Phys. Rev. Lett.* 84:5564 (2000).
41. R. Cauble et al., *Phys. Rev. Lett.* 70:210 (1993).
42. D. R. Farley et al., *Phys. Rev. Lett.* 83:1982 (1999).
43. R. Kauffman, X-ray radiation from laser plasma, in Ref. 2, p. 111.
44. R. Siegel, Laser generated intense thermal radiation, in Ref. 2, p. 163.
45. G. Jamelot, X-ray lasers, in: *Physics of Multiply Charged Ions*, D. Liesen, ed., Plenum, New York (1995) 291.
46. D. L. Matthews et al., *Phys. Rev. Lett.* 54:110 (1985).
47. C. Labaune et al., eds, *Inertial Fusion Science and Applications 99*, Elsevier, Paris (2000).

48. C. E. Max et al., *Phys. Fluids* 33:209 (1974).
49. G. A. Askar'yan, S. V. Bulanov, F. Pegoraro and A.M. Pukhov, *JETP Lett.* 60:251 (1994).
50. F. Califano, R. Prandi, F. Pegoraro and S.V. Bulanov, Nonlinear Weibel instability in the interaction of an underdense plasma with a "relativistic" laser pulse, in: Ref. 17, p. 123.
51. A. Pukhov and J. Meyer-ter-Vehn, *Phys. Rev. Lett* 76:3975 (1996).
52. M. Borghesi et al., *Phys. Rev. Lett.* 78:897 (1997).
53. M. H. Key et al., Progress in fast ignitor research with the NOVA Petawatt laser facility, in: *Fusion Energy 98*, IAEA, Vienna (1999), p. 1098.
54. P. A. Norreys et al, *Phys. Plasmas* 6:2150 (1999).
55. M. Borghesi et al, *Phys. Rev. Lett.* 83:4309 (1999).
56. T. Tajima and J. M. Dawson, *Phys. Rev. Lett.* 43:267 (1979).
57. A. Pukhov, Z.-M. Sheng and J. Meyer-ter-Vehn, *Phys. Fluids* 6:2847 (1999).
58. W. P. Leemans et al., *Phys. Fluids* 5:1615 (1998).
59. F. Amiranoff et al., *Plasma Phys. Controll. Fusion* 38:A295 (1996).
60. P. A. Norreys et al., *Phys. Fluids* 6:2150 (1999).
61. K. W. D. Ledingham et al., *Phys. Rev. Lett.* 84:899. (2000).
62. T. E. Cowan et al., *Phys. Rev. Lett.* 84:903.
63. G. Pretzler et al., *Phys. Rev. E* 58:1165 (1998).
64. J. H. Nuckolls, L. Wood, A. Thiessen and G. B. Zimmermann, *Nature (London)* 239:129 (1972).
65. J.J. Duderstadt and G. A. Moses, *Inertial Confinement Fusion*, John Wiley, New York, USA (1982).
66. S. Atzeni, *Selected Topics in ICF Target Design*, ILE, Osaka (1995).
67. S. Atzeni, *Jpn. J. Appl. Phys.* 34:1980 (1995).
68. J.D. Lawson, *Proc. Royal Soc. (London)* 70B:6 (1957).
69. J. Meyer-ter-Vehn, *Nucl. Fusion* 22: 561 (1982).
70. S. Atzeni and A. Caruso, *Nucl. Fusion* 23:1092 (1983).
71. R. L. McCrory and C. P. Verdon, Computer modeling and simulation in inertial confinement fusion, in *Inertial Confinement Fusion, Proceedings of the International School of Plasma Physics Pietro Caldirola*, A. Caruso and E. Sindoni, eds., Editrice Compositori, Bologna (1989), p. 89.
72. J. Meyer-ter-Vehn and C. Schalk, *Z. Naturforschung* 37A:955 (1982).
73. A. Caruso, High-gain radiation driven targets, in *Inertial Confinement Fusion, Proceedings of the International School of Plasma Physics Pietro Caldirola*, A. Caruso and E. Sindoni, eds., Editrice Compositori, Bologna (1989), p. 139.
74. S. Atzeni, Heavy-ion induced fusion, in: *Physics of Multiply Charged Ions*, D. Liesen, ed., Plenum, New York, USA (1995) 319.
75. S. Atzeni, *Europhys. Lett.* 11:639 (1990).
76. S. E. Bodner, Symmetry and stability in laser fusion, in Ref. 2, p. 247.
77. S. P. Ryan et al., *Phys. Plasmas* 6:2072 (1999).
78. N. D. Delamater et al. *Phys. Plasmas* 7:1609 (2000).
79. M. H. Emery et al., *Phys Fluids* B3:2640 (1991).
80. D. G. Colombant et al., *Phys. Plasmas* 7:2046 (2000).
81. V. N. Goncharov et al., *Phys. Plasmas* 7:2062 (2000).
82. G. I. Taylor, *Proc. Royal Soc. (London)* A201:192 (1950).
83. S. Chandrasekhar, *Hydrodynamic and Hydromagnetic Stability*, Oxford University Press, Glasgow, UK (1961).
84. D. L. Youngs, *Physica* 12D:32 (1984).
85. H. Takabe et al., *Phys. Fluids* 28:3676 (1985).
86. R. Betti et al., *Phys Fluids* 5:1446 (1998).
87. S. G. Glendinning et al., *Phys. Rev. Lett.* 78:3318 (1997).
88. K. S. Budil et al., *Phys. Rev. Lett.* 76:4536 (1996).
89. M. Marinak et al, *Phys. Plasmas* 3:2070 (1996).
90. M. D. Cable et al., *Phys. Rev. Lett.* 73:2316 (1994).
91. T. S. Dittrich et al., *Laser Part. Beams* 17 217 (1999).
92. S. E. Bodner et al., *Phys. Plasmas* 5:1901 (1998).
93. J. A. Paisner et al., *Laser Focus World* 30:75 (1994).
94. M. André, Laser Megajoule project status, in Ref.47 p. 32. Also: M. André, *Fusion Eng. Des.* 44:43 (1999).
95. L. Suter et al., *Phys. Plasmas* 7:2092 (2000).
96. S. Atzeni, *Phys Plasmas* 6:3316 (1999).

97. P. Mora and J.-C. Gauthier, eds., *Fourth International Workshop on Fast Ignition, April 3–5, 2000, Palaiseau*, Ecole Polytechnique, Palaiseau (2000).
98. K. Takahashi *et al.*, *Phys. Rev. Lett.* 84:2405 (2000).
99. M. Roth *et al.*, Fast ignition by intense laser-accelerated proton beams, submitted to *Phys. Rev. Lett.* (2000); see also: R. B. Stephens *et al.*, Update on fast ignition experiments at NOVA PetaWatt, paper IAEA-CN-77/IFP/10, 18th IAEA Fusion Fusion Energy Conference, Sorrento, Italy 4–10 October 2000; M. Roth *et al.*, Ion particle acceleration: a new approach to fast ignition?, in Ref. 97.
100. R. A. Snavely *et al.*, *Phys. Rev. Lett.* 85:2945 (2000).
101. S. Atzeni and M. L. Ciampi, *Nucl. Fusion* 37:1665 (1997); S. Atzeni, Improved studies on ignition requirements for DT and T-poor fast ignitors, in Ref. 47, p. 415.

EXPERIMENTAL STUDY OF PETAWATT LASER PRODUCED PLASMAS

Michael H. Key

University of California
Lawrence Livermore National Laboratory
PO Box 808
Livermore
CA 94550

INTRODUCTION

The operation of the first 1 petawatt (PW) laser at the Lawrence Livermore National Laboratory (LLNL) in 1996 provided a laser power which has not yet been equaled elsewhere and it enabled a range of novel scientific investigations which are reviewed here and briefly set in context with theoretical work and other related experiments.

A special feature of the physics of interaction of PW class lasers with matter at high focused intensities is that free electrons achieve strongly relativistic energies in the field of the laser beam. Much of the new science is associated with this relativistic laser plasma interaction which is defined by a ratio greater than unity of the transverse momentum of an electron to $m_e c$. Represented by the parameter $a = eE/m_e c$, the ratio can be expressed in practical units as $(I_{18}\lambda^2/1.37)^{1/2}$ where I_{18} is the intensity in units of 10^{18} Wcm^{-2} and λ is the wavelength in microns. Associated with the relativistic regime is a strong effect of the magnetic force $e\mathbf{v}\times\mathbf{B}$ relative to the electric force $e\mathbf{E}$ from the electromagnetic wave. The trajectory of an electron in a plane polarized wave therefore includes both transverse oscillation and forward motion, the latter twice per optical cycle due to the $e\mathbf{v}\times\mathbf{B}$ force. There is therefore a net zig-zag motion in the direction of the laser light. As the interaction become more strongly relativistic the forward motion becomes dominant relative to the transverse motion.

An intensity approaching 10^{21}Wcm^{-2} is characteristic of 1PW laser power focused strongly onto a target as discussed later. At this intensity $a = 27$ and the electric field of the laser is 100kV/nm scaling with intensity as $I^{1/2}$. The strong electric field of the light wave field ionizes bound electrons with binding energies up to 4keV[1]. The relativistic interaction of laser light with matter is therefore invariably with a plasma.

The relativistic mass increase of the electrons causes self-focusing of the laser light through the associated decrease in polarizability of the plasma.[2] Reduction of the negative refractive index of the free electrons and therefore a net increase in the refractive index causes the self-focusing. In the self-focused beam a powerful current of electrons is driven forward by the laser light through the zig-zag forward motion and the associated magnetic field causes a z-pinch effect. Some electrons are accelerated by coherent interactions with plasma waves, the magnetic field and the laser electric field [3] and reach energies many times greater than the ponderomotive potential[4] $\phi = m_e c^2 ((1+a^2)^{1/2}-1)$.

At the critical density the laser light is strongly absorbed and converted to a forward directed stream of relativistic electrons through several mechanisms, notably the Brunel mechanism[5] for oblique P polarized light and jxB absorption[6] at normal incidence. These processes transfer energy of the order of the ponderomotive potential ϕ to the electrons. At $10^{21} Wcm^{-2}$ $\phi = 13$ MeV and a Boltzmann-like energy spectrum with a mean energy or 'temperature' similar to ϕ is produced[4,7]. The electron current for one petawatt of power carried by 10 MeV electrons is 100MA. This is 300x the Alfven limit[8] (17γ kA, where γ is the relativistic Lorentz factor) at which the self magnetic field field field of the current creates a Larmor radius for the electrons which is smaller than the radius of the beam. Propagation of the beam is impossible above this limit. Almost 100% compensation by return current is therefore required for the relativistic electrons generated by a PW laser to penetrate a plasma.

The light pressure at the critical density is $p = I/c$ and is of magnitude 300Gbar at $10^{21} Wcm^{-2}$. This light pressure drives the critical density interface forward launching an optically drive shock wave with a velocity [9] approximately $(p/\rho)^{1/2}$ where ρ is the density. The forward velocity exceeds 10^9 cm/s driving protons to more than 1MeV energy.

The richness and complexity of the new physics of the relativistic laser matter interaction make it difficult to model. The most sophisticated particle in cell (PIC) codes can today model only a small space and time element of the complete process. Never the less these PIC models give us our best understanding of the relativistic interactions. Effects such as relativistic self focusing including the magnetic pinching of the self focused beam[2], particle trajectories and acceleration processes within the self focused beam[10] and absorption and hole boring at the critical density [9] have all been modeled with PIC codes.

At densities above critical the forward travelling relativistic electrons and the cold electron return current create the conditions for strong growth of the Weibel instability[11]. It causes filamentation of the electron flow on a microscopic scale of the order of the classical skin depth c/ω_p with the instability growing with a time constant of the order of the plasma frequency ω_p.[12]

PW CLASS LASERS

PW class lasers are based on the basic concept of Chirped Pulse Amplification (CPA) first demonstrated in 1985[13]. The first 1PW laser was brought into operation in

1996[14] at the Lawrence Livermore National Laboratory. It generated 500J, 500fs pulses from a Nd-glass laser amplifier system fed with a 1ns duration chirped pulse from a Ti-sapphire mode-locked laser. The output pulse was compressed to 500 fs using a pair of 1m diameter diffraction gratings. With a feedback controlled deformable mirror incorporated into the optical system, the beam when focused with a f/3 parabolic mirror, had a focal spot diameter of approximately 8μm at full width half maximum (fwhm) intensity. The focal spot pattern was not a simple Airy function but was much more complex with significant scattering of energy into the wings of the distribution. About 30% of the energy fell inside the first minimum of the intensity pattern. The peak intensity was about 5% of the diffraction limit, reaching 3×10^{20} Wcm^{-2} with a broad distribution function of intensity as shown in figure 2.

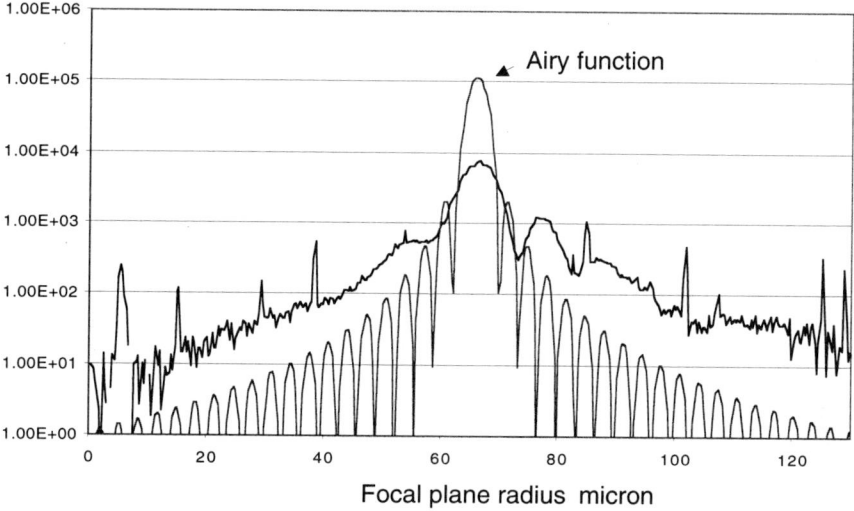

Figure 1. Measured intensity pattern of the focus of the 1PW laser beam (obtained from a 16 bit CCD camera recording an equivalent plane focal image). Shown for comparison is the corresponding diffraction limited Airy function.

Today the original PW laser has been shutdown after three years of operation, in preparation for the construction of the National Ignition Facility (NIF). Similar 1PW lasers are however under construction in Japan, France the U.K. and Germany and these lasers also currently operate at 0.1 to 0.2 PW. There is in addition a wider family of PW class lasers under active development using variants of the CPA architecture and technology. The development of large aperture transmission gratings offers the future possibility of multi-kJ multi-ps pulses for purposes such as fast ignition.[15] Nd glass laser pumped Ti-sapphire lasers have generated over 10J of energy in pulses down to 80 fs [16] providing a smaller scale technology for developing single pulse PW power. High rep-rate is obtained with Nd-YAG laser pumped Ti-sapphire lasers, producing up to a few Joules of energy in 20 fs pulses at 10 Hz.[17] The 10 Hz lasers promise to be important for many practical applications of ultra-high intensity. The possibility of optical parametric short pulse amplification offers prospects for 100J lasers and beyond with pulse lengths as short as 10 fs[18] opening the way to the next decades of power above 1PW.

PRE-FORMED PLASMA AND RELATIVISTIC SELF FOCUSING

A PW laser pulse focused onto a solid target interacts invariably with preformed plasma created by laser energy arriving before the main pulse. Amplified spontaneous emission (ASE) during the time that optical gates are open before the arrival of the main pulse, typically generates a few times 10^{-4} of the main pulse energy with a duration of one or two nanoseconds. Leakage of the mode locked pulse train of the pulse generator through closed electro-optical gates creates pre-pulses with typically 10^{-6} to 10^{-4} of the main pulse energy. Pre pulse levels can be reduced by increasing the contrast ratio of the electro-optical gating, by frequency conversion to the 2nd harmonic and by using OPCPA.

Measurements for the LLNL petawatt laser showed a pre-formed plasma extending as much as 200µm from the target surface with an axial scale length of about 40µm in a quasi- exponential density profile. Relativistic self-focusing in this plasma is evidenced by the x-ray images shown in Figure 2.[19] Here x-ray images of the focal spot on a solid Au target show the laser energy concentrated into a region of the order 20µm diameter (with indications of filamentary sub structure) when the focused beam is displaced up to 300 micron relative to the target such that in the absence of plasma, the irradiated area would be 100µm in diameter.

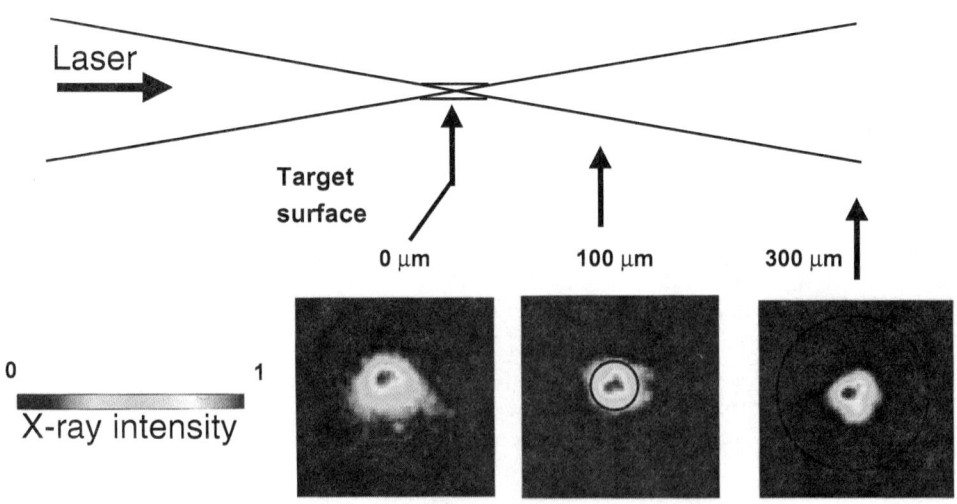

Figure 2. X-ray images of an Au target irradiated with the PW laser. The sketch denotes the focused beam with correct scale relative to the images. Rings show the irradiated area in the absence of any plasma. Images are 130 µm squares.

RELATIVISTIC ELECTRONS, BREMSSTRAHLUNG AND POSITRONS

Experiments illustrated in figure 3 have shown efficient conversion of PW laser power to relativistic electrons[20]. A magnetic electron spectrometer has measured the energy spectrum showing the presence of electrons up to 100MeV in quasi- Maxwellian energy spectrum with a mean energy of 5-10MeV.[21] Up to 40% conversion to relativistic electrons has been revealed by Kα fluorescence measurements [22].

Associated with the electrons and detected by the same the magnetic deflection spectrometer, are positrons (see figure 3) produced mainly by the e^+e^- pair creation absorption of >1MeV photons, the photons originating from Bremsstrahlung of the relativistic electrons in the solid target. Interesting is that an excess of positrons above the yield from absorption of photons, has been observed giving evidence for the first time in the laboratory, of the production of an electron-positron plasma by collisional processes[23]. This opens the door to laboratory simulation of a state of matter which is important in astrophysics in connection with gamma ray bursts.

The Bremsstrahlung process can be very efficient and measurements with the petawatt laser have shown as much as 11J or 2.9% of the laser energy converted to x-rays with photon energies exceeding 0.5MeV. Monte Carlo modeling of the experiment shows that this requires efficient conversion to relativistic electrons at a level of the order 40-50%.[24]

PHOTO- NUCLEAR PROCESSES

Bremsstrahlung produces photons with energies up to the maximum energies of the electrons. Photons of more than about 10MeV energy readily induce photo-nuclear interactions through the giant resonance when protons in the nucleus are driven at their resonant frequency with the respect to neutrons. The resulting photo-neutron production is readily detected from the associated nuclear activation and radioactive decay of elements in a solid target. In the petawatt experiments extensive photo-nuclear interactions have been observed[20,21,25]. These include the *(γ,xn)* processes in Au with the number x of ejected neutrons ranging from 1-7 [26]. A closely related process also observed[27], is the photo -fission of U^{238}, which is illustrated in Figure 4.

Nuclear activation can be used to estimate the angular spread of the relativistic electrons because Bremsstrahlung photons are emitted with a cone angle of *1/γ* and therefore mirror very closely the direction of the electrons, which produce them. An array of gold disks covering the forward hemisphere was used to determine the angular spread of photo-nuclear activation, which was found to be rather reproducibly, within a 100° cone but to have a variable off access direction spanning more than 100°. The latter may be associated with a hosing instability of the self-focused filament in the plasma.

The *(γ,xn)* photo-nuclear processes in Au have cross section peaks at energies ranging from 10MeV to 80MeV and enabled direct measurement [27] of the spectrum of the Bremsstrahlung photons from the yields of nuclear activation as shown in Figure 5.

The applicability of PW lasers as intense MeV x-ray sources for hard x-ray radiography has been studied. There are potential advantages in the forward directed

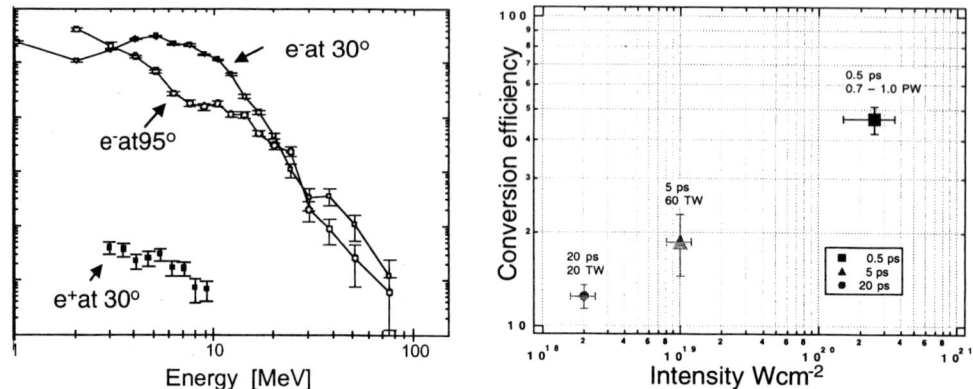

Figure 3. Left: electron and positron energy spectra recorded with magnetic spectrometers at 30° and 95° to the laser axis. Right: conversion efficiency of laser energy to relativistic electrons determined from Kα fluorescence in layered targets for 0.5 ps, 5ps and 20ps pulses, plotted at the corresponding intensities on the target.

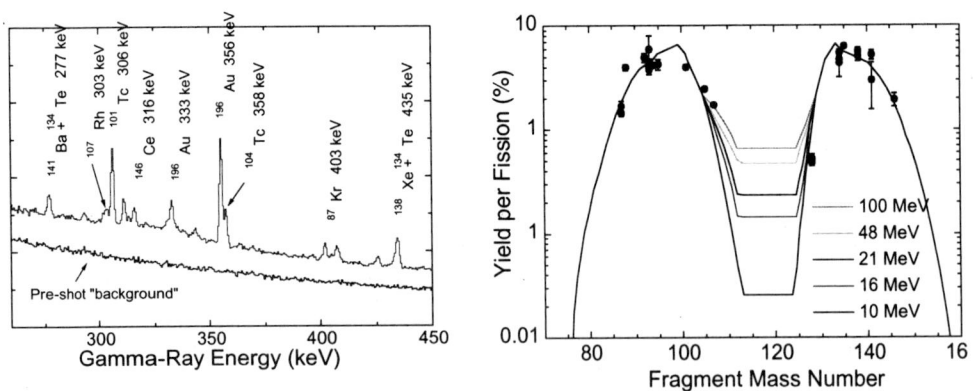

Figure 4. Left: gamma ray spectrum of nuclear activated target containing U^{238}. Right: Atomic fragment masses in laser induced photo-fission of U^{238}. Data are shown and solid lines give theoerical results for varying photon energies.

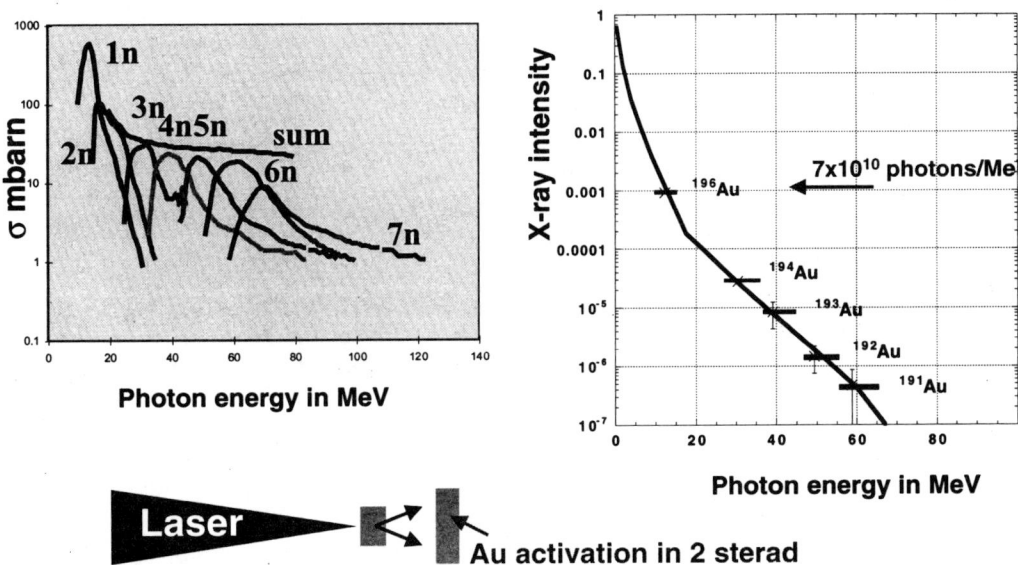

Figure 5. Left :photo-nuclear cross sections $^{197}Au(\gamma,xn)$ .Right: the spectrum of Bremsstrahlung inferred from the different channels of nuclear activation of an Au target.

emission of hard x-rays in the MeV region, the small area of the relativistic electron beam and therefore of the x-ray source, and the efficient conversion into x-rays[28]. An example is shown in Figure 6 where a heavy metal test object up to 10 cm thick is radiographed with sub mm spatial resolution and ps temporal resolution.

GENERATION OF HIGH-ENERGY PROTON BEAMS AND PROTON INDUCED NUCLEAR PHENOMENA

A very interesting outcome of experiments with the LLNL petawatt laser was the discovery of highly collimated very energetic proton beams emitted perpendicular to the rear surface of thin foil targets.[29] A novel diagnostic technique was used with the original intention of recording the angular distribution of relativistic electrons produced by the laser. It consisted of a conical assembly of thin sheets of radio-chromic film and Ta as shown in Figure 7. The radio-chromic film responds to ionizing radiation by changing color and the optical density of the film can be interpreted as dose of ionizing radiation. With assumptions as to the nature of the radiation, the dose can be related to the flux of ions, electrons or photons. The original discovery was in November 1998. The PW laser was focussed at 45° P polarized on a 125μm thick gold foil. The radio-chromic film showed an exceptionally intense beamed emission perpendicular to the back surface of the foil.

At first sight this was thought to be a beam of electrons.[30] From the attenuation of the 'electrons' through successive thickness of 100μm Ta foils in the detector, it was possible to estimate the energy spectrum and total current density of the beam. This revealed a paradox. The amount of charge in the 'electron' beam exceeded that which could escape from the target without charging the target to a potential higher than the energy of the electrons. The 'electron 'current also exceeded the Alfven limit. Resolution of this paradox was suggested by observations made with a nuclear track detector CR39. Etching of a thin sheet of this plastic material produces small holes in response to ionizing particles. The hole size associated with protons is distinctive and it was observed that protons penetrated through a 7mm thick Al frame holding the CR39 foil. The energy of protons penetrating the Al had to be more than 40 MeV. This result suggested that the beam seen in the radio-chromic film images through several 100μm sheets of Ta could be a proton beam.

Re-analysis of radiation dose in the radio-chromic film images assuming the beam to be protons, indicated a Boltzmann-like energy spectrum with a mean energy of 3MeV and a total energy content of about 10J. These numbers appeared plausible in the light of the measured energies of the hot electron source and well established relationships between accelerated fast ion energies and hot electron temperature[31]. More detailed examination of the radio-chromic film images revealed that underlying the sharply defined footprint of the proton beam there was a much weaker diffuse pattern of more penetrating radiation. Quantitative analysis assuming this to be electrons, indicated approximately 1J of electrons in a Boltzmann–like energy spectrum with a mean energy of 3 to 4 MeV. This was consistent with the known electron energies and the allowed escape of hot electrons from the target prior to the Coulomb potential exceeding the mean electron energy. The Alfven limit was also not violated The angular pattern of electron emission was found to be rather variable sometimes with a strong peak in the direction of the proton beam (for 45°P polarization), but more often diffuse over most of the forward hemisphere with sometimes multiple angular peaks.

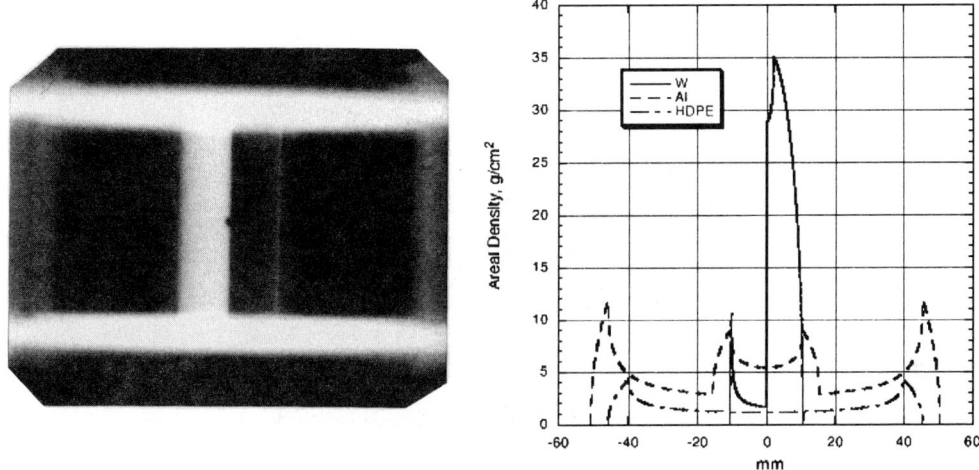

Figure 6. Left: Projection radiographic image. Right: plot of the areal density of the test object along a horizontal line through the center of the image.

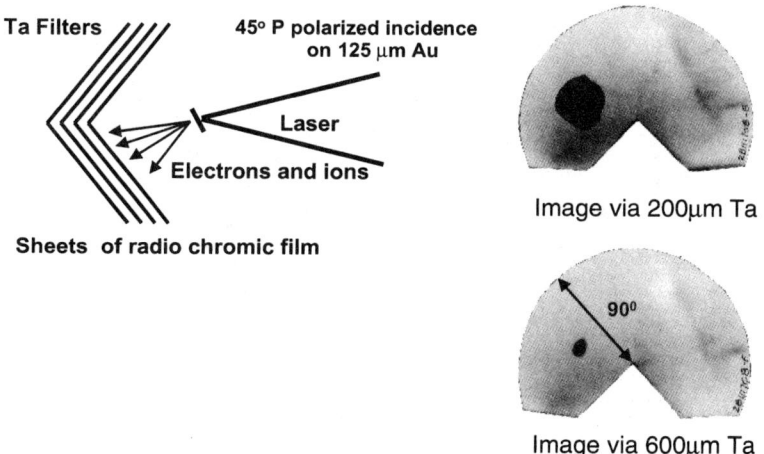

Figure 7. Radio-chromic film images showing beamed emission perpendicular to the target surface and a diffuse background.

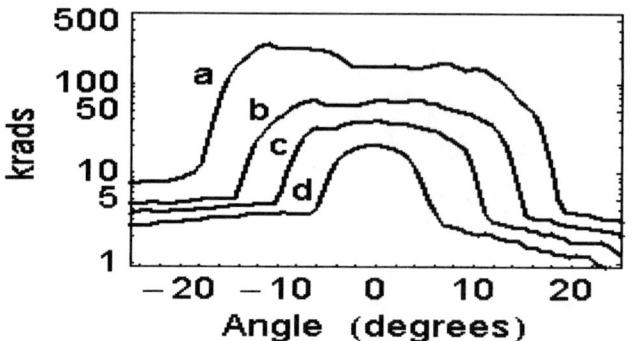

Figure 8. Proton beam angular pattern as a function of energy. Plots a to d have proton energies exceeding 17, 24, 30 and 35 MeV respectively.

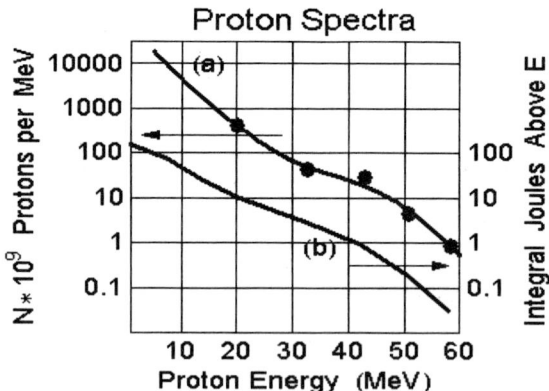

Figure 9. Energy spectrum and integrated energy of the proton beam from a 100 micron thick CH target.

For gold targets the proton beam had very sharply defined boundaries and was uniform within a roughly circular cross section. Its angular width became smaller at higher energies (selected by thicker Ta layers in front of the radio-chromic film) as shown in Figure 8. For CH targets there was irregular structure in the boundary and in the intensity pattern of the proton beam (see figure 10). CH targets gave a higher total proton emission than gold targets. For example 48J or 12% of the incident laser energy in a proton beam with energies >10 MeV was recorded as shown in figure 9 with a 100 μm thick CH target.[32]

An interesting feature of the proton beam was that its energy exceeded the threshold for proton induced nuclear processes. Accordingly an experiment was devised in which the beam was incident on multiple layers of 250μm thick Ti, radiochromic film and 1.2mm Be. The motivation was to observe neutron yield from (p, n) processes in Be and also the (p, n) nuclear activation process which created V^{48} from Ti^{48}. The results of this experiment showed the production of 3×10^{10} neutrons and 10^9 atoms of V^{48} in the Ti foils, the latter detected by gamma ray emission spectroscopy. Analysis showed that 3×10^{13} protons with 30J of energy, a maximum energy of 40 MeV and a mean energy in the Boltzmann- like spectrum of 6 MeV, were incident on the foils.[32]

The energy content and energy spectrum were deductions from Monte Carlo computer modeling, used to describe the response of the radiochromic film to protons (and electrons) and also the nuclear activation of Ti as shown in Figure 9. The response of radiochromic film to protons through different thicknesses of material is sharply peaked in energy due to the Bragg peak of energy deposition. The analogous modeling for nuclear activation of Ti gives slightly broader but similar response curves[32].

Interesting further information from the nuclear activation of Ti was from auto-radiographic images obtained by sandwiching the activated Ti foils between thin layers of Pb and photographic film. These auto-radiographic images were obtained by accumulating the radioactive decays over a period of a day and then developing the film to show the spatial pattern of the nuclear activation. When this was compared with the spatial pattern of the co-sandwiched layers of radio-chromic film there was a good match confirming that the radiographic film images corresponded to the proton beam as shown in figure 11.

The mechanism generating the proton beam was attributed to[24] the hot electron driven plasma blow-off from the rear surface of the target in which the leading component was protons either from absorbed layers of hydrocarbons on the gold targets or from the bulk material of the CH targets. Well established theory of hot electron driven expansion and fast ion production shows there is a separation of the ion components of the plasma with protons generally carrying the larger fraction of the energy and reaching the maximum energies[31]. Another possibility considered was the direct acceleration of ions by the light pressure incident at the front surface. It was of interest therefore to discriminate between front and rear surface acceleration and an experiment was devised in which a laser was incident on a wedged CH target. The result was striking in that two proton beams were observed perpendicular to each of the rear un-irradiated surfaces of the wedged target confirming very unambiguously the acceleration of the proton beam from the rear target surface.[32]

To compare proton beam emission from the front and back surfaces, cones of radio-chromic film were set-up covering most of the forward and rear hemispheres with a small entrance aperture for the laser beam. This showed that there was no well collimated proton beam emitted to the front and that protons with energy great than the detection threshold of the experiment which was 3MeV, when integrated over the whole forward hemisphere carried less than 5% of the energy of the well collimated proton beam to the rear.

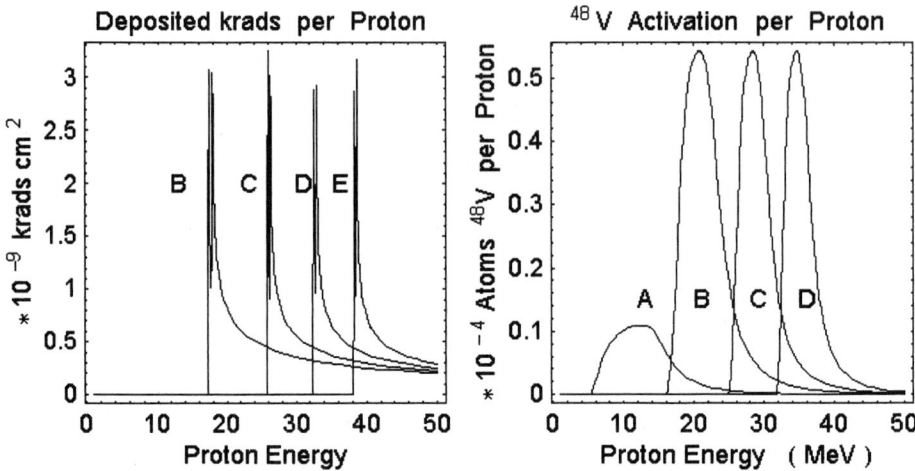

Figure 10. Monte Carlo modeling of the response to protons of radiochromic film (left) and nuclear activation of Ti (right) for the conditions of the experiment shown in figure 11.

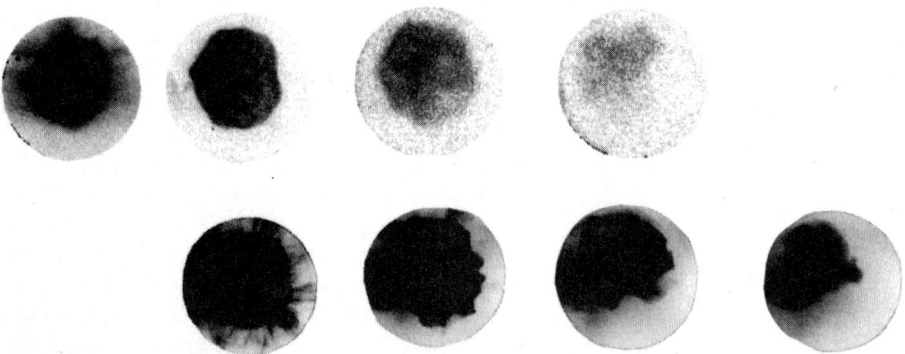

Figure 11. Autoradiography images from Ti foils (above) and corresponding radio-chromic images (below) with minimum proton energy increasing from left to right as the beam penetrates through more layers of the detector. Response curves A to D in figure 10 correspond to images here from left to right.

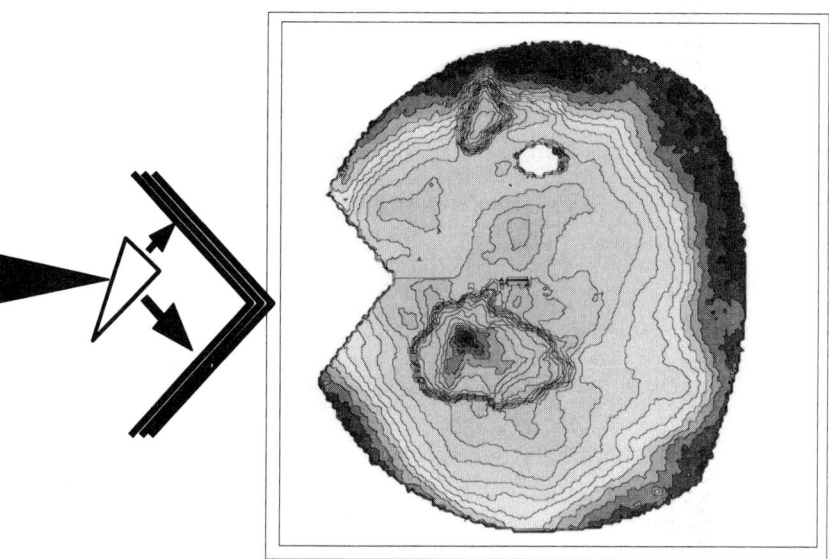

Figure 12. Proton beams from a wedged CH target. A contour plot of the image on radio-chromic film shows two beams perpendicular to the two rear surfaces of the wedge.

More detailed information on the energy spectrum of the protons[32] was obtained by adapting the magnetic spectrometer used to study electrons, for recording protons. The results showed an extremely sharp cut off in the energy spectrum of the protons, which was function of the angle relative to the rear surface of the target. The maximum proton energies were perpendicular to the rear surface and reached 60MeV. At 45° the energy dropped to 15 MeV. The spectra also showed that in the vicinity of the cut-off of the energy spectrum there were quasi-mono energetic features which theory[31] has shown to be associated with separation of the protons from other ionic components of the plasma.

The source size of the proton emission was of interest because this together with the angular divergence determines the emittance of the beam, which is important in considering it's potential applications. Limited information on the source size was obtained from the penumbral shadowing of the proton beam across the slit at the entrance to the magnetic spectrometer. This indicated a source size less than 400µm but was close to the resolution limit for the instrumental set-up. Better measurements were obtained in subsequent experiments at the 0.1PW, 1 ps Vulcan laser in the UK and the 0.1 PW, 100fs Janusp laser at LLNL, which indicated source sizes of the order of 30-50µm. Interesting from these data is that the normalized emittance of the proton beam was as small as 0.5π mm mrad. By comparison the CERN proton linear accelerator has a normalized emittance of 1.7. The laser produced proton beams are therefore of accelerator quality.

In the interpretation of experimental data on proton beams obtained at the Rutherford Appleton Laboratory, it has been postulated that the angular pattern and variation of energy with angle is due to protons accelerated at the front surface traveling through the solid target, and being deflected in angle by the self-generated magnetic field of the electron beam.[33] Studies of the variation of the proton energy with target thickness in another experiment from the University of Michigan[34] were also interpreted as due to a front surface mechanism. Contrary evidence is seen in analysis of neutron emission when solid CD_2 targets are irradiated with short laser pulses causing light pressure driven CD_2 ions to penetrate into the target producing beam target D-D fusion reactions[35]. Analysis of the angular spectrum of neutron energy shows that the mean energy of the light pressure driven ions is typically about 200keV and broadly consistent with the light pressure driven shock velocity[36]. These ions from the front surface are therefore of insufficient energy to explain the observed proton beams. Further evidence in favor of a rear surface acceleration mechanism was obtained in a Rutherford Laboratory experiment using an auxiliary laser beam to pre-form plasma on the rear surface of the target before the short pulse was incident. It was seen that this stopped the production of the proton beam and the interpretation was that this occurred through eliminating the sharp density gradient[37].

The mechanism of acceleration now seems well described[24] in terms of a hot electron source of temperature related to the ponderomotive potential of the laser at the front surface, the penetration of these relativistic electrons through the target, the formation of a Debye-sheath at the rear surface and acceleration of protons with a maximum accelerating field due to the Debye-sheath. This field can be expressed as $E=kT/eL$ where T is the relativistic electron temperature and L is the smaller of the scale length of the density gradient or the vacuum Debye length. A sharp density discontinuity at the rear surface of solid targets therefore maximizes the accelerating field which scales as $(N_e T_e)^{1/2}$, where N_e is the number density of relativistic electrons trapped in the target by the Coulomb potential. The power transfer to accelerated protons is greatest when the field is at its maximum. A weak density gradient such as produced by preformed plasma on the rear surface, reduces the accelerating field.

A qualitative picture of the variation of proton energy with angle is given by considering the radial diminution in hot electron density, which implies a radial reduction in the accelerating field. This causes a bell shaped development of the Debye sheath as the plasma expands from the rear surface. Numerical modeling with PIC techniques in 2 dimensions has given further insight in to this behavior[38]. The local accelerating field is on axis and strongest where the electron density is highest and weaker and more off axis to the margins. This structure can explain the maximum proton energies on axis and lower proton energies off axis in the observed proton beams.

A corollary of the fact that acceleration is initially perpendicular to the local surface is that a spherically concave rear surface can therefore produce a focused beam, the focal spot size being determined by the emittance. Such fast ion focussing was observed over 20 years ago in early studies of laser driven implosions of the exploding pusher type where an irradiated spherical shell was heated by hot electrons and exploded symmetrically inward and outward. When there was no gas fill in the shell, spherically focussed fast ions created an early implosion event seen in x-ray streak camera images[39].

FAST IGNITION

Fast ignition of thermo-nuclear fusion is an exciting concept[40] with major potential for inertial fusion energy (IFE). It relies on the production of compressed thermo-nuclear fuel by conventional laser or ion driven techniques. It has however a major difference from conventional inertially confined fusion (ICF)[41] in that the ignition is not in a central hot spot created as part of an implosion, but instead in a hot spot produced by fast external

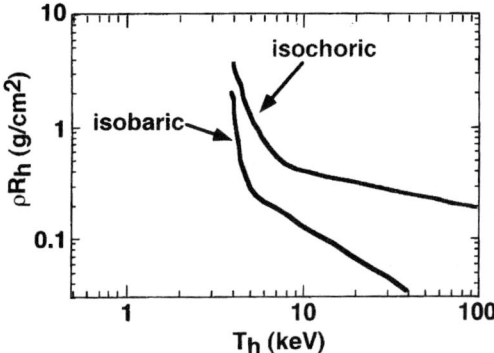

Figure 13. Conditions for isobaric and isochoric thermo-nuclear spark ignition.

injection of energy from a short pulse laser beam. Conventional ICF has an isobaric implosion core in which the hot spot is of lower density than the surrounding cold fuel. Fast ignition is isochoric with higher pressure in the hot spot, which is at equal density with the rest of the fuel. The importance of this difference is linked to the ignition spark energy requirements.

Many analyses have covered the conditions required to initiate a fusion burn.[42] For an isobaric spark the energy is minimized at a density radius product ρr of 0.3gcm^{-2} and a temperature of 5keV. For an isochoric spark these condition become 0.5gcm^{-2} and a temperature of 10keV as illustrated in figure 13.

At first sight the energy requirement (at constant density) is six times higher for the isochoric relative to the isobaric spark. However, the ρr dependence leads to the total spark energy scaling as ρ^{-2}. Fast ignition (isochoric) occurs at the full compressed fuel density whereas isobaric ignition occurs at typically a 20x lower density than the rest of the fuel.

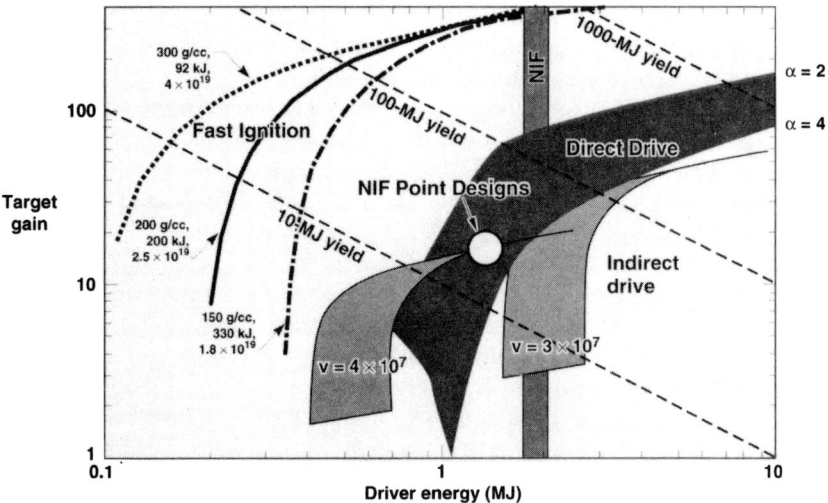

Figure 14. Thermonuclear gain comparisons for the NIF.

159

The net result is that the spark density can be higher and the ignition energy lower for fast ignition with lower fuel density and less energy used for compression. The energy gain of a fusion burn can therefore be higher. At the scale of the NIF fast ignition is predicted [43], to give gains as high as 300x relative to the 15x for the indirectly driven point design and typically 50x for directly driven central spark ignition. It is the high gain of fast ignition which makes it attractive for IFE because a driver of <10% electrical efficiency requires an energy yield of a few hundred times to give a cost effective power producing cycle.

Fast ignition at the scale of NIF with 300x gain might proceed for example, with the fuel compressed to 200g /cc in a sphere of the order 340m diameter requiring 1.2MJ of energy to compress the fuel at a ratio of the internal energy to the Fermi-degenerate limit of 2. The ignitor-pulse would require 200kJ of laser energy in 20 ps delivered from 20 of the NIF beams into a cluster of 25μm diameter focal spots. With critical density located 100μm from the dense fuel a two fold increase in area of the electron beam would deliver the energy to an optimum 50μm diameter ignition spark. 50% conversion of laser light to the second harmonic, 50% conversion to electrons and 30% loss in the transport process are assumed [43].

TRANSPORT OF ENERGY AND HEATING BY RELATIVISTIC ELECTRONS

Crucial to the question of how to ignite the target is the transport of energy from the deposition point to the dense fuel. The previous discussion assumes that the energy is transported by relativistic electrons with reasonable collimation. The energy transport by hot electrons is however strongly modified by electric and magnetic fields. The Weibel instability causes the forward and return currents to filament as discussed previously. This filamentation is on the spatial scale of the plasma skin depth, which is sub-micron. The time scale for this development is the inverse plasma frequency, which is sub-femtosecond. The total forward current greatly exceeds the Alfven limit and therefore it must be compensated by cold return current. The cold return current also causes Ohmic heating and a potential build up[44]. Electrons entering the target experience loss of energy through the opposing potential and through collisionless dissipation in the process of coalescence of filaments in the non-linear phase of the Weibel instability[45].

Theory has predicted that there will be a net forward current of the order of the Alfven limit giving a net azimuthal magnetic field which guides the electron transport in a well collimated overall pattern[46]. Perhaps the best description is obtained with new class of hybrid PIC codes, which treat hot electrons as particles but the background plasma as a fluid. The code PARIS is a good example [47]. This suggests that an initially uniform electron beam penetrating into the solid would first filament by the Weibel instability, and that coalescence of the Weibel filaments will occur to cause a large scale irregular structure and that there may be evolution to an annular pattern of collimated current transport.

Transport experiments need spatial resolution of the relativistic electron flux and heating. This has been accomplished by a variety of techniques using sensor layers or observations of the rear surface of a thin foil target. These experiments used 5 ps pulse duration and correspondingly one order of magnitude reduced power and intensity more appropriate to fast ignition.

Initial studies used a non-space resolved sensor, which was D-D thermo-nuclear fusion in electron heated solid CD_2 targets where there was a front layer of CH or Cu not

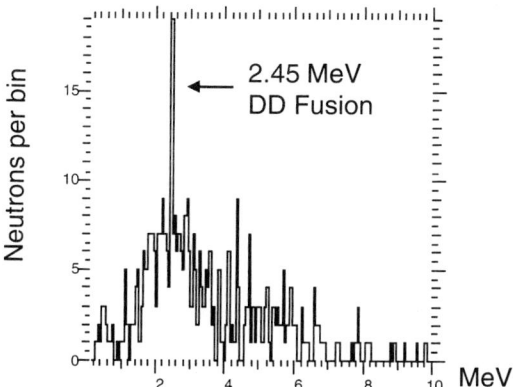

Figure 15. D-D thermonuclear fusion in relativistic electron heated solid CD_2, evidenced by the narrow peak in the neutron energy spectrum at 2.45 MeV .The total thermonuclear neutron yield is 6×10^4 and the inferred temperature is >500eV.

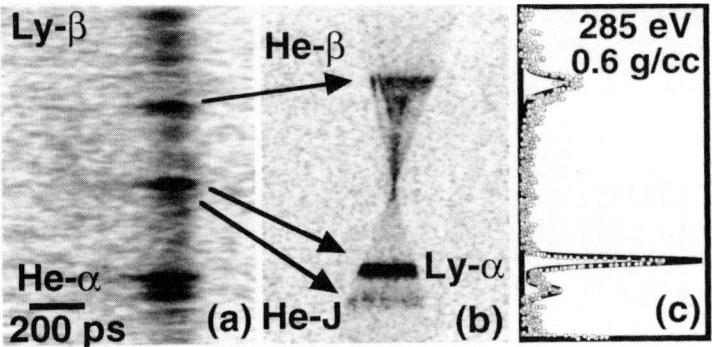

Figure 16. Left steak time resolved x-ray spectrum, centre, time integrated x-ray spectrum ,
Right, comparison of spectrum and modelling. 5 ps irradiation of a CH target with a 0.5 µm Al layer 15 µm below the CH surface.

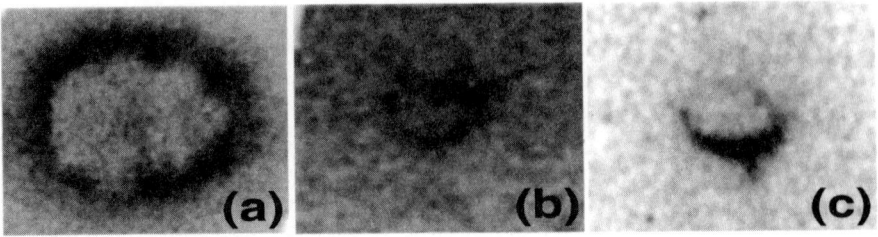

Figure 17. X-ray images of buried indicator layers in CH targets at depths of 15, 50 and 100 μm for a,b and c respectively.

containing deuterium. A narrow fusion peak at 2.45MeV was observed [43] in neutron energy spectra, where a broad underlying component is due to photo-nuclear and proton induced nuclear processes, and the narrow peak is due to thermo-nuclear fusion. The measurements were at the signal to noise limit for detecting the peak in the background but suggested temperatures > 0.5 keV produced somewhere in the solid target.

Buried aluminum tracer layers have been used for x-ray spectroscopy[48] to show the production of 300eV temperature at a density of 0.6g/cc as shown in figure 16. Images of the x-ray emission have a characteristic ring pattern suggestive of an annular current flow. This ring pattern has been observed to persist to depths of 100μm inside solid CH and is illustrated in figure 17 [49]. The energy transport is well collimated but in a rather large diameter annulus of about 80 μm diameter.

A full understanding of the development of the annular energy transport pattern will require advances in numerical modeling but preliminary indications from hybrid PIC codes suggest that this may be a consequence of the Weibel instability amplifying initially strong Fourier components arising from the edge shape of the electron beam pattern.[50]

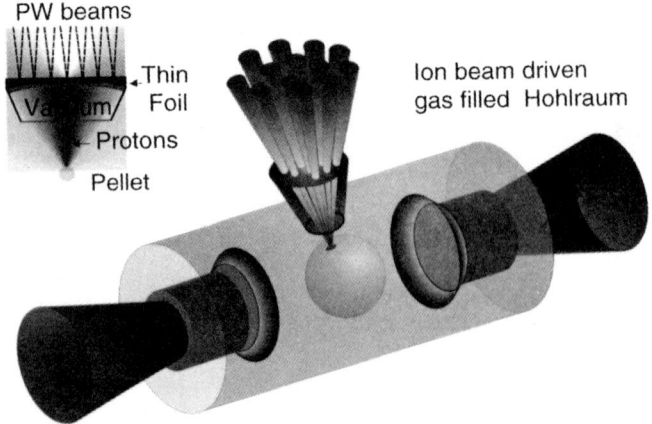

Figure 18. Schematic proton beam fast ignition in an ion beam heated hohlraum.

Other measurements (with 0.1 PW power and 0.5 ps pulses) at the rear surface of thin foil targets have shown from Planckian optical emission, a heating pattern which gets wider for increased target thickness and breaks-up into filamentary structure[51]. There is however good collimation in the early stages of the beam propagation up to 100μm depth. These shorter pulse experiments have Weibel growth and may therefore not evolve to a strong annular pattern.

The efficiency of energy transfer and its collimation is a crucial question for fast ignition and current experiments are beginning to give encouraging answers. More work is needed however to draw definite conclusions on the feasibility of electron induced fast ignition.

A radical alternative concept is to use proton beams with a spherically curved rear surface of a thin foil target giving ballistic focusing. Such proton beams might be generated at the side of a hohlraum in which fuel compression is accomplished through heating by ion beams. This involves indirect drive of the compression by intense thermal x-ray emission. The advantage of ballistic focusing of protons is lack of the many complicating effects in electron beam transport. The crucial issue for proton beams will be the possibility of an efficient transfer of laser energy into protons within a rather narrow energy spread from about 15-25MeV required to insure the short duration of the power impulse at the target[52]. The emittance of the beam from a large area thin foil source is another issue, which is unverified but the concept is at a very early stage and is novel and interesting.

CONCLUSIONS

The PW laser at LLNL has enabled many interesting experiments, which are part of a world wide effort to understand and exploit ultra-high intensity laser matter interactions. The science includes basic physics and links to astrophysics together with applied physics opportunities and are exciting and novel. Practical applications are being explored in areas such as radiography, ion accelerators and radio-nuclide sources, particularly using rep-rated ultra-short-pulse lasers. High-energy petawatt class lasers will be developed further for research in fusion energy where fast ignition has important potential advantages if basic energy transport issues can be resolved.

ACKNOWLEDGMENTS

Many colleagues and collaborators have contributed to the work I have discussed including E M. Campbell, T. Cowan, R. Freeman, S. Hatchett, E. Henry,
J. Koch, B. Langdon, B. Lasinski, A. MacKinnon, A. Offenberger, D. Pennington,
M. Perry, T. Phillips, M. Roth, C. Sangster, M. Singh, R. Snavely, R. Stephens,
M. Stoyer, S. Wilks, K. Yasuike. Work performed under the auspices of the U.S. Department of Energy by the Lawrence Livermore National Laboratory under Contract No. W-7405-ENG-48.

REFERENCES

[1] M. D. Perry and G. Mourou, Science **264**, 917 (1994)

[2] A. Pukhov and J. Meyer-ter-Vehn, Phys. Rev. Lett. **76**, 3975 (1996)

[3] A. Pukhov et al., Phys..Plasmas,. **6**, 2847 (1999)

[4] E Lefebvre, G Bonnaud,Phys.Rev.E.**55**,1011,(1997)

[5] F. Brunel, Phys. Rev. Letts.. **59**, 52 (1987)

[6] W L Kruer , K Estabrook Phys.Fluids **28**,430,(1985)

[7] S. C. Wilks, Phys. Fluids B **5**, 2603 (1993)

[8] H Alfven, Phys,Rev,Lett. **55**, 425,(1939)

[9] S C Wilks,WL Kruer,M Tabak, A B Langdon, Phys.Rev.Lett.**69**,1383 (1992)

[10] B. F. Lasinski, A. B. Langdon, S. P. Hatchett, M. H. Key, and M. Tabak, Phys. Plasmas **6**, 2041 (1999)

[11] E. W. Weibel, Phys. Rev Lett. **2**, 83 (1959).

[12] M. Honda, J. Meyer-ter-Vehn, A. Pukhov, Phys. Plasmas **7**, 1302 (2000)

[13] D Strickland and G Mourou,Opt.Comm.**56**, 219, (1985)

[14] M. D. Perry, D. Pennington, B. C. Stuart, G. Tietbohl, J. A. Britten, C. Brown, S. Herman, B. Golick, M. Kartz, J. Miller, H. T. Powell, M. Vergino, and V. Yanovsky, Opt. Lett. **24**, 160 (1999)

[15] M D Perry, private communication

[16] P Springer, private communication

[17] K. Yamakawa, M. Aoyama, et.al. Opt. Lett. **23**, 525

[18] I N Ross et al Appl. Opt. 39,2422,(2000)

[19] M. H. Key, E. M. Campbell, T. E. Cowan, S. P. Hatchett, et al. J. Fusion Energy **17**, 231 (1998)

[20] M. H. Key, K. Estabrook, B. Hammel, S. Hatchett, D. Hinkel, J. Kilkenny , J. Koch, et.al. Phys. Plasmas **5**, 1966 (1998)

[21] T E Cowan et al . Laser and Particle beams,17773,(1999)

[22] K. B. Wharton, S. P. Hatchett, S. C. Wilks, M. H. Key, J. D. Moody, V. Yanovsky, A. A. Offenberger, B. A. Hammel, M. D. Perry, and C. Joshi, Phys. Rev. Lett. **81**, 822 (1998)

[23] E P Liang,S C Wilks, M Tabak, Phys. Rev. Lett. **81**, 4887 (1998)

[24] S. P. Hatchett, C. G. Brown, T. E. Cowan, E. A. Henry, J. Johnson, M. H. Key, J. A. Koch, et.al, Phys. Plasmas **7**, 2076 (2000)

[25] T W Phillips, M D Cable, T E Cowan et.al. Rev. Sci. Instr. **70**, 1213 (1999)

[26] M. A. Stoyer et al. Rev. Sci. Instr. (in press) (2000)

[27] T. E. Cowan, A.W. Hunt, T.W. Phillips, et al. Phys. Rev. Lett. **84**, 903 (2000)

[28] M D Perry , J A Sefcik, T Cowan, S Hatchett et al. Rev.Sci. Instr. 70,265, (1999)

[29] M. H. Key, E.M. Campbell, T. E. Cowan, S. P. Hatchett et.al. in Inertial Fusion Sciences and Applications 99, Christine Labaune, William J. Hogan, and Kazuo A. Tanaka,Editors, publ. Elsevier (1999)

[30] M. H. Key, E. M. Campbell, R W Lee, M. Singh, T. Cowan, A. MacKinnon, R. Snavely. .et.al. Fusion Energy 1998, Publ IAEA Vienna , **3**, 1093 (1999)

[31] Y. Kishimoto, K. Mima, T. Watanabe, and K. Nishikawa, Phys. Fluids **26**, 2308 (1983).

[32] R. A. Snavely, M. H. Key, S. P. Hatchett, T. E. Cowan, M. Roth, T. W. Phillips, M. A. Stoyer, E. A. Henry, T. C. Sangster, M. S. Singh, S. C. Wilks, A. MacKinnon, A. Offenberger, D. M. Pennington, K. Yasuike, A. B. Landon, B. F. Lasinski, J. Johnson, M. D., Phys Rev. Lett. **85**, 2945, (2000)

[33] K. Krushelnick et al. Physics of Plasmas, **7**, 2055 (2000). E. L. Clark et al.. Phys. Rev. Lett. **84**, 670, (2000)

[34] A. Maksimchuk et al., Phys. Rev. Lett., **84**, 4108 (2000)

[35] L Disdier,J-P Garconnet, G Malka, J-L Miquel Phys Rev Lett, **82**, 1454(1999)

[36] C Toupin E Lefebvre,G Bonnaud. Phys Plasmas (in press) (2000)

[37] A J MacKinnon M Borghesi, S Hatchett, M H Key et. al. Phys.Rev.Lett. (in press) (2000)

[38] S.C. Wilks, A.B. Langdon, T.E. Cowan, M. Roth, M. Singh, S. Hatchett, M.H. Key, D. Pennington, A. MacKinnon, and RA Snavely. Phys. Plasmas. (in press) (2000)

[39] M H Key in Annual Report of the Central Laser Facility 1978 RL78-039. Publ. Rutherrford Laboratory, UK (1978)

[40] M. Tabak, J. Hammer, M. E. Glinsky, W. L. Kruer, S. C. Wilks, J. Woodworth, E. M. Campbell, M. D. Perry, and R. J. Mason, Phys. Plasmas **1**, 1626 (1994)

[41] J. D. Lindl, Phys. Plasmas **2**, 3933 (1995)

[42] S. Atzeni, Phys. Plasmas **6**, 3316 (1999)

[43] M. H. Key, E. M. Campbell, T. E. Cowan, S. P. Hatchett, et.al. J. Fusion Energy **17**, 231 (1998)

[44] A. R. Bell, J. R. Davies, S. Guérin, and H. Ruhl, Plasma Phys. Control. Fusion **39**, 653 (1997)

[45] M. Honda, J. Meyer-ter-Vehn, A. Pukhov, Phys. Rev. Lett. **85**, 2128 (2000)

[46] J. R. Davies, A. R. Bell, M. G. Haines, and S. M. Guérin, Phys. Rev. E **56**, 7193 (1997). A. R. Bell, J. R. Davies, and S. M. Guérin, Phys. Rev. E **58**, 2471 (1998).). J. R. Davies, A. R. Bell, and M. Tatarakis, Phys. Rev. E **59**, 6032 (1999)

[47] L. Gremillet, presented at the 4th International Workshop on the Fast Ignition of Fusion Targets, Paris, March 2000

[48] J. A. Koch, C. A. Back, C. Brown, K. Estabrook, B. A. Hammel, S. P. Hatchett, M. H. Key, et.al. Lasers and Particle Beams **16**, 225 (1998)

[49] J. A. Koch, S. P. Hatchett, M. H. Key, R. W. Lee, D. Pennington, R. B. Stephens, and M. Tabak, *Inertial Fusion Sciences and Applications 99*, Eds. C. Labaune, W. Hogan, K. Tanaka (Elsevier, Paris, 2000), pg. 463.

[50] J A Koch , R R Freeman, S P Hatchett, M H Key ,R.W.Lee, D Pennington, R .B Stephens, M Tabak. Phys Rev E (submitted) (2000)

[51] R. Kodama, presented at the 4th International Workshop on the Fast Ignition of Fusion Targets, Paris, March 2000

[52] M. Roth, T. Cowan, M. Key, S. Hatchett, C. Brown, M. Christl, W. Fountain, J. Johnson, T. Parnell, D. Pennington, M. D. Perry, T. W. Phillips, R. Snavely, S. C. Wilks, K. Yasuike, Phys. Rev. Lett. (in press), (2000)

RELATIVISTIC LASER PLASMA INTERACTION

J. Meyer-ter-Vehn, A. Pukhov, and Zh.-M. Sheng[1]

Max-Planck-Institut für Quantenoptik
D-85748 Garching, Germany

INTRODUCTION

The new technique of Chirped Pulse Amplification (CPA) allows to generate laser pulses with focussed intensities far above 10^{18}W/cm^2. It has opened a new branch of relativistic plasma physics.[1] At intensities beyond 10^{18}W/cm^2, now available from table-top systems, the laser pulses self-focus in plasma channels, a few wavelengths in diameter, and accelerate electrons up to relativistic energies along these channels. The channel currents reach 10 - 100 kA and generate magnetic field up to GigaGauss level. One of the most remarkable features is the formation of well collimated relativistic electron beams emerging from the channels with energies up to GeV. In dense matter, they trigger cascades of γ-rays, e$^+$e$^-$ pairs[2] and a host of nuclear and particle processes[3]. The present notes are mainly based on work at the Max-Planck-Institute for Quantum Optics (MPQ), including experiments at the ATLAS laser facility[4] and simulations. The tree-dimensional particle-in-cell code VLPL (Virtual Laser Plasma Laboratory), developed at MPQ,[5] has turned out as a superior tool to analyze real experiments and to explore the non-linear plasma kinetics at such intensities.

In this lecture, we start from a basic discussion of single-electron motion in an intense light field and then turn to the *cold plasma* approximation and an analytic description of relativistic selffocussing. The core part of the paper addresses the topic of electron acceleration in plasma channels. It became clear recently that it is *direct laser acceleration* by a resonance mechanism analogous to that in free-electron-lasers which drives the electrons to multi-MeV energies well above the ponderomotive energy.[6] At the end, a brief discussion of plasma wave excitation, wave breaking and wakefield acceleration is added. As it turns out, wakefield acceleration dominates for ultra-short pulses, shorter than a plasma wavelength. Another topic of high current interest is ion acceleration by laser-generated space charge fields. For space reasons, it could not be treated in the present lecture.

---

[1] Present address: Institute of Laser Engineering, Osaka University, Suita, Osaka, Japan

RELATIVISTIC ELECTRON MOTION IN A STRONG LASER FIELD

The Laser Field

We start this lecture with some basic definitions of the light field, its dimensionless amplitude a, and the relativistic intensity threshold $a = 1$. A plane light wave is described by its vector potential

$$\mathbf{A}(\mathbf{r}, t) = \mathrm{Re}\{\mathbf{A}_0 e^{i\psi}\},$$

where $\mathbf{A}_0 = A_0 \hat{e}_y$ for linear polarization (LP), $\mathbf{A}_0 = (A_0)(\hat{e}_y \pm i\hat{e}_z)$ for circular polarization (CP) (+ and - for right and left-circular polarization, respectively), and phase $\psi = \mathbf{k} \cdot \mathbf{r} - \omega t$. The dispersion relation in vacuum is $\omega = kc$ with $k = |\mathbf{k}| = 2\pi/\lambda$ and c the velocity of light. Electric and magnetic fields are then obtained as

$$\mathbf{E} = \mathrm{Re}\{-\frac{i\omega}{c}\mathbf{A}_0 e^{i\psi}\}, \quad \mathbf{B} = \mathrm{Re}\{i\mathbf{k} \times \mathbf{A}_0 e^{i\psi}\}, \tag{1}$$

and the Poynting vector is $\mathbf{S} = \frac{c}{4\pi}\mathbf{E} \times \mathbf{B}$. The intensity of the light is then

$$I = |\mathbf{S}| = \frac{\omega k}{4\pi} A_0^2 \times \begin{cases} \frac{1}{2}(1 + \sin 2\psi), & \text{for LP} \\ 1, & \text{for CP} \end{cases} \tag{2}$$

and the averaged intensity

$$I_0 \lambda^2 = \frac{\omega k \lambda^2}{8\pi} A_0^2 = \frac{\pi}{2} c A_0^2$$

for linear and twice this value for circular polarization.

The Relativistic Threshold

The relativistic threshold intensity is reached when electrons caught by the light wave acquire the velocity of light. For non-relativistic electrons with $|\mathbf{v}| \ll c$, the equation of motion

$$m\frac{d\mathbf{v}}{dt} = -e(\mathbf{E} + \frac{\mathbf{v}}{c} \times \mathbf{B}) \approx -e\mathbf{E}, \tag{3}$$

has the integrals

$$\mathbf{v} = \mathrm{Re}\{\frac{e\mathbf{E}}{im\omega}\} = -\frac{eA_0}{mc}\begin{cases} \hat{e}_y \cos\varphi, & \text{for LP} \\ (\hat{e}_y \cos\varphi \mp \hat{e}_z \sin\varphi), & \text{for CP} \end{cases} \tag{4}$$

$$\tag{5}$$

$$\mathbf{x} = \mathrm{Re}\{\frac{e\mathbf{E}}{m\omega^2}\} = \frac{eA_0}{mc}\begin{cases} \hat{e}_y \sin\varphi, & \text{for LP} \\ (\hat{e}_y \sin\varphi \mp \hat{e}_z \cos\varphi), & \text{for CP} \end{cases} \tag{6}$$

Introducing the dimensionless light amplitude

$$a_0 = \frac{eA_0}{mc^2},$$

we obtain the intensity in the form

$$I_0 \lambda^2 = \frac{\pi}{2} c A_0^2 = \frac{\pi}{2} P_0 a_0^2 = \left[1.37 \times 10^{18} \frac{W}{cm^2} \mu m^2\right] a_0^2. \tag{7}$$

The relativistic intensity threshold is reached at $a_0 = 1$, when the quiver velocity approaches c. Of course, the electron trajectories then differ from the simple transverse oscillation derived above. We shall derive them in detail below. Equation 7 involves the relativistic power unit

$$P_0 = \frac{mc^2}{e^2/mc^3} = \frac{mc^2}{e}\frac{mc^3}{e} = 8.67\text{GW}, \tag{8}$$

which may be written as the product of the voltage $mc^2/e = 511\text{kV}$ corresponding to the rest energy of the electron and the current unit $J_0 = mc^3/e = 17\text{kA}$, which is related to the Alfven current $J_A = J_0\beta\gamma$ with $\beta = v/c$ and $\gamma = 1/\sqrt{1-\beta^2}$; currents larger than J_A cannot be transported in vacuum due to magnetic selfinteraction.

Symmetries and Invariants

An exact analytic description is possible for single electrons in a plane light wave of arbitrary amplitude. For a simple derivation, it is important to make appropriate use of the symmetries and invariants of the problem. The relativistic Lagrange function of a particle with charge q moving in electromagnetic potentials $\mathbf{A}$ and ϕ is given by

$$L(\mathbf{r}, \mathbf{v}, t) = -mc^2\sqrt{1 - \frac{v^2}{c^2}} + \frac{q}{c}\mathbf{v}\cdot\mathbf{A} + q\phi. \tag{9}$$

From the Euler-Lagrange equation

$$\frac{d}{dt}\frac{\partial L}{\partial \mathbf{v}} - \frac{\partial L}{\partial \mathbf{r}} = 0, \tag{10}$$

we obtain the equation of motion

$$\frac{d\mathbf{p}}{dt} = q(\mathbf{E} + \frac{\mathbf{v}}{c}\times\mathbf{B}). \tag{11}$$

The canonical momentum $\mathbf{p}^{\text{can}} = \partial L/\partial\mathbf{v} = m\gamma\mathbf{v} + q\mathbf{A}/c = \mathbf{p} + q\mathbf{A}/c$, where $\gamma = 1/\sqrt{1-v^2/c^2}$. For a plane light wave, there exist two symmetries providing two constants of motion. Planar symmetry implies $\partial L/\partial\mathbf{r}_\perp = 0$ and therefore conservation of the canonical momentum in transverse direction

$$\partial L/\partial\mathbf{v}_\perp = \mathbf{p}_\perp + \frac{q}{c}\mathbf{A}_\perp = \text{constant}. \tag{12}$$

The second invariant derives from the wave form of $\mathbf{A}(t - x/c)$. Making use of the relation $dH/dt = -\partial L/\partial t$ for the Hamilton function $H(\mathbf{x},\mathbf{p},t) = E$, which expresses the time-dependent energy of the particle, one obtains

$$\frac{dE}{dt} = -\frac{\partial L}{\partial t} = c\frac{\partial L}{\partial x} = c\frac{d}{dt}\frac{\partial L}{\partial v_x} = c\frac{dp_x^{\text{can}}}{dt} = c\frac{dp_x}{dt},$$

where $A_x = 0$ for transverse light, and therefore

$$E - cp_x = \text{constant}. \tag{13}$$

For electrons initially at rest, this gives the kinetic energy

$$E_{kin} = E - mc^2 = p_x c, \tag{14}$$

and in combination with $E = mc^2\gamma = \sqrt{(mc^2)^2 + p_\perp^2 c^2 + p_x^2 c^2}$ after some algebra

$$E_{kin} = \frac{p_\perp^2}{2m} = p_x c = mc^2(\gamma - 1). \tag{15}$$

This is a remarkable relation which is relativistically exact, even though the term $p_\perp^2/2m$ looks like a non-relativistic kinetic energy. It holds approximately also for a finite laser beam. Applying it to the scattering of single electrons out of a laser focus, it implies

$$\tan^2\theta = \left(\frac{p_\perp}{p_x}\right)^2 = \frac{2mE_{kin}}{(E_{kin}/c)^2} = \frac{2}{\gamma-1}$$

relating the energy γ of scattered electrons to the angle θ, under which they emerge from the focus. This relation has been verified experimentally.[7]

Trajectory of a Relativistic Electron in a Plane Wave

For an electron in a plane light pulse of finite duration, the relativistic equation of motion can be integrated exactly. For an electron initially at rest, the invariants give

$$\mathbf{a} = \frac{e\mathbf{A}_\perp}{mc^2}, \tag{16}$$

$$\hat{\mathbf{p}}_\perp = \frac{\mathbf{p}_\perp}{mc} = \mathbf{a} = (0, a_y, a_z), \tag{17}$$

$$\hat{E}_{kin} = \frac{E_{kin}}{mc^2} = \gamma - 1 = \hat{p}_x = \frac{\hat{p}_\perp^2}{2} = \frac{a^2}{2}, \tag{18}$$

The equations of motion then read

$$\hat{p}_x = \gamma\beta_x = \frac{\gamma}{c}\frac{dx}{dt} = a^2/2, \tag{19}$$

$$\hat{p}_y = \gamma\beta_y = \frac{\gamma}{c}\frac{dy}{dt} = a_y, \tag{20}$$

$$\hat{p}_z = \gamma\beta_z = \frac{\gamma}{c}\frac{dz}{dt} = a_z, \tag{21}$$

Since $\gamma = 1 + a^2/2$, it is obvious that

$$\beta_x = \frac{a^2/2}{1+a^2/2} \to 1, \quad \beta_y = \frac{a_y}{1+a^2/2} \to 0, \quad \beta_z = \frac{a_z}{1+a^2/2} \to 0,$$

and $\tan\theta = p_\perp/p_\parallel \to 0$ when $a \gg 1$. This means that the motion of the electron, though transverse for low field strengths $|\mathbf{a}| \ll 1$, is more and more directed in the direction of light propagation in the relativistic regime $|\mathbf{a}| \gg 1$. Integration of eqs. 19 for a given light pulse $\mathbf{a}(t-x/c)$ is straightforward in the variable $\tau = t - x(t)/c$, for which

$$\gamma\frac{d}{dt} = \gamma\frac{d\tau}{dt}\frac{d}{d\tau} = \gamma(1-\frac{1}{c}\frac{dx}{d\tau})\frac{d}{d\tau} = (1+\frac{a^2}{2}-\frac{a^2}{2})\frac{d}{d\tau} = \frac{d}{d\tau}.$$

and therefore $d\tau = dt/\gamma$, such that the equations of motion obtain the simple form

$$\frac{dx}{d\tau} = c\frac{a^2}{2}, \tag{22}$$

$$\frac{dy}{d\tau} = ca_y, \tag{23}$$

$$\frac{dz}{d\tau} = ca_z. \tag{24}$$

For a box-shaped pulse of linear polarization with $a_y = a_0\cos(\omega\tau)$ for $0 < \tau < N(2\pi/\omega)$, $a_z \equiv 0$, $a^2 = a_y^2 = a_0^2\cos^2(\omega\tau)$, the trajectory of an electron at $x = y = z = 0$ at time

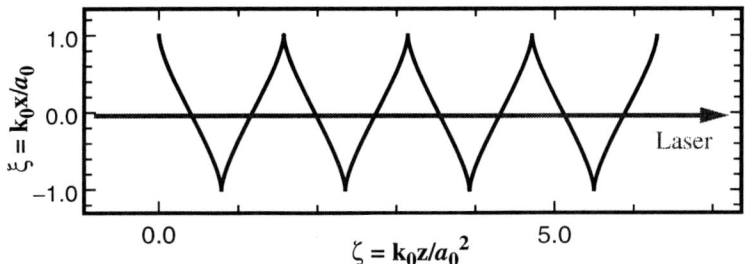

Figure 1: Electron trajectory in a plane electromagnetic wave.

$\tau = 0$ is obtained in the form

$$x(\tau) = \frac{ca_0^2}{2}\int_0^\tau \cos^2(\omega\tilde{\tau})d\tilde{\tau} = \frac{ca_0^2}{4}\left[\tau + \frac{1}{2\omega}\sin(2\omega\tau)\right], \tag{25}$$

$$y(\tau) = ca_0 \int_0^\tau \cos(\omega\tilde{\tau})d\tilde{\tau} = \frac{ca_0}{\omega}\sin(\omega\tau). \tag{26}$$

It consists of a drift motion in x-direction

$$x_d(t) = \frac{a_0^2}{a_0^2 + 4}ct \tag{27}$$

and superimposed a figure-8 trajectory in the drift frame:

$$ky = a_0 \sin(\omega\tau), \quad k(x - x_d) = \frac{a_0}{8}\sin(2\omega\tau). \tag{28}$$

The electron trajectory in normalized coordinates is shown in Fig. 1. The trajectory is self-similar in these coordinates. The transverse oscillations of the electron are proportional on the laser amplitude itself. The longitudinal motion, however, is proportional to the amplitude squared. Thus, at small laser amplitudes, $a \ll 1$, the electrons make mostly transverse oscillations. At ultra-relativistic laser amplitudes, $a \gg 1$, however, the electrons are strongly pushed forward.

If the laser pulse is an ideal plane wave, the electron remains at rest after the laser pulse is over. An electron trajectory in a finite laser pulse is shown in Fig. 2. Here, a Gaussian laser pulse, $a = a_0 \exp{-(t/\tau)^2}\cos(\omega_0 t - k_0 z)$ overtakes a resting electron. The laser pulse parameters are: $a_0 = 2$, $\omega_0\tau = 12.5$. We see that the electron oscillation become more and more longitudinal as the laser amplitude rises. After the laser pulse is over, the electron remains at rest. There is no net energy transfer between an electron and the laser pulse in vacuum, provided the laser pulse is a plane electromagnetic wave.

For circular polarization with

$$\mathbf{a}(\mathbf{r},t) = \mathrm{Re}\{a_0(\hat{e}_y \pm i\hat{e}_z)e^{-i\omega\tau}\},$$

one has $a^2 = a_y^2 + a_z^2 = a_0^2/2 = const$ and therefore a constant $\gamma = 1 + a_0^2/4$. The electron moves with constant velocity along the helical trajectory

$$x(\tau) = \frac{ca_0^2}{4}\tau = \frac{a_0^2}{a_0^2+4}ct, \tag{29}$$

$$y(\tau) = \frac{ca_0}{\sqrt{2}\omega}\sin(\omega t/\gamma), \tag{30}$$

$$z(\tau) = \mp\frac{ca_0}{\sqrt{2}\omega}\cos(\omega t/\gamma), \tag{31}$$

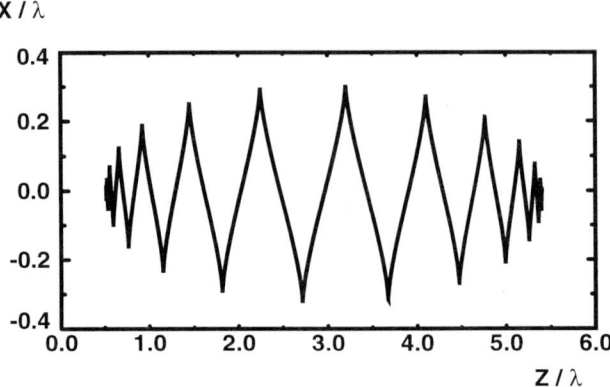

Figure 2: Electron trajectory in a finite duration laser pulse.

The drift velocity is the same as for linear polarisation, and the circular motion in the perpendicular plane is time-dilated by a factor γ. Note that the pulse overtakes the electron after some time and that the electron is finally again in its initial state. No net energy transfer takes place.

THE COLD PLASMA EQUATIONS

The *cold plasma* equations describe the plasma essentially as a zero temperature electron fluid for which density $N(\mathbf{x},t)$ and velocity $\mathbf{v}(\mathbf{x},t)$ are functions of space and time. Ions are assumed to form a uniform immobile background with density $N_i = N_0$ which neutralizes space charge on the average. The current density is then given by $\mathbf{J} = -eN\mathbf{v} = -eN\mathbf{P}/m\gamma$, where $\mathbf{P} = m\gamma\mathbf{v}$ is the momentum and $\gamma = \sqrt{1+p^2}$ the relativistic factor. Here and in the following, we use the dimensionless variables:

$$\mathbf{a} = \frac{e\mathbf{A}}{mc^2}, \quad \varphi = \frac{e\phi}{mc^2}, \quad \mathbf{p} = \frac{\mathbf{P}}{mc}, \quad n = \frac{N}{N_0}.$$

The wave equation describing the light propagating in plasma is then obtained from Maxwell's equations in the form

$$\nabla^2 \mathbf{a} - (1/c^2)(\partial^2 \mathbf{a}/\partial t^2) = (\partial/c\partial t)\nabla \cdot \varphi + (\omega_p^2/c^2)(n\mathbf{p}/\gamma),$$
$$\nabla^2 \varphi = (\omega_p^2/c^2)(n-1). \tag{32}$$

Here, the source term on the right-hand side consists of time-dependent space charge fields, expressed by the electrostatic potential φ, and currents.

The equation of motion for the cold plasma is taken as

$$\frac{d}{dt}\mathbf{P}(\mathbf{x},t) = -e(\mathbf{E} + \frac{\mathbf{v}}{c} \times \mathbf{B}). \tag{33}$$

where pressure terms proportional to plasma temperature are neglected. This is the essential point of the *cold plasma* approximation to describe relativistic laser plasma interaction, namely that thermal plasma forces are much smaller than electric and magnetic forces. They are therefore neglected. With $\mathbf{B} = \nabla \times \mathbf{A}$ and $\mathbf{E} = -\nabla\phi - (1/c)\partial\mathbf{A}/\partial t$, the Coulomb gauge $\nabla \cdot \mathbf{A} = 0$ and the time derivative $d/dt = \partial/\partial t + \mathbf{v}\cdot\nabla$, we find from Eq. 33

$$(\frac{\partial}{\partial t} + \mathbf{v}\cdot\nabla)\mathbf{P} = -e\left[-\frac{1}{c}\frac{\partial\mathbf{A}}{\partial t} - \nabla\phi + \frac{\mathbf{v}}{c}\times(\nabla\times\mathbf{A})\right]. \tag{34}$$

Making further use of the relations

$$\nabla\gamma = \nabla\sqrt{1+p^2} = \frac{1}{2\gamma}\nabla p^2, \quad \mathbf{v}\times\nabla\times\mathbf{p} = \nabla_p(\mathbf{v}\cdot\mathbf{p}) - (\mathbf{v}\cdot)\mathbf{p} = c\nabla\gamma - (\mathbf{v}\cdot\nabla)\mathbf{p} \tag{35}$$

and switching to dimensionless variables, we finally arrive at the central relativistic equation of motion

$$\frac{1}{c}\frac{\partial}{\partial t}(\mathbf{p}-\mathbf{a}) - \mathbf{v}\times\nabla\times(\mathbf{p}-\mathbf{a}) = \nabla(\varphi-\gamma). \tag{36}$$

A basic solution is $\mathbf{p} = \mathbf{a}$, for which $\nabla(\varphi - \gamma) = 0$. In this case the electrostatic force $\nabla\varphi$ just balances the ponderomotive force $\nabla\gamma = \nabla\sqrt{1+p^2} = \nabla\sqrt{1+a^2} = (2\gamma)^{-1}\nabla a^2$. For example, it describes how plasma electrons are pushed forward in front of a laser pulse exciting plasma waves or the pressure equilibrium in a self-focussed laser channel in which the ponderomotive force expels electrons building up a radial electrostatic field. The creation of strong electrostatic fields of same order as the laser electric field is a key feature of relativistic laser plasma interaction. These electrostatic fields accelerate electrons and in particular ions to multi-MeV energies. For the purpose of simple modelling, it is convenient to consider circularly polarized light beams

$$\mathbf{a} = Re\{(\hat{e}_x \pm i\hat{e}_y)a(r,z,t)e^{-i\psi}\}$$

with $r = \sqrt{x^2+y^2}$, $\psi = kz - \omega t$, and $|\mathbf{a}|^2 = |a(r,z,t)|^2/2$ varying slowly in time and space. Under these conditions and $\mathbf{p} = \mathbf{a}$, also $\gamma = \sqrt{1+p^2} = \sqrt{1+|a|^2/2}$ and φ vary slowly in space and time, just as the envelope function $|a|^2$. In this case, the wave equation (32) can be written in the form

$$\nabla^2\mathbf{a} - \frac{1}{c^2}\frac{\partial^2\mathbf{a}}{\partial t^2} = \frac{\omega_p^2}{c^2}\frac{n\mathbf{a}}{\gamma}. \tag{37}$$

This equation will be used to discuss relativistic self-focussing in the next chapter.

RELATIVISTIC SELF-FOCUSSING

Focus of a Gaussian Beam

Let us consider a light beam of the form

$$\mathbf{a}(\mathbf{r},t) = Re\{\mathbf{a}_0(\mathbf{r},t)\exp[i(\mathbf{kr}-\omega t)]\}, \tag{38}$$

and let us assume that the amplitude $\mathbf{a}(\mathbf{r},t)$ varies much less with $\mathbf{r}$ and t than the phase factor, i.e.

$$\partial a_0/\partial t| \ll |\omega a_0|, \quad \partial a_0/\partial z| \ll |ka_0|. \tag{39}$$

The wave equation in vacuum $(\nabla^2 - (1/c^2)\partial^2/\partial t^2)\mathbf{a} = 0$ can then be reduced to the envelope equation $(\nabla_\perp^2 + 2ik\partial/\partial z)a_0(r,z) = 0$, where we have neglected second derivative relative the first derivative terms and have used the dispersion relation $\omega^2 = k^2c^2$. Also we have restricted ourselves to a cylindrical beam with radius r, $\nabla_\perp^2 = \partial^2/\partial r^2 + (1/r)\partial/\partial r$, and the axial coordinate z chosen such that the envelope is independent of time.

For a Gaussian beam with $a_0(r,z) = \exp(P(z) - Q(z)(r/r_0)^2)$ it is straightforward to derive the functions $P(z)$ and $Q(z)$ satisfying the enevelope equation, and one obtains

$$a_0(r,z) = \frac{e^{-r^2/(r_0^2(1+z^2/L_r^2))}}{\sqrt{1+z^2/L_R^2}} \exp\left\{-i\arctan\left(\frac{z}{L_R}\right) + i\left(\frac{r}{r_0}\right)^2 \frac{z/L_R}{1+z^2/L_R^2}\right\} \tag{40}$$

Here, the front factor describes a hyperbolic envelope where r_0 is the focus radius at $z=0$ and $L_R = kr_0^2/2$ the Rayleigh length, determining the length of the focal waist. The phase factor describes spherical phase fronts of the incoming and outgoing wave with the phase jump of π when passing the focus.

Threshold Power For Relativistic Self-Focussing

We now switch to a beam focussed into a plasma and described by Eq. (37). For a low-intensity beam ($a \ll 1$), where a denotes the amplitude of a circularly polarized beam, the plasma density remains undisturbed ($n=1$) and the electrons are non-relativistic ($\gamma = 1$) such that Eq. 37 is linear in $\mathbf{a}$ and one obtains the usual plasma dispersion relation for electromagnetic waves

$$\omega^2 = \omega_p^2 + (kc)^2. \tag{41}$$

With increasing amplitude, but still $a^2 \ll 1$, non-linearity first arises through $1/\gamma = 1/\sqrt{1+a^2/2} \approx 1 - a^2/4$, while the density, determined by the Poisson equation in the form $n \approx 1 + (c/\omega_p)^2 \nabla^2 \gamma$, contributes only in higher order. Under these conditions, the wave equation in envelope approximation reads

$$\left[\nabla_\perp^2 + 2ik\frac{\partial}{\partial z}\right] a = -\frac{\omega_p^2}{c^2}\frac{|a|^2}{4}a. \tag{42}$$

and can be used to determine the threshold power for self-focussing. While the term $\nabla_\perp^2 a$ disperses the beam, the term $(\omega_p/c)^2|a|^2 a/4$ tends to focus the beam. At threshold both tendencies just balance each other, and we may consider the Gaussian envelope $a(r,z) \sim a_0(z)e^{-(r/R(z))^2}$ to test the balance. The corresponding beam power is

$$P = \frac{\pi R^2 I_0}{2} = \left(\frac{\omega^2}{16c^2}a_0^2 R^2\right) P_0, \tag{43}$$

where the relation (7) has been used involving the beam intensity on axis, I_0, and the power unit $P_0 = m^2 c^5/e^2 = 8.7$ GW. Going through an algebra, not given here for space reasons, one can derive

$$\frac{d^2 R(z)}{dz^2} = \frac{4}{k^2 r_0^3}\left[1 - \frac{1}{32}\frac{\omega_p^2}{c^2}|a_0|^2 r_0^2\right]. \tag{44}$$

for the beam radius $R(z)$.[8] Apparently, beam contraction sets in for $|a_0|^2 r_0^2 \geq 32(c^2/\omega_p^2)$. This gives the critical power for relativistic self-focussing

$$P_{\text{crit}} = 2.0(\omega/\omega_p)^2 P_0 \simeq 17.4\,\text{GW} \cdot (\omega/\omega_p)^2. \tag{45}$$

For $P > P_{\text{crit}}$, the beam contracts, expels electrons due to the ponderomotive force and forms a narrow plasma channel.

Plasma Channels and Electron Cavitation

For powers already slightly above P_{crit}, contraction of the beam leads to a strong increase of the light intensity on axis, and ponderomotive expulsion of electrons from the beam region sets in, forming a narrow plasma channel with a depleted electron density. Sun et al.[9] have described this situation by numerically solving the equations

$$\left[\nabla_\perp^2 + 2ik\frac{\partial}{\partial z} + \frac{\omega_p^2}{c^2}\sigma\right]a = \frac{\omega_p^2}{c^2}\frac{na}{\gamma}. \tag{46}$$

where $n = H(1+(c/\omega_p)^2 \nabla_\perp^2 \gamma)$ and $\gamma = \sqrt{1+a^2/2}$. Here, the step function $H(x) = (x+|x|)/2x$ accounts also for the case of empty channels, $n=0$. The parameter σ controls beam dispersion in the finite channel and is obtained as an eigenvalue when solving Eq. (46) for appropriate boundary conditions. Sun et al. find the self-focussing threshold for $\sigma=0$ and $P_{\text{crit}} = 17.2(\omega/\omega_p)^2\text{GW}$. The small difference compared to Eq. 45 is due to the assumption of a Gaussian beam made above, while the correct solution leads to Bessel functions for the radial profile.

For powers larger than P_{crit}, full collapse of the beam envelope is stopped by the action of the ponderomotive pressure, which expels electrons from the beam region and eventually leads to an electron-free channel. Complete cavitation is found for $\sigma = 0.8778$ and $P_{\text{cav}} = 1.06 P_{\text{crit}}$. One can estimate the size of cavitated channel as a function of a_0. Taking $\nabla_\perp^2 \gamma \sim -\gamma^2/R^2$ and $n=0$, one finds $R \approx (c/\omega_p)(1+a_0^2/2)^{1/4}$.

3D PIC simulations and experiments on relativistic self-focusing

The first 3D PIC simulations of relativistic laser self-focusing have been published by Pukhov and Meyer-ter-Vehn.[10] In that simulation, Fig. 3, a 10 TW laser pulse with $I_0 = 10^{19}$ W/cm^2 propagates through a slightly underdense plasma, $n_e/n_c = 0.36$. The laser power here is much higher than the threshold P_c. The laser pulse is subject to filamentory instability, because the power is sufficient to create many channels of self-focusing. This filamentary stage is clearly seen in the perspective view, shown in Fig. 3. Few well-defined channels are formed. These channels coalesce then creating a very narrow, $1-2\ \lambda$ wide, super-channel. The laser intensity in the channel rises by a factor 20 and reaches $I_{foc} = 2 \times 10^{20}$ W/cm^2.

The super-channeling effect has been observed experimentally as well.[11] The experiment has been done at Rutherford Appleton Laboratory with the 30 TW VULCAN laser pulse. The laser pulse was sent on a deliberately preformed plasma with exponential density profile of a scale length 40 μm. The channel has been observed in the second-harmonic emission, see Fig. 4. The laser pulse comes from the right side, self-focuses and channels over many diffraction lengths up the density gradient.

The 3D PIC simulations of this experiment, depicted in Fig. 5, show that the electron density in the channel is strongly depleted. In Fig. 5a, we see a transverse, $(Y-Z)$, cut of the electron density distribution, Fig. 5(a) and the line cut along Y-axis, Fig. 5(b). The electron density depression on the channel axis is about 50%. It is not a complete cavitation as would be expected from the standard "fluid" theory. In addition, the simulations reveal strong currents of very energetic electrons driven by the laser pulse.[12] Both self-focussed channels and collimated electron beams have been observed experimentally.[6] The measured electron energy spectrum extends to 12 MeV energies, much higher than ponderomotive energy, which was 1-2 MeV in this

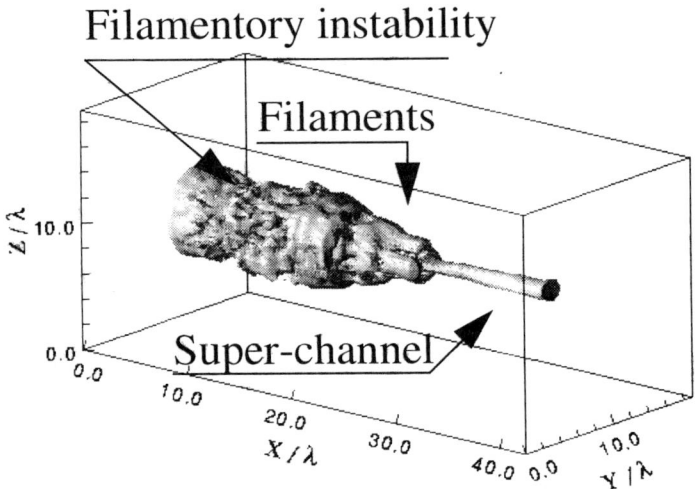

Figure 3: 3D perspective view of the self-focusing laser pulse, with $I_0 = 1.2 \times 10^{19}$ W/cm^2 and radius $r = 6$ μm, in a plasma with $n = 0.36 n_c$. It is the first 3D-PIC simulation of this phenomenon.[10]

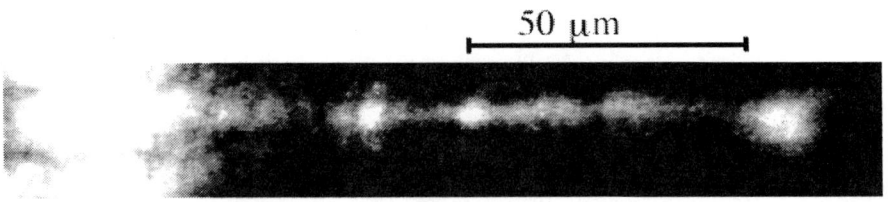

Figure 4: First clear experimental evidence of a relativistically channeling laser pulse in a preformed plasma, observed by M. Borghesi et al..[11]

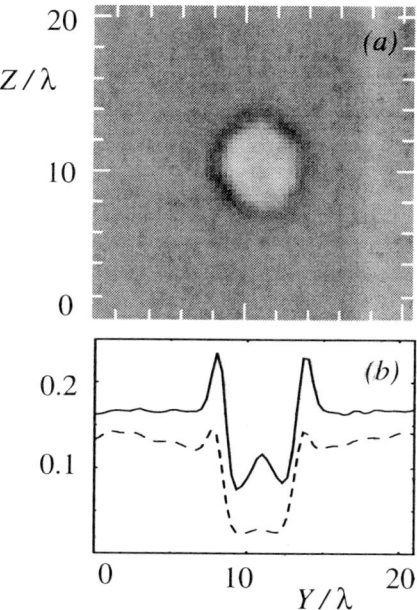

Figure 5: 3D PIC simulation of the experiment. Electron density cut (a) in the $(Y-Z)$-plane, perpendicular to the laser propagation axis; (b) the linear cut along the Y-axis. The solid line shows n_e/n_c and the broken line includes the relativistic correction: $n_e/<\gamma>n_c$.

case. The question arises about what drives these electrons. It is addressed in the next chapter.

DIRECT LASER ACCELERATION OF ELECTRONS IN PLASMA CHANNELS

In the past, electron acceleration by lasers in plasma has been discussed mainly in terms of plasma waves and wakefields, starting with the pioneering paper by Tajima and Dawson.[13] Here, we describe a different acceleration mechanism which turns out to be dominating in selffocussed plasma channels near critical density. Analyzing recent experiments by Gahn et al.,[6] performed at intensities of a few 10^{18}W/cm^2 and plasma densities around 10^{20}/cm^3, and corresponding 3D-PIC simulation, which describe these data almost quantitatively, we have found strong evidence that the electrons are accelerated in the channels directly by the transverse electric field of the laser wave. The physical mechanism is analogous to that occuring in free electron lasers (FEL), but with opposite phase. Here, we adopt the name *inverse free electron laser* (IFEL), which had been coined before by Courant et al..[14] The wiggler field which makes the relativistic electrons oscillate in a usal FEL is replaced by the quasi stationary electric and magnetic self-fields of the laser channel. The betatron frequency at which trapped electrons oscillate in the channel are discussed next.

Betatron Frequency of Electrons Trapped in a Plasma Channel

Let us consider a stationary cylindrical laser plasma channel with trapped electrons having a uniform density $n_e = f n_0$. Here f is the electron depletion factor ($0 \leq f \leq 1$), while the ion density is $n_i = n_0$. The net charge density $(1-f)n_0$ creates the radial electric field

$$-eE_r = (1-f)m\omega_p^2/2 \cdot r. \tag{47}$$

The light propagating in the channel drives the elecrons in forward direction at almost the velocity of light, producing a current density $-efn_0c$ which creates an azimuthal magnetic field

$$-eB_\varphi = fm\omega_p^2/2 \cdot r. \tag{48}$$

Solving the equation of radial motion of an electron in the channel, $m\gamma d^2r/dt^2 = -eE_r - -eB_\varphi = -m\omega_p^2/2 \cdot r$, we find that the oscillation (betatron) frequency is

$$\omega_\beta^2 = \omega_p^2/2\gamma, \tag{49}$$

independent of the depletion factor f. The parallel propagation of photons and relativistic electrons in the channel, with electrons making ω_β oscillations transverse to the channel axis, is analogous to the configuration of a free electron laser (FEL), for which the magnetic wiggler has been replaced by the electric and magnetic channel fields. As it is well known from FEL's, resonant energy exchange occurs between photons and electrons, depending on the relative phase. In the present case of relativistic laser plasma channels, incident photons transfer energy to plasma electrons. This mechanism has been identified by the authors of this paper as the dominant acceleration mechanism,[6] as soon as long laser pulses (longer than plasma wavelength) self-focus in a plasma.

Spectrum of electrons accelerated in plasma channel: 3D VLPL simulations

In order to gain more insight into the process of particle acceleration in plasma, we have performed a set of 3D PIC simulations.[15] The laser pulse is incident on the plasma with an exponential density profile $n(x) = n_0 \exp(-x/L_s)$, where L_s is the plasma scale length. The maximum plasma density is higher than the critical one for the particular laser intensity. This configuration was chosen because it is easy to be set up experimentally. Such a plasma can be formed on a surface of a solid state body with a prepuls, which is naturally present or can deliberately be created.

The laser pulse is Gaussian in space and time with dimensionless amplitude $a = eA/mc^2 = a_0 \exp(-t^2/\tau^2)\exp(-r^2/\sigma^2)$, where t is time, and $r = \sqrt{y^2 + z^2}$ is the distance from the axis. The time duration is set to $\tau = 150$ fs and the laser is assumed to be focused at the left boundary of a plasma box in a round focal spot with radius $\sigma = 6\lambda$. When this geometry is maintained, the maximum intensity of the laser pulse is defined by its power. For our parameters $I/I_{18} = 1.76P$ (TW), where $I_{18} = 10^{18}$ W/cm^2. The laser is linearly polarized in $z-$direction.

Fig. 6 shows spectra of fast electrons obtained at four different laser powers: 1 TW, 10 TW, 100 TW and 1 PW. The spectra indicate an approximately exponential roll-off at high energies

$$n_e(\varepsilon) \sim n_0 \exp(-\varepsilon/T_{eff}), \tag{50}$$

having an "effective temperature" T_{eff} growing with laser intensity. Fig. 7 illustrates the dependence of T_{eff} versus laser intensity I in a logarithmic plot. The results of the numerical simulations suggest that T_{eff} grows like the square root of the intensity:

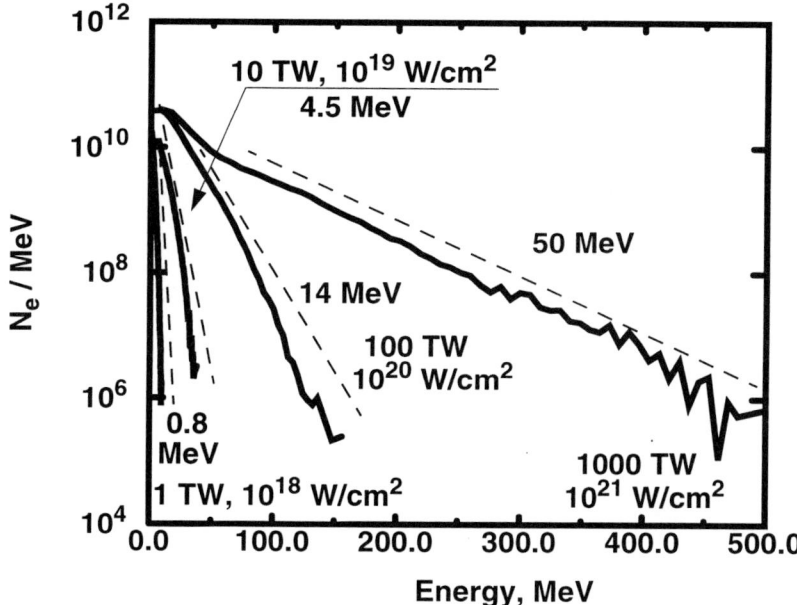

Figure 6: Electron energy spectra for different laser powers. The plasma scalelength is $L = 30$ μm.

$$T_{eff} \sim \alpha(I/I_{18})^{1/2}, \qquad (51)$$

where the coefficient is $\alpha \approx 1.5$ MeV. The "effective temperature" for the PetaWatt laser pulse appears to be as high as 50 MeV with the tail touching 0.5 GeV value. The density of electron energy distribution shows as much as 10^6 particles per MeV at this highest energy. One should notice that the exponential energy spectrum is cut off at even higher energies. The cut-off energy is set by the finite force acting on the particle and the finite acceleration length. In our present simulations, however, we may not have resolved this cut-off because the number of macroparticles, though huge (up to 10^9), may have been still too low. Each numerical macroparticle substitutes for about 10^5 physical electrons.

Electron trajectories in plasma channel: 2D test simulation

The 3D PIC simulations display very rich and complicated physics of the laser-plasma interaction. Because of this complexity some of the basic mechanisms of electron acceleration in relativistic channels are easier to analyse in 2D geometry. We have performed test PIC simulations with the 2D version of the code VLPL. A laser pulse with peak intensity of $I_0 = 10^{19}$ W/cm^2 and a transverse Gaussian profile with $\sigma = 2\lambda$ is incident on a 100λ thick slab of underdense plasma, $n_e = 0.16 n_c$. Ions in the plasma are kept immobile to prevent channel expansion which would complicate evaluation of the results. The laser is polarized in y-direction (p-polarisation). We have chosen the laser pulse to be rather narrow initially so that the laser pulse cannot focus further and change its intensity. Simulation results after 200 laser periods are presented in Fig. 8. The Figs. 8a-d show distributions of (a) electron density n_e/n_c; (b) intensity I/I_{18}; (c) self-generated magnetic field B_z/B_0, where $B_0 = mc\omega_0/e = 107$ MG for a laser beam

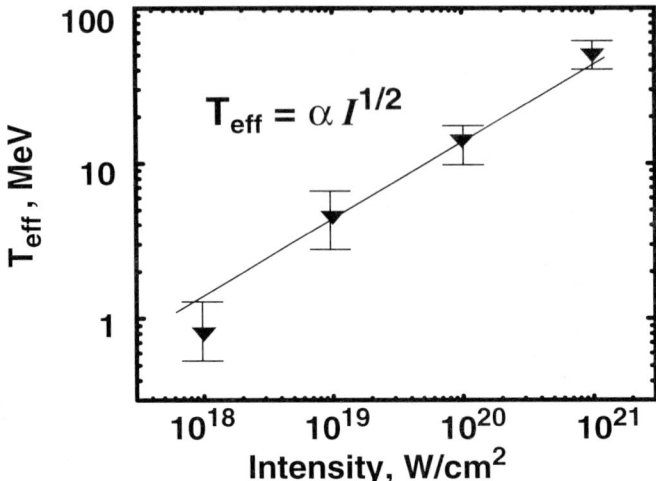

Figure 7: Effective temperature of the electron distributions for different laser intensities. The best fit gives the scaling $T_{eff} \sim \alpha(I/I_{18})^{1/2}$ with $\alpha \approx 1.5 MeV$.

with 1 μm wavelength; (d) static electric field $eE_y/mc\omega_0$ in the channel. In addition, we have drawn trajectories of some selected electrons in Fig. 8a. These trajectories represent different groups of particularly interesting electrons. They are marked by numbers.

Electrons with numbers '1', '2', and '4' make betatron oscillations in the self-generated magnetic and electrostatic fields of the channel. The electron with number '1' enters the channel at the mouth, while number '4' penetrates through the channel wall. The electron '2' has been scattered out of the channel after just a couple of betatron oscillations. Electron '3' experiences cyclotron rotation in the quasistatic magnetic field at the channel boundaries. Electron '5' contributes to the return current flowing around the channel. This electrons drifts from outside towards the channel wall, bounces, and is scattered away by the static magnetic field, which has the defocusing polarity for the return current.

We have drawn separately the trajectory of the most energetic electron as the white line in Fig. 8c. The electron originates in the background plasma outside of the channel. It slowly drifts to the channel mouth, catches the laser pulse at a favored phase and starts to gain energy continuously, making betatron oscillations in the channel. The corresponding trajectory in longitudinal phase space of the electron is plotted in Fig. 9. The frequency of the betatron oscillations seen in the p_y plot slowly decreases as the electron is accelerated in the forward direction (see p_x plot). In addition to the secular acceleration, the p_x dynamics shows also oscillations at twice the betatron frequency. Note that no plasma wave structure is observed in the channel. Therefore, the only possible explanation for the electron energy gain is direct acceleration by the laser pulse. We also observe that the oscillations of transverse momentum of the electron have 5 times larger amplitude than that of the dimensionless laser vector potential. We conclude that the electrons are resonantly driven by the laser field, making transverse oscillations in the channel field on top of the motion along the channel axis.

Additional strong evidence for the mechanism of betatron resonance acceleration

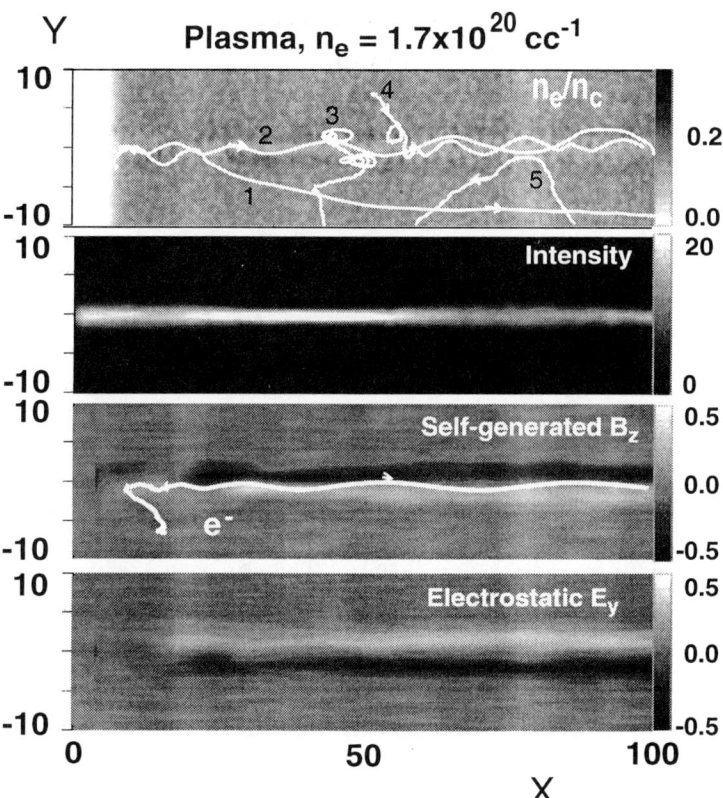

Figure 8: Test 2D PIC simulations of laser, $I = 10^{19}$ W/cm^2, channeling in underdense, $n = 0.16\ n_c$, plasma. (a) Electron density n_e/n_c; (b) intensity I in units of 10^{18}W/cm^2; (c) self-generated magnetic field $eB_z/mc\omega_0$; (d) static electric field $eE_y/mc\omega_0$. The black numbers mark white lines corresponding to trajectories of individual electrons. For more detail see the text.

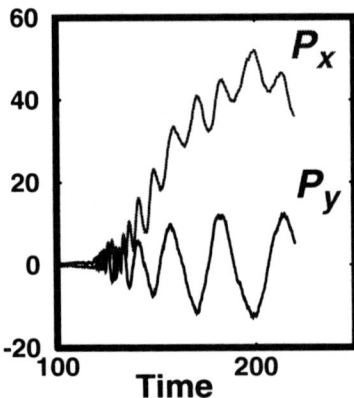

Figure 9: Phase space of the fastest electron in the 2D PIC simulation. The transverse momentum, P_y, shows regular betatron oscillations of the electron in the self-generated fields. The longitudinal momentum, P_x, grows on average while displaying oscillations at *twice* the betatron frequency.

we find from looking at the electron density inside the channel in Fig. 10. We observe an electron density modulation with a period of half the laser wavelength and a contrast $K = (n_{max} - n_{min})/(n_{max} + n_{min}) \approx 0.3$. These modulations are similar to the natural microbunching of the fast electrons in inverse free electron lasers. The difference here is that in a usual FEL electrons are bunched once per wavelength of the electromagnetic wave. This is because the electron motion is primarily governed by the wiggler. The resonant electrons then have a fixed phase shift between the laser field and oscillations in the wiggler field. As the wiggler phase is fixed in the space, the resonance can be achieved only once per laser wavelength. The situation is different in the plasma channel. The phase of the betatron oscillations is not fixed spatially, and, when in resonance, is attached to the laser phase only. As we are considering a linearly polarized laser beam, the transverse velocity of the resonant electrons $\beta_\perp = v_\perp/c$ oscillates with the laser period, while the longitudinal velocity $\beta_\parallel \approx 1 - \beta_\perp^2/2$ oscillates twice per laser period, leading to the electron bunching in space two times per laser wavelength. In simulations with a circularly polarized laser beam, such bunching is naturally absent.

Electron acceleration at betatron resonance

We now discuss electron acceleration by the IFEL mechanism analyticly. The presence of self-generated fields is the characteristic feature of a laser pulse channeling in plasma due to relativistic and ponderomotive effects. Ponderomotive expulsion of background plasma electrons from of the channel creates a radial electrostatic field, while the current of the accelerated electrons generates an azimuthal magnetic field. Both these fields depend about linearly on radius and reach their maxima at the channel boundaries. We have to take into account the influence of these fields on the electron dynamics to understand the mechanism of acceleration. Such an acceleration in the presence of self-generated magnetic field has been already considered by Pukhov al.,[12] where it was termed B-loop mechanism. In the present paper, we generalize our consideration by taking into account the electrostatic field and giving more insight into the details of the acceleration.

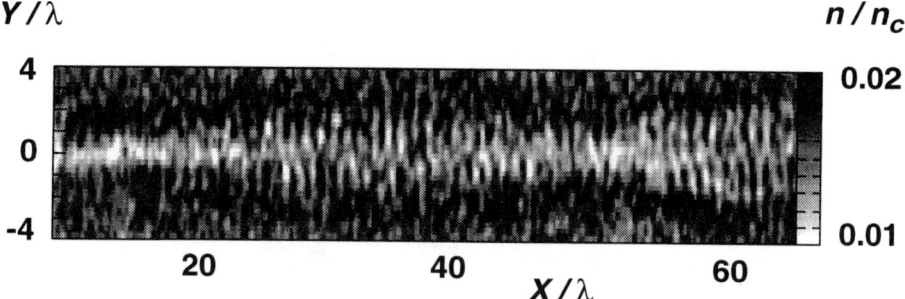

Figure 10: Electron density inside channel. Electron bunching at half the laser wavelength are seen.

Let us consider an electron moving in a plane electromagnetic wave described by $E_y^l = E_0 \cos \omega_0(t - x/v_{ph})$, $B_z = E_y/v_{ph}$ which runs with the phase velocity $v_{ph} > c$ in X-direction. In addition, we impose a static electric field $E_y^s = \kappa_E y$ and a magnetic field $B_z^s = -\kappa_B y$. For simplicity, we exploit plane 2D ($X-Y$) geometry. In the following discussion, we will use the dimensionless variables:

$$p \to p/mc; \quad v \to v/c; \quad t \to \omega_0 t;$$
$$x \to x\omega_0/c; \quad E \to eE/m\omega_0 c; \quad B \to eB/m\omega_0. \tag{52}$$

In these variables, one can write down the equations of particle motion:

$$\frac{dp_x}{dt} = -\frac{v_y}{v_{ph}} E^l - v_y B^s, \tag{53}$$

$$\frac{dp_y}{dt} = -\left(1 - \frac{v_x}{v_{ph}}\right) E^l - E^s + v_x B^s, \tag{54}$$

$$\frac{d\gamma}{dt} = -v_y(E^l + E^s). \tag{55}$$

Using the expressions $B^s = -\kappa_B y$ and $E^s = \kappa_E y$ for the static fields, we can immediately find from Eqs. (53), (55) the constant of motion

$$-v_{ph} p_x + W = W_0, \tag{56}$$

where

$$W = \gamma - 1 + (\kappa_E + v_{ph}\kappa_B) y^2/2 \tag{57}$$

is the sum of the kinetic $K = \gamma - 1$ and the potential $U = (\kappa_E + v_{ph}\kappa_B) y^2/2$ energy of the particle in the static fields. Here $\gamma = \sqrt{1 + p_x^2 + p_y^2}$ is the relativistic γ-factor.

An exact analytical solution of Eqs. (53)-(55) appears to be impossible in view of their non-linear nature. Nevertheless, these equations exhibit the mechanism of acceleration. To make it explicit, we rewrite the equation for transverse motion (54) as

$$\frac{d^2 y}{dt^2} + \omega_\beta^2 y = \left[\left(\frac{dy}{dt}\right)^2 - \left(1 - \frac{v_x}{v_{ph}}\right)\right] \frac{E^l}{\gamma} + \left(\frac{dy}{dt}\right)^2 \frac{\kappa_E y}{\gamma}. \tag{58}$$

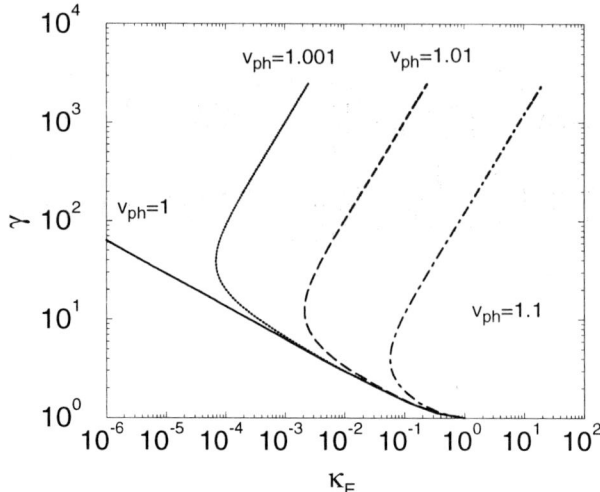

Figure 11: Dependence of the resonant electron energy γ on the electrostatic field parameter κ_E for different phase velocities of the electromagnetic wave. For $v_{ph} > 1$, the self-generated fields must be sufficiently strong for the resonance to appear.

This is indeed an equation of a driven oscillator with the eigenfrequency

$$\omega_\beta^2 = (\kappa_E + v_x \kappa_B)/\gamma, \tag{59}$$

corresponding to betatron oscillations in the static E^s and B^s fields. The driving force on the right-hand-side (RHS) of Eq. (58) hits the resonance when

$$\omega_\beta = 1 - v_x/v_{ph}. \tag{60}$$

This resonance condition states that when an electron makes one oscillation, the electromagnetic wave, which propagates with $v_{ph} > c$, overtakes it exactly by one period. It appears that, for the general case $v_{ph} > 1$, the resonant electron energy γ is not a monotonical function of κ_E and κ_B. This is shown in Fig. 11, where we have set $\kappa_B = 0$ for simplicity. In particular, for a given $v_{ph} > 1$, the resonance can only be achieved for a sufficiently strong self-generated electrostatic field satisfying the condition $\kappa_E > \gamma_0 [5 v_{ph}^2 + 4 - 3 v_{ph} (v_{ph}^2 + 8)^{1/2}]/(2 v_{ph}^2)$, where the resonant energy $\gamma_0 = \sqrt{2} [v_{ph}(v_{ph}^2 + 8)^{1/2} - v_{ph}^2 - 2]^{-1/2}$ (or $v_x = [-v_{ph} + (v_{ph}^2 + 8)^{1/2}]/2$). For very strong self-generated fields, i.e., higher parameters κ_E, the resonance can be achieved at two different energies, as seen in Fig. 11.

The physics relevant here is very similar to that of inverse free electron lasers (Courant et al., 1985). However, instead of being bent by a periodic wiggler, electrons make betatron oscillations in the self-generated fields. Equations describing the inverse free electron laser at the betatron resonance (β–IFEL) are much simpler in the ultra-relativistic limit $\gamma \gg 1$. In this case, ω_β changes slowly on the time scale of one betatron oscillation and for the transverse electron motion we may write:

$$\begin{aligned} p_y &= P_y \cos\theta_\beta, \\ d_t \theta_\beta &= \omega_\beta, \end{aligned} \tag{61}$$

where P_y is the magnitude of the oscillating transverse momentum. The maximum transverse displacement of the electron is $y_0 = P_y/\omega_\beta$. The laser electric field at the electron position is $E^l = E_0 \cos(t - x/v_{ph})$. Using (55) we write down the final set of β–IFEL equations:

$$d_x\left(\gamma + \frac{\kappa_E y^2}{2}\right) = -E_0 \frac{P_y}{2p_x}(\cos\Psi + \cos\psi), \qquad (62)$$

$$v_x d_x \Psi = \omega_\beta - \left(1 - \frac{v_x}{v_{ph}}\right), \qquad (63)$$

$$v_x d_x \psi = \omega_\beta + \left(1 - \frac{v_x}{v_{ph}}\right), \qquad (64)$$

where Ψ is the ponderomotive phase (slow oscillatory) of an electron in the bucket produced by the laser wave, while the fast phase ψ rotates as $2\omega_\beta t$. These oscillations at twice the betatron frequency are clearly seen in the p_x plot of Fig. 9.

As it follows from the gain equation (62), electrons are accelerated if their ponderomotive phase satisfies $\pi/2 < \Psi < 3\pi/2$, and are decelerated otherwise. The maximum acceleration is achieved when the betatron oscillations are exactly in counterphase with the laser electric field (because of the negative electron charge). This means, that when the electron moves with its highest transverse velocity at the channel axis, the electric field of the laser pulse is also at its maximum and has the accelerating direction. When the electron reaches the turning point at the channel boundary, the laser field vanishes. As the electron reverses its transverse motion, the direction of the laser electric field reverses either, and the electron continues gaining energy.

The continuous growth of γ leads to a corresponding decrease of ω_β and detuning according to Eq. (63). The electron dephases and acceleration eventually stops. To describe this effect quantitatively, we use Eq. (62)-(63) and obtain for the ponderomotive phase Ψ the usual FEL nonlinear pendulum equation:

$$d_t^2 \Psi = E_0 \frac{\omega_\beta}{4\gamma} \sin\alpha \cos\Psi. \qquad (65)$$

Here, α is the average angle of electron propagation direction with respect to the channel axis, so that $\sin\alpha = P_y/\gamma$. We neglected smaller terms $\sim \gamma^{-3}$. After many synchrotron rotations in the bucket, the phase space is mixed. Also, different electrons run at different angles α to the channel axis, and the electron energy spectrum thermalizes. As we see from Eq. (62), the electron energy gain is proportional to the laser pulse electric field E_0, which in turn is $\sim I^{1/2}$. Thus, we expect the effective electron "temperature" to depend on the laser intensity in the same way; an accurate derivation of the dependence is still pending.

Equation (65) tells us that the electrons are making synchrotron oscillations in the ponderomotive bucket with the bouncing frequency $\Omega_B = (E_0 \omega_\beta \sin\alpha/4\gamma)^{1/2}$. The bucket height is $2\Omega_B$. Electrons are becoming trapped in the bucket if they enter it with $|d_t \Psi| < 2\Omega_B$. At high laser intensities and strong static fields, the bucket is sufficiently deep to trap even low energy background electrons. One may estimate the threshold to be

$$E_0^2(\kappa_E + v_x \kappa_B) > 1. \qquad (66)$$

When the trapping condition (66) is not satisfied, the electrons can be pre-accelerated by the plasma wakefield and then be trapped.

In the discussion above, we supposed the channel to be infinite in transverse direction. This is not the case in reality. As the electrons are making transverse oscillations,

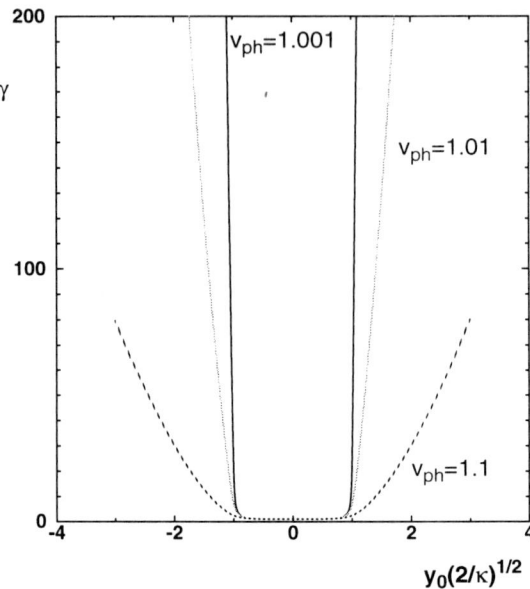

Figure 12: Relativistic γ-factor of an electron vs transverse extent y_0 of betatron oscillations for different phase velocities v_{ph} of the laser pulse.

they may overtake the potential barrier and leave the channel transversely. Thus, the potential well associated with the channel must be sufficiently deep and wide to trap the fast electrons. Some interesting estimations can be obtained from the conservation law (57). It connects the particle energy with its transverse position and the channel parameters. The expression is greatly simplified for an electron initially at rest at the channel axis, so that $W_0 = 0$. Furthermore, we can use the fact that the transverse momentum is $p_y = 0$ at the maximum displacement from the axis and obtain

$$\gamma = \gamma_{ph}^2 \left[\frac{\kappa y_0^2}{2} - 1 + v_{ph} \sqrt{\gamma_{ph}^{-2} + \left(\frac{\kappa y_0^2}{2} - 1 \right)^2} \right], \tag{67}$$

where $\gamma_{ph}^2 = 1/(v_{ph}^2 - 1)$, and $\kappa = \kappa_E + v_{ph}\kappa_B$. Fig. 12 shows how γ depends on the oscillation amplitude y_0. The behavior is threshold-like. For small oscillations nearby the channel axis the particle energy gain is small. As soon as the amplitude y_0 becomes larger than the critical radius $\rho_\beta = \sqrt{2/\kappa}$, the energy gain takes off steeply as $\gamma \sim \gamma_{ph}^2(1 + v_{ph})\kappa y_0^2/2$. The closer the phase velocity v_{ph} is to the speed of light, the steeper is the dependence. This allows us to draw two conclusions. (i) The hot electrons oscillate in the channel within the radius ρ_β. (ii) The background plasma electrons are accelerated by the betatron resonance mechanism only if the self-generated fields in the channel are sufficiently strong,

$$\rho(E_{max} + B_{max}) > 2, \tag{68}$$

where E_{max} and B_{max} are the maximum static fields at the channel boundary at radius ρ.

The condition (68) can be expressed in terms of laser intensity or power. For simplicity, we neglect the magnetic field and consider the electrostatic fields only. This assumption is valid at the beginning of the interaction, when the plasma is still cold and most of the background electrons are expelled from the channel due to the ponderomotive force. Balancing the electrostatic field with the ponderomotive force, we may estimate $E_{max} \sim \sqrt{1 + a_0^2/2}/\rho$, where a_0 is the laser amplitude at the axis and ρ is the radius of the laser filament. Substituting E_{max} in Eq.(68), we find the condition $a_0^2 > 2$ for the laser amplitude, or for the intensity $I\lambda^2 > 2.74 \times 10^{18}$ Wμm^2/cm^2. According to the results of Sun (1987), this condition is satisfied as soon as the laser power significantly exceeds the critical power 45 for self-focusing and a cavitated region appears.

3D PIC simulations do suggest that this mechanism of electron acceleration acts in relativistic channels in an underdense plasma when the laser power is significantly higher than P_c. The simulations indicate that the threshold power P_{th} is about $6P_c$.

The acceleration at the betatron resonance effectively works when the laser power overcomes significantly the critical power for self-focusing. It is also one of the most effective mechanisms of laser energy absorption in relativistic channels.

PLASMA WAVES AND WAKEFIELDS

In this final chapter of the lecture, we briefly outline the theory of non-linear plasma waves and wave breaking in one-dimensional geometry. The basic idea of using plasma waves for particle acceleration is touched at the end.

Electron Plasma Waves

Plasma waves are longitudinal: $\mathbf{E}||\mathbf{k}$. It can be described by combining Poisson's equation with the equation of motion and the continunity equation for the electron fluid:

$$\nabla \cdot \mathbf{E} = -4\pi e(n - n_0), \tag{69}$$

$$m\frac{d\mathbf{u}}{dt} = -e\mathbf{E}, \tag{70}$$

$$\frac{\partial n}{\partial t} + \nabla \cdot (n\mathbf{u}) = 0. \tag{71}$$

This is for a plane wave in the x-direction.

Linearizing these equations for small density perturbations $|n - n_0| \ll n_0$ around the uniform background density n_0, we find linearizing

$$n(x,t) = n_0 + n_1(x,t), \tag{72}$$
$$\mathbf{u}(x,t) = u_1(x,t)\hat{e}_x, \tag{73}$$
$$\mathbf{E}(x,t) = E_1(x,t)\hat{e}_x, \tag{74}$$

where n_1, v_1 and E_1 vary in $\propto \exp[i(kx - \omega t)]$. The linearized equations

$$ikE_1 = -4\pi e n_1, \tag{75}$$
$$-i\omega m u_1 = -eE_1, \tag{76}$$
$$-i\omega n_1 + ikn_0 u_1 = 0, \tag{77}$$

give
$$\omega^2 = \frac{4\pi e^2 n_0}{m} \equiv \omega_p^2, \tag{78}$$

For *warm* plasma, including the thermal pressue $P_e = nk_B T_e$ in the equation of motion with $T_e = mv_{th}^2$ being the temperature of the electron fluid, one has
$$\omega^2 = \omega_p^2 + 3k^2 v_{th}^2. \tag{79}$$

Nonlinear Plasma Waves

The following description of one-dimensional plasma waves and wavebreaking is taken from Sheng et al..[16] The one-dimensional fluid equations are
$$\frac{\partial n}{\partial t} + \frac{\partial}{\partial x}(nu) = 0, \tag{80}$$
$$\frac{\partial u}{\partial t} + u\frac{\partial}{\partial x}u = -\frac{e}{m}E, \tag{81}$$
$$\frac{\partial E}{\partial x} = 4\pi e(n_0 - n). \tag{82}$$

We look for wave solutions of $n(x,t)$, $u(x,t)$, and $E(x,t)$ that depend only on $\tau = \omega_p(t - x/v_{ph})$. Since $\partial/\partial t = \omega_p d/d\tau$, $\partial/\partial x = -(\omega_p/v_{ph})d/d\tau$, one finds
$$\hat{n} = 1/(1 - \hat{u}), \tag{83}$$
$$\frac{d}{d\tau}(\hat{u} - \hat{u}^2/2) = -\hat{E}, \tag{84}$$
$$\frac{d\hat{E}}{d\tau} = \frac{\hat{u}}{1 - \hat{u}} \equiv \hat{J}, \tag{85}$$

where the dimensionless variables are $\hat{n} = n/n_0$, $\hat{u} = u/v_{ph}$, $\hat{E} = E/(m\omega_p v_{ph}/e)$. For $F(\hat{u}) \equiv \hat{u} - \hat{u}^2/2$ one has
$$\frac{d^2 F}{d\tau^2} = -\frac{\hat{u}}{1 - \hat{u}}, \tag{86}$$

The first integral is
$$\frac{1}{2}\frac{d}{d\tau}\left(\frac{dF}{d\tau}\right)^2 = -\frac{\hat{u}}{1-\hat{u}}\frac{dF}{d\tau} = -\frac{\hat{u}}{1-\hat{u}}\frac{dF}{d\hat{u}}\frac{d\hat{u}}{d\tau} = -\hat{u}\frac{d\hat{u}}{d\tau},$$

$$\frac{1}{2}\left(\frac{dF}{d\tau}\right)^2 = C - \hat{u}^2/2 = \frac{1}{2}\hat{E}^2 \tag{87}$$

with $\hat{E}^2 = 2C - \hat{u}^2$. The maximum electric field is $\hat{E}_{max} = \sqrt{2C}$ for $\hat{u} = 0$, and $\hat{u}_{max} = \sqrt{2C}$ for $\hat{E} = 0$. The second integral
$$(1 - \hat{u})\frac{d\hat{u}}{d\tau} = -\hat{E} = \pm\sqrt{\hat{E}_{max} - \hat{u}^2}, \tag{88}$$

$$\tau - \tau_0 = \int^{\hat{u}/\hat{E}_{max}} \frac{d\xi}{\sqrt{1-\xi^2}} - \hat{E}_{max}\int^{\hat{u}/\hat{E}_{max}} \frac{d\xi^2}{2\sqrt{1-\xi^2}}$$
$$= \arcsin(\hat{u}/\hat{E}_{max}) + \hat{E}_{max}\sqrt{1 - \hat{u}/\hat{E}_{max}} \tag{89}$$

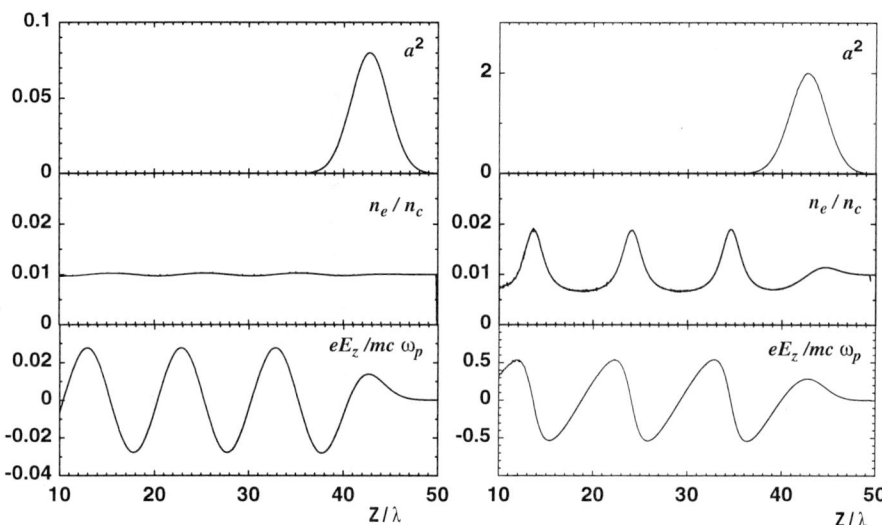

Figure 13: Wake field created by short laser pulses at two different amplitudes. The left column: $a_1 = 0.2$, the right column: $a_2 = 1$. The frames show the normalized laser intensity, a^2, the electron density n_e/n_c, and the accelerating plasma field $eE_z/mc\omega_p$.

provides an implicit representation of the non-linear wave. For $\hat{E}_{max} \ll 1$, it can be written in the explicit form

$$\hat{u} = \hat{E}_{max} \sin(\tau - \tau_0), \tag{90}$$

$$\hat{n} = \frac{1}{1-\hat{u}} = 1 + \hat{E}_{max} \sin(\tau - \tau_0), \tag{91}$$

$$\hat{E} = \sqrt{\hat{E}_{max} - \hat{u}} = \hat{E}_{max} \cos(\tau - \tau_0), \tag{92}$$

As a numerical example, we show one-dimensional VLPL simulations of wake field generation by a short laser pulse with circular polarization for two different intensities, Fig. 13. The first laser pulse has amplitude $a_1 = 0.2$ and the second $a_2 = 1$. The laser pulses have a duration of 15 fs, the plasma density is $n_e = 0.01 \, n_c$. The first laser pulse with $a_1 = 0.2$ produces a nearly linear wakefield, see Fig. 13 (left column). The density perturbation $\delta n_e/n_0$ is small and the electric field is harmonic. At the higher laser amplitude, $a_2 = 1$, the plasma wave becomes nonlinear. The electron density perturbation $\delta n_e/n_0 \approx 1$ and the electric field has a saw-tooth form. This plasma wave is close to the wavebreaking limit.

Wave-Breaking

For $u = v_{ph}$, the electron density diverges. The maximum electric field is obtained at $\hat{u} = 1$ and gives $\hat{E}_{max} = 1$, correspondin in dimensional units to

$$E_{max} = \frac{m\omega_p v_{ph}}{e}. \tag{93}$$

Plasma waves with larger amplitudes break, i.e. a new plasma regime evolves which cannot be described within the hydrodynamic approximation. It involves groups of electrons of different velocity ranges which interpenetrate and require a kinetic description.

In the present context, we are only concerned with the threshold value for wavebreaking and how it is modified by relativistic and thermal pressure effects.

Relativistic and Thermal Pressure Corrections

The relativistic equation of motion including thermal pressure is

$$\frac{d}{dt}(m\gamma u) = -eE - \frac{1}{n}\frac{\partial P_e}{\partial x}, \quad (94)$$

where

$$\frac{d}{dt} = \frac{\partial}{\partial t} + u\frac{\partial}{\partial x} = \left(\omega_p - \frac{\omega_p}{v_{ph}}u\right)\frac{d}{d\tau} = \omega_p(1-\hat{u})\frac{d}{d\tau}.$$

The relativistic momentum is

$$\frac{d}{dt}(m\gamma u) = m\omega_p v_{ph}(1-\hat{u})\frac{d}{d\tau}(\gamma\hat{u})$$

and $\gamma = 1/\sqrt{1-u^2/c^2} = 1/\sqrt{1-(\hat{u}\beta_{ph})^2}$ with $\beta_{ph} = v_{ph}/c$. Using the adiabatic equation $P_e/P_0 \approx \gamma(n/n_0)^\Gamma$ and $P_0 = n_0 k_B T_0$, $v_{th}^2 = k_B T_0/m$, $\alpha = (v_{th}/v_{ph})^2$, $\hat{n} = 1/(1-\hat{u})$, $d\hat{n}^\Gamma/d\hat{u} = \Gamma/(1-\hat{u})^\Gamma$, the thermal pressure reads

$$-\frac{1}{n}\frac{\partial P_e}{\partial x} = -\frac{P_0}{n_0}\frac{n_0}{n}\left(-\frac{\omega_p}{v_{ph}}\right)\frac{d}{d\tau}\left[\gamma\left(\frac{n}{n_0}\right)^\Gamma\right],$$

$$= m\frac{k_B T_0}{m}\frac{\omega_p}{v_{ph}}(1-\hat{u})\frac{d}{d\tau}(\gamma\hat{n}^\Gamma)$$

$$= m\omega_p v_{ph}(1-\hat{u})\frac{d}{d\tau}(\alpha\gamma\hat{n}^\Gamma).$$

Finally, we obtain

$$(1-\hat{u})\frac{d}{d\tau}[\gamma(\hat{u} - \alpha\hat{n}^\Gamma)] = -\hat{E}. \quad (95)$$

Defining the function F by

$$\frac{dF}{d\tau} = \frac{dF}{d\hat{u}}\frac{d\hat{u}}{d\tau}, \quad \frac{dF}{d\hat{u}} = (1-\hat{u})\frac{d}{d\hat{u}}[\gamma(\hat{u} - \alpha\hat{n}^\Gamma)],$$

we can perform the integration

$$\frac{1}{2}\hat{E}^2 = C - \Delta(\hat{u}), \quad (96)$$

$$\Delta(\hat{u}) = \int_0^{\hat{u}} \frac{\tilde{u}}{1-\tilde{u}}\frac{dF}{d\tilde{u}}d\tilde{u} = \int_0^{\hat{u}} \tilde{u}\frac{d}{d\tilde{u}}[\gamma(\tilde{u}-\alpha\hat{n}^\Gamma)]d\tilde{u}$$

$$= \gamma\hat{u}(\hat{u}-\alpha\hat{n}^\Gamma) - \int_0^{\hat{u}} \gamma(\tilde{u}-\alpha\hat{n}^\Gamma)d\tilde{u}$$

$$= \gamma\hat{u}\left(\hat{u} - \frac{\alpha}{(1-\hat{u})^\Gamma}\right) - \frac{1}{\beta_{ph}^2}\left(\frac{1}{\gamma}-1\right) + \frac{\alpha}{\Gamma-1}\left(\frac{1}{(1-\hat{u})^{\Gamma-1}}-1\right), \quad (97)$$

From this general result, we can the different limiting values for the wavebreaking amplitude:

- the classical result valid for cold non-relativistic plasma

$$eE_{max}/m\omega_p v_{ph} = 1; \quad (98)$$

- the cold relativistic limit with $\gamma_{ph} = (1 - v_{ph}^2/c^2)^{-1/2}$:

$$eE_{max}/m\omega_p c = \sqrt{2}(\gamma_{ph} - 1)^{1/2}, \tag{99}$$

- the non-relativistic wavebreaking formula, first given by Coffee, accounting for thermal pressure with $\mu = 3T/mv_{ph}^2$:

$$eE_{max}/m\omega_p v_{ph} = (1 - \frac{1}{3}\mu - \frac{8}{3}\mu^{1/4} + 2\mu^{1/2})^{1/2}. \tag{100}$$

- and finally the results for a relativistic warm plasma for $v_{ph} = c$

$$eE_{max}/m\omega_p c \approx (\frac{4}{27}\frac{mc^2}{T})^{1/4}. \tag{101}$$

Laser Accelerators

For a phase velocity near the velocity of light $v_{ph} \approx c$, the maximum eletric field amounts to (in the classical limit):

$$eE_{max} = \frac{mc^2 \omega_p}{c} = 30 \left(\frac{n_e}{10^{21}\mathrm{cm}^{-3}}\right)^{1/2} \mathrm{GeV/cm}.$$

Laser accelerator schemes explore possibilities to use this field to accelerate particles (Tajima and Dawson 1979). The plasma field exceeds those possible in conventional accelerators (0.1 —1 MeV/cm) by several orders of magnitude and might allow for the development of very compact accelerators.

In small-amplitude approximation, the plasma wave is linear, $E_Z = E_p \sin \omega_p(z/v_p - t)$. An electron can be accelerated in z-direction by the wave. As the electron is accelerated, its velocity v_Z increases and approaches the speed of light, $v_z -> c$. The phase velocity of the plasma wave v_p equals the group velocity of the laser pulse v_g and thus is smaller than the vacuum speed of light, $v_p < c$. Consequently, the electron will outrun the plasma wave and leave the accelerating regime. This electron phase detuning sets limits on the maximum energy gain of the electron in the plasma wave. The detuning length L_d corresponds to the distance the electron must go before its phase slips by one half of a plasma wave. For an ultra-relativistic electron, $\gamma \gg 1$, the detuning time t_d is given by $\omega_p(c/v_p - 1)t_d = \pi$. We get

$$L_d = ct_d = \gamma_p^2 \lambda_p,$$

where $\gamma_p = 1/\sqrt{1 - (v_p/c)^2} \approx \omega_0/\omega_p$ is the plasma-wave γ-factor. The maximum energy gain after the dephasing limit is then

$$W_{max} = eE_p L_d \approx 2\pi(n_c/n_e)(E_p/E_{max})mc^2.$$

The maximum energy gain is larger for tenuous plasmas, but the rate of acceleration decreases. An advanced particle accelerator based on plasma waves will, most probably, include several accelerating stages.

Particle acceleration by strong laser waves in plasma is currently of high interest. In view of the fast progress presently made in generating ultra-short (few-cycle) laser pulses with high intensity (tera-watt and beyond), this area of wakefield accelerator research is expected to become very important in the next future and would deserve a broader discussion. Because of the limited space, we could not give it here.

REFERENCES

1. G.A. Mourou, C.P.J. Barty, M.D. Perry, Ultrahigh-intensity lasers: physics of the extreme on the tabletop, *Physics Today* 51:22 (1998).

2. C. Gahn, G.D. Tsakiris, G. Pretzler, K.J. Witte, C. Delfin, C.-G. Wahlstrm, D. Habs, Generating positrons with femtosecond-laser pulses, *Appl. Phys. Lett.* 77:2662 (2000).

3. S. Karsch, D. Habs, T. Schtz, U. Schramm, P.G. Thierolf, J. Meyer-ter-Vehn, A. Pukhov, A., Particle physiocs with peta-watt class lasers, *Laser & Particle Beams* 17:565 (1999).

4. K.J. Witte, C. Gahn, J. Meyer-ter-Vehn, G. Pretzler, A. Pukhov, G. Tsakiris, Physics of ultra-intense laser-plasma interaction, *Plasma Phys. Contr. Fusion* 41B:221 (1999).

5. A. Pukhov, Three-dimensional electromagnetic relativistic particle-in-cell code VLPL (Virtual Laser Plasma Laboratory), *J. Plasma Physics* 61:425 (1999).

6. C. Gahn, G.D. Tsakiris, A. Pukhov, J. Meyer-ter-Vehn, G. Pretzler, P. Thirolf, D. Habs, K.J. Witte, Multi-MeV electron beam generation by direct laser acceleration in high-density plasma channels, *Phys. Rev. Lett.* 83:4772 (1999).

7. C.I. Moore, J.P. Knauer, D.D. Meyerhofer, Observation of the transition from Thomson to Compton scattering in multi-photon interactions with low-energy electrons, *Phys. Rev. Lett.* 74:2439 (1995).

8. G. Shvets, private communication (1998).

9. G. Sun, E. Ott, Y.C. Lee, P. Guzdar, Selffocussing of short intense pulses in plasmas, *Phys. Fluids* 30:526 (1987).

10. A. Pukhov and J. Meyer-ter-Vehn, Relativistic magnetic self-channeling of light in near-critical plasma: three-dimensional particle-in-cell simulation, *Phys. Rev. Lett.* 76:3975 (1996).

11. M. Borghesi, A.J. Mackinnon, L. Barringer, R. Gaillard, L. Gizzi, C. Meyer, O. Willi, A. Pukhov, and J. Meyer-ter-Vehn, Experimental observation of the relativistic self-channeling of an intense laser pulse in a preformed plasma,*Phys. Rev. Lett.* 78:879 (1997).

12. A. Pukhov and J. Meyer-ter-Vehn, Relativistic laser-plasma imteraction by multi-dimensional particle-in-cell simulation, *Phys. Plasmas* 5:1880 (1998).

13. T. Tajima and J.M. Dawson, Laser electron accelerator. *Phys. Plasmas* 43:267 (1979).

14. E.D. Courant, C. Pellegrini, W. Zakowicz, High-energy inverse free-electron-laser accelerator, *Phys. Review* 32A:2813 (1985).

15. A. Pukhov, Zh.-M. Sheng and J. Meyer-ter-Vehn, Particle acceleration in relativistic laser channels *Phys. Plasmas* 6:2847 (1999).

16. Zh.-M. Sheng and J. Meyer-ter-Vehn, Relativistic wave-breaking in warm plasma, *Phys. Plasmas* 4:493 (1997).

DENSE ULTRAFAST PLASMAS

Jean-Claude Gauthier

LULI, UMR 7605 CNRS-CEA-Ecole polytechnique-Université Paris VI
Ecole polytechnique, 91128 Palaiseau (France)

INTRODUCTION

The interaction of high power lasers with plasmas has been an active field of research for over 30 years. Motivated primarily by inertial confinement fusion[1] and x-ray laser research,[2] most studies have used nanosecond pulse gas lasers – CO_2 lasers in the early days, then excimer (KrF and XeCl) lasers – and solid state Nd:glass lasers, eventually frequency up-converted to 0.53μm ($2\omega_0$), 0.34μm ($3\omega_0$), and 0.26μm ($4\omega_0$). Laser-plasma interaction physics has been studied extensively to explore the efficiency of collisional absorption by inverse bremsstrahlung, parametric instability growth rates,[3,4] filamentation, x-ray conversion efficiency and hydrodynamic instabilities.[5] Numerous other applications of laser-produced plasmas do exist.[6] The development of bright incoherent short wavelength radiation[7] is a major research topic that is being developed for numerous purposes,[8] including x-ray sources for x-ray lithography around 10nm wavelength and hard ($h\nu > 10$keV) x-ray sources for material characterization and solid-state physics. Laser-driven shock wave research for equation of state measurements has also important developments in planetary research, geophysics and astrophysics.

In the 80s, advances in laser science and optical technology have opened new possibilities for the study of high energy density plasmas, notably dense laser-produced plasmas. One of these advances is laser beam smoothing exploiting coherence control techniques, both in the spatial and frequency domains.[9] These techniques help in suppressing filamentation of hot spots, thereby allowing to generate more uniform plasmas with better controlled characteristics. Combined with laser frequency-quadrupling, which allow collisional absorption at high densities, typical signatures of dense plasma effects such as molecular satellites in hydrogen-like ion emission could be observed for the first time.[10]

The second breakthrough is the implementation of the Chirped Pulse Amplification (CPA) technique[11] which has opened new opportunities in the domain of ultra-intense physics. Plasma physics has become one of the major relevant scientific fields for use of these methods. It is now possible to generate the highest electromagnetic fields ever produced in the laboratory.[12] Indeed, the interactions of such high-intensity and ultra-short duration laser pulses with plasmas have opened the field of optics in relativistic plasmas. In addition, the time duration of these pulses being less than 10^{-13}s, this is shorter than the time-scale of significant hydrodynamic motion of ions or solid target surfaces. Consequently, solid-density matter may be heated from room temperature to several hundreds of electronvolts without the usual change in density that accompanies long-pulse irradiation.[13]

Ultrafast laser-based plasma studies will play a pivotal role in the development of experimental dense plasma physics in the coming years.[14,15] Ultrafast plasmas will also have important applications in material processing,[16] thin film growth using ultrafast pulsed-laser deposition,[17] and ultrashort pulse x-ray sources.[18] Moreover, in the future, femtosecond x-ray pulses will have tremendous applications in broad areas of science, including condensed matter physics, chemistry, biology, and engineering because the ability to time-resolve atomic motions by x-ray diffraction (XRD) could open entirely new fields of scientific research.[19] The potentially most rewarding, but also most demanding application of femtosecond XRD will be the characterization of ultrafast structural processes in complex,[20,21,22] eventually biological, molecules.

In these lectures, theoretical and experimental research on dense plasmas produced with short-pulse, high-intensity lasers will be described in a tutorial overview. In the next Section, basics of dense plasma physics, dense plasma parameters, and the laser technology which has driven this new field will be quickly reviewed. Basic models of femtosecond laser-solid interaction will be described in the third Section. Energy transport (particles, photons) will be studied in the fourth Section. Atomic physics and spectroscopy of dense plasmas will be reviewed in the fifth Section. Finally, applications in the area of ultrafast x-ray generation and new dense plasma diagnostics will be examined in the last Section.

DENSE PLASMAS

Basics of dense plasma physics and thermodynamics

In the context of the interaction of plasmas with electromagnetic fields, it is difficult to classify plasmas according to their density. It is more appropriate to consider whether or not their response to the field is sensitive to particle interactions. Weakly interacting plasmas are called ideal plasmas. Accordingly, dense but very hot plasmas belong to the category of ideal plasmas. Plasmas produced by laser irradiation of solid targets are non-ideal because of their high density, partial ionization and great optical depth. Ideal plasmas are typical of conditions reached in magnetic-confinement fusion, e.g. an electron density of about 10^{14}cm^{-3} and a temperature of a few keV. The plasma is completely ionized so that its average charge Q is equal to the nuclear charge Z. Electric interactions are weak so that $Z^2e^2/\lambda_D < kT$, where $\lambda_D = (kT/4\pi n_e e^2)^{1/2}$ is the Debye length, T the temperature and n_e and n_i the electron and ion densities, respectively. This implies that a large number of particles participate in the atomic potential screening, the number of electrons in a Debye sphere being $4\pi/3\lambda_D^3 n_i \gg 1$.

Because ideal plasmas are optically thin due to their relatively low densities, radiation produced by bremsstrahlung, bound-bound, and bound-free transitions escape freely, a feature which is very useful for plasma diagnostics.[23] Transport coefficients of ideal plasmas are strongly dependent on the temperature. The electron conductivity $\sigma \propto T^{3/2}$ and the thermal conductivity $\kappa \propto T^{5/2}$ have large variations with temperature because of the dominance of small-angle Coulomb deflections[24] (cross-section $\propto v^{-3}$). Ideal plasmas obey the perfect gas law with the electron and ion plasma pressures given by $p_e = Zn_i kT_e$ and $p_i = n_i kT_i$, respectively. Figure 1 gives a schematic density-temperature diagram showing the ideal and non-ideal plasma regions. At high densities and low temperatures, plasmas are degenerate i.e. are characterized by a temperature which is lower than the Fermi temperature $kT_F = \hbar^2/2m(3\pi^2 n_e)^{2/3}$. At solid density, the Fermi temperature of, say, aluminum is about 5 eV.

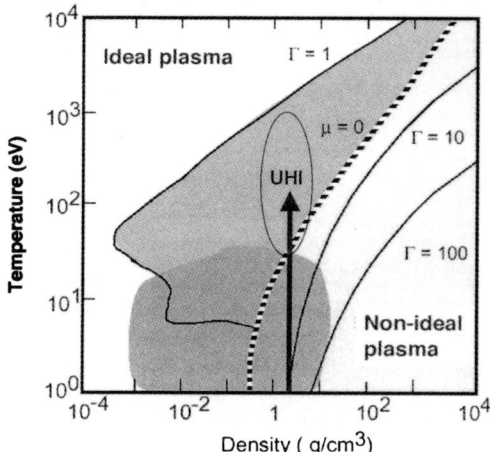

Figure 1. Density-temperature diagram showing the regions of ideal and non-ideal plasmas together with the values taken by the coupling parameter Γ. Degenerate plasmas are in the temperature region below the Fermi temperature. The gray region around 10eV delimits the so-called "warm dense matter" region. The arrow shows a path of isochoric heating of solid-density matter by a ultra-high intensity (UHI) laser.

For degenerate plasmas, the Maxwell distribution $f(\epsilon) \propto \exp(-\epsilon/kT)$ is replaced by the Fermi-Dirac distribution,[25] which writes for fermions:

$$f(\epsilon) \propto 1/(1 + \exp((\epsilon - \mu)/kT)) \qquad (1)$$

where μ is the chemical potential. The radical change in the temperature scaling of the pressure and electronic transport coefficients occurring in degenerate plasmas is summarized in Table 1.

Table 1. Scaling of the pressure p, electrical conductivity σ, and thermal conductivity κ for degenerate and non-degenerate plasmas

	non-degenerate	degenerate
p	$n_e T$	$n_e T_F \propto n_e^{5/3}$
σ	$T^{3/2}$	$T_F^{3/2}$
κ	$T^{5/2}$	$T T_F^{3/2}$

The Debye temperature partitions classical from quantum behavior of a given material. Classical properties occur in the domain $T \gg \Theta_D$ and quantum properties in the $T \leq \Theta_D$ interval. Albeit most of the interaction studies of dense plasmas with intense electromagnetic fields ignore quantum mechanics, quantum-kinetic theories of correlated charged-particle systems do exist[26] to investigate the influence of the electromagnetic field on two-particle scattering processes.

In non-ideal plasmas, a distinction has to be made between strongly-coupled plasmas and highly-correlated plasmas. Strongly-coupled plasmas are characterized by their Coulomb energy being very much greater than their thermal energy. The strong coupling parameter, Γ, is the ratio of the ion-ion potential energy to the thermal energy:

$$\Gamma = Z^2 e^2 / (R_0 kT) \qquad (2)$$

where the interparticle spacing $R_0 = (3/4\pi n_i)^{1/3}$ is also called the ion-sphere radius. Significant values of Γ are shown in Fig. 1. The $\Gamma = 1$ line indicates where the region of strong coupling begins. Obviously, for large values of Γ, the ion screening length becomes the ion sphere radius. Because strongly-coupled plasmas are usually dense, reabsorption of radiation is important, the photon population builds up toward the black-body distribution. Correlation effects are important when the emission properties of any ion inside the plasma are perturbed by the presence of the nearest neighbor ion,[27] plasmas are then named "highly-correlated".

Relativistic plasma conditions are reached whenever the temperature T or the ponderomotive energy $E_p = e^2 E_0^2 / 4 m_e \omega^2$ are approaching the electron rest mass energy $m_e c^2$. Here, we have considered a wave of amplitude E_0 and angular frequency ω. This is a regime which, as we will see in the present lectures and in the lecture by Pr. M. Key, is easily reached nowadays with high-intensity, short-pulse lasers.

In the Section on laser-solid interaction, we will see an intermediate state of plasma conditions dubbed "warm dense matter". This is the state of matter making a transition between solid state physics and plasma physics, like, for example, in dense metallic and sub-metallic hydrogen. In a condensed matter system, the issue becomes the temperature of the system relative to the Fermi energy. In the case where $T < E_{Fermi}$ the normal methods of condensed matter theory will work, as only a few valence/conduction bands need to be taken into account. However, when $T > E_{Fermi}$ the number of bands that are required to describe the system become enormous. Further, one must include both excited cores and ionized states in the description, as multiple species are required. Finally, a very important quantity in plasma physics is the plasma frequency $\omega_p = (4\pi n_e e^2 / m_e)^{1/2}$ which is the highest frequency which can propagate in a plasma of electron density n_e.

Thermodynamic quantities

Pressure and internal energy directly govern the hydrodynamic motion of plasmas. Once one knows the amount of laser energy deposited in the plasma, then one gets the temperature by the relation $E(\rho, T)$ and the pressure by the relation $P(\rho, T)$, where ρ is the matter density. Shock wave strength in plasmas is governed by the driving pressure, together with the sound speed $c_s^2 = (\partial P/\partial \rho)_S$. In thermodynamic equilibrium, all thermodynamic quantities are calculated from the free energy $F(\rho, T)$:

$$P = \rho^2 \partial F / \partial \rho$$
$$S = -\partial F / \partial T$$
$$E = F + TS$$

where S is the entropy. Combining these relations, one gets a condition of thermodynamic consistency which turns out to be very important if one uses data tables from different sources

$$\rho^2 \partial E / \partial \rho = P - T \partial P / \partial T$$

As we have already seen before, the simplest equation of state is the non-interacting ideal-plasma approximation, in which

$$P_e = Q\rho kT_e/AM_p$$
$$E_e = (\frac{3}{2}Q\rho kT_e + E_i(Z,Q))/AM_p$$

where A is the atomic number and M_p the proton mass. Ionization energies $E_i(Z,Q)$ enter directly in the equation of state (EOS) and are very important for the value of all thermodynamic quantities. We will concentrate later on various means of calculating the ionization energies. Substituting the electron EOS into the thermodynamic consistency relation, we get an approximate version of the Saha equilibrium:

$$\frac{T(\partial Q/\partial T)}{\rho(\partial Q/\partial \rho)} = 3/2 + I(Q)/kT; \quad I(Q) = \partial E_i/\partial Q$$

There is an obvious solution, a being a scaling constant:

$$Q = a\frac{T^{3/2}}{\rho}\exp(-I(Q)/kT).$$

A correct description of ionization requires an accurate treatment of thermal excitation (e.g. by collisions, or radiation from the external photon bath) and a proper account of the specific heat of the excited states (i.e. their populations). In dense plasmas, degeneracy of the free electrons and continuum lowering resulting from the electrostatic potentials of neighbor ions have to be included.[28] These corrections are increasingly important at higher densities. The important point is that the equation for pressure and energy has also to be improved to ensure thermodynamic consistency.[29]

Two-temperature equation of state

Up to now, we have not explicitly considered the possibility that the electrons and ions have different temperatures. Indeed, thermodynamic equilibrium can be reached over time scales which are usually much longer than the typical laser pulse duration of a fraction of a picosecond. Electrical (screening) energy is exchanged very rapidly between electron and ions, in a time of the order of the ion plasma frequency (<10fs). Heat is exchanged at the collisional time-scale which can be of several picoseconds. As a result, it is relatively easy to reach a non-equilibrium state with $T_e \neq T_i$ when using femtosecond laser pulses. Fortunately, one has, most of the time $T_i \ll M_iT_e/m_e$. This allows one to treat a system of electrons at temperature T_e as if they were interacting with fixed ions at constant positions $\{R_i\}$. The probability distribution for classically defined ion positions $\{R_i\}$ and moments $\{P_j\}$ writes:

$$P(\{R_i\},\{P_j\}) = 1/\mathcal{Z} \exp - \left[\sum_j \frac{P_j^2}{2M_i} + F_e(\{R_i\},T_e)\right]/kT_i$$

where F_e, the electron free energy, is calculated for fixed ion positions (the so-called Born-Oppenheimer approximation) and $\mathcal{Z}$ is the partition function. Using a quantum-statistical calculation with the hamiltonian:

$$H = \sum_i \frac{p_j^2}{2m_e} + \frac{1}{2}\sum_{ij} \frac{e^2}{|r_i - r_j|} - \sum_{ij} \frac{Qe^2}{|r_i - R_j|} + \frac{1}{2}\sum_{ij} \frac{(Qe)^2}{|R_i - R_j|} \quad (3)$$

the electron free energy is obtained from:

$$\exp(-F_e/kT_e) = \mathcal{Z}_e = Trace[e^{-H/kT_e}].$$

Obviously F_e cannot be calculated exactly but for many purposes, a reasonable approximation (but marginal for dense plasmas) assumes:

$$F_e \approx F_0(\rho, T_e) + \frac{1}{2}\frac{(Qe)^2}{|R_i - R_j|}e^{-|R_i-R_j|/\lambda_D(\rho,T_e)}$$

where λ_D is the Debye electron screening length and F_0 is the ideal-plasma electron free energy. The total free energy allows to recover the thermodynamic functions

$$F = -kT_i \log Z(T_e, T_i, \rho).$$

Electron and ion entropies and electron energy are thus defined as:

$$S_{e,i} = -(\partial F/\partial T_{e,i})_{\rho,T_{e,i}}, \quad E_e = F_e + T_e S_e.$$

For changes of the energy at constant density, we have:

$$dQ_e = T_e dS_e = C_{ee}dT_e + C_{ei}dT_i,$$
$$dQ_i = T_i dS_i = C_{ie}dT_e + C_{ii}dT_i,$$

where the C_{kl} form a matrix including all kinds of collision pairs between electrons and ions. The matrix nature of the specific heat is natural because when a sudden heat pulse is applied to the dense plasma, the laser energy is absorbed by the electrons and raises their temperature. The temperature rise induces an immediate change of the electron screening length making the ion interaction stronger. The weaker ion repulsion must be compensated; ions readjust their spatial correlations (pair distribution function) to reach equilibrium with stronger forces. This arise in a few ion plasma oscillations and release thermal energy. As before, collisional heat transfer occur on a much longer time scale.

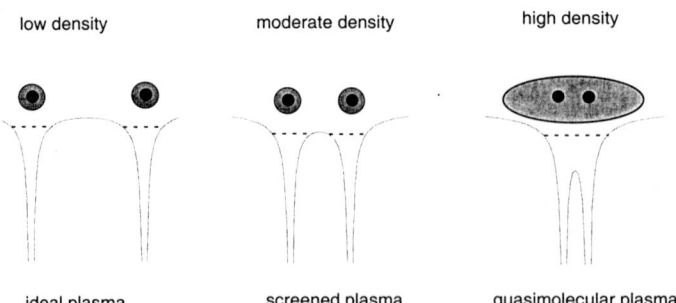

Figure 2. From ideal to quasimolecular plasmas by changing the distance between ions.

Dense plasma models

All the models of ionic potential in plasmas share the concept of screening of the nuclear electrostatic potential by free and bound electrons. They are all designed to answer the question: how does the potential relate to the distribution of atoms? To illustrate that point, the potential can be split into a Coulomb part and a screening part:

$$V(r) = \frac{Ze}{r}S(r),$$

e.g. in the Debye-Hückel model, one assumes $S(r) = \exp(-r/\lambda_D)$ where λ_D is the Debye length. Schematically, the basic models of the ionic potential are depicted in Fig. 2 to represent the transition from an ideal plasma situation to a dense plasma situation, when the interionic distance varies. Differences between the various ionic potential models are focused on the accuracy to which the model treats bound electrons, to what statistical method is used to describe the influence of electrons (Boltzmann, or Fermi-Dirac, or a full quantum mechanical treatment), and to what distance are the neighboring ions treated as separate objects.[30] In most of the models, the use of the Poisson equation to relate the potential variations with the spatial charge distributions reflects the electrostatic nature of plasma particle interactions:

$$\nabla^2 V(\vec{r}) = -4\pi e(\sum_{Q=0}^{Z} Q N_Q(\vec{r}) - n_e(\vec{r}))$$

Debye-Hückel model. In the Debye-Hückel model, Boltzmann radial distributions n_e and N_Q are used: this limits the validity to low density plasmas (say more than 10 electrons in a Debye sphere) and $\Gamma_{ii} = \frac{Z^2 e^2}{R_i T} \ll 1$.

Thomas-Fermi model. This model is widely used in solid-state physics, astrophysics and EOS calculations.[31] It considers a nucleus of charge Z at $r = 0$ and Z electrons (bound and free, with no excited states) confined to the ion-sphere. The total potential is zero on the boundary and outside of the ion sphere where a continuous background of electrons neutralizes the ion. The potential is divided in a nuclear part $V_N(r) = Ze/r$ and an electronic part which satisfies the above-mentioned Poisson equation. It solves the coupled equations

$$V(r) = V_N(r) + V_e(r)$$
$$n_e(r) = \int dp f_e(r,p) = \frac{1}{2\pi^2}(\frac{2mc^2 T_e}{(\hbar c)^2})^{3/2} F_{1/2}(\frac{\mu + eV(r)}{T_e})$$

where μ is the chemical potential determined from the charge neutrality requirement $Z = \int_0^{R_i} n_e(r) d^3r$ and $F_{1/2}$ is one of the Fermi integrals. The ionization Q_{eff} is found from the number of electrons with positive energies. R. More has found a convenient algorithm to calculate Q_{eff} for any element.[32]

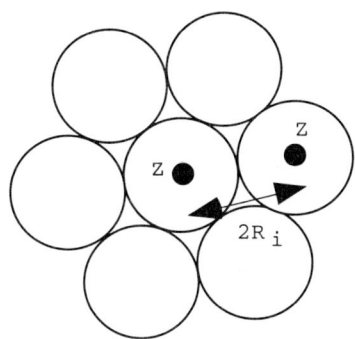

Figure 3. Schematic representation of the tight binding of ions at high densities. For hydrogenic ions, we have $R_i = (\frac{3(Z-1)}{4\pi N_e})^{1/3}$.

Over the years, many corrections and improvements to this model have been applied.[33] They include i) adding the exchange energy to antisymmetrize the electron wavefunctions which can then be calculated through the Dirac equation instead of the Schrödinger equation (the so-called TFD model), ii) adding a gradient correction because the gradient of $n_e(r)$ changes the electron energy (electrons are not "point-like"), and iii) removing the divergence of the electron density at $r = 0$.

Ion-sphere models. This is a generic term for models which assume charge neutrality inside an ion sphere. Bound electrons are treated quantum-mechanically with the Schrödinger or Dirac equation in a potential generated by the nucleus and bound and free electrons. The Pauli exclusion principle is accounted for by a Slater exchange potential. Free electrons are treated statistically by use of the Fermi-Dirac distribution. Closely-packed ion spheres are often used to describe non-degenerate plasmas at extreme densities (see Fig. 3).

Quasi-molecular models. Dicenter quasi-molecular features appear when the extent of the wavefunction of a bound electron is comparable to the interionic distance. The effect of the nearest neighbor is the most important,[34] as shown in the left part of Fig. 4. Spherical symmetry is no more useful to describe the electrostatic interactions, and because only ion pair correlations are important, cylindrical symmetry along the internuclear axis is appropriate. Bound electrons are treated in the framework of the Born-Oppenheimer approximation and elliptical coordinates are used. Electronic states depend only on the instantaneous positions of the nuclei. Schrödinger equation is solved in the potential of electrons and ions. A confinement volume is defined by $E_r = -[d\Phi(r)/dr]_{r=R_i} = 0$. This is equivalent to maintain global charge neutrality at the boundary.

The distribution of neighbors is evaluated by molecular dynamic calculations in which the positions of a (small) number of ions is followed by solution of equation of motion in a screened potential (screening distance can be R_i for $\Gamma_{ii} \approx 1$). Periodic boundary conditions are assumed. Potential curves are calculated self-consistently with the density of perturbing free electrons (see the right part of Fig. 4).

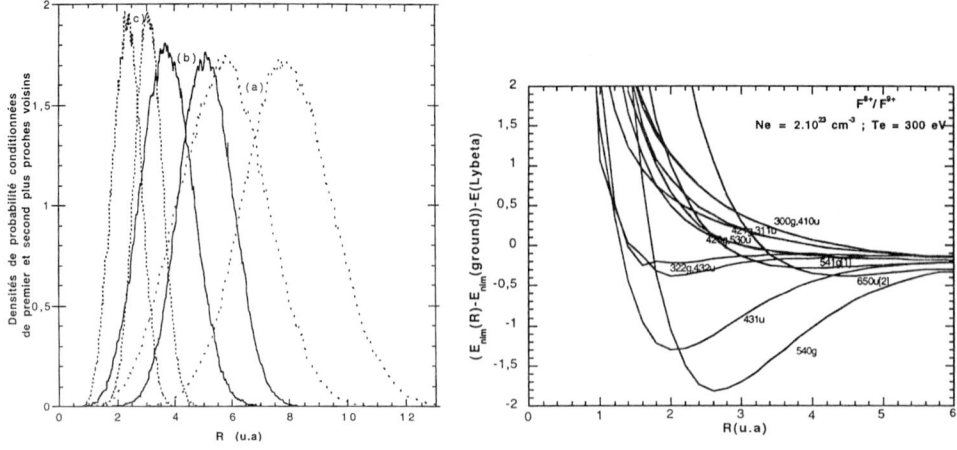

Figure 4. Distribution of nearest neighbors (left) a) $\Gamma_{ii} = 1.1$; b) $\Gamma_{ii} = 1.8$; c) $\Gamma_{ii} = 3.0$ and transition energies (right) around the Ly$_\beta$ line of F^{8+} as a function of internuclear distance between F^{8+} and F^{9+}.

Table 2. Electric field amplitude E, quiver velocity v_{osc}, excursion length δ_y, and hot electron energy W for Nd:glass and Ti:Al$_2$O$_3$ lasers

I	E	λ=1056 nm			λ=800 nm		
Wcm^{-2}	Vm^{-1}	v_{osc}/c	$\delta_y(nm)$	W(keV)	v_{osc}/c	$\delta_y(nm)$	W(keV)
10^{16}	3×10^{11}	0,06	15	1	0,048	8,7	0,59
10^{18}	3×10^{12}	0,536	128	96	0,433	78	56
10^{20}	3×10^{13}	0,988	235	2775	0,979	176	2002

How fast is fast and dense is dense?

The interaction of solids with high-intensity ultrafast lasers produce dense plasma conditions which are difficult to find together in other experimental situations. The atomic field (acting on an electron in the first Bohr orbit of the hydrogen atom) is of the order of $E_a = e/r_b^2 = m_e^2 e^5/\hbar^4$ V/cm= 5.1×10^9 V/cm for a laser irradiance of $I_a = 3.4 \times 10^{16}$ W/cm^2. Such irradiances can be reached very easily with commercial femtosecond lasers. For instance, a modest energy of 100mJ in a 100fs pulse duration corresponds to a power of 1 TW (10^{12} W). When focused to a diameter of 10μm, it allows to reach an intensity close to 3×10^{18} W/cm^2, the threshold of relativistic effects which occur when the oscillatory energy of the electrons is comparable to their rest energy $m_e c^2$. Table 2 gives the variation of two other interesting quantities which are the quiver velocity $eE_0/m_e\omega$ and the excursion amplitude $eE_0/m_e\omega^2$. Finally, the light pressure on the surface of the solid target is $P_0 = 3.3 \times 10^{-2} I/10^{12}$ Mbar, i.e. for $I_r = 10^{19}$ W/cm^2, we have $P_0 = 3.3$Gbar. Other time and spatial scales are specific to ultrafast plasma physics. For a 800nm wavelength laser of pulse duration 10-300fs, the photon energy is 1.55eV, as compared to the $\sim$ 10-15eV energy corresponding to plasma frequency at solid densities. For a metal target, the skin depth (the length over which the evanescent electromagnetic field penetrates) is about 10nm. The electron thermal diffusion depth is slightly much larger than at $\sim$ 100-200nm. Plasma expansion distances are quite small during the laser pulse duration. Typical values of 1Å/fs are reached at moderate intensities. Amazingly, this corresponds to huge accelerations[35] of $> 2 \times 10^{17}$g. In perspective of these numbers, the dynamics of the structural properties of condensed matter, i.e. the evolution of phase transitions and melting occur on the time scale of atom motions with typical atom oscillations of 10fs, onset of substrate disorder of 100fs, and thermal melting time $>$1ps.

Laser technology

It is now possible to find commercial lasers with a peak power of 1 TW fitting on one or two optical tables. This explains the frequently-used expression of "tabletop lasers" associated with ultrafast lasers.[12] This has been due largely to the new ability to manipulate short pulses[36, 37] and amplify them in a wide variety of broadband materials, the most popular being Titane:sapphire and Neodymium:glass which have saturation fluences of typically 1 to 10J/cm^2. The saturation fluence of an amplifying laser material denotes our ability to efficiently extract energy from an amplifier. Avoiding catastrophic self-focusing and small scale filamentation implies the pulse intensity to be kept at the GW/cm^2 level in the amplifiers. This can be done only if the pulse is of nanosecond duration. Therefore femtosecond pulses have to be stretched by a factor 10^3-10^4. This is the whole idea of CPA which was initiated and implemented by G. Mourou and his

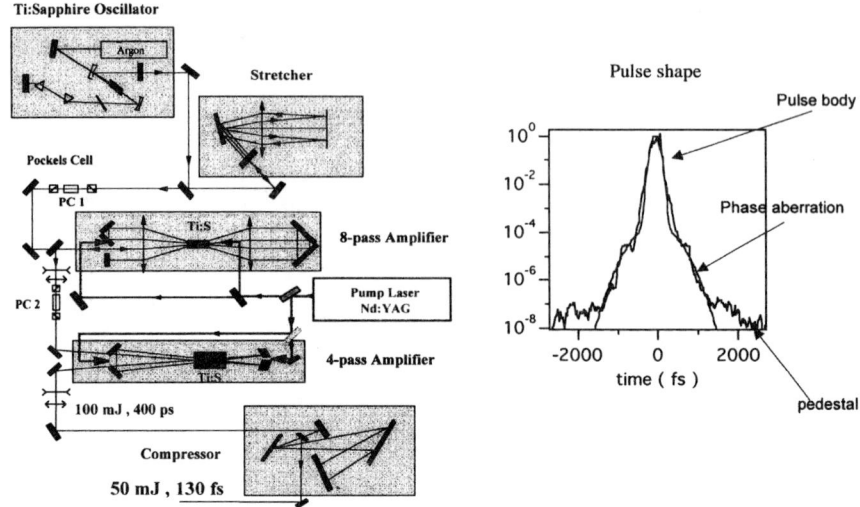

Figure 5. Schematic diagram of the Laboratoire d'Optique Appliquée laser in "salle verte". The right insert shows a typical pulse shape from this laser.

co-workers (see the lecture by G. Mourou in this book). The recompression of the pulse is the difficult point of CPA because it needs very large size gratings which are quite expensive and difficult to fabricate. In addition, the amplification process may severely distort the stretched pulse so that compression may not be entirely effective and higher order pulse aberrations may not be fully compensated for.

Numerous CPA systems have been developed in the recent few years. Figure 5 presents such an all-Titane:sapphire system which has been built at LOA and is now implemented in the so-called *salle-verte*.[38] It is made of a series of amplifiers pumped by the second harmonic of a Q-switched Nd-YAG laser. A Kerr-lens mode-locked oscillator delivers pulses of 100fs duration and 10nJ energy. The pulses are stretched to 450ps and amplified first in an eight-pass system, then in a four-pass amplifier before being compressed back to nearly 130fs. A similar system has been built at Lund University with TW-power capability.[39]

In Fig. 5, the right inset shows a typical pulse shape of a Titane:sapphire laser, measured with a third-order autocorrelation system, illustrating the difficulty in maintaining a good intensity contrast ratio between the peak of the pulse and its background. The figure clearly illustrates the three contributions to the pulse width: i) the pedestal at times larger than 2ps before the peak is due to amplified spontaneous emission (ASE), ii) improperly compensated phase aberrations between the stretcher and the compressor broaden the pulse around 1ps before the peak, and iii) the peak width itself is governed by the oscillator duration and the bandwidth of the amplifying material. One sees that for a nominal 10^{18} W/cm^2 irradiation, an intensity contrast ratio of 10^{-8} gives an irradiance of 10^{10} W/cm^2 before the pulse, a value more than enough to create a preplasma and perturb the physics under study.

To go to much higher peak powers, in the petawatt (10^{15} W) range, much shorter pulses and higher energy have to be delivered.[40] The short pulse limitation arises mainly from the aberrations induced in the telescope of the stretcher. It has been demonstrated that going to reflective optics improves greatly this limit, and 25fs pulses have been amplified up to the >50mJ regime.[41, 42] To increase the level of delivered energy, the implementation of CPA technology to large Nd:glass power chain has been performed, such as in the CEA Limeil center with the hybrid Titane:Sapphire and Nd:glass P102 laser.[43] However, the pulse duration is limited to ∼500fs by the amplifying media bandwidth.

Table 3. Multi-terawatt ultra-short pulse laser systems worldwide. Laser systems dismantled[‡] or in upgrade[↑] or construction are indicated

Laboratory	Name	Type	λ (nm)	Power (TW)	Pulse length (fs)	Rep. rate (Hz)
LLNL, USA	Petawatt‡	glass	1053	1000	500	-
Limeil, FR	P102‡	glass	1053	55	300	-
LULI, FR	PICO2000↑	glass	1053	100-1000	500	-
RAL, UK	VULCAN↑	glass	1053	100-1000	500	-
RAL, UK	ASTRA	Ti:Sa	800	10	50	10
LOA, FR	Yellow↑	Ti:Sa	800	10-100	20	5-10
GSI, DE	PHELIX↑	glass	1053	1000	500	-
UCSD, USA	-	Ti:Sa	800	10-100	20	10
Saclay, FR	UHI	Ti:Sa	800	10	50	10
ILE, JP	↑	glass	1053	100-1000	500	-
Kansai, JP	-	Ti:Sa	800	<100	20-50	few Hz
CELIA, FR	-	Ti:Sa	800	1	20	1000
MPQ, DE	ATLAS	Ti:Sa	800	1-8	130	10
CUOS, USA	↑	glass	800	1-8	500	-
KAIST, Korea	-	Ti:Sa	800	3	20	-

By mixing phosphate and silicate glass the pulse duration could be decreased[44] from 450fs down to 270fs. This approach has allowed in the recent years to get subpicosecond pulses into the kilojoule level, the so-called petawatt laser at Lawrence Livermore National Laboratory.[45, 46] We should note that one of the main technological barriers, the limited size of the gratings and their damage threshold, is now technically resolved by using multidielectric coatings. In Table 3, one can find a list, certainly incomplete, of existing or upgrading lasers with their current or expected performances. At least four petawatt-class laser systems are in construction or planned in Japan and Europe. The wealth of this field is clearly illustrated by the myriad of 1-10TW Titane:sapphire lasers which are operated over the world, very often as user facilities.

BASICS OF SHORT PULSE LASER INTERACTION WITH SOLIDS

Laser absorption

At normal incidence, simple scaling laws for the interaction of femtosecond high power lasers with solid targets have been soon derived,[47] just after the outcome of the first experimental results.[48] As an example, suppose that we use a laser with 5mJ energy, 160fs pulse duration, $0.8\mu m$ wavelength focused on a $20\mu m$ focal spot on a Si target. This corresponds to a modest laser irradiance of 10^{16} W/cm^2. Measured absorption is $\approx 30\%$ which corresponds to 500J/cm^2 absorbed fluence. This is deposited at solid electron density $(2 \times 10^{23} cm^{-3})$ on a skin layer of ≈ 100Å depth. We will see that the thermal conduction front propagates over a depth of the order of 1000Å during the pulse width. This gives about 2×10^7 J/gm energy deposition. Assuming ideal gas specific heat (but taking ionization into account as we have seen in the previous Section), we get an electron temperature of about 250 eV. The acoustic speed is of order 10^7 cm/s so there is only 100Å hydrodynamic expansion. Table 4 gives scaling laws for the electron temperature and thermal penetration scale length for a model laser pulse

203

Table 4. Simple scaling laws for a model laser pulse shape and normal incidence. C_l and C_T are scaling constants

	during pulse	after pulse
scale length	$C_l E^{5/9} t_0^{2/9} (t/t_0)^{7/9}$	$C_l E^{5/9} t_0^{2/9} (t/t_0)^{2/9}$
temperature	$C_T E^{4/9} t_0^{-2/9} (t/t_0)^{2/9}$	$C_T E^{4/9} t_0^{-2/9} (t/t_0)^{-2/9}$

shape showing a ramp increase from 0 to E in a time t_0 followed by a constant value of E. We assume a plasma at a temperature high enough for Spitzer's electron transport coefficients to be valid. During the rising edge of the pulse, the thermal penetration front spreads almost linearly whereas the temperature rises sharply and flattens around the peak of the pulse. These results compare quite favorably with more sophisticated hydrodynamics simulation, that we will describe below. In simple metals below 1eV, the thermal diffusivity $\chi = \kappa/C_e$ is approximately constant, the heat penetration depth scales[49] as $l_T = (2\chi t)^{1/2}$ and the velocity of the heat front decreases as $v_T = (\chi/t)^{1/2}$. This feature explains the interest of short pulses in laser treatment of materials, such as marking and drilling, because the "heat affected zone" is of very small extent.

Electrical conduction and electromagnetic field propagation. A basic ingredient in the interaction of ultrashort laser pulses by solids is the electrical conductivity of the material. Indeed, the conductivity σ and the dielectric constant ϵ are related through $\epsilon = 1 - i\sigma/\omega\epsilon_0$. A metal like aluminum is highly conducting and present a low absorption coefficient. To show simply how one can calculate σ, we start from the electron kinetic equation in the relaxation time approximation (the so-called Krook approximation):

$$\frac{\partial f}{\partial t} + v_z \frac{\partial f}{\partial z} - e/m_e(\mathbf{E} + \mathbf{v} \times \mathbf{B})\frac{\partial f}{\partial \mathbf{v}} = -\nu(f - F_0)$$

with

$$F_0 = n_e \Phi_M(|\mathbf{v}|)$$

the equilibrium distribution, which is a Maxwellian distribution or a Fermi distribution, according to the material density. In the above kinetic equation, $\delta f = f - F_0$ is a *small* deviation. Linearizing and Fourier transforming and assuming that the laser field varies in $\exp(-i\omega t)$, we get:

$$\delta f_k = -i \frac{n_e e}{m_e v_T^2} \frac{\Phi_M}{(\omega - k v_z + i\nu)} \mathbf{v} \cdot \mathbf{E}_k$$

The Fourier-transformed electrical current and conductivity obey:

$$J_k = -e \int d^3v\, \mathbf{v}\, \delta f_k = \sigma(\omega,k)\mathbf{E}_k$$

with the final result for the electrical conductivity:

$$\sigma(\omega,k) = i\omega_p^2/4\pi \int_{-\infty}^{\infty} dv_z \frac{1}{(\omega - k v_z + i\nu)} \frac{1}{\sqrt{2\pi} v_T} \times \exp(-v_z^2/2v_T^2)$$

The electrical conductivity depends on the laser frequency. It contains the non local term kv_z responsible for collisionless absorption and relies on the effective collision frequency $\nu(\omega)$ that we will discuss later.

Table 5. From collisional to non-collisional absorption: regimes of validity

NSE	ASE	HFSE	SIB
$\nu \gg \omega, k v_z \approx v_T/l_s$	$\omega \gg v_T/l_s \gg \nu$	$\nu \ll \omega, k v_z \approx v_T/l_s$	$v_T/l_s \gg \omega \gg \nu$
$l_s \gg v_T/\omega \gg v_T/\nu$	$l_s \ll v_T/\omega \ll v_T/\nu$	$l_s \gg v_T/\nu \gg v_T/\omega$	$v_T/\omega \ll l_s \ll v_T/\nu$

Starting from the Maxwell equation to get the propagation equation of the field, both in vacuum and in the solid, we recover the Helmholtz formulas ($\mathbf{B} = B(z,\omega)\hat{y}$, $\mathbf{E} = E(z,\omega)\hat{x}$):

$$\frac{\partial E}{\partial z} = i\omega/cB; \qquad \frac{\partial B}{\partial z} = -4\pi/cJ$$

Then, by combining these two expressions, we have (0_+ is inside the plasma):

$$\frac{d^2 E}{dz^2} + i4\pi\omega/c^2 J = 2i\omega/cB(0_+)\delta(z)$$

Fourier transforming, with the help of the Ohm's law, we get:

$$E(z) = 2i\omega/cB(0_+) \int_{-\infty}^{\infty} dk \frac{\exp(ikz)}{-k^2 + i(4\pi\omega/c^2)\sigma(|\mathbf{k}|,\omega)}$$

By using the concept of surface impedance $\mathcal{Z} = \frac{E(0_+)}{B(0_+)}$, the absorbed power P due to Joule heating is calculated and the absorption is:

$$A = \frac{P}{I_0} = 4\Re(\mathcal{Z})/|1+\mathcal{Z}|^2; \qquad I_0 = c|B_0|^2/8\pi$$

where $\Re$ denotes the real part of the complex expression and I_0 is the incident laser intensity.

Absorption in a skin layer. The dominant absorption mechanisms[50] are related to the ordering between three parameters: the skin layer depth l_s, the electron excursion length v_T/ω and the electron mean-free path v_T/ν. In the literature,[51, 52, 53] it is common to distinguish between the normal skin effect (NSE), the high-frequency skin effect (HFSE), the anomalous skin effect (ASE), and the sheath inverse bremsstrahlung (SIB). The ordering of the frequencies and spatial scales for each of the above-mentioned regimes is given in Table 5. In the normal skin effect regime, the dominant term in the electrical conductivity is $\sigma \approx \omega_p^2/4\pi\nu(0)$ and the absorption coefficient and skin depth are:

$$A_{NSE} = 2\frac{\sqrt{2\omega\nu(0)}}{\omega_p}; \qquad l_s = c/\omega_p\sqrt{\frac{2\nu(0)}{\omega}}$$

In the case of the high frequency skin effect, the collision frequency is still large but the plasma polarization screens the electric field, so that the skin depth is independent of the collision frequency:

$$A_{HFSE} = 2\frac{\nu(\omega)}{\omega_p}; \qquad l_s = c/\omega_p$$

With increasing laser intensity and electron temperature the role of collisions decrease. The sheath inverse bremsstrahlung regime is related to the effect of inelastic electron collisions on the electron trajectories,[54] at a sharp plasma interface over a sheath of thickness λ_D. In the anomalous skin effect regime, the electron crosses the skin depth within one laser oscillation because the electron mean-free-path is much larger than the skin depth. This regime is analogous to "delocalized" electron thermal conduction, when the electron temperature gradient is steeper than the mean-free-path.[55]

There is a variant of ASE which is called vacuum heating.[56] In this regime, which has been verified experimentally for sub-50fs pulses,[57] electrons are dragged out of the surface during the first-half period of the laser pulse and pushed back towards the target during the second half. Because the electrons are shielded from the effects of the electromagnetic field inside the target, they continue their trajectories and deposit their energy by inelastic collisions. The absorption coefficient for the sheath inverse bremsstrahlung can be written:

$$A_{SIB} = 8(2/\pi)^{1/2}(v_T/c)^3(\omega_p/\omega)^2; \qquad l_s = c/\omega_p,$$

and for the anomalous skin effect:[58]

$$A_{ASE} = 8/3(2/\pi)^{1/6}(v_T/c)^{1/3}(\omega/\omega_p)^{2/3}$$
$$l_s = (2/\pi)^{1/6}(c/\omega_p)^{2/3}(v_T/\omega)^{1/3}.$$

Figure 6 shows the relative contributions of the above-mentioned absorption processes.

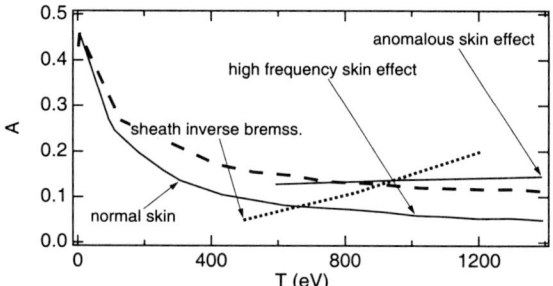

Figure 6. Different contributions to the absorption coefficients in the collisional regime (low temperature) and the non-collisional regime (high temperature).

The normal skin effect is dominant at low electron temperatures, when electron collisions play a role. The anomalous skin effect and sheath inverse bremsstrahlung, both non-collisional, are weakly dependent on the temperature and their absolute value is about 10%. They clearly dominate, at high temperatures, over the collisional processes. Vacuum heating could be significant only for ultrashort (< 50fs) laser pulse interaction.

Arbitrary incidence. Using Maxwell equations and eliminating either the electric or the magnetic field, one can get (with ϵ being the dielectric constant of either vacuum or the dense plasma):

$$\vec{\nabla} \wedge \vec{\nabla} \wedge \vec{E} + \frac{\omega^2}{c^2}\epsilon \vec{E} = 0$$

$$\vec{\nabla} \wedge \vec{\nabla} \wedge \vec{B} - \frac{\vec{\nabla}\epsilon}{\epsilon} \wedge \vec{\nabla} \wedge \vec{B} + \frac{\omega^2}{c^2}\epsilon \vec{B} = 0$$

For an angle of incidence θ with respect to the normal of the target, in S polarization (electric field perpendicular to the plane of incidence), we have $\vec{\nabla} \wedge \vec{E} = -\frac{\partial \vec{B}}{\partial t}$ and the propagation equation writes:

$$\frac{d^2\mathfrak{E}_x}{dz^2} + \frac{\omega^2}{c^2}(\epsilon(\omega, z) - \sin^2\theta)\mathfrak{E}_x = 0 \tag{4}$$

where $\mathfrak{E}_x$ is the component of the electric field perpendicular to the incidence plane. In P polarization (magnetic field perpendicular to the plane of incidence), we have $\vec{\nabla} \wedge \vec{B} = \frac{i\omega\epsilon}{c^2}\vec{E}$ and the propagation equation writes:

$$\frac{d^2\mathfrak{B}_x}{dz^2} - \frac{d\epsilon}{dz}\frac{1}{\epsilon(\omega, z)}\frac{d\mathfrak{B}_x}{dz} + \frac{\omega^2}{c^2}(\epsilon(\omega, z) - \sin^2\theta)\mathfrak{B}_x = 0 \tag{5}$$

where $\mathfrak{B}_x$ is the component of the magnetic field perpendicular to the incidence plane.

These equations can be solved by standard ordinary differential equation procedures (like the Runge-Kutta[59] method) or by using the matrix method used also to calculate thin film multilayer properties.[60] The P-polarization case is special because the dielectric constant goes to zero (assuming no collisions) at critical density ($\omega = \omega_p$). We will discuss with some detail two limiting cases which are important for pump/probe dense plasmas diagnostic purposes.[61] Fresnel formulas should be valid for very steep gradients (or equivalently shorter pulse duration) and the long laser pulse limit, making use of the WKB approximation,[4] should be recovered for long gradient scale lengths.

The Fresnel limit $\lambda \gg l_{n_e}$. The Fresnel limit is usually valid for UV wavelengths and/or very short (sub-100fs) laser pulses. In that case, the electron density gradient scale length is very much shorter than the wavelength. We define the electron density gradient scale length as $l_{n_e} = n_e/(dn_e/dz)$. In the case of S-polarized light, the field reflection coefficient $r_\perp$ and the intensity reflection coefficient R_S are given by (ϵ is the dielectric constant of the plasma):

$$r_\perp = \frac{\cos\theta - (\epsilon - \sin^2\theta)^{1/2}}{\cos\theta + (\epsilon - \sin^2\theta)^{1/2}}; \qquad R_S = |r_\perp|^2$$

The phase upon reflection $\exp(i\Phi_S) = r_\perp/|r_\perp|$ is negative for dielectrics and change sign for metals. The skin depth inside the material is $\delta = (k_0\Im[(\epsilon - \sin^2\theta)^{1/2}])^{-1}$ for arbitrary angle of incidence.

In the case of P-polarized light, the field reflection coefficient $r_\parallel$ and the intensity reflection coefficient R_P are given by:

$$r_\parallel = \frac{\epsilon\cos\theta - (\epsilon - \sin^2\theta)^{1/2}}{\epsilon\cos\theta + (\epsilon - \sin^2\theta)^{1/2}}; \qquad R_P = |r_\parallel|^2$$

The phase upon reflection for the magnetic field is $\exp(i\Phi_P) = r_\parallel/|r_\parallel|$. There is a phase difference of π between the reflected electric and magnetic field phases. With a dielectric target material, we recover the Brewster angle when $\tan\theta_B = \sqrt{\epsilon}$.

For polished metals, the absorption coefficient is easily deduced from the reflection coefficient because scattering is negligibly small. For dielectrics, transmission formulas are similar to reflection formulas. Metals and dielectrics have very different absorption coefficients at low intensities,[62] metals showing usually low absorption. Several attempts of increasing laser absorption by structuring the target surface have been made in the context of producing an efficient source of short pulse x-rays. They involved modifying the electromagnetic field-surface interaction by volume deposition in "black foam" targets,[63] surface plasmon absorption by a grating,[64] using "velvet-like"

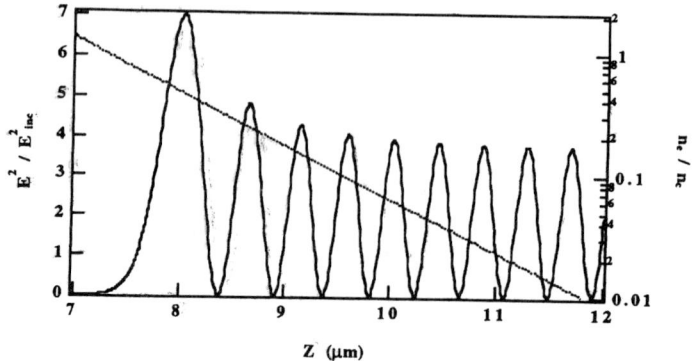

Figure 7. The shape of the field in a long scale length density gradient. The straight line gives the exponential (right scale) density gradient. Laser is coming from the right.

targets,[65] or, for oblique incidence, using a controlled prepulse to shape the electron density gradient.[66, 67]

In a series of seminal experiments, the group of Göttingen has measured[68, 69] the absorption of 250fs KrF laser pulses incident on solid targets of aluminum and gold as a function of polarization and angle of incidence for the intensity range of 10^{14}–2.5×10^{15} Wcm^{-2}. Maximum absorption of over 60% occurs for P-polarized radiation at angles of incidence in the range of 48°–57°. The measured results are in agreement with absorption on a steep density gradient.

The WKB limit $\lambda \leq l_{n_e}$. The dielectric function of the plasma varies "slowly" with space. Seeking a solution of Eq. 4 or Eq. 5 of the form:

$$\mathcal{E}_x(z) = \mathcal{E}_0 \exp(\pm \frac{i\omega}{c} \int^z \Psi(z') dz')$$

we find a solution, not valid close to the critical density (n_c) turning point at $n_c \cos^2 \theta$:

$$\mathcal{E}_x(z) = \frac{C}{k_0^{1/2}(\epsilon(\omega, z) - \sin^2 \theta)^{1/4}} \exp \pm i k_0 \int^z \sqrt{(\epsilon(\omega, z') - \sin^2 \theta)} dz' \qquad (6)$$

The phase shift of the reflected wave is:

$$\Phi(z) = \int_{z_{ref}}^{z} \sqrt{(\epsilon(\omega, z') - \sin^2 \theta)} dz' - \frac{\pi}{2}.$$

In a linear density gradient, an exact solution of Eq. 6 can be found in form of an Airy function. Figure 7 shows the spatial variations of the electric field (left scale) and of the electron density in an exponential density gradient for 45° incidence. The electric field grows closer to the turning point and becomes evanescent above critical density. This "resonant" behavior, called resonance absorption, will be described now; it is responsible for the optimum absorption found in experiments for different incidence angles.[66]

General case and resonance absorption. Resonance absorption and its damping mechanisms are physical processes which have been studied for many years in the context of laser fusion. The interested reader can find a full description of resonance absorption in the book of Kruer[4]. In short, in P-polarization, the electric field at the

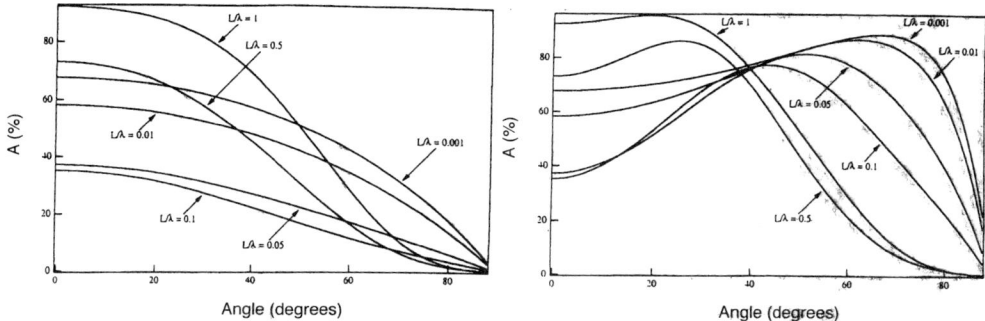

Figure 8. Calculated absorption as a function of the electron density gradient scalelength. Left: S-polarization, right: P-polarization.

turning point is perpendicular to the surface. This launch exponentially growing plasma waves towards the inside of the target and the field amplitude at resonance could be very much larger than the propagating incident field amplitude: this is very good for absorption. In the long scale length gradient regime, the field is damped by collisions and convection. In the steep gradient regime, the field is damped by wave breaking and collisions.[70] The plasma waves accelerate electrons which deposit their energy inside the target. Figures 8 and 9 show the numerical solution of Eqs. 4 and 5 in an exponential electron density gradient. We use the Drude model for the dielectric constant:

$$\epsilon = n_r^2 = 1 - n_e^*/(1 + i\nu^*) \tag{7}$$

in which $n_e^* = n_e/n_c$ and $\nu^* = \nu/\omega$ are the scaled electron density and collision frequency, respectively. Figure 8 show the calculated absorption as a function of the incidence angle for different electron density gradient scale lengths. For short scale lengths, we recover the Fresnel results. In S-polarization, absorption decreases monotonously with the angle. On the contrary, there is a clear absorption maximum in P-polarized light. The peak of P-polarized light absorption clearly shifts towards normal incidence when the scale length increases, in good agreement with the observations[68, 69] of Fedosejevs et al.

Figure 9 shows the numerically calculated absorption as a function of the plasma scale length for different dense plasma collisionality parameters. Here, $\theta=40°$. The scaled collision frequency $\nu^* = \nu/\omega$ is found to be the dominant parameter to "control" absorption at a fixed incidence angle.

To support the approximate Drude model for the plasma dielectric constant, the resistivity of nearly solid density aluminum has been measured as a function of temperature over 4 orders of magnitude above ambient by observing the self-reflection of an

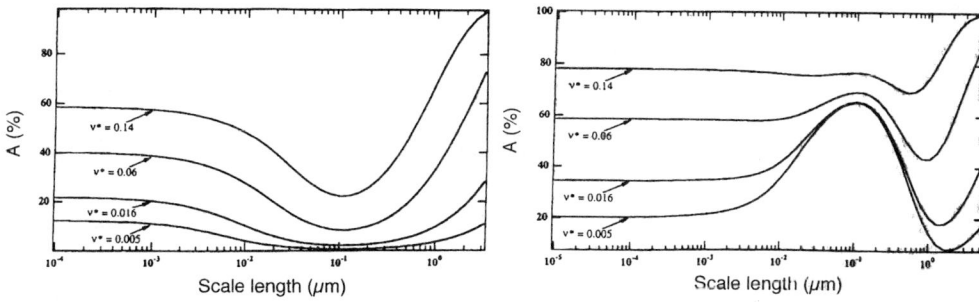

Figure 9. Calculated absorption as a function of the dense plasma collisionality. Left: S-polarization, right: P-polarization.

intense, subpicosecond, 308-nm-light pulse incident on a planar Al target.[71] As an increasing function of electron temperature, the resistivity is observed initially to increase, reaching a maximum value which is relatively constant over an extended temperature range, then decrease at the highest temperatures. The broad maximum is interpreted as "resistivity saturation," a condition in which the mean free path of the conduction electrons reaches a minimum value as a function of temperature, regardless of the extent of any further disorder in the material. These measurements of laser absorption for high-contrast ultrashort pulses have been extended on a variety of solid targets over an intensity range of 10^{13} to 10^{18} W/cm^2. These data[62] give an indirect measure of dense plasma electrical conductivity, in rather good agreement with an ultrashort pulse laser absorption model.[72] At high intensity, all target materials reach a "universal plasma mirror" state and reflect about 90% of the incident light.

Ponderomotive force

A light beam of flux density (intensity) I impinging normally upon a surface of reflectivity R exerts the pressure $P_r = (1+R)\frac{I}{c}$, also called "radiation pressure". At sufficiently high intensity, in a gradient of electric field, electrons are submitted to the oscillating electric field $\vec{E}$, which is a linear force and to the effects of the transverse magnetic field $\vec{v} \wedge \vec{B}$, which, in contrary, is a nonlinear force. As a result, electrons are expelled of the regions of high fields. The ponderomotive force arises whenever the radiation changes in space. It has an oscillatory component and a secular component (the radiation pressure). This force is responsible for many phenomena in laser-matter interaction such as parametric instabilities (speckles), self-focusing, filamentation and can have very interesting applications[73] such as beatwave[74] and wakefield[75] acceleration of electrons. Albeit a more correct formulation uses a tensorial approach, the expression for the ponderomotive force can be written in 1-D:

$$\vec{F} = -\frac{\epsilon_0}{2}\frac{n_e}{n_c}\vec{\nabla}\langle E_h^2 \rangle$$

where E_h is the oscillatory amplitude of the field and the $\langle \rangle$ denote the average over one field period. Ponderomotive pressure is larger than thermal pressure at laser intensities above 10^{16} W/cm^2. A steepening of the electron density gradient due to the ponderomotive force do occur around critical density, resulting in a "shelf" appearing in front of the target. Accordingly, there is a modest increase of density inside the target. This was used to explain anomalously large Stark broadening[76] of x-ray resonance line emissions of massive targets irradiated by an intense short pulse laser. Competition between the ponderomotive force and thermal forces in short-scale-length laser plasmas have been studied by X. Liu and D. Umstadter[77] using Doppler pump/probe measurements.

At extreme intensities, above 10^{19} W/cm^2, the manifestations of the ponderomotive force have been studied mostly numerically,[78,79,80,81] using particle-in-cell codes. Basically, the super intense laser beam impinging on a dense plasma creates a bow shock travelling into the material and leaving a rarefaction wave at velocity $v_S \approx I^{1/2}\rho_0^{-1/2}$ behind it. The target surface is bored by the laser pressure, with Rayleigh-Taylor-like instabilities and electron jet formation towards the inside of the target. Large magnetic fields are created. Laser hole-boring is one of the key ingredients of the fast igniter scheme for fusion,[82] for which more details can be found in the lecture notes of Pr. M. Key.

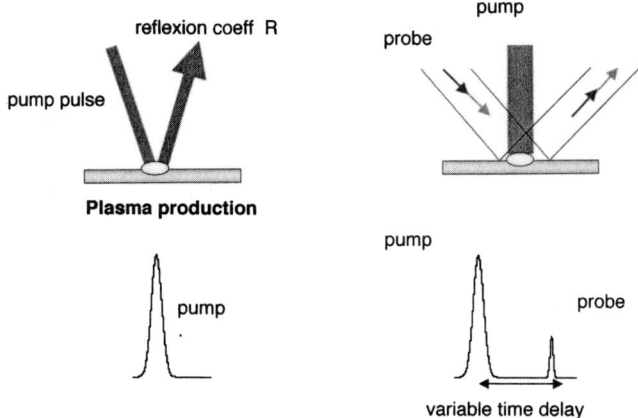

Figure 10. Basics of pump-probe optical techniques.

Pump/probe experiments

Due to the short duration of the pulse standard time-resolving methods, like fast photodiodes and the optical or x-ray streak cameras, do not have the time resolution to follow ultrafast dense plasma dynamics. Pump/probe optical techniques have to be used instead.[83] Fig. 10 shows the difference between time-integrated (eventually space-resolved) optical diagnostics and the pump/probe technique. It can be used in connection with several optical arrangements like shadowgraphy, reflectometry, and interferometry (spatial and spectral). Pump/probe techniques have had a great success in the study of fast phase transitions in semiconductors and metals by pump/probe reflectometry using multi-wavelength, quadrature polarization[84, 85, 86] detection, of spectral blue-shifting[87] in gases ("photon acceleration"), and of plasmas.[88] A more recent technique sensitive to the phase change of a probe pulse has been proposed recently (see below).

Indeed, optical methods giving access to the phase shift of a probe beam in transmission or reflection have shown their usefulness in solid state physics,[89, 90, 91] plasma physics,[92, 93, 94] and material damage studies.[95] Second harmonic generation from a probe beam gives a wealth of new information on the plasma dynamics, due to the non-linear nature of the interaction.[96] Pump/probe techniques in transmission and reflection have been elegantly exploited for the study of thermal electron transport.[97] In the experiment, targets are glass slides covered by carbon to favor absorption and the 100fs laser irradiates the target at 5×10^{14} W/cm^2 intensity. During the laser pulse, the probe pulse transmission is quenched by the increase of the electron density. Complementary Doppler shift measurements give the speed, which is found to be supersonic, of the electron thermal wave. By fitting the reflectivity data with the solution of Helmholtz equations within the approximation of the Drude model, a collision frequency can be recovered. Results point towards resistively saturated metallic character of the plasma. Electron scattering lengths comparable with the average interionic spacing (~ 3Å) imply localization of thermal electrons.

Phase measurements with pump/probe methods

Frequency-domain interferometry enables to measure the amplitude and the phase shift difference induced by the index of refraction of the plasma between a pair of femtosecond probe pulses with simultaneously high spatial and temporal resolution.

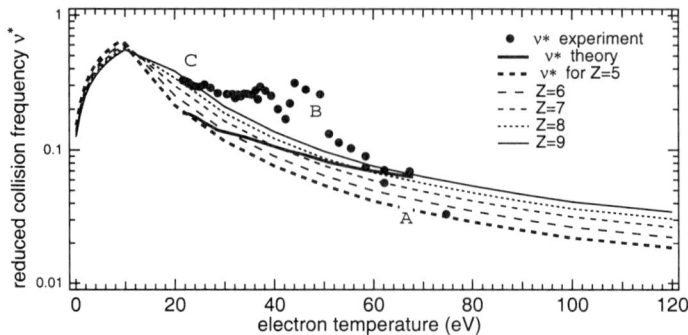

Figure 11. Collision frequency ν^* (dots) deduced from the measured S reflectivity as a function of electron temperature. Results of Lee and More's model are shown for comparison. Regions A, B, and C correspond to decreasing electron densities of $n_{reflection} \approx 10^{23}$ cm^{-3}, $n_{reflection} \approx 10^{22}$ cm^{-3}, and $n_{reflection} \approx 3 \times 10^{21}$ cm^{-3}, respectively.

The theory of operation of the interferometer is straightforward.[98] After one of the two probe pulse interacts with the plasma, the twin pulses are recombined in a spectrometer. Let us consider a point along the entrance slit of the spectrometer which is the image of a point along the diameter perpendicular to the plane of incidence of the probe beams. We assume that the probe pulse spectral widths are Fourier-transform limited so that $\Delta\omega \times \Delta t \approx 2\pi$. The expression for the reference pulse field is $\overrightarrow{E_0(t)} = E_0(t)e^{i\omega_0 t}$. After being reflected from the plasma surface, the probe pulse delayed by Δt from the reference pulse undergo a phase change and its intensity is reduced by a factor R (describing an effective reflection coefficient). The probe pulse field is now $\overrightarrow{E_1(t)} = E_0(t - \Delta t)\sqrt{R}e^{i(\omega_0(t-\Delta t)+\Delta\Phi)}$. The intensity measured by the camera is the Fourier transform of the sum of these two fields which can be developed to give the expression

$$I(\omega) = I_0(\omega)(1 + R + 2\sqrt{R}\cos(\omega\Delta t + \Delta\Phi))$$

The measured signal is sensitive to both the amplitude (through R) and the phase (through $\Delta\Phi$). It has been shown[92] that the probe wave reflected by a dense plasma is phase-shifted by three contributions. The first contribution corresponds to the propagation phase of the incoming and outgoing probe beams. At the wave turning point (when $n_e = n_c \cos^2\theta$), there is a contribution from the Doppler motion of the plasma. There is also a reflection phase, which is the extra phase from the WKB contribution discussed earlier. The important point is that the phase contribution most sensitive to probe polarization is the reflection phase. Moreover, the phase shift difference between S- and P-polarized light is a weakly varying function of the electron collisionality ν^* and a strongly varying function of the electron density gradient scale length. This is due to the specific effect of resonance absorption which maximizes its effect whenever the electron density gradient scale length $l_{n_e} \approx (0.8/\sin\theta)^3 \lambda/2\pi$.[4] Optical probing allows to test theoretical models of conductivities and electron-ion equilibration.[99]

In our group, using the technique of frequency-domain interferometry, we have studied recently the dynamics of ultrafast laser-produced plasmas at moderate laser irradiances.[100] From the complex reflection coefficient measured with the two in-quadrature probe pulse polarizations, we have determined the peak electron temperature during the pump laser pulse, the gradient scale length at critical density, the variation of the real and imaginary part of the dielectric constant of the plasma, and, using the Drude model, the collisionality of the plasma as a function of time and temperature.

In Fig. 11 we have plotted the measured values of the reduced collision frequency ν^* as a function of the measured temperature together with the results of the collision frequency obtained by Lee and More[101] for various values of the average charge Z^*. The heavy solid line takes into account the variation of the electron temperature with time, the corresponding Z^* being calculated under local thermal equilibrium (dense plasma) conditions. Agreement between experiment and theory is satisfactory at the higher densities (marked by label A in Fig. 11) reached early in time, when the electron density gradient is steep. The comparison quickly gets worse when the electron density at the reflection point decreases to critical density. However, the discrepancy between theory and experiment is never larger than a factor of two.

ENERGY TRANSPORT BY PHOTONS, AND PARTICLES

Transport theory

As we have seen earlier, the high amount of energy brought by the laser has to be dissipated by electron and photon transport because for ultrashort pulses the plasma hydrodynamic expansion is reduced. For electrons, the relevant quantities are the transport coefficients[102] for electronic (electrical conductivity) and thermal conduction. Electron current and magnetic field generation are governed by the electrical conductivity. Indeed, the relevant quantity is the collisionality of the thermal electrons. Material conductivity also plays a role in electron transport inhibition.[103] Heat is transported towards the bulk by electron thermal conduction. The relevant quantity is the ratio of the electron-electron mean-free-path to the electron temperature gradient scale length. This define the boundary between "local" and "non-local" thermal transport. Suprathermal electron transport becomes important at high irradiances.[104, 105] Poor conductivity inhibits return currents and limit fast electron transport. High currents, higher than the Alfven limit,[106] are channeled by the self-generated magnetic field.[78, 107, 108] Energy is also transported by x-ray photons. The relevant quantities to calculate energy transport are then the x-ray emission conversion efficiency and the material opacities.[109, 110]

Transport coefficients in the transition regime between solid state and plasma physics. Here we will derive, very simply, the expressions for the electrical and thermal transport coefficients by using kinetic theory. A more comprehensive review on electron transport can be found in Ref. 111. We start from the Boltzmann equation:

$$\frac{\partial f}{\partial t} + \vec{v} \cdot \frac{\partial f}{\partial \vec{r}} - e\vec{E} \cdot \frac{\partial f}{\partial \vec{p}} = K_c[f]$$

We use the Krook's approximation for the relaxation part of the kinetic equation $K_c[f] = -\frac{(f-f_0)}{\tau_c}$ with $\nu_{ei} = 1/\tau_c$. The equilibrium distribution is the Fermi-Dirac distribution normalized by the condition:

$$\int f_0(\mathbf{p}) d^3\mathbf{p} = n_e$$

and $\nu_{ei} = n_i \sigma_{tr} v$ is the velocity-dependent collision frequency. Lee and More[101] have recalled that the electrical and thermal conductivities can be written:

$$\sigma_{ei} = e^2 K_0 \tag{8}$$

$$\kappa_{ei} = \frac{1}{T_e}(K_2 - K_1^2/K_0) \text{ with}$$

$$K_n = -\int \frac{v^2 \epsilon^n \partial f_0}{3\nu_{ei} \partial \epsilon} d^3 \mathbf{p}$$

The real and imaginary parts of the dielectric permittivity, using its relation with conductivity $\epsilon = 1 - i4\pi\sigma_{ei}/\omega$, are:

$$\epsilon_r(\omega) = 1 + 4\pi e^2 \int \frac{v^2}{3} \frac{1}{\omega^2 + \nu_{ei}^2} \frac{\partial f_0}{\partial \epsilon} d^3 \mathbf{p} \tag{9}$$

$$\epsilon_i(\omega) = -4\pi e^2/\omega \int \frac{v^2}{3} \frac{\nu_{ei}}{\omega^2 + \nu_{ei}^2} \frac{\partial f_0}{\partial \epsilon} d^3 \mathbf{p}. \tag{10}$$

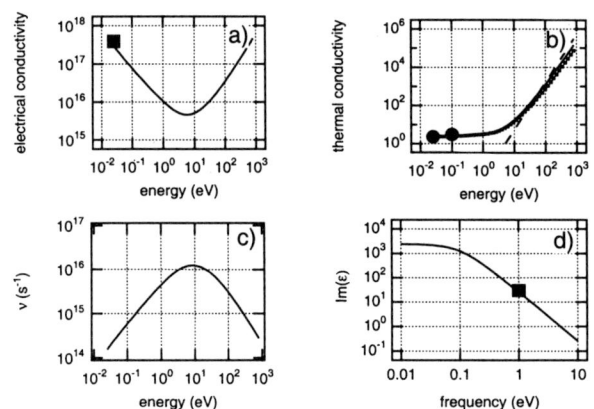

Figure 12. Calculations of (a) the electrical conductivity, (b) thermal conductivity (solid line), Spitzer (dashed line), (c) collision frequency as a function of electron temperature. (d) Calculation of the imaginary part of the dielectric constant ϵ as a function of frequency. Dots and squares are measured quantities.

We should note here a common error done in using the Drude model for conductivity. Because of the integral definition of σ_{ei} (see Eq. 8), the correct expression for σ is:

$$\sigma(\omega) = \frac{ne^2}{m} \langle \frac{\tau}{1 - i\omega\tau} \rangle \neq \frac{ne^2}{m} \frac{\langle \tau \rangle}{1 - i\omega\langle \tau \rangle},$$

where $\langle \rangle$ denotes the integral over the distribution function. However, $\langle \tau \rangle$ is often taken as an *ad-hoc* collisional parameter, but one has to be careful in comparing it to a transport coefficient from theory. The velocity-dependent rate of Coulomb collisions is:

$$\nu_{ei} = \frac{4\pi n_i Z_i^2 e^4}{m^2 v^3} L_{ei}$$

where $L_{ei} = \ln \Lambda_{ei}$ is the Coulomb logarithm and $\Lambda_{ei} = q_{max}/q_{min}$, where q_{max} and q_{min} are the maximum and minimum momentum transfer of an electron scattered off an ion of charge Z_i, respectively. For a weakly non-ideal plasma, we have $\Lambda_{ei} = \frac{2mv}{\hbar D_{ei}}$, where D_{ei} is a modified expression of the Debye length:

$$D_{ei}^{-2} = \frac{4\pi n_i Z_i^2 e^2}{T_i} + \frac{4\pi n_e e^2}{[T_e^2 + (2\epsilon_F/3)^2]^{1/2}}$$

where the effect of electron degeneracy is incorporated through ϵ_F. At T_i above the Debye temperature, electron waves are scattered off phonons and $\nu_{ei} \propto e^2 T_i/\hbar^2 v$ and $\ln \Lambda_{ei} \propto \Lambda_{ei}^2$.

To bridge the gap between the theories of low- and high-temperature electrical conductivity, in other words between solid state and plasma physics, Basko et al. worked out[112] a subtle procedure, in the limit of the Lorentz (no $e-e$ collisions) model, where the Coulomb logarithm can be written:

$$L_{ei} = \ln(1 + \frac{\Lambda_{ei}}{1 + g_{ei}/\Lambda_{ei}})$$

Figure 12 gives the results of the calculation of the collision frequency, the electrical and thermal conductivity, and the imaginary part of the dielectric constant for aluminum at solid density, using the prescriptions given by Basko. Another approach is the one developed by Lee and More[101] for DC conductivity (including the effects of magnetic fields). This is a piecewise model in which different approximations for the collision frequency are linked together.

Thermal electron transport: dense plasma effects At solid and higher densities; the collision time is limited by small-angle interference scattering. In the calculation of the collision time, which one way or the other, enters the conductivity formula (see Eqs. 9 and 10) the important items are the ionic cross-section $\frac{d\sigma}{d\Omega}$ and the ion structure factor $S(q)$, so that one can write:

$$1/\tau = \pi n_i v \int \frac{d\sigma}{d\Omega} S(q) \frac{q^3 dq}{k^4}.$$

The Ziman formula[113, 114] is used at near-solid densities. Because of the strong ion correlations (lattice effects), the conductivity rises at low temperature ($\sigma \propto 1/T$). The important physical effect is that degeneracy inhibit e-e collisions. Indeed, e-e collisions do not directly reduce the current J but e-e collisions alter $f(\epsilon)$.

Suprathermal electron transport

Because Coulomb scattering involves a v^{-4} variation of the differential cross section with velocity, electrons with an energy above the Fermi energy (in a solid) or above the maximum of the Maxwell distribution (in a LTE plasma) have very low collisionality. Accordingly, these electrons, dubbed "suprathermal" or "hot", play a particular role in energy transport. In solid state physics, such "ballistic" electrons are important in thermal transport.[115] One of the first manifestation of the role of suprathermal electrons in dense laser-produced plasmas is the delocalized thermal transport, when their mean-free-path is larger than the thermal gradient scale length.

At very high laser intensities, we have already assessed the dominant role of nonlinear collisionless processes in the production of "hot" electrons which deposit their energy deeply in the bulk of the target. Resonant absorption,[4] vacuum heating,[56] and $J \times B$ heating[116] are the most significant hot electron sources. Typical electron energy distributions show a "thermal", Maxwellian body and a high energy tail, often made of at least two exponential functions with characteristic hot electron "temperatures" A convenient scaling for the hot electron "temperature" is:

$$T_{hot} \approx (I\lambda^2)^{1/3-1/2},$$

where I is the laser intensity. Typically, for $1\mu m$ wavelength and 10^{17} W/cm^2 intensity, the hot electron temperature is about 150keV.

The stopping power of charged particles in solid matter under the influence of Coulomb forces is usually calculated by including the combined effect of elastic and inelastic collisions on nuclei, and bound and free electrons of the target.[117] The stopping power of electrons in matter takes the general form:

$$\frac{dE}{dx} = -\frac{4\pi e^4}{E}[n_b L_b + n_f(L_f + L_w)]$$

where n_b and n_f are the number densities of bound and free electrons, respectively. L_w is a small term originating from the stopping by plasma waves. Stopping numbers $L_{f,b,w}$ are dominated by logarithmic terms which vary weakly with density. For convenience, several codes have been written in the past, mostly to evaluate radiation doses behind shielding materials. One of the most popular is ITS.[118] In such codes, the 3-D trajectory of a single electron (or any other charged particle) interacting with the target through elastic (screened Rutherford cross sections) and inelastic (Bethe stopping power cross sections) scattering is calculted. By weighting the results obtained at several electron energies, one can predict the penetration characteristics of a more complex electron distribution function.[119] The energy deposition is evaluated as a function of position inside the target. The temperature rise is calculated using the equation of state of the material. The effect of particle heating on the energy deposition itself is neglected in such codes. Fluorescence efficiencies incorporating Auger decay branching ratios are used to calculate the bremsstrahlung and K_α line emission at different target depths. The optical depth from the emitting point to the target surface is taken into account.

A recent experiment[120] has addressed the problem of the variation of the stopping power in shock compressed matter, i.e. at matter densities above solid density. The probed plasma is dense ($\Gamma \approx 6$), and degenerate ($E_F/T \approx 4$) Unexpectedly, because the areal mass does not change on compression, the experiment show larger penetration for compressed matter. This was explained[121] by the heating effect which changed both the uncompressed and compressed matter energy release. Transport inhibition was also invoked because, for large currents, electron transport is magnetized and a low energy electron return current neutralize the hot electrons. As we have seen before, neutralization processes depend on the dense plasma bulk conductivity σ. This explains the longer penetration depth for compressed matter because the current heats the plasma and the electron density is larger so that the higher conductivity reduces the transport inhibiting electric fields. From this highly significant experiment, the important point to keep in mind is that hot electron transport play a significant role in high intensity short pulse laser matter interaction.

Energy transport by photons

Energy transport by photons in dense plasmas obey the radiative transfer equation. It is written in simple form here neglecting time dependent effects and using 1-D geometry:

$$\frac{dI_\nu(\mathbf{r})}{d\mathbf{r}} = j_\nu(\mathbf{r}) - k_\nu(\mathbf{r})I_\nu(\mathbf{r})$$

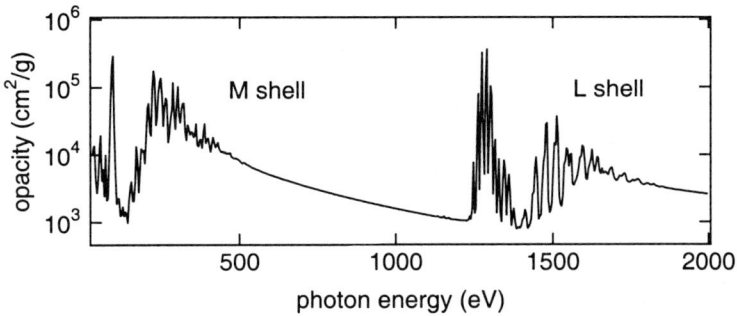

Figure 13. LTE opacity of Germanium at $T = 50 eV$, $\rho = 10^{-2} g/cc$.

Opacities k_ν and emissivities j_ν are related by the source function $S_\nu = j_\nu/k_\nu$ which is the Planck formula at LTE. The calculation of the ionization state Z^* is crucial in determining photon transport. Indeed, dense plasma effects, like pressure ionization, have to be taken into account to predict the charge distribution accurately. Opacities and emissivities are built from the contributions of bound-bound, bound-free, and free-free (bremsstrahlung) transitions. Figure 13 shows the calculation of the opacity of germanium under LTE conditions which can be found in the expanding plume of a nanosecond laser-created plasma.[122] One can clearly see contributions of line radiation from the M-shell ($n = 3 \to n = 4$) and the L-shell ($n = 2 \to n = 3$) bound-bound transitions, together with the background due to bound-free and free-free radiation. In the energy budget of ultrafast dense plasmas, conversion to thermal x-rays is a small factor ($\ll 1\%$).[65, 123, 124, 125, 126] However, because of the high energy involved this could be sufficient to preheat the target to a few eV,[127] or, at more extreme intensities, to a few tens or hundreds of eV.[128]

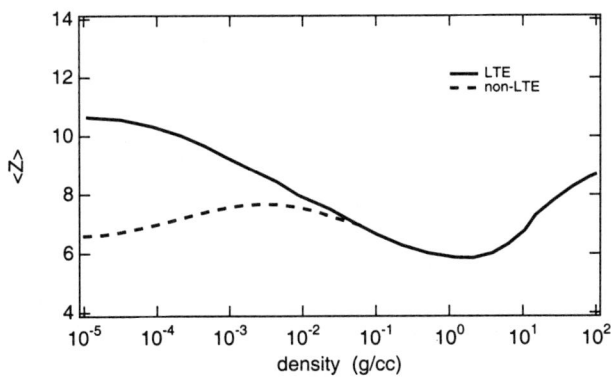

Figure 14. LTE and non-LTE ionization of aluminum at 50 eV as a function of density.

ATOMIC PHYSICS AND SPECTROSCOPY OF DENSE PLASMAS

Basics of multicharged ions physics in dense plasmas

The problem of the *ab-initio* calculation of atomic structure under dense plasma conditions is formidable.[129] Only very limited-size self-consistent field molecular dynamic approaches have been already attempted and, significantly, on "simple" atomic systems, e.g. helium. Accordingly, most atomic physics calculations in dense plasmas have to rely on the average atom model which results from generalizing the usual atomic self-consistent field theory to finite density and temperature. The spherical-cell model has a long history.[130] This model is particularly suitable for describing ionization and atomic properties of dense plasmas ($\Gamma \geq 1$). To describe it simply, let us assume a continuum of positive charges outside a ion sphere of radius $4\pi/3 R_0^3 = n_i^{-1}$ where n_i is the ion density. One imagines the plasma to be made up of these ion spheres closely packed together. Coulomb repulsion ensures that the nearest neighbor is to be located at $\approx 1.7 R_0$. The cell itself contains a *nonuniform* electron gas and one nucleus at center. The self-consistent field method solves four basic equations which are used to calculate the potential with s as a set of quantum numbers:

$$-\frac{\hbar^2}{2m}\nabla^2 \Psi_s - eV(r)\Psi_s = \epsilon_s \Psi_s$$

$$\nabla^2 V = 4\pi e n(r)$$

$$p_s = \frac{D_s}{1 + \exp(\epsilon_s - \mu)/kT}$$

$$n(r) = \sum_s p_s |\Psi_s(r)|^2$$

with the assumption that the potential and the field vanish at R_0. Figure 14 shows the variation of the average charge Z^* for aluminium as a function of temperature at a density of 1 g/cm^3, calculated with such a model. The increase of ionization due to pressure ionization is clearly visible at densities above solid density.

Several improvements have been brought to this model over the years to incorporate fine structure effects such as spin-orbit coupling, relativistic effects, and quasi-molecular states. The use of the Dirac equation for one-electron wavefunctions instead of the Schrödinger equation takes care of relativity. Regarding the accuracy of the potential, local-density exchange potential and correlation potential have been introduced. The external plasma has been described more accurately by a radial distribution function $g(r)$ obtained from a fluid-structure theory or Monte Carlo simulations.[131] One of the drawback of the self-consistent model is that it represents the effect of neighbor ions by a smooth distribution, instead of point charges. Because the microfield generated by point charges is not properly taken into account, Stark effect and the Inglis-Teller merging of lines into the continuum are not well described. In addition, this model does not include the possibility for the plasma to be degenerate.

To handle non-LTE situations, a screened hydrogenic average atom model has been settled. In order to describe correctly this regime, one must consider both radiative and collisional phenomena at once. The usual approach to solve efficiently – with little computer time – the collisional-radiative (CR) equations is within the framework of the hydrogenic "average atom" (AA) with screened charges.[132, 133, 134] In this model, the collisional-radiative equations of an average-ion are solved and the resulting screened charges are used to reconstruct the one-electron atomic potential. The total ion energy $E_{ion} = -\sum_{nl} P_{nl} Q_{nl}^2/(2n^2)$, the screened charges $Q_{nl} = Z - \sum_{n'l'} \sigma_{nl;n'l'} P_{n'l'}$ depend on a set of screening coefficients[135] and the populations P_{nl} of the shells. Average-ion wave functions, oscillator strengths and Slater integrals are calculated by quantum mechanics in the reconstructed potential. Dipolar matrix elements and Slater integrals have improved values with respect to hydrogenic formulas, especially for the orbitals close to the ion core. This improvement in Slater integrals results in more accurate line shapes which are very important for calculating opacities.[122, 136] Figure 14 shows the non-LTE results which, as they should, depart strongly from the superimposed LTE results at low densities.

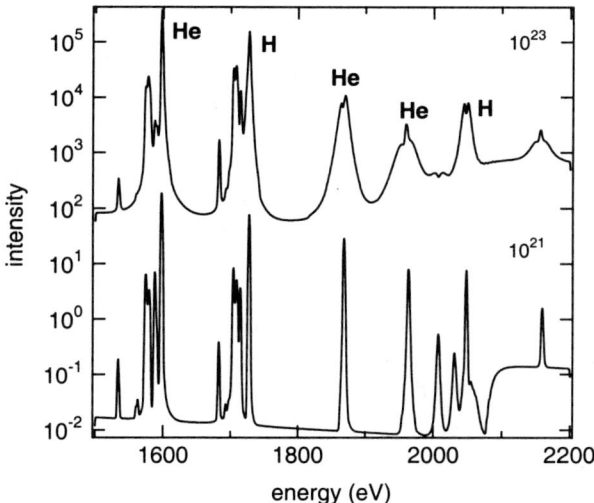

Figure 15. Stark broadening of Al He-like and H-like resonance lines at a temperature of 600 eV for two electron densities.

Basics of line broadening in dense plasmas

Line broadening is a diagnostic of choice to probe dense plasma effects through the Stark effect.[137] There are several other sources of line broadening in plasmas. From the definition of the dipolar moment of an ionic transition, one derives the Einstein coefficients:

$$B_{ji} = \frac{8\pi^2}{3c\hbar^2} \frac{1}{g_j} \sum_{M_i M_j} |d_{ij}|^2; \qquad A_{ji} = \frac{8\pi h \nu_{ij}^3}{c^3} B_{ji}.$$

Natural broadening manifests itself by a Lorentz profile of width $\Gamma_i + \Gamma_j$ with $\Gamma_j^{-1} \sim \sum_i A_{ji}$. At high temperatures or high velocities, Doppler broadening comes into play with a Gaussian profile of width $\Delta\omega/\omega \propto v/c$ where v is either a hydrodynamic velocity (motional) or a thermal speed v_{th}. For dense plasmas, collisional broadening has to be calculated from a statistical average over many perturbers. The single event can be thought of as a disruption of the ion emission by a (collisional) perturbation. In the impact approximation, which is valid near the center of the lines, the collision time is much smaller than the time between 2 collisions. The profile is Lorentzian, its width being related to the collision cross-section. In the wings of the spectral lines, the characteristic time is small ($\Delta\omega \times \Delta t \sim \hbar/2$) and during that short time, ion perturbers are almost immobile. The corresponding quasistatic approximation relates the broadening parameters to the strength of the microfield generated by the quasi-immobile (over the time of interest of emission) ions. Ion dynamics effects, in the intermediate "time of interest" region, show their effects much closer to line center. Stark broadening calculations at high densities involve molecular dynamics calculations of the microfield and full quantum mechanical calculations of the atomic structure and level mixing for a range of fields. In a dense plasma, on top of all the preceding effects, self-absorption play a very important role in the spectral line shape. Figure 15 shows the Stark broadening of resonance lines of aluminum calculated with the code FLY[138] of

R.W. Lee. Satellite lines are visible on the "red" side of the $n = 2 \to n = 1$ transitions of He-like and H-like ions. They are due to $n = 2 \to n = 1$ transitions in Li-like and He-like ions with one spectator electron in $n \geq 2$, respectively. One can see also that the line width due to the Stark effect increases with the electron density and the quantum number n.

Density effects manifest themselves also on bound-free transitions due to pressure ionization and continuum lowering. Pressure ionization occurs because the strong compression of the ion raises electron kinetic energies relative to potential energies. High pressure "liberates" electrons as it crushes the ions into smaller volumes. At lower densities, pressure destroys excited states (electrons belong to several cores). A similar manifestation on spectra occurs from field-induced overlapping of Rydberg levels, the so-called Inglis-Teller effect. Pressure ionization is often understood as a continuum lowering.[139] In models, these effects are approximately taken into account by changing smoothly the ionization potentials or the statistical weights for increasing densities. There has been many different formulas for the evaluation of the continuum lowering,[140] we summarize below the results:

$$\text{Sphere ion-cell (IC):} \quad \Delta E_{IC} = \frac{Ze^2}{r_i}, r_i = \left(\frac{3}{4\pi N_i}\right)^{1/3}$$

$$\text{Debye-Hückel (DH):} \quad \Delta E_{DH} = \frac{Ze^2}{r_D}, r_D = \left(\frac{kT}{4\pi e^2 (Z^* + 1) N_e}\right)^{1/2}$$

$$\text{Stewart-Pyatt (SP):} \quad \Delta E_{SP} = \frac{([(r_i/r_D)^3 + 1]^{2/3} - 1)kT}{2(Z^* + 1)}$$

$$\text{Inglis-Teller limit (IT):} \quad E_{n+1} - E_n = r_n \left(\frac{Ze^2}{r_i^2}\right)$$

DH is to be used for weakly coupled plasmas ($\Gamma \ll 1$), IC for "strongly coupled" plasmas ($\Gamma > 1$), and SP for intermediate cases ($\Gamma \approx 1$). Recent experiments[141, 142] have tried to disentangle the various expressions for the shift and the broadening of the continuum edges with the difficulty of obtaining independently measured values for the plasma parameters n_e and T_e, so that this remains an open problem.

Hydrocode and ionization dynamics codes

The various hydrocodes which have been developed for laser-induced fusion (the most famous among them being LASNEX[143]) have all been rapidly adapted to treat the case of ultrashort pulse interaction. Different flavors of hydrocodes (MULTI-fs,[144, 145] FILM-fs,[146] UBC,[99, 147] etc) are now available. The set of fluid equations to be solved is:

$$\frac{\partial \rho}{\partial t} + \rho^2 \frac{\partial u}{\partial m} = 0$$

$$\frac{\partial u}{\partial t} + \frac{\partial (p_e + p_i)}{\partial m} = 0$$

$$\frac{\partial \epsilon_e}{\partial t} + p_e \frac{\partial u}{\partial m} = -\frac{\partial S_e}{\partial m} - \sum_k \frac{\partial S_k}{\partial m} + \frac{\partial S_L}{\partial m} - \gamma(T_e - T_i)$$

$$\frac{\partial \epsilon_i}{\partial t} + p_i \frac{\partial u}{\partial m} = \gamma(T_e - T_i)$$

The system is closed by the EOS which relates the electron ϵ_e, p_e and ion ϵ_i, p_i energies and pressures to the density ρ and temperatures T_e and T_i. S_k is the radiative flow of photon group k with energies $h\nu_k \leq h\nu \leq h\nu_{k+1}$. S_e is the electron heat flow. S_L is the energy deposited by the laser. It is calculated by self-consistent solution of the Helmholtz equations using the Drude approximation to the dielectric constant. FILM-fs[146] incorporates the effects of the ponderomotive pressure (see above). None of these codes can treat the relativistic regime $I \gg 10^{17}$ W/cm².

Within the framework of the approximations made, such as the Drude model for electron conduction, there are a few adjustable parameters (within a restricted range governed by comparison to various experiments) to match the isochoric solid/warm plasma transition. These parameters are: the momentum transfer collision frequency which enters the dielectric constant formula (see Eq. 7), the electron-ion collision frequency for energy transfer entering the parameter γ in the hydrodynamic equations above, and the correct expressions of the electron and thermal conductivities. Figure 16

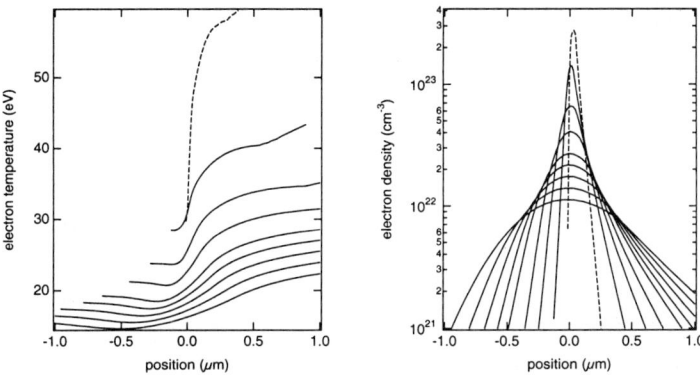

Figure 16. Electron temperature (left) and density (right) obtained with the FILM-fs code from 0.7ps (dashed curve) to 8.7ps after the peak of the laser pulse, for an aluminum foil of 80nm thickness irradiated at 10^{15} W/cm² with the 300fs, 0.53μm LULI laser pulse.

shows the results of hydrocode simulation, using FILM-fs, of the expansion of a thin foil. We should note at this point that most of the ultrashort pulse hydrodynamic codes are 1-D. This is justified by the ratio of the size of the focal spot to the longitudinal gradient scale length $L/\lambda \ll 1$. As an experimental proof of this finding, Doppler velocities (phase shifts) have been found in perfect agreement with the measured radial profile of laser intensities at the rear of a 400nm Al layer irradiated by a 10^{14} W/cm² laser pulse.[93]

Spectral line formation and collisional-radiative calculations

Under LTE conditions, the ionic populations are given by the Saha-Boltzmann equation:

$$\frac{N_{X+1}}{N_X} = \frac{G_{X+1}}{G_X} \exp(\xi - \frac{I_X}{kT})$$

with $\xi = -\mu_e/kT$ is of order 10 for $N_e = 10^{21} - 10^{23}$ cm⁻³ and $T_e = 300 - 2000 eV$. In fact, LTE is seldom found in short pulse laser plasmas because of time-dependent effects and steep spatial gradients. Under conditions of collisional equilibrium, the ratio of two adjacent ions is:

$$\frac{N_{X+1}}{N_X} = \frac{Q_X}{R_{X+1} + W_{X+1}}$$

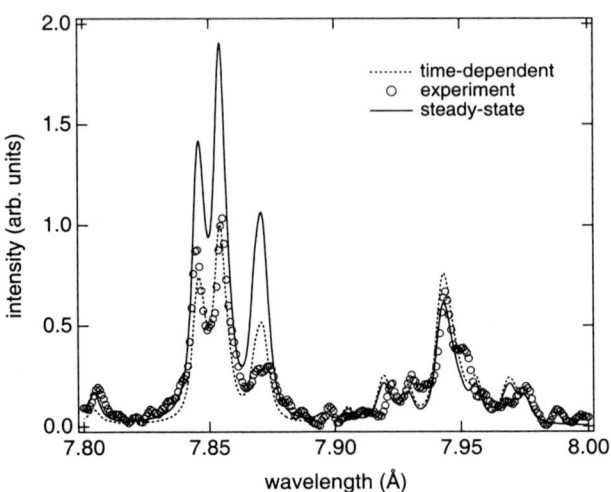

Figure 17. Comparison of time-dependent (broken curve), steady state (full curve) simulations and experiments (circles) for a laser intensity of $10^{16} W/cm^2$.

where Q_X, R_{X+1}, W_{X+1} are the ionization and recombination rates. Using hydrogenic values, one finds:

$$\xi = -\log(CX(kT)^{3/2}(I_X/kT)^{5/2}) \approx 3 - 4$$

Otherwise, one solves the full master equations:

$$\frac{d\{N\}}{dt} = \{A\}\{N\}$$

where $\{A\}$ is the matrix of collisional and radiative excitation and de-excitation rate constants. Figure 17 shows the simulation of a medium intensity experiment of the interaction of a 100fs laser pulse with an aluminum target. Experimental results of Li-like and Be-like satellite emissions are compared with full-blown hydrodynamic simulations post-processed by time-dependent (dashed line) and steady-state excited level populations deduced from a collisional model. As we can see, the stationary calculations largely overestimate the emission from the lithium-like fraction.[148]

X-RAY SOURCE APPLICATIONS OF DENSE ULTRAFAST PLASMAS

Ultra-short pulse x-rays by thermal and suprathermal electrons

As we have seen before, laser-plasmas have a number of characteristics that make them valuable as short pulse x-ray sources for applications such as time-resolved x-ray diffraction and absorption experiments: i) a wide range of pulse durations from a few hundred femtoseconds to hundreds of nanoseconds, ii) very bright x-ray sources can be generated with source sizes as small as a few microns, and iii) laser-plasma x-ray sources can be accurately synchronized with other events that can be driven, triggered or stimulated by the same laser light.

Thermal x-ray emission from laser-produced plasmas in the subpicosecond regime occurs with pulse durations of the order of a few tens of picoseconds,[149, 150, 151, 152] measured by a streak camera. Conversion efficiencies around 200eV (i.e. the ratio of x-ray energy to laser energy) are about 0.1% for a 20% bandwidth.[151, 153] Peak brilliance

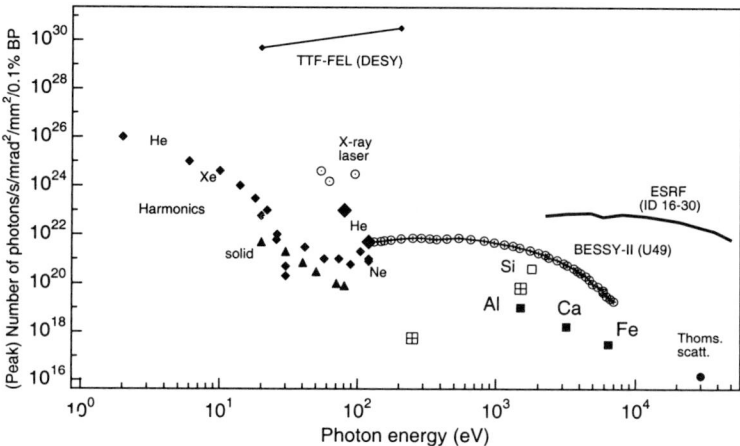

Figure 18. Peak spectral brilliance in $photons/(s.mrd.^2mm^20.1\%BW)$ as a function of photon energy. Diamonds: harmonic generation in gases; triangles: harmonic generation on solids; open circles: x-ray laser; solid lines: synchrotron radiation insertion devices; squares: laser plasma characteristics radiation; circle: Thomson scattering.

results for laser-produced plasma thermal x-ray sources are given in Fig. 18 for a laser pulse duration of 600ps (this is also roughly the duration of the x-ray pulse) near 200eV and for a streak-camera-measured x-ray pulse duration of 4ps at 1500eV (open crossed-squares).[154] Hydrodynamic simulations of laser-plasma interaction predict shorter x-ray pulses with thin foils instead of massive targets. This feature is still the subject of active experimental confirmation.

Suprathermal electrons produced by non-collisional absorption mechanisms have proved to be a convenient way of generating x-rays in the photon energy range above one keV[67, 119] and in the MeV region.[155] In the 1-10keV energy range, efficient production of K_α radiation in aluminum, calcium, and iron has been demonstrated.[67] The x-ray throughput was controlled by varying the energy contrast ratio between the main ultrashort pulse and its nanosecond pedestal. Results are shown in Fig. 18 (filled and open squares) for several target materials. Peak brilliance is comparable to BESSY2 synchrotron radiation but the streak-measured pulse duration is below 2ps, instead of ≈ 50ps. The measured source diameter is about $10\mu m$, the repetition rate is $10Hz$ and the total number of photons per shot is about 10^{10}.

High-order harmonic generation from dense plasmas and applications

Other sources of high harmonics[156, 157, 158, 159, 160] are produced from dense plasma surfaces. In solids, the very large pressure is responsible for the generation of harmonics. Pushing on the steep vacuum-solid interface at the laser frequency modulates the reflected beam at the same frequency. This "oscillating mirror" gives rise to a series of sidebands[161] on the reflected spectrum, separated by ω or 2ω where ω is the fundamental laser pulsation. Accordingly, all harmonic orders are emitted, contrary to the gas phase where only odd orders are symmetry permitted. In addition, contrary to gas harmonics, the emission is diffuse (into 2π steradians). Figure 18 summarizes the results obtained at the Rutherford Appleton Laboratory[162] (triangles). Recent experiments[163] have shown the strong dependence of the harmonic conversion efficiency on the temporal pulse shape and the resulting density scale length L/λ of the preformed plasma. There is some analogy here with the optimization of K_α radiation with the electron density gradient scale length.[66, 164] The highest conversion efficiencies are achieved for short density scale lengths $L/\lambda < 0.4$ which result from high-contrast ratio pulse interactions.

Very interesting applications of high harmonic generation from gases[165] and solid targets targets[166, 167] using the transmission of high order harmonics have already been implemented. This technique has the advantage of providing direct access to ultrahigh electron densities, temperature, and the dielectric constant of the plasma.[168]

Future applications of short pulse x-rays

Many fundamental processes in physics, materials science, chemistry, and biology occur on the ultrafast (picosecond or subpicosecond) time scale. Some of these processes can be initiated by transient optical excitation, and followed in their time evolution by ultrafast infrared, visible and ultraviolet spectroscopy. Pump-probe optical techniques are sensitive to electronic excitations, whereas extension to the sub-nanometer wavelength range should make possible the direct monitoring of atomic positions.

The most recent experimental results[21, 169, 170, 171, 172] show that the present technical capabilities permit great advances in the understanding of the response of solids to impulsive heating. We expect that in the next few years, phenomena such as ultrafast melting, ordinary laser-induced melting and propagation of strain waves will be characterized in detail. Beyond this stage awaits the challenge of detecting many different reflections in order to reconstruct the directions of atomic motions – in other words, the development of true femtosecond crystallography.

Acknowledgments

It is a pleasure to acknowledge the strong support of my colleagues Jean-Paul Geindre, Patrick Audebert, Claude Chenais-Popovics, François Amiranoff, Michel Koenig, Arnold Migus and Elizabeth Leboucher-Dalimier from Laboratoire d'Utilisation des Lasers Intenses, Antoine Rousse from Laboratoire d'Optique Appliquée, Klaus Eidmann from Max Planck Institut for Quantum Optics and many helpful discussions with students and other lecturers attending the Erice school.

REFERENCES

1. J. Lindl, Development of the indirect-drive approach to inertial confinement fusion and the target physics basis for ignition and gain, *Phys. Plasmas* 2:3933 (1995).
2. R.C. Elton, "X-Ray Lasers," (Academic Press Inc., San Diego, 1990).
3. See several contributions in "Laser Plasma Interaction," edited by R. Balian and J.-C. Adam (North-Holland, Amsterdam, 1980).
4. W.L. Kruer, "The Physics of Laser Plasma Interactions", (Addison-Wesley, Redwood-City, 1988).
5. E.M. Campbell, The physics of megajoule, large scale, and ultrafast short-scale laser plasmas, *Phys. Fluids* B 4:3781 (1992).
6. L.J. Radziemski and D.A. Cremers, "Laser Induced Plasmas and Applications," (Dekker, Basel, 1989).
7. I.C. Turcu and J.B. Dance, "X-Rays From Laser Plasmas," (John Wiley and Sons, Chichester, 1999).
8. See several contributions in "New Ultrafast X-Ray Sources and Applications", edited by J.-C. Gauthier, *C. R. Acad. Sci. Paris, Séries IV*, vol. 1, (Elsevier, Paris, 2000).
9. S.H. Glenzer, L.J. Suter, R.L. Berger, K.G. Estabrook, B.A. Hammel, R.L. Kauffman, R.K. Kirkwood, B.J. MacGowan, J.D. Moody, J.E. Rothenberg, and R.E. Turner, Hohlraum energetics with smoothed laser beams, *Phys. Plasmas* 7:2585 (2000).

10. E. Leboucher-Dalimier, A. Poquérusse, and P. Angelo, Space-resolved x-ray emission from the densest part of laser plasmas: molecular satellite features and asymmetrical wings, *Phys. Rev. E* 47:R1467 (1993).
11. P. Maine, D. Strickland, M. Pessot, J. Squier, P. Bado, G. Mourou, and D. Harter, "Chirped Pulse Amplification: Present and Future," in *Ultrafast Phenomena VI*, edited by T. Yajima, K. Yoshihara, C. B. Harris, and S. Shionoya (Springer-Verlag, New York, 1988), p. 205, and references therein.
12. G.A. Mourou, C.P. Barty, and M.D. Perry, Ultra-high intensity lasers: physics of the extreme on a tabletop, *Physics Today* January 98:22 (1998).
13. M.M. Murnane, H.C. Kaypten, and R.W. Falcone, High density plasmas produced by ultrafast laser pulses, *Phys. Rev. Lett.* 62:155 (1989).
14. J.-C. Gauthier, "Short Pulse Laser Interaction With Solid Targets," in *Laser Interaction with Matter*, edited by S. Rose, (Institute of Physics Publishing, Bristol, 1995).
15. P. Gibbon and E. Förster, Short-pulse laser plasma interactions, *Plasma Phys. Control. Fusion* 38:769 (1996).
16. X. Liu, D. Du, and G. Mourou, Laser ablation and micromachining with ultrashort laser pulses, *IEEE J. Quantum Electron.* 33:1706 (1997).
17. P.P. Pronko, P.A. VanRompay, Z. Zhang, and J.A. Nees, Isotope enrichment in laser-ablation plumes and commensurately deposited thin films, *Phys. Rev. Lett.* 83:2596 (1999).
18. A. Rousse, Les flashes X prennent les atomes de vitesse, *La Recherche* 315:46 (1998).
19. Several contributions in "Time-Resolved Diffraction," edited by J.R. Helliwell and P.M. Rentzepis, *Oxford Science Publications*, (Clarendon, Oxford, 1997).
20. F. Ráksi, K. Wilson, Z. Jian, A. Ikhlef, C.Y. Côté, and J.-C. Kieffer, Ultrafast x-ray absorption probing of a chemical reaction, *J. Chem. Phys.* 104:6066 (1996).
21. A. Rousse, C. Rischel, I. Uschmann, P.A. Albouy, J.-P. Geindre, P. Audebert, J.-C. Gauthier, E. Förster, J.L. Martin, A. Antonetti, Ultrafast x-ray time-resolved surface structural dynamics of optically excited organic materials, *Journal of Applied Crystallography* 32:977 (1999).
22. T. Guo, C. Rose-Petruck, R. Jimenez, F. Ràski, J.A. Squier, B.C. Walker, K. Wilsonand, and C.P.J. Barty, Picosecond-Milliangstroem Resolution Dynamics by Ultrafast X-Ray Diffraction, in proceedings of "Applications of X-Rays generated from Lasers and Other Bright Sources," edited by A. Kyrala and J. C. Gauthier (SPIE, San Diego, 1997), 3157:84-92.
23. H. R. Griem, "Plasma Spectroscopy," (Mc Graw Hill, New-York, 1964).
24. L. Spitzer, Jr. "Physics of Fully Ionized Gases", (John Wiley and Sons, New-York, 1963).
25. R. L. Liboff, "Kinetic Theory," (John Wiley, New-York, 1998).
26. M. Bonitz, Th. Bornath, D. Kremp, M. Schlanges, and W.D. Kraeft, Quantum kinetic theory for laser plasmas. Dynamic screening in strong fields, *Contrib. Plasma Phys.* 39:229 (1999), and references therein.
27. P. Gauthier, S.J. Rose, P. Sauvan, P. Angelo, E. Leboucher-Dalimier, A. Calisti and B. Talin, Modeling the radiative properties of dense plasmas, *Phys. Rev. E* 58:942 (1998).
28. F.J. Rogers, Ionization equiibrium and equation of state in strongly coupled plasmas, *Phys. Plasmas* 7:51 (2000).
29. R. M. More, "Plasma Processes in Non-Ideal Plasma", in *35th Scottish Universities Summer School in Physics: Laser-plasma interactions 3* (IOP Publishing, Bristol, 1988).
30. D.A. Baiko, A.D. Kaminker, A.Y. Potekhin, and D.G. Yakovlev, Ion structure factors and electron transports in dense coulomb plasmas, *Phys. Rev. Lett.* 81:5556 (1998).
31. R. M. More, "Semi-Classical Calculations of Atomic Processes," in *Laser Interaction with Atoms, Solids, and Plasmas*, edited by R. M. More, NATO ASI Series, vol. 327 (Plenum, New-York and London, 1994).
32. R.M. More, "Atoms in Dense Plasmas," in *Atoms in Unusual Situations*, edited by J.-P. Briand NATO ASI Series B: Physics Vol. 143 (Plenum, New-York, 1986), p. 155.
33. D. Salzman, "Atomic Physics in Hot Plasmas," (Oxford University Press, New-York, 1998).
34. P. Gauthier, Etude des structures dicentriques formées transitoirement dans les plasmas denses et chauds," PhD Thesis (University Paris VI, 1996). Avalable upon request to the author.
35. R. Sauerbrey, Acceleration in femtosecond laser-produced plasmas, *Phys. Plasmas* 3:4712 (1996).
36. A recent review on modern ultrafast optics can be found in *IEEE J. Select Topics in Quantum Electron.* 4:157-459 (1998).
37. A. Migus, "Femtosecond and Subpicosecond Ultra-Intense Lasers", in *45th Scottish Universities Summer School in Physics: Laser-plasma interactions 5: Inertial Confinement Fusion* (IOP Publishing, Bristol, 1994).

38. C. Le Blanc, G. Grillon, J.-P. Chambaret, A. Migus, and A. Antonetti, Compact and efficient multipass Ti:Sapphire system for femtosecond chirped-pulse amplification at the terawatt level, *Optics Lett.* 18:140 (1993).
39. S. Svanberg, J. Larsson, A. Persson, and C.G. Wahlstroem, Lund high-power laser facility, *Phys. Scr.* 49:187 (1994).
40. C. P. J. Barty, T. Guo, C. Le Blanc, F. Raski, C. Rose-Petruck, J. Squier, K.R. Wilson, V.V. Yakovlev and K. Yamakawa, Generation of 18fs multiterawatt pulses regenerative pulse shaping and chirped pulse amplification, *Optics Lett.* 21:668 (1996).
41. J.-P. Chambaret, C. Le Blanc, A. Antonetti, G. Chériaux, P.F. Curley, G. Darpentigny and F. Salin, Generation of 25 TW,32 fs pulses at 10 Hz, *Optics Lett.* 21:1921 (1996).
42. A. Antonetti, F. Blasco, J. P. Chambaret, G. Chériaux, G. Darpentigny, C. Le Blanc, P. Rousseau, S. Ranc, G. Rey, and F. Salin, A laser system producing 5×10^{19} W/cm^2 at 10 Hz. *Appl. Phys. B* 65:197 (1997).
43. C. Rouyer E. Mazataud, I. Allais, A. Pierre, S. Seznec, C. Sauteret, G. Mourou, and A. Migus, Generation of 50 TW femtosecond pulses in a Ti:sapphire/Nd:glass chain, *Optics Lett.* 18:214 (1993).
44. D. Descamps, "Développement d'une chaîne laser Ti:saphir - Nd:verres capable de délivrer une puissance de 200 TW et étude d'un amplificateur a verres dopés au néodyme pompé par laser," PhD thesis (University Paris VI, 1997). Available upon request to the author.
45. M.D. Perry and G. Mourou, Terawatt to Petawatt Subpicosecond Lasers, *Science* 264:917 (1994).
46. M.D. Perry, D. Pennington, B. Stuart, G. Tietbolh, J. Britten *et al.* , Petawatt Laser Pulses, *Optics Lett.* 24:160 (1999).
47. M.D. Rosen, "Scaling laws for femtosecond laser-plasma interactions", in *Femtosecond to Nanosecond High-Intensity Lasers and Applications*, SPIE 1229:160 (1990).
48. J.-C. Kieffer, P. Audebert, M. Chaker, J.-P. Matte, H. Pépin, T.W. Johnston, P. Maine, D. Meyerhoffer, J. Delettrez, D. Strickland, P. Bado, and G. Mourou, Short pulse laser absorption in very steep plasma density gradients, *Phys. Rev. Lett.* 62:760 (1989).
49. A.P. Kanavin, I.V. Smetanin, V.A. Isakov, Yu.V. Afanasiev, B.N. Chickov, B. Wellegehausen, S. Nolte, C. Momma, and A. Tünnermann, "Heat transport in metals irradiated by ultrashort pulses", *Phys. Rev. B* 57:14698 (1998).
50. E. Lefebvre, "Mécanismes d'absorption et d'émission dans l'interaction d'une impulsion laser ultra-intense avec une cible surcritique", PhD thesis, (University of Orsay, 1996). Available upon request to the author.
51. F. Cornolti, P. Mulser, and M. Hahn, Absorption of ultrashort laser pulses in solid targets, *Laser and Particle Beams* 9:465 (1991).
52. W. Rozmus and V.T. Tikhonchuk, Skin effect and interaction of short laser pulses with dense plasmas, *Phys. Rev. A* 42:7401 (1990).
53. T.-Y. Brian Yang, W.L. Kruer, R.M. More and A.B. Langdon, Absorption of laser light in overdense plasmas by sheath inverse bremsstrahlung, *Phys. Plasmas* 2:3146 (1995).
54. P.J. Catto and R.M. More, Sheath inverse bremsstrahlung in laser produced plasmas, *Phys. Fluids* 20:704 (1977).
55. F.J. Luciani, P. Mora, and A. Bendib, Magnetic field and non-local transport in laser-created plasmas, *Phys. Rev. Lett.* 55:2421 (1985), and references therein.
56. F. Brunel, Not-so-resonant resonant absorption, *Phys. Rev. Lett.* 59:152 (1987).
57. M.K. Grimes, Y.-S. Lee, A.R. Rundquist and M.C. Downer, Experimental identification of vacuum heating at femtosecond-laser-irradiated metal surfaces, *Phys. Rev. Lett.* 82, 4010-4013 (1999).
58. A.A. Andreev, K.Yu. Platonov, and J.-C. Gauthier, Skin effect in stronly inhomogeneous laser plasmas with weakly anisotropic temperature distribution, *Phys. Rev. E* 58:2424 (1998).
59. W.H. Press, W.T. Vetterling, S.A. Teukolsky, and B.P. Falnnery, "Numerical recipes," (Cambridge University Press, New-York, 1992).
60. M. Born and E. Wolf, "Principles of Optics," (Pergamon, Oxford, 1998).
61. H.M. Milchberg and R.R. Freeman, Light absorption in ultrashort scale length plasmas, *J.O.S.A. B* 6:1351 (1989).
62. D.F. Price, R.M. More, R.S. Walling, G. Guethlein, R.L. Shepherd, R.E. Stewart, and W.E. White, Absorption of ultrashort laser pulses by solid targets heated rapidly to temperatures 11000 eV, *Phys. Rev. Lett.* 75:252 (1995).
63. S.P. Gordon, T. Donnelly, A. Sullivan, H. Hamster and R.W. Falcone, X-rays from microstructured targets heated by femtosecond lasers, *Opt. Lett.* 19:484 (1994).

64. J.-C. Gauthier, S. Bastiani, P. Audebert, J.P. Geindre, K. Neuman, T. Donnelly, M. Hoffer, R.W. Falcone, R. Sheperd, D. Price and B. White, Femtosecond laser-produced plasma x-rays from periodically modulated surface targets, *Proc. SPIE* 2523:242 (1995).
65. G. Kulcsár, D. AlMawlawi, F.W. Budnik, P.R. Herman, M. Moskovits, L. Zhao, and R.S. Marjoribanks, Intense picosecond x-ray pulses from laser plasmas by use of nanostructured velvet targets, *Phys. Rev. Lett.* 84:5149 (2000).
66. S. Bastiani, A. Rousse, J.-P. Geindre, P. Audebert, C. Quoix, G. Hamoniaux, A. Antonetti, and J.-C. Gauthier, Experimental study of the interaction of subpicosecond laser pulses with solid targets of varying initial scale lengths, *Phys. Rev. E* 56:7179 (1997).
67. J.-C. Gauthier, S. Bastiani, P. Audebert, J.-P. Geindre, A. Rousse, C. Quoix, G. Grillon, A. Mysyrowicz, A. Antonetti, R. Mancini, et A. Shlyaptseva, Characterization of a femtosecond laser-produced plasma X-ray source by electronic, optical and X-ray diagnostic techniques, *Proc. SPIE* 3157:52 (1997).
68. R. Fedosejevs, R. Ottmann, R. Sigel, G. Kuhnle, S. Szatmari, and F.P. Schäfer, Absorption of femtosecond laser pulses in high-density plasma, *Phys. Rev. Lett.* 64:1250 (1990).
69. R. Fedosejevs, R. Ottmann, R. Sigel, G. Kuhnle, S. Szatmari, and F.P. Schäfer, Absorption of subpicosecond ultraviolet laser pulses in hih-density plasma, *Appl. Phys. B* 50:79 (1990).
70. P. Mulser, "Theory of Plasma Wave Absorption," in *High Power Laser-Matter Interaction* (Springer, Heidelberg, 1995).
71. H.M. Milchberg, R.R. Freeman, S.C. Davey, and R.M. More, Resistivity of a simple metal from room temperature to $10^6 K$, *Phys. Rev. Lett.* 61:2364 (1988).
72. W. Rozmus, V.T. Tikonchuk, and R. Cauble, A model for ultrashort laser pulse absorption in solid targets, *Phys. Plasmas* 3:360 (1996).
73. T. Tajima and J. Dawson, Laser electron accelerator, *Phys. Rev. Lett.* 43:267 (1979).
74. F. Amiranoff, D. Bernard, B. Cros, F. Jacquet, G. Matthieussent, J.R. Marquès, P. Miné, P. Mora, A. Modena, J. Morillo, F. Moulin, Z. Najmudin, A.E. Specka, and C. Stenz, A summary of the beatwave experiments at Ecole Polytechnique, *IEEE Trans. Plas. Sci.* 24:296 (1996)
75. F. Amiranoff, S. Baton, D. Bernard, B. Cros, D. Descamps, F. Dorchies, F. Jacquet, V. Malka, J. R. Marquès, G. Matthieussent, P. Miné, A. Modena, P. Mora, J. Morillo, and Z. Najmudin, Observation of laser wakefield acceleration of electrons, *Phys. Rev. Lett.* 81:995 (1998).
76. O. Peyrusse, M. Busquet, J.-C. Kieffer, Z. Jiang, and C.Y. Côté, Generation of hot solid-density plasmas by laser radiation pressure confinement, *Phys. Rev. Lett.* 75:3862 (1995).
77. X. Liu and D. Umstadter, Competition between ponderomotive and thermal pressures in short-scale-length laser-plasmas, *Phys. Rev. Lett.* 69:1935 (1992).
78. A. Pukhov and J. Meyer-ter-Vehn, Relativistic magnetic self-channeling of light in near-critical plasma: three-dimensional particle-in-cell simulation, *Phys. Rev. Lett.* 76:3975 (1996).
79. S.C. Wilks, Simulations of ultraintense laser-plasma interactions, *Phys. Fluids B* 5:2603 (1993).
80. J. C. Adam, A. Héron, S. Guérin, G. Laval, P. Mora, and B. Quesnel, Anomalous absorption of very high-intensity laser pulses propagating through moderately dense plasmas, *Phys. Rev. Lett.* 78:4765 (1997).
81. K. Nagashima, Y. Kishimoto, and H. Takuma, Propagation of a relativistic ultrashort laser pulse in a near-critical-density plasma layer, *Phys. Rev. E* 58:4937 (1998).
82. M. Tabak, J. Hammer, M. Glinsky, W. Kruer, S. Wilks, J. Woodworth, and E.M. Campbell, Ignition and high gain with ultra powerful lasers, *Phys. Plasmas* 1:1626 (1994).
83. One can find the most recent advances in pump/probe techniques in the Proceedings of "Ultrafast Phenomena XI", edited by T. Elsaesser, J.G. Fujimoto, D.A. Wiersma, and W. Zinth, vol. 63, *Springer Series in Chemical Physics* (Springer Verlag, Berlin, 1998) and in volumes 38, 46, 48, 53, 55, and 60.
84. M.C. Downer and C.V. Shank, Ultrafast heating of silicon-on-sapphire by femtosecond optical pulses, *Phys. Rev. Lett.* 56:761 (1986).
85. X.Y. Wang and M.C. Downer, Femtosecond time-resolved reflectivity of hydrodynamically expanding metal surfaces, *Optics Lett.* 17:1450 (1992).
86. E. Mazur, "Interaction of Ultrashort Laser Pulses With Solids", in *Spectroscopy and Dynamics of Collective Excitations in Solids*, edited by B. Di Bartolo, NATO ASI Series (Plenum Press, New-York, 1996).
87. W.M. Wood, C.W. Siders, and M.C. Downer, Femtosecond growth dynamics of an underdense ionization front measured by spectral blueshifting, *IEEE Transactions on Plasma Science* 21:20 (1993).
88. D. von der Linde, Laser-Plasma Interaction in the Femtosecond Time Regime, in *Photon and Electron Collisions with Atoms and Molecules*, edited by P.G. Burke and C.J. Joachain, 279-295 (Plenum Press, New York, 1997).

89. P. Martin, S. Guizard, Ph. Daguzan, G. Petite, P. D' Oliveira, P. Meynadier, and M. Perdrix, Subpicosecond study of carrier trapping dynamics in wide-band-gap crystals, *Phys. Rev.* B 55:5799 (1997).
90. L. Lepetit, G. Chériaux, and M. Joffre, Linear techniques of phase-measurement by femtosecond spectral interferometry for applications in spectroscopy, *J.O.S.A.* B 12:2467 (1995).
91. P. Audebert, P. Daguzan, A. Dos Santos, J.-C. Gauthier, J.-P. Geindre, S. Guizard, G. Hamoniaux, K. Krastev, P. Martin, G. Petite, and A. Antonetti, Space-time observation of an electron gas in SiO_2, *Phys. Rev. Lett.* 73:1990 (1994).
92. P. Blanc, P. Audebert, F. Falliès, J.-P. Geindre, J.-C. Gauthier, A. Dos Santos, A. Mysyrowicz, and A. Antonetti, Phase dynamics of reflected probe pulses from sub-100fs laser-produced plasmas, *J.O.S.A.* B 13:118 (1996).
93. R. Evans, A.D. Badger, F. Falliès, M. Mahdieh, T. Hall, P. Audebert, J.-P. Geindre, J.-C. Gauthier, A. Mysyrowicz, G. Grillon, and A. Antonetti, Time- and space-resolved optical probing of femtosecond-laser-driven shock waves in aluminum, *Phys. Rev. Lett.* 77:3359 (1996).
94. J.R. Marquès, J.-P. Geindre, F. Amiranoff, P. Audebert, J.-C. Gauthier, A. Antonetti, and G. Grillon, Temporal and spatial measurement of the electron density perturbation produced in the wake of an ultrashort pulse, *Phys. Rev. Lett.* 76:3566 (1996).
95. C. Quoix, G. Grillon, A. Antonetti, J.-P. Geindre, P. Audebert, and J.-C. Gauthier, Time-resolved studies of short pulse laser-produced plasmas in silicon dioxide near breakdown threshold, *EPJ: Applied Physics* 5:163 (1999).
96. D. von der Linde, "Second Harmonic Production From Solid Targets," in *Laser Interaction with Atoms, Solids, and Plasmas*, edited by R. M. More, NATO ASI Series, vol. 327 (Plenum, New-York and London, 1994).
97. B.-T.V. Vu, O.L. Landen, and A. Szoke, Time-resolved probing of femtosecond-laser-produced plasmas in transparent solids by electron thermal transport, *Physics of Plasmas* 2:476 (1995).
98. [4] E. Tokunaga, A. Terasaki and T. Kobayashi, Frequency-domain interferometer for femtosecond time-resolved phase spectroscopy, *Opt. Lett.* 17:1131 (1992).
99. A. Ng, A. Forsman, and G. Chiu, Electron thermal conduction waves in a two-temperature dense plasma, *Phys. Rev. Lett.* 81:2914 (1998).
100. C. Quoix, G. Hamoniaux, A. Antonetti, J.-C. Gauthier, J.-P. Geindre, and P. Audebert, Ultrafast plasma studies by phase and amplitude measurements with femtosecond spectral interferometry, *JQSRT* 65:455 (2000).
101. Y.T. Lee and R.M. More, An electron conductivity model, *Phys. Fluids* 27:1273 (1984).
102. H.M. Milchberg, I. Lyubomirsky, and C.G. Durfee. III, Factors controlling the x-ray pulse emission from an intense femtosecond laser heated solid, *Phys. Rev. Lett.* 67:2654 (1991).
103. A.R. Bell, J.R. Davies, S. Guérin, and H. Ruhl, Fast-electron transport in high-intensity short-pulse laser - solid experiments, *Plasma Phys. and Cont. Fusion* 39:653 (1997).
104. F. Beg, A.R. Bell, A.E. Dangor, C.N. Danson, A.P. Fews, M.E. Glinsky, B.A. Hammel, P. Lee, P.A. Norreys, and M. Tatarakis, A study of picosecond lasersolid interactions up to 10^{19} Wcm^2, *Phys. Plasmas* 4:447 (1997).
105. M.H. Key, M.D. Cable, T.E. Cowan, K.G. Estabrook, B.A. Hammel, S.P. Hatchett, E.A. Henry, D.E. Hinkel *et al.* , Hot electron production and heating by hot electrons in fast ignitor research, *Phys. Plasmas* 5:1966 (1998).
106. S. Humphries, "Charged Particle Beams" (Wiley Interscience, New York, 1990).
107. L. Gremillet, F. Amiranoff, S. D. Baton, J.-C. Gauthier, M. Koenig, E. Martinolli, F. Pisani, G. Bonnaud, C. Lebourg, C. Rousseaux, C. Toupin, A. Antonicci, D. Batani, A. Bernardinello, T. Hall, D. Scott, P. Norreys, H. Bandulet and H. Pépin, Time-resolved observation of ultra-high intensity laser-produced electron jets propagating through transparent solid targets, *Phys. Rev. Lett.* 83:5015 (1999).
108. M. Borghesi, A.H. Mackinnon, A.R. Bell, G. Malka, C. Vickers, O. Willi, J.R. Davies, A. Pukhov, and J. Meyer-ter-Vehn, Observation of collimated ionization channels in aluminum-coated glass targets irradiated by ultraintense laser pulses, *Phys. Rev. Lett.* 83:4309 (1999).
109. J. Oxenius, "Kinetic Theory of Particles and Photons" (Springer-Verlag, Heidelberg, 1986).
110. H.F. Beyer, H.-J. Kluge, and V.P. Shevelko, "X-Ray Radiation of Highly Charge Ions" (Springer-Verlag, Berlin, 1997).
111. I.P. Schkarowsky, T.W. Johnston, and Bachinsky, "The Particle Kinetics of Plasmas," (Addison-Wesley, Reading, 1966).
112. M. Basko, Th. Löwer, V.N. Kondrashov, A. Kendl, R. Sigel, and J. Meyer-ter-Vehn, Optical probing of laser-induced indirectly driven shock waves in aluminum, *Phys. Rev.* E 56:1019 (1997).

113. J.M. Ziman, "Principles of Theory of Solids," (Cambridge Univ. Press, Cambridge, 1969).
114. W. Ebeling, A. Foerster, V.E. Fortov, V.K. Gryaznov, A.Ya. Polishchuk, "Thermophysical Properties of Hot Plasmas," (B.G. Teubner Verlag, Stuttgart, 1991).
115. S.D. Brorson, J.G. Fujimoto, and E.P. Ippen, Femtosecond electronic heat-transport dynamics in thin gold films, *Phys. Rev. Lett.* 59:1962 (1987).
116. W.L. Kruer and K. Estabrook, $J \times B$ heating by very intense laser light, *Phys. Fluids* 28:430 (1985).
117. J.D. Jackson, "Classical Electrodynamics", (John Wiley and Sons, New-York, 1975).
118. J.A. Halbleib and T.A. Mehlhorn, "ITS: The Integrated TIGER Series of Coupled Electron/Photon Monte Carlo Transport Codes", Sandia Natinal Laboratories, Albuquerque, New Mexico, Sandia Report SAND84-0573, 1984.
119. A. Rousse, P. Audebert, J. P. Geindre, F. Falliès, J.-C. Gauthier, A. Mysyrowicz, G. Grillon, and A. Antonetti, Efficient $K\alpha$ x-ray source from femtosecond laser-produced plasmas, *Phys. Rev. E* 50:2200 (1994).
120. T.A. Hall, S. Ellwi, D. Batani, A. Bernardinello, V. Masella, M. Koenig, A. Benuzzi, J. Krishnan, F. Pisani, A. Djaoui, P. Norreys, D. Neely, S. Rose, M.H. Key and P. Fews, Fast electron deposition in laser shock compressed plastic targets, *Phys. Rev. Lett.* 81:1003 (1998).
121. D. Batani, J.R. Davies, A. Bernardinello, F. Pisani, M. Koenig, T.A. Hall, S. Ellwi, P. Norreys, S. Rose, A. Djaoui, and D. Neely, Explanations for the observed increase in fast electron penetration in laser shock compressed materials, *Phys. Rev. E* 61:5725 (2000).
122. A. Mirone, J.-C. Gauthier, F. Gilleron, and C. Chenais-Popovics, Non-LTE opacity calculations with nl splitting for radiative hydrodynamic codes, *JQSRT* 58:791 (1997).
123. Z. Jiang, J.-C. Kieffer, J.P. Matte, M. Chaker, O. Peyrusse, D. Gilles, G. Korn, A. Maksimchuk, S. Coe, and G. Mourou, X-ray spectroscopy of hot solid density plasmas produced by subpicosecond high contrast laser pulses at 10^{18}–10^{19} W/cm^2, *Phys. Plasmas* 2:1702 (1995).
124. See several contribution in "Applications of X-rays Generated from Lasers and Other Bright Sources", edited by J.-C. Gauthier and G. Kirala, Proc. SPIE 3157 (SPIE, Bellingham, 1997).
125. C.Y. Côté, J.-C. Kieffer, Z. Jiang, A. Ikhlef, and H. Pépin, KeV x-ray emission produced by a sub-picosecond laser interacting with a controlled preformed plasma, *J. Phys. B: At. Mol. Opt. Phys.* 31:L883 (1998).
126. J. Yu, Z. Jiang, J.-C. Kieffer, and A. Krol, Hard x-ray emission in high intensity femtosecond laser-target interaction, *Phys. Plasmas* 6:1318 (1999).
127. E.T. Gumbrell, R.A. Smith, T. Ditmire, A. Djaoui, S.J. Rose, and M.H.R. Hutchinson, Picosecond optical probing of ultrafast energy transport in short pulse laser solid target interaction experiments, *Phys. Plasmas* 5:3714 (1998).
128. J.A. Koch, S.P. Hatchett, M.H. Key, R.W. Lee, D. Pennington, R.B. Stephens, and M. Tabak, Measurements of deep heating generated by ultra-intense laser-plasma interactions, in Inertial Fusion Sciences and Applications, edited by C. Labaune, B. Hogan, and K. Tanaka (Elsevier, Paris, 2000), p. 463-466.
129. G. Massacrier, Self-consistent shemes for the calculation of ionice structures and populations in dense plasmas, *JQSRT* 51:221 (1994).
130. R.M. More, "Theory of Atoms in Dense Plasmas", Lawrence Livermore National Laboratory, Livermore, LLNL Report UCRL-100885 (LLNL, unpublished, 1989).
131. F. Perrot, Model for atomic species in a dense plasma: Description and applications, *Phys. Rev. A* 35:1235 (1987).
132. W.A. Lokke and W.H. Grassberger, "A non-LTE Emission and Absorption Coefficient Subroutine", Report UCRL-5227, (LLNL, unpublished, 1977).
133. R.M. More, "Physics of Dense Plasmas", Report UCRL-84991 (LLNL, unpublished, 1981).
134. Shi-Chang Li, Guo-Xing Han, and Ze-Qing Wu, Opacity calculations for non-LTE systems, *JQSRT* 54:257 (1995).
135. G. Faussurier, "Traitement Statistique des Propriétés Spectrales des Plasmas à l'Equilibre Thermodynamique Local dans le Cadre du Modèle Hydrogénique Écranté", PhD Thesis (Ecole polytechnique, 1996). Available upon request to the author.
136. A. Mirone, F. Gilleron, and J.-C. Gauthier, Statistical mechanics of average ions in non local thermodynamic equilibrium plasmas, *JQSRT* 60:551 (1998).
137. H. Griem, "Spectral Line Broadening by Plasmas", (Academic Press, New-York, 1974).
138. FLY is the time-dependent version of RATION described in R.W. Lee, B.L. Whitten and R.E. Strout, *JQSRT* 54:81 (1995).
139. G. Chiu and A. Ng, Pressure ionization in dense plasmas, *Phys. Rev. E* 59:1024 (1999).

140. J. Davies and M. Blaha, "Problems in Line Broadening and Ionization Lowering", in *Atomic Processes in Plasmas*, edited by Y-K Kim and R.C. Elton, AIP Conference Proceedings 206 (American Institute of Physics, New-York, 1989).
141. O. Renner, P. Sondhauss, D. Salzmann, A. Djaoui, M. Koenig, and E. Förster, Measurement of the polarization shift in hot and dense aluminum plasma, *JQSRT* 58:851 (1997).
142. M. Nantel, G. Ma, S. Gu, C. Y. Côté, J. Itatani, and D. Umstadter, Pressure ionization and line merging in strongly coupled plasmas produced by 100-fs laser pulses, *Phys. Rev. Lett.* 80:4442 (1998).
143. G.B. Zimmerman and W.L. Kruer, Numerical simulation of laser initiated fusion, *Comm. in Plasma Physics* 11:51 (1975).
144. R. Ramis, R. Schmalz, and J. Meyer-ter-Vehn, MULTI a radiative-hydrodynamic computer code for laser-plasma interaction, *Comput. Phys. Commun.* 49:475 (1988).
145. K. Eidmann, J. Meyer-ter-Vehn, T. Schlegel, and S. Hüller, Hydrodynamic simulation of subpicosecond laser interaction with solid-density matter, *Phys. Rev. E* 62:1202 (2000).
146. U. Teubner, P. Gibbon, E. Förster, F. Falliès, P. Audebert, J.-P. Geindre, and J.-C. Gauthier, *Phys. Plasmas* 3:2679 (1996).
147. A. Forsman, A. Ng, G. Chiu, and R.M. More, Interaction of femtosecond laser pulses with ultrathin foils, *Phys. Rev. E* 58:R1248 (1998).
148. P. Audebert, J.-P. Geindre, A. Rousse, F. Fallis, J.-C. Gauthier, A Mysyrowicz, G. Grillon, et A. Antonetti, K-shell emission dynamics of Be-like to He-like ions from a 100 fs laser-produced aluminium plasma, *J. Phys. B: Atom. Optical Mol. Phys.* 27:3303 (1994).
149. J.-C. Kieffer, Z. Jiang, A. Ikhlef, and C.Y. Côté, Picosecond dynamics of hot solid-density plasma, *J.O.S.A.* B 13:132 (1996).
150. J. Workman A. Maksimchuk, X. Liu, U. Ellenberger, J.S. Coe, C.Y. Chien, and D. Umstadter, Control of bright picosecond X-ray emission from intense subpicosecond laser-plasma interactions, *Phys. Rev. Lett.* 75:2324 (1995).
151. J.F. Pelletier, M. Chaker, and J.C. Kieffer, Picosecond soft x-ray pulses from a high intensity laser-plasma source, *Optics Lett.* 21:1040 (1996).
152. J. Workman, A. Maksimchuk, X. Liu, U. Ellenberger, J.S. Coe, X.-Y. Chien, and D. Umstadter, Picosecond soft x-ray source from subpicosecond laser-produced plasmas, *J.O.S.A.* B 13:125 (1996).
153. D. Alterbernd, U. Teubner, P. Gibbon, E. Förster, P. Audebert, J.P. Geindre, J.C. Gauthier, G. Grillon, and A. Antonetti, *J. Phys. B: Atom. Mol. Opt. Phys.* 30:3969 (1997).
154. C.Y. Côté, "Etude de la dynamique d'un plasma chauffé par une impulsion laser subpicoseconde par la spectroscopie keV avec résolution temporelle," PhD Thesis, University of Montreal (1996).
155. J.D. Kmetec, Ultrafast laser generation of hard x-rays, *Journal of Quantum Electronics* 28:2382 (1992).
156. W. Theobald, R. Häßner, and R. Sauerbrey, Temporally Resolved Measurement of Electron Densities ($> 10^{23} cm^{-3}$) with High Harmonics, *Phys. Rev. Lett.* 77:298 (1996).
157. D. von der Linde and K. Rzàzewski, High-order optical harmonic generation from solid surfaces, *Appl. Phys. B* 63:499 (1996).
158. P. Gibbon, Harmonic generation by femtosecond laser-solid interaction: a coherent water-window light source?, *Phys. Rev. Lett.* 76:50 (1996).
159. D. von der Linde, T. Engers, G. Jenke, P. Agostini, G. Grillon, E. Nibbering, A. Mysyrowicz, and A. Antonetti, Generation of high-order harmonics from solid surfaces by intense femtosecond laser pulses, *Phys. Rev. A* 52, R25 (1995).
160. P. Norreys, M. Zepf, S. Moustaizis, A.P. Fews, J. Zhang, P. Lee, M. Bakarezos, C.N. Danson, A. Dyson, P. Gibbon, P. Loukakos, D. Neely, F.N. Walsh, J.S. Wark, and A.E. Dangor, Efficient extreme UV harmonics generated from picosecond laser pulse interactions with solid targets, *Phys. Rev. Lett.* 76:1832 (1996).
161. R. Lichters, J. Meyer-ter-Vehn and A. Pukhov, Short-pulse laser harmonics from oscillating plasma surfaces driven at relativistic intensities, *Phys. Plasmas* 3:3425 (1996).
162. M.H. Key, T.W. Barbee, et al. , in *X-ray Lasers 1996*, Eds. S. Svanberg and C.G. Wahlström, (Institute of Physics Publishing, Oxford, 1996), p. 9.
163. M. Zepf, G.D. Tsakiris, G. Pretzler, I. Watts, D.M. Chambers, P.A. Norreys, U. Andiel, A.E. Dangor, K. Eidmann, C. Gahn, A. Machacek, J.S. Wark, and K. Witte, Role of the plasma scale length in the harmonic generation from solid targets, *Phys. Rev. E* 58:R5253 (1998).
164. Ch. Reich, P. Gibbon, I. Uschmann, and E. Förster, Yield optimization and time-structure of femtosecond laser plasma K_α sources, *Phys. Rev. Lett.* 84:4846 (2000).

165. P. Salières, L. Le Déroff, T. Auguste, P. Monot, P. d'Oliveira, D. Campo, J.-F. Hergott, H. Merdji, and B. Carré, Frequency-domain interferometry in the xuv with high-order harmonics, *Phys. Rev. Lett.* 83:5483 (1999).
166. R. Hässner, W. Theobald, S. Niedermeir, H. Schillinger, and R. Sauerbrey, High-order harmonics from solid targets as a probe to high-density plasmas, *Optics Lett.* 22:1491 (1997).
167. P. Gibbon, D. Altenbernd, U. Teubner, E. Förster, P. Audebert, J.-P. Geindre, J.-C. Gauthier, and A. Mysyrowicz, Plasma density determination by transmission of laser-generated surface harmonics, *Phys. Rev.* E 55:R6352 (1997).
168. W. Theobald, R. Hässner, R. Kingham, R. Sauerbrey, R. Fehr, D.O. Gericke, M. Schlanges, W.D. Kraeft, and K. Ishikawa, Electron densities, temperatures, and the dielectric function of femtosecond laser-produced plasmas, *Phys. Rev.* E 59:3544 (1999).
169. P. Chen, V. Tomov and P.M. Rentzepis, Time resolved heat propagation in a gold crystal by means of picosecond x-ray diffraction, *J. Chem. Phys.* 104:10001 (1996).
170. A.H. Chin, R.W. Schoenlein, T.E. Glover, P. Balling, W.P. Leemans and C.V. Shank, Ultrafast structural dynamics in InSb probed by time-resolved x-ray diffraction, *Phys. Rev. Lett.* 83:336 (1999).
171. J. Larsson, P.A. Heimann, A.M. Lindenberg, P.J. Schuck, P.H. Bucksbaum, R.W. Lee, H.A. Padmore, J.S. Wark and R.W. Falcone, Ultrafast structural changes measured by time-resolved X-ray diffraction, *Applied Physics* A 66:587 (1998).
172. C. Rose-Petruck, R. Jimenez, T. Guo, A. Cavalleri, C.W. Siders, F. Ràski, J.A. Squier, B.C. Walker, K. Wilson and C.P.J. Barty, Picosecond-milliangstrom lattice dynamics measured by ultrafast X-ray diffraction, *Nature* 398:310 (1999).

MAGNETIC FIELDS AND SOLITONS IN RELATIVISTIC PLASMAS

F. Pegoraro,[1] S. Bulanov,[2] F. Califano,[3] T. Esirkepov,[4]
M. Lontano,[5] N. Naumova,[2] V. Vshivkov[6]

[1] Physics Department, Pisa University and INFM, Pisa, Italy
[2] General Physics Institute, RAS, Moscow, Russia
[3] INFM, Pisa, Italy
[4] Moscow Institute for Physics and Technology, Dolgoprudny, Russia
[5] Institute for Plasma Physics - CNR, Milan, Italy
[6] Institute of Computation Technology, SD RAS, Novosibirsk, Russia

INTRODUCTION

In this paper we show that relativistic nonlinearities in a plasma interacting with ultra short high intensity laser pulses, lead to the formation of long-lived, slow-propagating coherent structures such as magnetic vortices and solitons. These structures are part of the complex nonlinear interaction between the laser pulse and the underdense plasma and represent the basic ingredients of the long time electron behaviour in the wake of the laser pulse.

A relativistic plasma interacting with an ultrashort, ultraintense laser pulse exhibits new phenomena where the nonlinearity of the relativistic particle kinematics and the nonlinearity of the magnetic part of the Lorentz force become dominant. In this sense relativistic plasmas can provide an arena in which to test and extend our understanding of nonlinear phenomena in continuous systems and to explore regimes where, on the time scale of interest, dissipative effects are in general negligible with respect to nonlinear effects.

In the investigation of relativistic plasmas a new basic tool, besides experiments and analytical modelling, has taken a major role: multi-dimensional, fully relativistic Particle in Cell (PIC) numerical codes have made it possible[1] to reproduce, in many important plasma regimes, the kinetic plasma behaviour that shapes the plasma and the laser pulse dynamics. PIC simulations indeed are not only used for validating analytical models or for reproducing experimental results, but can also play the vital role of an investigative tool for discovering new phenomena and new interaction regimes.

An important result to be stressed from these analyses is that, notwithstanding the richness and variety of the nonlinear relativistic electrodynamics of the laser plasma interaction, basic building blocks can be identified that are common throughout the realm of nonlinear continuous system dynamics and that represent elementary nonlinear excitations of these systems.

In this paper some recent results will be reviewed on two such basic components of the behaviour of nonlinear systems, vortices and solitons, and it will be described how they enter the interaction between an ultrashort ultraintense laser pulse and a relativistic plasma. Vortices appear in the more general context of the generation of a quasi-static (i.e., low frequency) magnetic field in the plasma which, together with its inverse process, magnetic field annihilation due to field line reconnection, represents one of the most important problems for both laboratory and astrophysical plasmas. Superstrong quasi-static magnetic fields in a laser plasma has been studied extensively over many years: they are observed in laser produced plasmas and can affect the plasma dynamics.

Solitons appear in the form of stable structures where low frequency electromagnetic radiation is trapped and, together with magnetic vortices and high energy particles, represent an important channel of conversion of the electromagnetic pulse energy into plasma energy. In a homogeneous plasma their propagation velocity is very small but, in an inhomogeneous plasma they are accelerated against the density gradient and their electromagnetic energy can thus be extracted and detected experimentally.

This paper is organized in two parts. The first part deals with magnetic field generation, dynamics and with magnetic vortices. The second part deals with solitons. Although most of the results described are obtained from 2-D and 3-D PIC simulations, no figures representing these numerical results will be shown here as these figures are available in the papers that are quoted in the text. In this context we mention in particular Ref.2, where the animations of these numerical results are presented, and all the articles published in the same electronic issue of J. Plasma Fusion Res on a CD.

PLASMA DYNAMICS CONSTRAINTS ON THE MAGNETIC FIELD GENERATION

A direct link between the particle and the magnetic field dynamics in a plasma, which can be applied to a variety of different plasma regimes, is obtained by combining Faraday's law

$$\nabla \times \mathbf{E} = -\frac{1}{c}\frac{\partial \mathbf{B}}{\partial t} \tag{1}$$

and the mean electron momentum equation

$$m_e n \left[\frac{\partial \mathbf{u}_e}{\partial t} + (\mathbf{u}_e \cdot \nabla)\mathbf{u}_e\right] = -\nabla \cdot \mathbf{\Pi}_e - ne\left[\mathbf{E} + \frac{\mathbf{u}_e}{c} \times \mathbf{B}\right] + \mathcal{C}, \tag{2}$$

where $\mathbf{\Pi}_e$ is the effective electron "pressure" tensor and $\mathcal{C}$ stands generically for collisional effects (such as electron viscosity and, most important, electron resistivity). For the sake of simplicity in this section we use nonrelativistic equations.

As long as the form and dependencies of the effective "pressure" tensor $\mathbf{\Pi}_e$ are not specified, Eq.(2) is general (aside for the dissipative term $\mathcal{C}$, Eq.(2) corresponds to the first velocity moment of Vlasov's equation for the electron distribution function $f_e(\mathbf{x},\mathbf{v},t)$) and is not based on any fluid model. Kinetic effects enter the expression of the pressure tensor which is defined in terms of the electron distribution function by

$$\Pi_{e,jk} \equiv \int d\mathbf{v}\, f_e(v_j - u_{e,j})(v_k - u_{e,k}), \quad \text{where} \quad \mathbf{u}_{e,j} \equiv (\int d\mathbf{v}\, f_e v_j)/(\int d\mathbf{v}\, f_e). \tag{3}$$

In the absence of a fluid closure, the expression of $\mathbf{\Pi}_e$ in Eq.(2) must be determined independently, e.g., from the electron distribution function $f_e(\mathbf{x},\mathbf{v},t)$ as obtained from the solution of the Vlasov equation.

In the case of low frequency, large scale phenomena in a magnetized plasma described by the Magnetohydrodynamic equations, we can identify the electron mean velocity $\mathbf{u}_e$ with the plasma fluid velocity $\mathbf{u}$ and we can assume that the pressure is isotropic, $\mathbf{\Pi}_e \to p_e \mathbf{I}$, and that it obeys a polytropic closure of the form $p_e = p_e(n)$. For these low frequency phenomena, the effect of electron inertia on the l.h.s. of Eq.(2) and of electron viscosity in the collisional term $\mathcal{C}$ can be neglected in most cases. Then, from Eqs.(1,2) we obtain

$$\nabla \times \left[\mathbf{E} + \frac{\mathbf{u}}{c} \times \mathbf{B}\right] = \nabla \times \left(\frac{\eta c}{4\pi} \nabla \times \mathbf{B}\right), \qquad (4)$$

where η is the electric resistivity of the plasma. In the ideal limit $\eta \to 0$, Eq.(4) reduces to the well known magnetic flux conservation theorem (see, e.g., Ref.3, chapter 3)

$$\frac{d\Phi}{dt} = 0, \qquad (5)$$

where Φ is the magnetic flux through a surface moving together with the plasma, i.e. with the plasma fluid velocity $\mathbf{u}$. The flux conservation expressed by Eq.(5) is generally referred to as the "freezing" of the magnetic field in the plasma.

In the case of fast phenomena that occur on times scales much shorter then the ion dynamical time, we can assume that the ions remain at rest. Again, if we assume an isotropic electron pressure with a polytropic closure and neglect collisional effects, from Eqs.(1,2) we obtain

$$\nabla \times \left[\mathbf{E}_e + \frac{\mathbf{u}_e}{c} \times \mathbf{B}_e\right] = 0, \qquad (6)$$

where the "generalized" electric and magnetic fields $\mathbf{E}_e, \mathbf{B}_e$ include the effect of electron inertia, obey Faraday's equation and are defined by

$$\mathbf{E}_e \equiv \mathbf{E} - \frac{m_e c}{e}\frac{\partial \mathbf{u}_e}{\partial t} \equiv -\frac{1}{c}\frac{\partial \mathbf{A}_e}{\partial t}, \qquad \mathbf{B}_e \equiv \mathbf{B} + \frac{m_e c}{e}\nabla \times \mathbf{u}_e \equiv \nabla \times \mathbf{A}_e. \qquad (7)$$

Here the generalized vector potential $\mathbf{A}_e \equiv \mathbf{A} + (m_e c/e)\mathbf{u}_e$ is related to the electron canonical momentum. Equation (6) expresses the freezing of the generalized magnetic field $\mathbf{B}_e$ (often called "generalized vorticity", in contrast to the standard fluid vorticity $\nabla \times \mathbf{u}_e$) in the electron fluid. In a uniform density plasma the generalized vector potential $\mathbf{A}_e$ can be written as $\mathbf{A} - d_e^2 \nabla^2 \mathbf{A}$, where $d_e \equiv c/\omega_{pe}$, is the collisionless electron skin depth and ω_{pe} is the plasma frequency. Thus, in the case of phenomena characterized by spatial scales larger than d_e, the generalized vector potential $\mathbf{A}_e$ and the fields $\mathbf{E}_e$ and $\mathbf{B}_e$ reduce to $\mathbf{A}$, $\mathbf{E}$ and $\mathbf{B}$ respectively. In this limit, if the assumptions mentioned above Eq.(6) apply, the magnetic field $\mathbf{B}$ is frozen in the electron fluid[4].

These flux conservation theorems are widely used both in astrophysical and laboratory plasmas as they are very convenient when describing the plasma behaviour on space- and time-scales where dissipative effects are unimportant. However they are based on two strong assumptions that, as is well known, can be easily violated in a real plasma in particular when kinetic effects become important:

a) that the effective pressure tensor Π_e in Eq.(2) is isotropic (we recall that, in the case of an anisotropic pressure tensor, $\nabla \times (\nabla \cdot \Pi)$ does not vanish so that, in general, an anisotropic effective pressure tensor violates magnetic flux conservation) and, if $\Pi = p\mathbf{I}$,

b) that the scalar pressure p satisfies a polytropic closure ($\nabla \times [(1/n)(\nabla p)]$ does not vanish unless $p = p(n)$). This implies that the magnetic flux conservation is violated if the electron temperature gradient is not parallel to the density gradient (baroclynic effect).

Indeed all the mechanisms of magnetic flux generation that have been introduced in the literature or investigated experimentally, see e.g., Refs.5,6 and more recently Refs.7-16, can be viewed as violations either of condition a) (non-potential ponderomotive force, electron anisotropy etc.) or of condition b) (baroclynic effect). In the next section the explicit case of the magnetic field generation by the current filamentation instability will be examined. In this case the anisotropy of the effective pressure tensor Π_e arises from the relative motion of the two counterstreaming (cold) electron populations.

PRESSURE ANISOTROPY, REPULSION OF OPPOSITE CURRENTS AND MAGNETIC FIELD GENERATION

An anisotropic electron distribution function leads according to Eq.(3) to an anisotropic pressure tensor Π_e which can generate a magnetic field due to the development of a Weibel-type instability[17]. In this section we will briefly recall the physical mechanism that is at the basis of the Weibel instability and derive the dispersion relation of the closely related "electromagnetic current filamentation instability" (ECFI) that is of direct interest for explaining the generation of a quasistatic magnetic field in the wake of an ultra short ultraintense laser pulse propagating in a plasma[11,18–20].

The electromagnetic current filamentation instability occurs in the case of two counterstreaming electron populations (with zero net total current) and develops perpendicularly to the direction of the electron streams leading, because opposite currents repel each other, to their spatial separation and to the generation of a magnetic field. The connection between the Weibel instability and the current filamentation instability can be seen by observing that, in the framework of the mean electron momentum equation (2), the effect of the relative velocity between the two counterstreaming electron populations appears as a contribution to the effective pressure tensor Π_e. This can be understood by referring, e.g., to an anisotropic electron distribution with temperature T_x along the x-direction larger than the temperature $T_\perp$ in the y, z directions. By interpreting the portions of this distribution function with positive and with negative velocity along x as corresponding to two different electron populations with non zero, oppositely directed, net stream velocities, we can draw the analogy with a distribution function which consists of two separate populations with isotropic temperature equal to $T_\perp$ and velocity separation $\delta u_x \sim 2(2T_x/m_e)^{1/2}$. Clearly this analogy is meaningful only if T_x is sufficiently larger than $T_\perp$.

In the case of two counter propagating electron populations, the transverse electromagnetic current filamentation instability is coupled to the two stream electrostatic instability that develops along the x direction. For the sake of illustration we will recall here the linear dispersion relation of these coupled instabilities in the two electron fluid approximation following the analysis of Refs.21,22. (This analysis has been extended to a kinetic treatment[23]. More recently it has been shown[24] that the saturation mechanism of these coupled instabilities is related to the formation of coherent vortex-like structures in phase space).

Assuming the ions to be at rest and to provide a uniform neutralizing background, the linear dispersion relation can be obtained by linearizing the relativistic equations for the two counter-streaming cold electron populations together with Maxwell's equations:

$$\frac{\partial n_\alpha}{\partial t} = \nabla \cdot \mathbf{j}_\alpha, \qquad \frac{\partial \mathbf{p}_\alpha}{\partial t} = -\mathbf{u}_\alpha \cdot \nabla \mathbf{p}_\alpha - (\mathbf{E} + \mathbf{u}_\alpha \times \mathbf{B}), \qquad (8)$$

$$\frac{\partial \mathbf{B}}{\partial t} = -\nabla \times \mathbf{E}, \qquad \frac{\partial \mathbf{E}}{\partial t} = \nabla \times \mathbf{B} - \sum_\alpha \mathbf{j}_\alpha, \qquad (9)$$

with $\mathbf{u}_\alpha = \mathbf{p}_\alpha/(1+p_\alpha^2)^{1/2}$, and $\mathbf{j}_\alpha = -n_\alpha \mathbf{u}_\alpha$, $\alpha = 1, 2$. In Eqs.(8,9) quantities are normalized on a characteristic density $\bar{n}$, on the speed of light c and on the plasma frequency $\bar{\omega}_{pe} = (4\pi\bar{n}e^2/m)^{1/2}$. We consider a homogeneous plasma with velocities along the x direction $u_{0,\alpha}$, such that the net current density is zero $\sum_\alpha n_{0,\alpha} u_{0,\alpha} = 0$, and a perturbation with frequency ω and wavevector $\mathbf{k} = (k_x, k_y)$, such that the perturbed magnetic field, arising from the separation along y of the oppositely directed currents along x, is in the z direction. Defining $\Omega_\alpha = \omega - k_x u_{0,\alpha}$ and $\Gamma_\alpha = (1 - u_{0,\alpha}^2)^{-1/2}$, the linear dispersion relation reads:

$$(1 - \Omega_2^{-2}) \left[k_x^2 (1 + \Omega_4^{-2}) - \omega^2 (1 - \Omega_1^{-2}) - 2\omega k_x \Omega_3^{-2} \right] \tag{10}$$

$$+ k_y^2 \left[(1 - \Omega_1^{-2})(1 + \Omega_4^{-2}) + \Omega_3^{-4} \right] = 0,$$

with

$$\Omega_1^{-2} = \sum_\alpha \frac{n_{0,\alpha}}{\Gamma_\alpha \Omega_\alpha^2}, \quad \Omega_2^{-2} = \sum_\alpha \frac{n_{0,\alpha}}{\Gamma_\alpha^3 \Omega_\alpha^2}, \quad \Omega_3^{-2} = \sum_\alpha \frac{n_{0,\alpha} u_{0,\alpha}}{\Gamma_\alpha \Omega_\alpha^2}, \quad \Omega_4^{-2} = \sum_\alpha \frac{n_{0,\alpha} u_{0,\alpha}^2}{\Gamma_\alpha \Omega_\alpha^2}. \tag{11}$$

When the perturbation propagates parallel to the mean electron streams, i.e. $k_y = 0$, the electrostatic two-stream instability amplifies the electric field E_x with a growth rate obtained by solving the equation $1 - \Omega_2^{-2} = 0$. No magnetic field is produced in this case. In the opposite limit, $k_x = 0$, the dispersion relation reduces to[18]

$$\omega^2 (1 - \Omega_2^{-2})(1 - \Omega_1^{-2}) - k_y^2 \left[(1 - \Omega_1^{-2})(1 + \Omega_4^{-2}) + \Omega_3^{-4} \right] = 0, \tag{12}$$

which contains two oscillatory solutions and one purely growing electromagnetic instability (the current filamentation instability) which amplifies the magnetic field B_z with a growth rate that is linear on k_y for $k_y d_e < 1$ (in dimensional units) and becomes approximately constant and of order ω_{pe} for $k_y d_e > 1$ when the velocity on the two counterstreaming beams is close to the velocity of light.

The fact that in the relativistic case the ECFI growth rate is of the order of of the Langmuir frequency indicates that this mechanism of magnetic field generation can indeed be effective in the case of the interaction of an ultrashort, ultraintense laser pulse with a plasma where most phenomena occur on timescales of the order of the electron dynamical time ω_{pe}^{-1}. In this framework the two counterstreaming electron populations consist of a smaller population of fast (relativistic) electrons, accelerated by the laser pulse interacting with the plasma, and by a larger population of slow electrons that provide the return current needed in order to maintain charge neutrality in the plasma.

In Ref.11 this model was introduced in the case of a laser pulse undergoing relativistic self-focussing and filamentation while propagating in an underdense plasma. In this case the acceleration of the fast electrons was thought to result from the break of the Langmuir wake wave produced behind the laser pulse. In such a configuration, both the fast electrons and the slow electrons in the return current are correctly described as collisionless populations moving against a background of immobile neutralizing ions. The physics of the fast electron beam propagation and the dynamics of the return current are somewhat more involved in the case of an overdense plasma where additional effects are important, such as the strong spatial inhomogeneity of the system and in particular the effect of resistivity on the return current, which can affect the fast electron penetration range, as discussed recently (see Refs.25-34 and references therein) in the context for example of the Fast Ignitor scheme[10] of Inertial Fusion.

It is clear however that the linear stability analysis in a homogeneous plasma sketched above is not sufficient in order either to determine the efficiency of the conversion of the kinetic energy of the fast electrons into magnetic energy, which require a nonlinear saturation analysis, or the spatial structure of the magnetic field generated in the wake of a laser pulse. Besides, it is of interest to determine not only how the magnetic field is created at given energy source (i.e., at given fast electron beam), but also the self-consistent evolution of the laser pulse and of the fast electron population in the presence of the magnetic field. The dynamical effect of the magnetic field on the laser pulse propagation will be discussed in the next section. A further question related to the spatial structure of the generated magnetic field is its evolution on time scales longer than ω_{pe}^{-1}.

A rough estimate of the magnitude of the generated magnetic field can be obtained by observing that the ECFI growth rate reaches its maximum value for wavenumbers of the order of the inverse collisionless electron skin depth $d_e^{-1} \equiv \omega_{pe}^{-1}/c$. Thus we may expect that the characteristic transverse size of the current channels produced by the nonlinear evolution of the ECFI be of the order of d_e. Since the maximum current density in the current channel is given by $J_{max} \sim -enc$, in the quasistatic approximation we obtain for the maximum dimensionless value of the generated magnetic field

$$\frac{eB}{m_e c \omega} \sim \frac{\omega_{pe}}{\omega}. \tag{13}$$

Here ω is the carrier frequency of the laser pulse and the normalization is chosen so as to follow the one generally adopted for the dimensionless amplitude a of the laser pulse

$$a \equiv \left|\frac{eA}{m_e c^2}\right| \equiv \left|\frac{eE}{m_e c \omega}\right|, \tag{14}$$

where A and E are the amplitudes of the vector potential and of the electric field in the pulse. In deriving this estimate we have disregarded correction factors arising from the relativistic modification of the electron mass (which depend on the laser pulse amplitude, see Refs.11, 18-20), from the difference between the fast electron and the slow electron densities that affect the channel width and the instability growth rate and from the density depletion in the channel. For a relativistic laser pulse laser pulse, $a > 1$, with wavelength $\lambda \sim 1\mu m$ propagating in a plasma with density, e.g., half its critical value, the amplitude B of the generated quasistatic magnetic field given by Eq.(13) is extremely large, being of the order of $100 MG$.

A similar estimate can be obtained from energy considerations, by requiring that the magnetic energy density be at most of the order of the kinetic energy density of the fast electrons. Taking this latter to be roughly of order $nm_e c^2$, we obtain

$$\frac{e^2 B^2}{m_e^2 c^2} \equiv \Omega_{ce} < \omega_{pe}^2, \tag{15}$$

in agreement with Eq.(13). This estimate is also consistent with the simplest form (relativistic factors are not properly included in these simplified reasonings) of the so-called Alfvèn limit[35] for the total current inside the current channel. This limit states that the width of the current channel (in our case d_e) should not exceed the characteristic orbit size (in our case taken to be of order c/Ω_{ce}) in the magnetic field generated by the current in the channel.

A more detailed estimate of the magnitude of the magnetic field and of the efficiency of the conversion from kinetic to magnetic energy can be obtained by studying the kinetic saturation of the ECFI as done, for the parameters of interest for the laser pulse plasma interaction, in Ref.23. The overall result is that, in the case of two symmetric oppositely propagating fast beams, the conversion efficiency can be rather large, leading to approximate equipartition between kinetic and magnetic energy, in agreement with Eq.(15). On the other hand, when the beams are non-symmetric, as is the case of interest here where the velocity of the electrons in the return current is much smaller than that of the fast electrons, the conversion efficiency drops significantly below energy equipartition.

In the case of a self-focused laser pulse propagating in an underdense plasma, the fast electron beam is strongly localized in the plane perpendicular to its direction of propagation and the separation between the fast electron current and the return current is expected to lead to a strongly inhomogeneous magnetic field. In a two-dimensional (2-D) model, where the fast electrons beam propagates along x and is localized in y, the magnetic field that arises from the nonlinear development of the ECFI is directed along z and is essentially dipolar; it consists of two "ribbons" of opposite polarities and vanishes at the fast beam axis at $y = 0$. These two ribbons can be seen as the intersection with the $z = 0$ plane of the cylindrical magnetic sheet that would be produced in a 3-D configuration by a cylindrical laser pulse. In this configuration, in the simple case on an azimuthally symmetric instability, field lines are circles in the y-z plane. Multiple dipolar magnetic structures are formed in the case when the laser pulse filaments causing the formation of multiple fast electron beams. The effect of the finite transverse with of the beam was investigated in 2-D in Ref.22, where it was shown that the magnetic field inhomogeneity along y is enhanced by the fact that the ECFI has a resonant-type behaviour in the inhomogeneity direction that leads to the formation of a (fluid) spatial singularity.

If the laser pulse interacts with an overdense plasma, e.g., with a solid target preceded by a strongly inhomogeneous plasma produced by nonlinear effects or simply by the interaction with the prepulse, the fast electrons are expected to be produced at the so called "resonant surface" and to propagate in a strongly inhomogeneous medium. In this case a model for the ECFI must include a density inhomogeneity along the direction of propagation of the fast beam, in addition to the transverse inhomogeneity discussed above. This analysis has not yet been performed in the framework of the application of the ECFI to the magnetic field generation in an overdense plasma and may be affected by (electrostatic) surface phenomena at the solid plasma interface which may imprint the further development of the ECFI.

MAGNETIC INTERACTION

Quasistatic magnetic fields of such high intensities that they can bend the electron trajectories on scales as short as the electron skin depth must have have an important dynamical effect of the propagation of the charged particles in the plasma (see e.g., Refs.11,18,27,28,36). These magnetic fields can focalize and channel charged particles and even provide a longitudinal electron acceleration mechanism in the transverse laser fields[37]. In an underdense plasma, with a density not too far below the critical density, these effects on the energetic particles are transferred to the laser propagation itself, leading to the magnetic interaction between different channels of laser radiation in the plasma. The basic physical mechanism relies on the fact that, in such conditions, fast electrons can overcome the laser pulse and bring ahead of the pulse information about the magnetic field generated at the back of, or inside, the pulse, thus generating a positive feed-back. The action on the laser pulse propagation is due to the modification in the dielectric constant caused by the "heavier" fast electrons that replace the "lighter" ambient electrons in front of the pulse. This mechanism of magnetic interaction has been proposed[11] in order to explain the coalescence between different channels of a filamented relativistic pulse in an underdense plasma

In an overdense plasma the magnetic field is expected to play an important role in the electron transport[27], and in the plasma hole-boring, possibly providing the electron focalization that is required in the fast Ignitor scheme (see e.g., Refs.10,28,38,39). Indeed the question is still open whether magnetic guiding and magnetic coalescence will suffice in producing a (single) well collimated electron beam that propagates deeply into the target.

As far as ions are concerned, in the context of particle acceleration in the petawatt power regime, the self-generated magnetic field has been shown in 2-D and 3-D numerical simulations reported in Ref.36 to focalize a narrow plasma filament inside the channel bored by the pulse in a plasma slab. The ions in this filaments have been shown[36] to be accelerated up to relativistic energies by the anisotropic "Coulomb explosion"[40] that occurs when the electron expand in vacuum after the pulse has bored through the plasma slab accelerating the ions behind. In this case also the strong magnetic field expands in vacuum, being to first approximation frozen in the expanding plasma as described above in the second section and collimates the ion beam[36].

MAGNETIC VORTICES

The long time behaviour of the bipolar quasistatic magnetic field in the wake of an ultrashort laser pulse in an underdense plasma has been shown in Refs.20,41 to develop a vortex structure: the bipolar magnetic ribbon develops an instability which tends to bend it and to produce an electron velocity pattern similar to the von Kàrmàn row in hydrodynamics. The resulting configuration of the magnetic field corresponds that of an antisymmetric vortex street in the electron fluid velocity, where the oppositely polarized vortices are shifted, one with respect to the other, along the two chains.

After the ECFI has saturated and the counterstreaming electron populations have "thermalized", the electrons can satisfactorily be described as a single cold population. As discussed in the second section, in the absence of dissipation and of source terms and for relatively fast phenomena such that the ions can be taken as immobile, the generalized vorticity $\nabla \times [\mathbf{p} - (e/c)\mathbf{A}]$, i.e. of the rotation of the canonical electron momentum field, is frozen in the (cold) electron flow (see Eqs.(6,7)). If we further assume that the electron motion that sustains the quasistatic magnetic field is slow compared to the Langmuir time and that its velocity is much smaller than the speed of light c, the electron fluid can be regarded as incompressible and the electron fluid velocity is related to the magnetic field as $\mathbf{v} = -(c/4\pi e n)\nabla \times \mathbf{B}$. Thus the domains where the magnetic field is stronger correspond to vortices in the electron fluid motion.

In a two dimensional configuration where the plasma currents flow in the x-y plane, taking $\mathbf{B}$ to be along the z-axis, we obtain for $\mathbf{B} \equiv B\mathbf{e}_z$

$$(\partial/\partial t + \hat{z} \times \nabla B \cdot \nabla)(\Delta B - B) = 0, \qquad (16)$$

where the time and space units are the inverse cyclotron frequency in the generated magnetic field Ω_{Ce}^{-1}, and $d_e = c/\omega_p$, the collisionless electron skin depth. Equation (16) is known as the Hasegawa-Mima equation in the limit of zero drift velocity[42]. A more general type of vortex structure, where charge neutrality is relaxed, has been considered recently in this context in Ref.43.

Equation (16) admits point-like vortex solutions. These solutions provide a convenient tool for representing the antisymmetric vortex street in the wake of the laser pulse, and for showing that its propagation velocity is slow compared to the speed of light c when the distance between the vortices becomes larger than d_e, as consistent with the numerical results presented in Ref.20. A stability analysis can also be performed and indicates that a stability domain exists for values of the geometrical parameters of the configuration in agreement with the simulations.

We can summarize these results by saying that in the nonlinear interaction between an ultraintense, ultrashort laser pulse and an underdense plasma, a finite fraction of the pulse energy is transformed into quasistationary magnetic structures that survive as long as electron momentum transfer to the ions does not become important. These structures remain for a long time in the plasma because of their slow propagation velocity. Thus vortex structures, which are a basic ingredient of fluid turbulence, turn out to be also important in the fast phenomena of relativistic plasmas.

SUBCYCLE RELATIVISTIC SOLITONS

Solitons are another basic ingredient of fluid turbulence that numerical simulations indicate are realized in the nonlinear laser plasma interaction[44]. Relativistic solitons are self-trapped, finite size, electromagnetic waves that propagate without diffraction spreading. Self-trapping appears because the electromagnetic wave modifies the local refractive index through the relativistic increase of the electron mass and the redistribution of the electron density under the pondermotive pressure of the radiation. On the basis of the indications provided by the numerical results on the soliton formation[44], we will concentrate on sub-cycle low frequency solitons[45]. As in the case of magnetic vortices, these low frequency solitons have low propagation velocities and remain long in the plasma.

In the rest of this section we shall recall some of the mathematical results that have been recently obtained in the case of 1-D relativistic solitons in plasmas[45,46]. In the following section we present a model description[44] of s- and p-polarized solitons in 2-D and some remarks about their structure in 3-D. The soliton generation mechanism and their dynamics in an inhomogeneous plasma will be discussed in the following sections. An extension of the 1-D results to the case where an external magnetic field is present has been recently given in Ref.47.

First we study[46] finite amplitude circularly polarized waves in a cold collisionless unbounded relativistic plasma described by Maxwell's equations and by the hydrodynamic equations of an electron fluid in a fixed ion background with ion density $n_i(x)$:

$$\Delta \mathbf{A} - \frac{1}{c^2}\partial_{tt}\mathbf{A} - \frac{1}{c}\nabla\partial_t\varphi - \frac{4\pi e n_e}{m_e c^2 \gamma}(\mathcal{P} + \frac{e}{c}\mathbf{A}) = 0, \qquad (17)$$

$$n_e = n_i(x) + \frac{1}{4\pi e}\Delta\varphi, \qquad (18)$$

$$\partial_t \mathcal{P} = \nabla(e\varphi - m_e c^2 \gamma) + \frac{1}{\gamma}(\mathcal{P} + \frac{e}{c}\mathbf{A}) \times \mathrm{rot}\mathcal{P}. \qquad (19)$$

The continuity equation is implied by Eqs.(17,18). These equations are written in the Coulomb gauge div $\mathbf{A} = 0$, $\mathcal{P} \equiv \mathbf{p} - e\mathbf{A}/c$ is the canonical electron momentum and the relativistic Lorentz factor is $\gamma = (1 + (\mathcal{P} + e\mathbf{A}/c)^2/(m_e c^2))^{1/2}$. We consider 1-D solutions where $\partial_y = \partial_z = 0$. This implies $A_x = 0$ and $\mathcal{P}_y = \mathcal{P}_z = 0$.

Assuming the electromagnetic wave to be circularly polarized, we introduce the new coordinates $X = x - v_s t$, and $\tau = t$, and look for solutions of the form:

$$\mathbf{A}_\perp = A_y + iA_z = A(X)\exp(i\omega((1-\beta_s^2)\tau - v_s X/c^2)), \quad p_{\parallel}/m_e c = \beta_s b(X). \qquad (20)$$

Inserting Eq.(20) into Eqs.(17-19) and assuming the ion density to be homogeneous, we obtain the system of coupled equations

$$\left(\gamma - \beta_s^2 b\right)'' = \frac{\omega_{pe}^2 b}{(\gamma - b)c^2}, \qquad a'' + \frac{\omega^2}{c^2}a = \frac{\omega_{pe}^2 \gamma_s^2}{(\gamma - b)c^2}a. \qquad (21)$$

Here $\gamma = (1 + a^2 + \beta_s^2 b^2)^{1/2}$, $\gamma_s = (1 - \beta_s^2)^{-1/2}$, $\beta_s = v_s/c$, $a = eA/m_e c^2$ and a prime denotes a differentiation with respect to the variable X. This system of equations corresponds to the Hamiltonian motion of a "particle" in a two dimensional potential field[49,52]. For a purely transverse electromagnetic wave, from Eqs.(21) with $b = 0$, we find that a is constant and that its frequency is [56]

$$\omega = \omega_{pe}\gamma_s/(1+a^2)^{1/4}. \tag{22}$$

In the x,t-coordinates Eq.(22) corresponds to a frequency ω given by $\omega^2 = k^2c^2 + \omega_{pe}^2/(1+a^2)^{1/2}$ with wave-number $k = v_s\omega/c^2$. In this case v_s is the group velocity of the wave and the wave phase velocity is $\omega/k = c^2/v_s$.

With boundary conditions $a(\infty) = b(\infty) = 0$, Eqs.(21) describe a one-dimensional relativistic electromagnetic soliton propagating through a cold collisionless plasma. For a small but finite amplitude, the soliton is described by the well known hyperbolic secant expression:

$$a = \frac{2(1-(\omega/\omega_{pe}\gamma_s)^2)^{1/2}\exp(i\omega((1-\beta_s^2)\tau - v_sX/c^2))}{\cosh\left(k_p^2 X\left(1-(\omega/\omega_{pe}\gamma_s)^2\right)^{1/2}\right)} \tag{23}$$

with frequency $\omega \approx \omega_{pe}\gamma_s(1-a_m^2/8)$, and amplitude $a_m = a(0,0)$. This solution corresponds to an isolated envelope soliton. Envelope solitons have been discussed in Refs.48-55.

Another exact solution[57,45] can be found in the limit of a soliton with zero propagation velocity. In the case $\beta_s = 0$, b vanishes and Eqs.(21) reduce to

$$a'' + k_p^2[(\omega/\omega_{pe})^2 - (1+k_p^2\gamma'')/\gamma]a = 0, \tag{24}$$

from which we obtain a soliton solution of the form

$$a(X,\tau) = \frac{2(1-(\omega/\omega_{pe})^2)^{1/2}\cosh\left(k_p^2 X\left(1-(\omega/\omega_{pe})^2\right)^{1/2}\right)\exp(i\omega\tau)}{\cosh^2\left(k_p^2 X\left(1-(\omega/\omega_{pe})^2\right)^{1/2}\right)+1-(\omega/\omega_{pe})^2}. \tag{25}$$

The soliton frequency ω depends on the soliton amplitude a_m as

$$a_m = 2\omega_{pe}(\omega_{pe}^2-\omega^2)^{1/2}/\omega^2. \tag{26}$$

This soliton solution is stable provided the electron density inside it does not vanish. This imposes the constraints $a_m < \sqrt{3}$ and $1 > (\omega/\omega_{pe}) > \sqrt{2/3}$. These properties have been verified in the PIC simulations reported in Ref.45.

HIGHER DIMENSION RELATIVISTIC SOLITONS

In two and three dimensions exact solutions of the type described above are not as yet available. Numerical simulations[44] however indicate that long-lasting coherent soliton-like structures, where electromagnetic energy from the laser pulse is self trapped, do occur. These structures have very small propagation velocity in a homogeneous plasma, oscillation frequency below the ambient plasma frequency and characteristic scale-length of the order of the electron skin depth.

In the 2-D case, the analogues of Eq.(24) can be obtained in the limit of weak nonlinearity for linearly polarized s- and p- solitons. In s-polarized solitons the z-component of the electric field and the azimuthal component of the magnetic field oscillate in time, while the electron density distribution remains constant. Assuming for the vector potential a time dependence of the form $a(r)\exp[-i(\omega_{pe} - \delta\omega)t]$, we find

$$\Delta_\perp a - k_p^2[2\delta\omega/\omega_{pe} - |a|^2]a = 0, \tag{27}$$

where $\Delta_\perp = r^{-1}\partial_r r \partial_r$ and $\delta\omega$ is the nonlinear frequency shift. The properties of this equation are well known in the theory of wave collapse[58] and self-focusing[59]. It describes localized 2-D solitons with frequency shift $\delta\omega$ and radius r_0 which depend on the soliton amplitude as $\delta\omega \sim a^2$ and $r_0 \sim 1/a$. This scaling agrees with the dependence of the frequency on the soliton amplitude in the case of a planar circularly polarized 1-D soliton. In p-polarized solitons the electric field is azimuthal and the magnetic field is directed along z. Below we give the analogue of Eq.(27) for the case of solitons with mixed s- and p-polarization where the vector potential has a z component and an azimuthal of the form $[a_z(r_\perp, \varphi) \, ; \, a_\varphi(r_\perp)] \exp[-i(\omega_{pe} - \delta\omega)t]$, where $\delta\omega$ is the nonlinear frequency shift:

$$r^{-1}\partial_r r \partial_r a_z + r^{-2}\partial_{\varphi\varphi} a_z - k_p^2[2\delta\omega/\omega_{pe} - (|a_z|^2 + |a_\varphi|^2)]a_z = 0, \tag{28}$$

$$r^{-1}\partial_r r \partial_r a_\varphi - r^{-2}a_\varphi - k_p^2[2\delta\omega/\omega_{pe} - (|a_z|^2 + |a_\varphi|^2)]a_\varphi = 0, \tag{29}$$

Taking $a_z = \varepsilon_s a_\varphi(r)\cos(\varphi)$, with $\varepsilon_s \ll 1$ a constant and neglecting $|a_z|^2$ compared to $|a_\varphi|^2$ in Eqs.(28,29), we obtain a single equation for the common amplitude $a_\varphi(r)$

$$r^{-1}\partial_r r \partial_r a_\varphi - r^{-2}a_\varphi - k_p^2(2\delta\omega/\omega_{pe} - |a|^2)a_\varphi = 0. \tag{30}$$

The typical form of the s- and p-solitons obtained by numerical integration of Eqs.(28,29) is presented in Ref.44.

In the three dimensional case the analog of the s-solitons (p-solitons) are the TE-solitons (TM-solitons). The mixed TE-TM mode corresponds to a toroidal soliton with orthogonal magnetic and electric field lines wound on the surface of a torus, The analogue of Eq.(27) for TE-solitons with vector potential $\vec{a}_\varphi = a_\varphi(r,\theta)\vec{e}_\varphi$ is

$$\Delta \vec{a}_\varphi - k_p^2[2\delta\omega/\omega_{pe} - |a_\varphi|^2]\,\vec{a}_\varphi = 0. \tag{31}$$

We expand $\vec{a}_\varphi$ in "toroidal modes" as

$$\vec{a}_\varphi = \sum_l a_l(r)\,[\partial_\theta Y_l^o(\theta)]\,\vec{e}_\varphi, \tag{32}$$

where Y_l^m are spherical harmonics, and take $m = 0$. The 3-D Laplacian operator Δ acting on the 3-D vector $\vec{a}_\varphi$ is diagonal on the above basis with

$$\Delta\,[a_l(r)\,[\partial_\theta Y_l^o(\theta)]\,\vec{e}_\varphi] = r^{-2}\left[\partial_r r^2 \partial_r a_l(r) - l(l+1)\,a_l(r)\right]\,[\partial_\theta Y_l^o(\theta)]\,\vec{e}_\varphi. \tag{33}$$

The cubic term in Eq.(31) couples all odd $l = 1, 3, 5...$ (and all even $l = 2, 4, 6...$) harmonics, leading to an an infinite set of coupled nonlinear differential equations. If we consider odd harmonics and, for the sake of illustration, truncate the series and take only $l = 1$, we obtain

$$r^{-2}\partial_r r^2 \partial_r a_1(r) - 2r^{-2}a_1(r) - k_p^2[2\delta\omega/\omega_{pe} - 3a_1^2/4]\,a_1(r) = 0 \tag{34}$$

instead of Eq.(27).

SUBCYCLE SOLITON GENERATION AND INTERACTION

The physical mechanism that produces these sub-cycle solitons, as seen e.g., in the simulations reported in Ref.44 is different from the standard process where the nonlinear steepening of the wave is counterbalanced by the effect of dispersion. In our case the dispersion effects come into play because of the frequency downshift of the laser pulse. Interacting with the underdense plasma the laser pulse loses its energy as it generates electrostatic and magnetostatic wake fields behind it. The frequency of these fields is much lower than the carrier frequency of the laser pulse and the laser-plasma interaction is adiabatic. In this case the ratio between the electromagnetic energy density and frequency is adiabatically conserved and the decrease of the laser pulse energy must be accompanied by the downshift of its carrier frequency. On the other hand the group velocity of the laser pulse decreases as the carrier frequency is downshifted and the pulse group velocity tends to zero at $\omega \to \omega_{\mathrm{pe}}$. Thus the depleted portions of the pulse lag behind and convert their energy into soliton energy with almost zero group velocity. The effectiveness of this mechanism can be verified by correlating the location of the soliton formation in the PIC simulations with the pulse depletion length. For a wide laser pulse the latter is equal[60] to $l_{\mathrm{depl}} \approx l_{\mathrm{p}}(\omega/\omega_{\mathrm{pe}})^2$, while a narrow laser beam decays after propagating over the length $\approx al_{\mathrm{p}}(\omega/\omega_{\mathrm{pe}})^2$.

We notice that in our 1D PIC simulations[45] we have observed solitary waves for about 1000 oscillation periods. In the 2D PIC simulations[44] we can observe solitons for more than 25 periods without appreciable changes in the amplitude and the frequency. This limit on time is imposed by the required simulation time, as ion effects can be expected to affect the solitons dynamics on time scales related to the inverse of the ion plasma frequency.

The collision between two 2-D s-solitons has also been observed[46] in 2-D simulations and shown to differ from that presented in Ref.45 for circularly polarized 1-D solitons where each soliton maintains its amplitude and propagation speed while the phases change. In 2-D, some of the characteristic properties of solitons are not recovered, and we have observed that the soliton interaction leads to merging of two s-solitons. Before merging there is a synchronization of the oscillations inside each soliton.

SOLITON EXTRACTION AND BURSTS OF SOLITON ENERGY

In a non-uniform plasma the propagation of the sub-cycle solitons is strongly affected by the inhomogeneity of the medium which can be used in order to extract and detect the solitons. The propagation in inhomogeneous media of solitons described by the nonlinear Schrödinger equation was discussed in Ref.61 and it was shown that this problem admits multisoliton solutions. These exact solutions show that the solitons are accelerated toward the plasma vacuum interface with an acceleration proportional to the gradient of the plasma density. A simple model of the soliton motion can be derived by referring to the propagation of a wave packet in a non-uniform dispersive medium. The geometric optics equations show that the wave packet is accelerated towards the low density side of the plasma and, if a density channel is formed e.g., by the laser pulse, the soliton oscillates in the transverse direction. As a result, the soliton moves towards the plasma vacuum interface, where it radiates its energy away in the form of low frequency electromagnetic waves, during its non-adiabatic interaction with the plasma boundary[44,46]. In the case of a laser pulse propagating in a plasma inhomogeneous in the transverse direction, the soliton moves in the direction perpendicular to the laser beam and abandons the region behind the laser pulse, where the wake field and electron vortices are localized, as shown by the numerical simulations presented in Refs.44,46. We see that the trapping of the electromagnetic energy becomes weaker as the solitons moves towards regions of lower density until they local oscillation frequency becomes larger than the ambient plasma frequency and they burst into low frequency electromagnetic radiation that can be used in principle in order to detect subcycle solitons.

CONCLUSIONS

Analytical and numerical investigations show that in the complex physics of the interaction of high intensity ultrashort laser pulses with plasmas, fundamental physical mechanisms can be identified that form the basic blocks of the nonlinear physics of continuous media such as vortices and solitons. The basic features of the vortex and soliton formation and dynamics have been described.

An interesting open question is the possibility of detecting such low frequency coherent structures experimentally. In the case of vortices their measurement is related to that of ultraintense low-frequency magnetic fields on spatial scales of a few μm and with polarity varying on the same spatial scale in plasmas with a strongly perturbed density profile.

In the case of solitons, the possibility of extracting them in an inhomogeneous plasma may indeed allow a indirect detection in the form of bursts of low frequency electromagnetic radiation. The knowledge of the frequency spectrum of the bursting radiation can be provided[44] by PIC simulations and can be an effective tool in discriminating the soliton events.

ACKNOWLEDGMENTS

We are pleased to acknowledge the use of the Origin SGI 2000 supercomputer at "Scuola Normale Superiore", Pisa Italy, and of the Cray T3E at "Cineca", Bologna Italy, under the INFM Parallel Computing Initiative.

REFERENCES

1. J.M. Dawson, Phys. Plasmas, **6**, 4436 (1999).
2. S.V. Bulanov, T.Zh. Esirkepov, *et al.*, J. Plasma Fusion Res., **75**, No. 5 (1999).
3. N. Krall, A. Trivelpiece, *Principles of Plasma Physics*, (McGraw-Hill, New York, 1978).
4. A.S. Kingsep, K.V. Chukbar, V.V. Yan'kov, 1990, *Reviews of Plasma Physics*, ed. by B. Kadomtsev, (Consultants Bureau, New York, N.Y.) **16**, 243.
5. J. Stamper, K. Papadopoulos, *et al.*, Phys. Rev. Lett., **26**,1012 (1971); J.A. Stamper, Laser Part. Beams 9, 841 (1991).
6. M.G. Haines, Can. J. Phys., **64**, 912 (1986).
7. S.C. Wilks, W.C. Kruer, M. Tabak, A.B. Langdon, Phys. Rev. Lett., **69**, 1383 (1992).
8. R.N. Sudan, Phys. Rev. Lett., **70**, 3075 (1993).
9. V.Yu. Bychenkov, V.I. Demin, V.T. Tikhonchuk, Sov. Phys. JETP, **105**, 118 (1994).
10. M. Tabak, Y. Hammer, *et al.*, Phys. Plasmas, **1**, 1626 (1994); R.J. Mason, M. Tabak, Phys. Rev. Lett., **80**, 524 (1998).
11. G.A. Askar'yan, S.V. Bulanov, F. Pegoraro, A.M. Pukhov, JETP Letters, **60**, 240 (1994).
12. L. Gorbunov, P. Mora, T.M. Antonsen, Jr. Phys. Rev. Lett. **76,** 2495 (1996).
13. A. Pukhov, J. Meyer-ter-Vehn, Phys. Rev. Lett. **76,** 3975 (1996).
14. L. Gorbunov, R. Ramazashvili, Sov. Phys. JETP, **87**, 461 (1998).
15. M. Borghesi, A.J. Mackinnon, *et al.*, Phys. Rev. Lett. **80**, 5137 (1998); M. Borghesi, A.J. Mackinnon, *et. al.*, Phys. Rev. Lett. **81**, 112 (1998).
16. Y. Sentoku. K. Mima, S-i. Kojima, H. Ruhl, Phys. Plasmas, **7**, 689 (2000).
17. E.W. Weibel, Phys. Rev. Lett., **2**, 83, (1959).

18. G.A. Askar'yan, S.V. Bulanov, et. al., Comm. Plasma Physics Contr. Fusion, **17**, 35, (1995); G.A. Askar'yan, S.V. Bulanov, et. al., Plasma Physics Reports, **21**, 835, (1995); S.V. Bulanov, T.Zh. Esirkepov, et. al., Physica Scripta, **T 63**, 280 (1996); F. Pegoraro, S.V. Bulanov, et. al., Physica Scripta, **T 63**, 262 (1996); F. Pegoraro, S.V. Bulanov, et al., in *Superstrong Fields in Plasma*, AIP Conf. Proc. No. 426, p. 113, M. Lontano, G. Mourou, F. Pegoraro, E. Sindoni eds., (AIP, New York, 1998).

19. G.A. Askar'yan, S.V. Bulanov, et. al., Plasma Phys. Contr. Fusion, **39**, 137 (1997); F. Califano, R. Prandi, et. al., in *Superstrong Fields in Plasma*, AIP Conf. Proc. No. 426, p. 123, M. Lontano, G. Mourou, F. Pegoraro, E. Sindoni eds., (AIP, New York, 1998).

20. S.V. Bulanov, T.Zh. Esirkepov,et al., Phys. Rev. Lett., **76**, 3562, (1996).

21. A. D. Steiger, C. H. Woods, Phys. Rev. A, **5**, 1467 (1971); V. Yu. Bychenkov, V. P. Silin, V. T. Tikhonchuk, Sov. Phys. JETP, **98**, 1269 (1990).

22. F. Califano, F. Pegoraro, S.V. Bulanov, Phys. Rev.,**E 56**, 963 (1997); F. Califano, R. Prandi, et. al., Phys. Rev. **E 58**, 7837 (1998); F. Califano, R. Prandi, et al., Journal of Plasma Physics **60**, 331 (1998) and references therein.

23. F. Califano, F. Pegoraro, S.V. Bulanov, A. Mangeney, Phys. Rev., **E 57**, 7048 (1998).

24. F. Califano, F. Pegoraro, S.V. Bulanov, Phys. Rev. Lett. **84**, 3602 (2000).

25. P. Gibbon, Phys. Rev. Lett., **73**, 664 (1994).

26. S.V. Bulanov, N.M. Naumova, F. Pegoraro, Phys. Plasmas, **1**, 745 (1994).

27. A.R. Bell, Phys. Plasmas, **1**, 1643 (1994); J.R. Davies, A. R. Bell, et. al., Phys. Rev. **E 56**, 7193 (1997); A.R. Bell, J.R. Davies S.M. Guerin, Phys. Rev. **E 58**, 2471 (1998); J.R. Davies, A.R. Bell, M. Tatarakis, Phys. Rev. **E 59**, 6032 (1999).

28. A. Pukhov, J. Meyer-ter-Vehn, Phys. Rev. Lett. **79**, 2686 (1997); M. Honda, J. Meyer-ter-Vehn, A. Pukhov, Phys. Rev. Lett. **85**, 2128 (2000).

29. T.A. Hall, S. Ellwi, et al., Phys. Rev. Lett. **81** , 1003 (1998).

30. A. Norreys, M. Santala, et al., Phys. Plasmas **6**, 2150 (1999).

31. H. Ruhl, A. Macchi, et al., Phys. Rev. Lett. **82**, 2095 (1999).

32. An-C. Tien, S. Backus, et al., Phys. Rev. Lett. **82**, 3883 (1999).

33. L. Gremillet, F. Amiranoff, et al., Phys. Rev. Lett. **83**, 5015 (1999).

34. E.L. Clark, K. Krushelnick, et al., Phys. Rev. Lett. **84**, 670 (2000).

35. H. Alfvén, Phys. Rev. **55**, 425 (1939); J.D. Lawson, J. Nucl. Energy, **C 1**, 31 (1959); M. Honda, Phys. Plasmas, **7**, 1606 (2000).

36. T. Esirkepov, Y. Sentoku, et al., JETP Letters, **70** , 82 (1999); S. Bulanov, F. Califano, et al., Plasma Physics Reports **25**, 468 (1999); S. Bulanov, N. Naumova, et al., JETP Letters, **71**, 407 (2000); Y. Sentoku, T.V. Liseikina, et al., Phys. Rev. **E**, (2000).

37. A. Pukhov, Z-M, Sheng, J. Meyer-ter-Vehn, Phys. Plasmas, **6** 2847 (1999).

38. P.A. Norreys, R. Allott, et al., Phys. Plasmas, **7**, 3721 (2000).

39. M. H. Key, et al., in *Proceedings of First Conference of Inertial Fusion Sciences and Applications*, Bordeaux, France, 1999, (Elsevier, Paris, 2000).

40. G. Sarkisov, V. Bychenkov, et al., JETP Letters, **66**, 828 (1997); G. Sarkisov, et al., Phys. Rev. **E 59**, 7042 (1999).

41. S.V. Bulanov, T. Zh. Esirkepov, et al., Plasma Physics Reports **23**, 284, (1997).

42. A. Hasegawa, K. Mima, Phys. Rev. Lett. **39**, 205 (1997); A. Hasegawa, K. Mima, Phys. Fluids 21, **87** (1978).

43. A.V. Gordeev, S.V. Levchenko, JETP Letters, **67**, 482 (1998); A.V. Gordeev, S.V. Levchenko, JETP Letters,**70**, 648 (1999).

44. S.V. Bulanov, T.Zh. Esirkepov, et al., Phys. Rev. Lett., **82**, 3440 (1999); Y.Sentoku, T.Zh. Esirkepov, et al., Phys. Rev. Lett., **83** , 3434 (1999).

45. T.Zh. Esirkepov, F.F. Kamenets, et al., JETP Letters, **68**, 36 (1998).

46. S.V. Bulanov, F. Califano, et al., Physica D, in press (2000).   F. Pegoraro, S.V. Bulanov, et al., Physica Scripta **T 84**, 89 (2000).

47. D. Farina, M Lontano, S.V. Bulanov, Phys. Rev., **E 62**, 4146 (2000).

48. N L. Tsintsadze, D.D. Tskhakaya, Sov. Phys. JETP, **45**, 252 (1977); N.L. Tsintsadze, et al., Phys. Rev., **E 58**, 4890 (1998).

49. V.A. Kozlov, A.G. Litvak, E.V. Suvorov, Sov. Phys. JETP, **76**, 148 (1979).

50. P.K. Shukla, et al., Phys. Fluids, **27**, 327 (1984).

51. K. Mima, T. Ohsuga, H. Takabe, et al., Phys. Rev. Lett., **57**, 1421 (1986).

52. P.K. Kaw, A. Sen, T. Katsouleas, Phys. Rev. Lett., **68**, 3172 (1992).

53. H.H. Kuehl, et al., Phys. Rev., **E 47**, 1249 (1993).

54. V.I. Berezhiani, S.M. Mahajan, Phys. Rev. Lett., **73**, 1110 (1994).

55. R.N. Sudan, Ya.S. Dimant, O.B. Shiryaev, Phys. Plasmas, **4**, 1489 (1997).

56. A.I. Akhiezer, R.V. Polovin, Sov. Phys. JETP **30**, 915 (1956).

57. T. Kurki-Suonio, P.J. Morrison, T. Tajima, Phys. Rev., **A 40**, 3230 (1982).

58. V. E. Zakharov, Sov. Phys. JETP, **35**, 908 (1972);   E. A. Kuznetsov, A. M. Rubenchik, V. E. Zakharov, Phys. Rep., **142**, 105 (1986); E. A. Kuznetsov, *Chaos* **6**, 381 (1996).

59. G.-Z. Sun, E. Ott, Y. C. Lee, P. Guzdar, Phys. Fluids, **30**, 526 (1987).

60. S.V. Bulanov, I.N. Inovenkov, V.I. Kirsanov, et al., Phys. Fluids **B 4**, 1935 (1992).

61. H. Chen, C. Liu, Phys. Rev. Lett. **37**, 693 (1976).

R-MATRIX-FLOQUET THEORY OF TWO-ELECTRON ATOMS IN INTENSE LASER FIELDS

Martin Dörr

Max-Born-Institut, D-12489 Berlin
mailto: doerr@mbi-berlin.de
http://mitarbeiter.mbi-berlin.de/doerr

INTRODUCTION

The purpose of this lecture is to discuss the R-matrix-Floquet (RMF) theory with a main focus on applications in two-electron atomic systems, such as the helium atom (He), the magnesium atom (Mg), and the negative ions of hydrogen (H$^-$) and of lithium (Li$^-$). Although Mg or Li$^-$ contain more than two electrons, in certain cases the inner shell can be taken as inert, providing simply an effective "core" potential, and thus Mg and Li$^-$ are effective two-electron systems. RMF theory is by no means restricted to effective two-electron systems, but these constitute the simplest multielectron atoms and are thus the most instructive for an introductory discussion.

The motivation for the discussion of RMF theory in the present school lies in the versatility and accuracy of the RMF method. A brief introduction has been given by N. Kylstra at this school. In fact, the RMF theory has been applied to multiphoton ionization (and detachment), to laser-assisted electron-atom scattering and to harmonic generation in atoms. It has also been applied to molecular multiphoton ionization [28]. All of these laser-driven or laser-assisted processes can thus be treated within a unique framework, which has been tested and perfected within many years of experience with field-free electron-atom scattering and photoionization computations. Thus, a suitable basis for accurate representation of atomic properties (e.g. polarizabilities, doubly excited resonant states, and generally atomic level structure) for any desired atomic or ionic species can be obtained. The emphasis is on a highly accurate and thus realistic description of the atomic species in question, including all states necessary for the description of the dynamics and therefore for ionization processes necessarily also involving the continuum. A collection of important papers on the R-matrix method and applications is given in the book [2].

The RMF theory yields precise total and partial cross-sections for multiphoton processes in the regime where non-perturbative dynamics, including atomic resonances is important. Laser field intensities range from the perturbative (single- or multi-photon) to the atomic order of magnitude. The intensity of a linearly polarized laser field with peak electric field of $\mathcal{E}_0 = 1$ a.u. $= 5.21 \times 10^9$ V/cm is 3.51×10^{16} W/cm^2. In applications of RMF theory, laser frequencies have usually been chosen of the order of magnitude of the relevant energy differences of the atomic system under study, such that in the perturbative limit multiphoton couplings of the order 1 to about 5 photons are relevant. At larger intensity, nonperturbative effects become apparent which evidently involve a large number of photons.

The outline of this paper and the corresponding lecture is as follows.

1.) Some basic notions on two-electron atoms are recalled, concerning the usual angular basis in spherical coordinates and the configuration interaction basis method.

2.) The basic equations of R-matrix-Floquet theory are presented. A brief resume of the Floquet method, its uses and limitations is given. The R-matrix-Floquet Hamiltonian matrix in the inner region is sketched. The asymptotic boundary conditions for (single) ionization or for electron atom scattering are discussed. The system of coupled equations for a single electron in the outer region is defined. A connection is drawn to the single active electron (frozen core, SAE) approximation. There have been many applications of RMF theory for atoms with more than two active electrons and some brief remarks on the differences with respect to two electron systems will be made.

3.) Some results on multiphoton ionization and detachment for He, Mg, H$^-$ and Li$^-$ are presented. An application to e-H scattering is also given.

4.) In the outlook, some further applications and future developments of RMF theory are mentioned. These comprise double ionization, new methods and algorithms, and further extensions of the R-matrix approach to time-dependent phenomena or heavier atoms.

Finally, a list of references concludes this paper. These contain a complete list of all applications of RMF theory to two-active-electron systems to-date and also some review articles containing resumes of RMF theory and results. The article [5] is the main source for references on RMF theory and related work up to 1999; some more focused reviews of RMF theory are given in [7, 6]. The review on two-electron atoms [8] contains also some discussions of the RMF method and results.

Atomic units ($e = m_e = \hbar = 1$) are used throughout the text unless otherwise indicated.

1. TWO ELECTRON ATOMS

The time-dependent Schrödinger equation (TDSE) for a $N+1$-electron atomic system in a laser field reads

$$i\frac{\partial}{\partial t}\Psi(\underline{\boldsymbol{X}}, t) = [H_{\text{atom}} + H_{\text{int}}(t)]\Psi(\underline{\boldsymbol{X}}, t). \tag{1}$$

Here $\underline{\boldsymbol{X}}$ denotes the spatial and spin coordinates of the two electrons, $\{\boldsymbol{x}_1, \boldsymbol{x}_2\}$, with $\boldsymbol{x}_i = \{\boldsymbol{r}_i, \chi_i\}$. The electron-field interaction is

$$H_{\text{int}}(t) = \frac{1}{c}\boldsymbol{A}(t)\cdot\boldsymbol{p} + \frac{2}{2c^2}\boldsymbol{A}^2(t), \tag{2}$$

where $\boldsymbol{p} = \sum_{i=1}^{2}\boldsymbol{p}_i$ is the sum of the electrons' momenta. The quantity $\boldsymbol{A}(t)$ is the vector potential of the external, classical field. The time-independent atomic Hamiltonian H_{atom} is

$$H_{\text{atom}} = -\sum_{i=1}^{2}\left[\frac{1}{2}\nabla_i^2 + \frac{Z}{r_i} - V_C(r_i)\right] + \frac{1}{|\boldsymbol{r}_i - \boldsymbol{r}_j|} \tag{3}$$

The atomic Hamiltonian is written in the nonrelativistic approximation and an infinitely heavy nucleus is taken.

The extra potential term $V_C(r_i)$ can account for a radial effective core potential, due to extra inert core electrons. More complicated forms of such a core potential, notably nonlocal forms incorporating angular dependence have been considered. In fact, in the R-matrix approach, generally the inner inactive electrons lead to more complex nonlocal forms of such a core potential.

Since the operators of permutation of the two electrons, P_{12}, of total angular momentum, L^2, and of spin, S^2, commute with the Hamiltonian (3) (note that we use the non-relativistic Hamiltonian, ignoring e.g. spin-orbit coupling), the atomic eigenstates are conveniently labeled by the eigenvalues with respect to these operators. For a two-electron system the total spin is either 0, leading to the singlet or para states, with spatial part of the wavefunction wavefunction even under P_{12}, or the total spin is 1, leading to the triplet or ortho states, with odd spatial wavefunction. The value of the total angular momentum is designated with the usual spectroscopic notation $L = 1$: S, $L = 2$: P, $L = 3$: D, etc. A leading superscript denotes the value of S (total spin) and finally a number is given to count the main quantum number. Thus, the singlet $L = 0$ series contains the levels 1^1S, 2^1S, 3^1S, etc. A total angular momentum L can be obtained by coupling two non-interacting angular momenta l_1 and l_2: $L \in \{l_1 + l_2, l_1 + l_2 - 1, ..., |l_1 - l_2|\}$. In an independent particle approach, one could identify an atomic state by its two-electron configuration. For example, the lowest He eigenstate, 1^1S, would have an $1s^2=1s1s$ configuration, the first excited would be 1s2s, where s designates $l_j = 0$ and the numbers give the main quantum number of the single-electron state. However, since the electrons interact, this constitutes only an approximation. The radial part of the single-electron wavefunctions is changed due to their interaction $1/|\mathbf{r}_1 - \mathbf{r}_2|$ and also the angular part is not given by a single sum of angular momenta. A better approximation, still retaining some of the computational simplicity of single-particle basis orbitals, consists in superposing several configurations, which must be properly antisymmetrized and with proper coupling of the angular momenta. This approach, called configuration interaction (CI), is followed in the implementation of the RMF method.

2. R-MATRIX-FLOQUET THEORY

The R-matrix approach [4] is based on the observation that, for a scattering process from a multielectron target, configuration space can be partitioned into an inner and and outer region. The *inner region* is defined by $r_i < a \ \forall \ i \in \{1, ..., N+1\}$, r_i being the i-th electron's radial coordinate and a the boundary of the inner region. In this region a quantum multiparticle interaction process must be solved. Outside of the sphere a there is only a single particle. The *outer region* is thus defined by one of the r_i being greater than a, while all the others are smaller. In a scattering context the residual N-electron system is called the target and the electron in the outer region is called the projectile. For the present two-electron case, $N = 1$.

Let us first consider the Floquet theory, which can generally be applied also outside the context of the R-matrix approach. The Floquet equations will be presented in more detail in the contribution of N. Kylstra to the present volume.

2.1 Floquet theory

In principle, the atomic dynamics in a short laser pulse should be obtained by solving the TDSE. The Floquet approach attempts to simplify this problem, relying on an adiabatic approximation, separating a short timescale, namely the laser field period, during which the laser field intensity does not change appreciably, from the longer timescale of variation of the laser pulse envelope (and possibly also phase or frequency chirp). Thus, for a definite intensity and frequency, the Floquet quasistationary eigenstates constitute the generalization of the field-free eigenstates of the atomic system, such as bound states or autoionizing states. After computation of the Floquet eigenstates for a range of field intensities, the temporal field intensity envelope variation can be included in a subsequent step, which will not be discussed here [32, 33].

The time dependent term in the Hamiltonian, $H_{int}(t)$, is due to the action of the external laser field. If this field is assumed to be perfectly monochromatic, then, following the Floquet theorem, solutions of the corresponding TDSE can be found which have the same periodicity, apart from a single eigenenergy phase factor.

Let us therefore assume that the classical external laser field is monochromatic, linearly polarized and spatially homogeneous. It is given by the vector potential $\boldsymbol{A}(t) = \hat{\boldsymbol{e}} A_0 \cos \omega t$, which is related to the electric field by $\boldsymbol{E}(t) = -(1/c)(d\boldsymbol{A}/dt)$. The Floquet ansatz for the wavefunction is then

$$\Psi(\underline{\boldsymbol{X}}, t) = e^{-iEt/\hbar} \sum_{n=-\infty}^{+\infty} e^{-in\omega t} F_n(\underline{\boldsymbol{X}}) \tag{4}$$

where the time-independent quantity E is called the quasi-energy, and the functions $F_n(\underline{\boldsymbol{X}})$ are called the Floquet components. If we substitute the Floquet ansatz into the Schrödinger equation (1), we obtain a system of time-independent coupled differential equations. These equations, together with appropriate boundary conditions, form an eigenvalue problem for the quasi-energies, which we can write as

$$(\mathbf{H}_F - E)\, \mathbf{F} = 0 \tag{5}$$

where the Floquet Hamiltonian $\mathbf{H}_F$ is an infinite matrix of operators. In the velocity or the length gauges, the Floquet Hamiltonian $\mathbf{H}_F$ is a tridiagonal matrix of operators, becoming a block-tridiagonal matrix when the operators are expressed on some atomic basis. Each diagonal block contains H_{atom} on the atomic basis plus $n\omega$, where n is its Floquet index. The off-diagonal blocks contain the dipole matrix elements between the atomic states, multiplied by the field strength.

$$\mathbf{H}_F = \begin{pmatrix} \cdots & & & & \\ D^\dagger & H_a + \omega & D^\dagger & & \\ & D & H_a & D & \\ & & D^\dagger & H_a - \omega & D^\dagger \\ & & & D & H_a - 2\omega & D \\ & & & & & \cdots \end{pmatrix}. \tag{6}$$

In actual calculations, the expansion (4) must be truncated at finite n, $-n_{emission} \leq n \leq n_{absorption}$. As the field becomes very strong, or the frequency becomes very low, many Floquet blocks are necessary for a converged calculation. This fact is setting a practical lower limit on the frequencies that can be considered. The limit on pulse duration is not so severe: numerical comparison between Floquet and fully-time-dependent results shows that a multimode Floquet theory can describe laser pulses as short as 4 cycles (fwhm) [33].

Finally, it must be noted that generalizations of the Floquet method to multimode (primarily bichromatic, commensurable or not) laser fields, that had been introduced earlier, have been incorporated within the RMF method [19, 22].

2.2 Inner region

Expanding the wavefunction Floquet components $F_n(\mathbf{X})$ on a basis, which in our case involves a close-coupling CI approach and thus an expansion in angular momenta, leads to a matrix representation for the operators in equation (6). The resulting Hamiltonian matrix consists of Floquet blocks. The field couples the angular momenta, and therefore the operator $\mathbf{L}^2$ is no longer a conserved quantity. Thus, in principle all L are involved, but this expansion must necessarily be truncated. To ensure convergence of the results, tests must be done on their stability with an increase of the number of channels retained in the calculation.

The wavefunction in the inner region is expressed on a basis, just as for the usual R-matrix calculations, involving a set of optimized single-particle orbitals $\varphi_j(r)$ which have decayed exponentially practically to zero at $r = a$ and a set of basis functions $u_{il}^\Lambda(r)$ which extend up to $r = a$. The latter are sometimes called "continuum basis functions" since they represent the outgoing electron. The Floquet component n of eigenfunction k in the configuration interaction approximation is then

$$\psi_{kn}(\mathbf{X}_{N+1}) = \mathcal{A} \sum_{\Lambda il} \bar{\phi}_i^\Lambda(\boldsymbol{x}_1, ..., \boldsymbol{x}_N, \hat{\boldsymbol{r}}_{N+1}, \sigma_{N+1}) \, r_{N+1}^{-1} \, u_{il}^\Lambda(r_{N+1}) \, a_{ilkn}^\Lambda, \qquad (7)$$

where $\mathcal{A}$ is the antisymmetrization operator and the a_{ilkn}^Λ are the expansion coefficients. For a general N-electron target, the channel functions $\bar{\phi}_i^\Lambda(\boldsymbol{x}_1, ..., \boldsymbol{x}_N, \hat{\boldsymbol{r}}_{N+1}, \sigma_{N+1})$ are formed by coupling the configuration interaction N-electron 'target' wavefunctions $\phi_i(\boldsymbol{x}_1, ..., \boldsymbol{x}_N)$ with the spin-angle functions of the scattered or ejected electron (which depend on $\hat{\boldsymbol{r}}_{N+1}$ and σ_{N+1}) to give a state with quantum numbers $\Lambda \equiv \lambda L S M_L M_S \pi$. Here L is the total orbital angular momentum quantum number, S the total spin quantum number, M_L and M_S are the corresponding magnetic quantum numbers and π is the parity of the $(N+1)$-electron system. The quantity λ indicates the remaining quantum numbers required to completely define the channel Λ. In the present case, $N = 1$, and the channel functions can simply be written as $\bar{\phi}_i^\Lambda(\boldsymbol{x}_1, \hat{\boldsymbol{r}}_2, \sigma_2)$, being composed just of a single $N = 1$ "target" orbital $\varphi_j(r)$, together with the spin-angle part of the "outer" electron.

Therefore, before the RMF calculation proper is started, there is a first step which consists in finding the basis orbitals φ_j and in defining the combinations of these orbitals which make up the configurations to be retained in order to build the wavefunctions ϕ_i. Which states are important depends on the problem to be studied and thus there is no universally useful CI basis for a given atomic system.

The surface terms in the resulting eigenvalue problem in the inner region are removed by means of a Bloch operator and thus one obtains a Hermitian eigenvalue problem defining a set of eigenvalues and eigenfunctions in the internal region. From the eigenvalues and their eigenvectors, the spectral decomposition of the R-matrix can be obtained [3], since the R-matrix $\mathbf{R}$ is just the resolvent of the Hamiltonian, with solution matrix $\mathbf{F}(r_{N+1})$:

$$\mathbf{F}(a) = \mathbf{R}(E) \left[r_{N+1} \frac{d\mathbf{F}}{dr_{N+1}} \right]_{r_{N+1}=a}, \qquad (8)$$

evaluated at the boundary a. There are as many linearly independent solutions $F(r_{N+1})$, compatible with the physical boundary conditions at small r, as there are channels in the problem: the matrices $\mathbf{F}$ and $\mathbf{R}$ are therefore square. The R-matrix element ij is calculated as

$$R_{ij}(E) = \frac{1}{2a} \sum_k \frac{w_{ik} \, w_{jk}}{E_k - E}, \qquad (9)$$

with the amplitudes on the surface $r_{N+1} = a$ defined as

$$w_{ik} = a \langle \bar{\phi}_i(\boldsymbol{x}_1, ..., \boldsymbol{x}_N, \hat{r}_{N+1}, \sigma_{N+1}) | \psi_{kn} \rangle |_{r_{N+1}=a} . \tag{10}$$

The index k denotes the different solutions obtained from the diagonalization of the Hermitian eigenvalue problem in the inner region.

2.3 Outer region

In the outer region only a one-electron problem must be solved. The outer electron is subject to both the external laser field and to the multipole potential from the core, including the nuclear Coulomb potential, the effective core potential and the interaction with the residual other active electron in the inner region. In the following, we abbreviate the outer electron's radius simply by r.

The coupled differential equations to be solved in the outer region can be written in matrix form as

$$\left[\frac{d^2}{dr^2} + \mathbf{P} \frac{d}{dr} - \mathbf{V}(r) + 2E \right] \mathbf{F}(r) = 0, \tag{11}$$

where $\mathbf{P}$ is an antihermitian matrix proportional to the field strength and $\mathbf{F}(r)$ is the solution matrix, all matrices being square.

These equations are solved at intermediate r, $a < r < a'$, by propagating the R-matrix using standard propagators. Within the close-coupling approach, the asymptotic channels of the outer electron are given by its angular momentum and its energy, which are coupled to the final state of the residual "target". Boundary conditions must be set in a frame in which the channels decouple asymptotically, which for the present case is the so-called acceleration or Kramers-Henneberger frame [32]. Then total and partial cross-sections into the various channels can be computed. Since we are considering processes taking place fully inside the field, the asymptotic states consist on the one hand of the field-dressed free electron, which has a well-defined average direction and energy, and which goes over, as the field is turned off, to a free electron with well-defined momentum. The residual 'target', on the other hand, is also dressed by the field. At large r, $r = a'$ the equations are therefore solved by an asymptotic expansion, incorporating the appropriate physical (resonance or scattering) boundary conditions. The frame transformation to the acceleration frame is performed formally at $r = \infty$ and it is incorporated in the asymptotic expansion.

At $r = a'$ the solution from the inner region is matched to the solution from the asymptotic expansion to obtain either the K-matrix for a scattering calculation or to obtain the quasienergy E in an ionization calculation. In the latter case, since we are using Gamow-Siegert boundary conditions, the energy must be found by an iterative procedure searching for a zero in the determinant of the matching matrix $\mathbf{B}$

$$\mathbf{B} = \mathbf{R}_V(E) \left[r \frac{d\mathbf{G}(E)}{dr} + \frac{1}{2} \mathbf{P} \mathbf{G}(E) \right] - \mathbf{G}(E), \tag{12}$$

with $\mathbf{G}(E)$ the solution matrix fulfilling the asymptotic boundary conditions for the energy E and all quantities evaluated at the matching radius $r = a'$. This direct specification of boundary conditions (as opposed to complex rotation calculations) allows the computation of virtual states or of shadow poles, as long as the propagation is not performed over a too large distance.

3. RESULTS

3.1 Ionization of and harmonic generation in He and Mg

Results for ionization of He have been presented in [9, 21, 20]. The results of [21], giving the ionization rate as a function of laser intensity for a 248 nm linearly polarized laser, compare to very good accuracy with the solution of the time-dependent Schrödinger equation [20, 28].

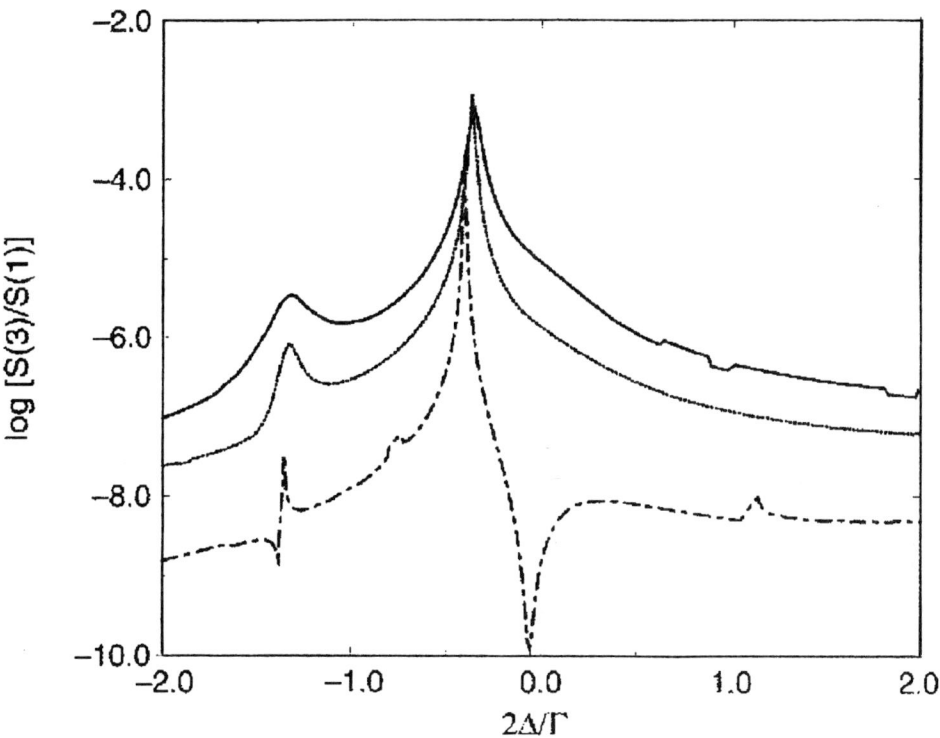

Figure 1: Third harmonic generation profile in Mg, for various laser intensities, namely (from bottom to top) 10^{10}, 5×10^{10}, and 10^{11} W/cm^2 for driving laser frequency near the three-photon resonance with the 3p3d ^{1}P^o autoionizing state. From Gebarowski et al, J Phys B **30**, 2505 (1997) [23].

Harmonic generation has been calculated with the RMF approach [23]. A detailed study of third-harmonic generation in magnesium has been performed, investigating the role of an autoionizing state resonance. The aim of these RMF calculations is *not* the computation of high-order harmonic generation (see the contribution of P Salieres to this school) but rather the determination of the influence of atomic structure on *low*-order harmonics. At resonances between bound and autoionizing states, the Floquet approach exhibits complex-energy degeneracies, which lead to interesting structure in the total and partial rates and shifts [11]. These have been termed "laser-induced degenerate states" or LIDS, in analogy with "laser-induced continuum structure" (LICS), to which they are closely related, for two-color experiments.

The generalization of the RMF approach to two frequencies, either incommensurable [19] or commensurable, has allowed the experimentally important study of double-resonances [22] and of coherent interaction between the fundamental frequency and one of its harmonics, for the commensurable case.

3.2 Multiphoton detachment from negative ions

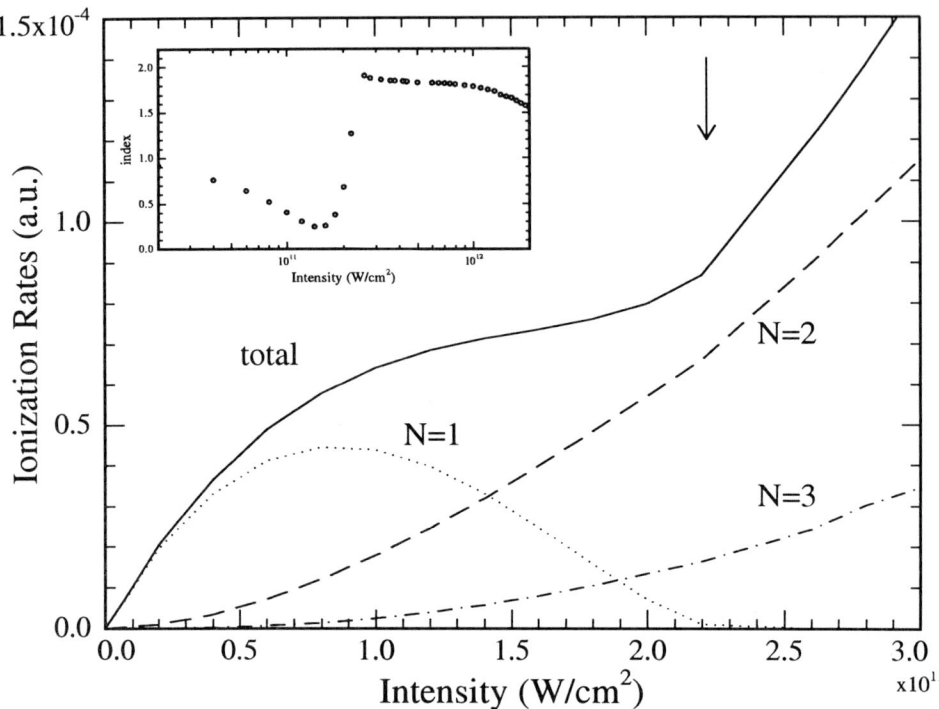

Figure 2: Ionization rates (total and partial) of the negative hydrogen ion in a laser field of frequency $\omega = 0.03$ a.u. Although the frequency is larger than the binding energy, due to electron correlation the shift is negative, resulting in channel closing (arrow). See [9] and [10].

Negatively charged ions have usually a rather weakly bound 'outer' electron and are notoriously difficult to describe accurately. The RMF theory has been applied extensively to the negative hydrogen ion H$^-$ [9, 10, 11, 14, 15], exploring total and partial rates. At low frequency, there is very good agreement with other perturbative approaches. It was found to be very useful to semi-empirically adjust the binding energy to the exact binding energy in order to obtain precise multiphoton ionization rates.

The perturbative regime for the field intensity depends quite sensitively on the frequency. Generally, at larger frequency, larger intensity can be reached within the perturbative regime. Near resonances, however, also the ones involving autoionizing states, perturbation theory breaks down at much smaller intensities [14].

Multiphoton detachment results for Li$^-$ have been reported in [16] and the non-perturbative effect of resonances has been investigated.

3.3 Electron-H scattering

The R-matrix approach being in fact primarily used for scattering calculations, "the electron-hydrogen problem" denotes generally the problem of two electrons with a proton nucleus, containing therefore also the H^- detachment problem. The e-H scattering problem in a low-frequency laser field has been solved [17], illuminating some controversial experimental and theoretical activity concerning the magnitude of the scattering cross-section. Although the net number of photons involved is quite small (1 or 2), it is very hard to reach converged values for this problem since many partial waves interfere destructively.

4. OUTLOOK AND FUTURE DEVELOPMENTS

A general basis in the inner region can in principle be obtained using any complete set of basis functions. Such an approach appears feasible only for two-active electron systems outside a closed (inactive) shell and a promising basis is given by the B-splines [12]. Within the context of field-free scattering, but with possible applications in RMF theory, pseudostate methods are presented in [26]. In [27], the e-2e process has been calculated using a two-outer-electron propagator method. A convergent close coupling method for two electrons, with possible extensions to more than two electrons has been presented [35].

These developments all have in view the question of correlated double multiphoton ionization [34]. This it is known as the "three (or more) body Coulomb problem", which even in the absence of a laser field is far from trivial. The question of nonsequential *versus* sequential multiple ionization is since long a key discussion on the way towards understanding correlated electron dynamics [34]. Our present RMF approach is restricted to a single outgoing electron and can thus only describe sequential multiple ionization. The domain of two-electron atomic systems is now well within reach of methods solving the TDSE directly [20, 31]. A review of the field has been given in [8]. Thus, the RMF method loses some of its importance for truly two-electron systems. It is still the only method capable of treating multiphoton processes in atomic systems involving three or more electrons. In any case, also for two- or one-electron systems, the Floquet results are an important tool for the interpretation of the fully time dependent results: the Floquet states give insight into the dynamics of the time dependent process, exhibiting the important atomic states involved and their intensity (or frequency) dependence, energy shifts and ionization widths.

At low frequencies, it is generally recognized that Single-Active-Electron model potential calculations can give a good approximation to the evolution of an atom in a laser field: at higher frequencies, however, when for example doubly excited resonances are important, these model potentials will no longer be appropriate.

The results presented here were all obtained for light atoms (having not more than three closed shells). For heavier atoms, the existing R-matrix programs incorporating the relativistic effects due to the large nuclear charge can be adapted to the RMF approach.

Time-dependent calculations for complex atoms modeled within a CI-'target' approach, thus within a basis similar to our R-matrix basis and with a single outgoing electron have been performed by the authors of [8]. It is desirable to produce a fully time-dependent R-matrix code, which has been proposed, with an application to a 1D model, in [30]. For this case, the inner region basis can be retained in its present form, however with time-dependent coefficients, while in the outer region the boundary conditions must be reformulated into fully absorbing or transparent boundary conditions, which again must be set in the accelerated frame.

Acknowledgments. Since almost ten years the author has been involved in a long, fruitful and pleasant collaboration on the development and implementation of the RMF method, involving the groups of Brussels (C J Joachain), Belfast (P G Burke), Daresbury (C J Noble) and Rennes (M Dunseath). The work presented here draws much on the input of many individuals who have contributed to the project. MD is supported by the Deutsche Forschungs-Gemeinschaft.

References

General, Review and Introductory:
[1] B. H. Bransden and C. J. Joachain, "Physics of Atoms and Molecules", Longman (London, 1983)
[2] P. G. Burke and K. A. Berrington eds., "Atomic and Molecular Processes: an R-matrix Approach", Institute of Physics (Bristol, 1993)
[3] P.G. Burke, A. Hibbert and W.D. Robb J. Phys. B **4**, 153 (1971) "Electron scattering by complex atoms"
[4] E. P. Wigner and L. Eisenbud, Phys. Rev. **72**, 29 (1947)
[5] C. J. Joachain, M. Dörr and N. J. Kylstra, "High intensity laser-atom physics", Adv. At. Mol. Opt. Phys. **42**, 225-286 (2000)
[6] C. J. Joachain, M. Dörr and N. J. Kylstra, "R-matrix-Floquet theory of multiphoton processes", Comm. At. Mol. Phys. **33**, 247 (1997)
[7] M. Dörr, "R-matrix-Floquet theory of multiphoton processes", p.191, in *Photon and Electron Collisions with Atoms and Molecules*, P G Burke and C J Joachain eds., Plenum Press (New York, 1997)
[8] P Lambropoulos, P Maragakis and J Zhang, Phys Rep **305**, 203–293 (1998) "Two Electron Atoms in Intense Fields"

Applications of RMF theory to two-electron systems:
[9] J. Purvis, M. Dörr, M. Terao-Dunseath, C. J. Joachain, P. G. Burke and C. J. Noble, Phys. Rev. Lett. **71**, 3943 (1993) "Multiphoton ionization of H^- and He in intense laser fields"
[10] M. Dörr, J. Purvis, M. Terao-Dunseath, P. G. Burke, C. J. Joachain, and C. J. Noble, J. Phys. B **28**, 4481 (1995) "RMF 5: multiphoton detachment of H^-"
[11] O. Latinne, N. J. Kylstra, M. Dörr, J. Purvis, M. Terao-Dunseath, C. J. Joachain, P. G. Burke and C. J. Noble, Phys. Rev. Lett. **74**, 46 (1995) "Laser-induced degeneracies involving autoionizing states in complex atoms"
[12] Hugo W van der Hart, J Phys B 30, 453 (1997) "B-spline methods in R-matrix theory for scattering in two-electron systems "
[13] D.H. Glass, P.G. Burke, H.W. van der Hart and C.J. Noble, J. Phys. B **30**, 3801 (1997) "RMF 8: a linear equations method"
[14] H.W. van der Hart and A. S. Fearnside, J. Phys. B **30**, 5657 (1997) "RMF 13: Resonances in H^- multiphoton detachment"
[15] A. S. Fearnside, J. Phys. B **31**, 275 (1998) "Intensity dependence of resonances profiles in multiphoton partial detachment rates of H^-
[16] D.H. Glass, P.G. Burke, C.J. Noble and G.B. Wöste, J. Phys. B **31**, L667 (1998) "RMF 14: Multiphoton detachment of Li^-"
[17] D. Charlo, M. Terao-Dunseath, K.M. Dunseath and J.M. Launay, J. Phys. B **31**, L539 (1998) "Angular distributions for e-H scattering in a CO_2 laser field"
[18] N.J. Kylstra, E. Paspalakis and P.L. Knight, J. Phys. B **31**, L719 (1998) "LICS in He: ab-initio non perturbative calculations"

[19] H.W. van der Hart, J. Phys. B **29**, 2217 (1996) "RMF 7: a two-colour approach"
[20] Parker, J.S.; Glass, D.H.; Moore, L.R.; Smyth, E.S.; Taylor, K.T.; Burke, P.G. J Phys B (2000) 33, L239-47 "Time-dependent and time-independent methods applied to multiphoton ionization of helium."
[21] Glass, D.H.; Burke, P.G. J Phys B (2000) 33, 407-19 "Resonances in multiphoton ionization of He at the KrF laser wavelength."
[22] Kylstra, N.J.; van der Hart, H.W.; Burke, P.G.; Joachain, C.J. J Phys B (1998) 31, 3089 "Singly, doubly and triply resonant multiphoton processes involving autoionizing states in magnesium."
[23] Gebarowski, R.; Taylor, K.T.; Burke, P.G. J Phys B (1997) 30, 2505 "RMF 12: Harmonic generation in magnesium"
[24] H. W. van der Hart, J. Phys. B **33**, 1789 (2000)
[25] M. Plummer and C. J. Noble, J. Phys. B **32**, L345 (1999)

Other R-matrix work:
[26] K. Bartschat, E.T. Hudson, M.P. Scott, P.G. Burke and V.M. Burke, *J. Phys.* B **29**, 115 (1996) "Electron-atom scattering at low and intermediate energies using an R-matrix basis with pseudo-states"
[27] K.M. Dunseath, M. LeDourneuf, M. Terao-Dunseath and J.M. Launay, Phys. Rev. A **54**,561(1996) "Two-dimensional R-matrix propagator: application to e-H scattering"
[28] See the contribution of P G Burke to this school
[29] Burke, P.G.; Colgan, J.; Glass, D.H.; Higgins, K. J Phys B (2000) 33, 143 "R-matrix-Floquet theory of molecular multiphoton processes."
[30] P. G. Burke and V. M. Burke, "Time-dependent R-matrix theory of multiphoton processes", J. Phys. B 30, L383 (1997)

Some selected other related work:
[31] R. Hasbani, E. Cormier and H. Bachau, J Phys B 33, 2101 (2000) "Resonant and non-resonant ionization of helium by XUV ultrashort and intense laser pulses"
[32] See the contribution of M Gavrila to this school
[33] H C Day, B Piraux and R M Potvliege, *Phys. Rev.* A **61**, 031402 (2000) "Multistate non-Hermitian Floquet dynamics in short laser pulses"
[34] This topic will be presented by F H M Faisal at this school
[35] See e. g. I Bray, J Phys B 33, 581 (2000) and references therein
[36] C W McCurdy, T N Rescigno and D Byrum, Phys Rev A 56, 1958 (1997)
[37] A "multi-electron multi-photon" method is being pursued by the group of T. Mercouris and C.A. Nicolaides, *e.g.* Phys. Rev. A **61**, 013407 (2000).

INTENSE-FIELD MANY-BODY S-MATRIX THEORY: APPLICATION TO RECOIL-MOMENTUM DISTRIBUTIONS FOR LASER-INDUCED DOUBLE IONIZATION

F. H. M. Faisal

Fakultät für Physik
Universität Bielefeld
D-33615 Bielefeld, Germany

INTRODUCTION

In recent years very high intensity short-pulse lasers with high repitition rates have been available in the laboratories. This has led to investigations of laser-atom interaction processes at hitherto unaccessable domains of double ionization signals vs. field intensity. Furthermore, new detection techniques have opend up the possibility of measurements of momentum distributions of the emitted electrons and ions in coincidence. Thus, for example, double ionization yields of He atom interacting with intense short pulse lasers have been measured (e.g., [1, 2, 3, 4, 5, 6]). More recently the momentum distributions of the two electrons emitted in double ionization of noble gas atoms have been determined via the measurement of the recoil momentum of the doubly charged ions in coincidence with an electron [7, 8]. These results provide important new tests of the mechanism of double ionization process in intense ultrashort laser pulses.

Here we shall briefly discuss the theoretical challenge posed by these observations and the progress made using the recently developed so-called 'intense-field many-body S-matrix theory (IMST)' to this end [9, 10, 11, 12, 13]. The theoretical difficulties for intense-field laser-atom interaction arise from a combibnation of
(a) highly non-perturbative light interaction
(b) quantum many-body problem, and
(c) non-separable Coulomb correlation interaction, $\sum_{i \neq j} 1/r_{ij}$.

One approach of attacking such problems is to try to integrate the Schrödinger equation of the system (atom coupled to the laser field) numerically. This approach is essentially exact but is limited by the computing power of the state-of-the-art computers to solve high dimensional partial differential equations over realistically large spece-time grids. Much progress have been made in treating one-electron systems (hydrogenic systems, or single-active-electron models) in 3-dimensions in the recent past. Two-electron systems in laser fields, involving 6-D partial differential equations, reaches the limit of simulation techniques for these problems using today's supercomputers (e.g. [14]).

An alternative approach to the numerical simulation, is a systematic approximation method. The recently introduced 'intense-field many-body S-matrix theory' (IMST) is designed to this end [9, 10]. In this theory the expansion of the S-matrix (or the probability amplitude) of the ionization process is rearranged in such a way that the dominant features of the process can appear in the leading terms of the asymptotic series of the S-matrix. Such a rearrangement has been shown to be possible [9] by introducing three partitioning of the total Hamiltonian of the interacting system, namely, an initial, a final *and* an intermediate partitioning. At the same time the laser interaction with the free electron (whether in the final or in the virtual intermediate states) is taken into account to all orders *non*-perturbatively by using the Volkov solution (i.e. the exact solution of a free elctron in a laser field, e.g. [15], p. 11) and the Volkov propagator [16, 17]. As will be seen below, introduction of the intermediate partition permits a remarkable flexibility to the S-matrix expansion that is not present in other versions, such as the usual 'prior' or 'post' [18, 19] form of the S-matrix (or evolution operator) series. This novel expansion allows one, in particular, to take into account, already in the leading terms of the theory, the virtual fragment-states of a *many*-body system, through which the reaction of interest may proceed predominantly. This can be done in IMST *without* affecting the orthogonal projections to the desired intial and final states of the transition of interest. In contrast, the usual 'prior' or 'post' expansions would have to be carried out to large (if not to infinite) orders, in order to account for such intermediate fragment-states. This is because the 'prior' or the 'post' expansion is restricted to use the *same* reference propagator that is appropriate only for the initial or the final-state of the process.

INTENSE-FIELD MANY-BODY S-MATRIX THEORY (IMST)

The essential steps for the derivation of IMST consists, first, in rewriting the time-dependent Schrödinger equation of the system as a time-dependent generalization of the Lippmann-Schwinger equation satisfying the initial condition. Thus, we partition the total Hamiltonian of the system $H(t)$, as

$$H(t) = H_i^0 + V_i(t). \tag{1}$$

i.e., as a sum of the initial reference Hamiltonian H_i^0 (of the unperturbed system) and write

$$[(i\hbar \frac{d}{dt} - H_i^0) - V_i(t)]|\Psi(t)>= 0 \tag{2}$$

as an integral equation

$$|\Psi(t)>= |\phi_i(t)> + \int_{t_i}^{t} dt_1 G(t,t_1) V_i(t_1)|\phi_i(t_1)> \tag{3}$$

where we have introduced the total Green's function (or propagator) G(t,t') by the definition

$$[i\hbar \frac{d}{dt} - H(t)]G(t,t') = \delta(t-t') \tag{4}$$

And, $\phi_i(t)$ is the solution of the initial unperturbed Hamiltonian, H_i^0 (e.g. He atom),

$$[i\hbar \frac{d}{dt} - H_i^0]|\phi_i(t)>= 0 \tag{5}$$

that is chosen to satisfy the initial condition. The initial interaction is $V_i(t)$(e.g. interation of the two electrons with the vector potential of the laser-field,

$$V_i(t) = \sum_{j=1,2} [-\frac{e}{mc}\mathbf{p}_j \cdot \mathbf{A}(t) + \frac{e^2}{2mc^2}A^2(t)] \tag{6}$$

Second, we introduce the final-state partition of the total Hamiltonian

$$H(t) = H_f^0(t) + V_f(t), \tag{7}$$

where, $H_f^{(0)}(t)$ is the final reference Hamiltonian (e.g. the two-electron Volkov Hamiltonian [16, 17]), and $V_f(t)$ is the final state residual interaction (e.g. the Coulomb ineraction, including the electron-electron correlation). The associated Green's function satisfies

$$[i\hbar\frac{d}{dt} - H_f^0(t)]G_f^0(t,t') = \delta(t - t') \tag{8}$$

Third, we expand the total Green's function in terms of the final Green's function in the form

$$G(t,t') = G_f^0(t,t') + \int_{t_1}^{\infty} dt_1 G_f^0(t,t_1)V_f(t_1)G(t_1,t') \tag{9}$$

and insert it in the expression for the total wavefunction $|\Psi(t)>$ above, Eq. (3), to get

$$\begin{aligned}|\Psi(t)> &= |\phi_i(t)> + \int_{t_i}^{t} dt_1 G_f^0(t,t_1)V_i(t_1)|\phi_i(t_1)> \\ &+ \int_{t_i}^{t}\int_{t_i}^{t_2} dt_2 dt_1 G_f^0(t,t_2)V_f(t_2)G(t_2,t_1)V_i(t_1)|\phi_i(t_1)>.\end{aligned} \tag{10}$$

Note that in this way we have arrived at a useful form for the wavefunction that not only satisfies the initial-state condition but also is well arranged for computing the transition amplitude by projection to any state of the final reference Hamiltonian, even when the latter is *not* identical to the intial reference Hamiltonian. This is because the projection is orthogonal and hence extracts only*one* term from the final reference Green's function (given in the properstate representation). Such a formulation is, therefore, useful not only in the present context but also for many other processes, e.g., charge exchange and chemical reactions (which invariably involve unequal reference Hamiltonians for the input and output channels). Thus, projecting onto the final state $|\phi_f(t)>$, belonging to $H_f^0(t)$, we write the transition amplitude for the i $\to$ f transition as:

$$\begin{aligned}(S-1)_{fi}(t) &= \int_{t_i}^{t} dt_1 <\phi_f(t_1)|V_i(t_1)|\phi_i(t_1)> + \\ &+ \int_{t_i}^{t}\int_{t_i}^{t_2} dt_2 dt_1 <\phi_f(t_2)|V_f(t_2)G(t_2,t_1)V_i(t_1)|\phi_i(t_1)>\end{aligned} \tag{11}$$

This is an exact *master* equation for the S-matrix, whose most important features, as indicated above, are (a) that the orthogonal projections for the initial and the final states are carried out independently of equal *or* unequal reference Hamiltonians, and (b) that the total Green's function appears *between* the initial and the final interactions. This is unlike the appearence of the Green's function at one end of the usual 'prior' or 'post' form (see e.g.,[18, 19])) expansion of the S-matrix theory (or the Dyson series of the evolution operator). As mentioned, it is this flexibility of Eq. (11) that permits the introduction of any *virtual* fragments propagator of interest, already in the leading terms of the IMST. Note that, inclusion of the effects of such virtual fragments-channels in the usual 'prior' and 'post' expansions may only be obtained indirectly, if at all, by summing them to very high orders, if not to infinite orders. It is this difference with the other forms of S-matrix (or evolution operator) expansions that makes IMST more useful for many-body reaction problems. Thus, we make the third intermediate partition of the total Hamiltonian

$$H(t) = H_0(t) + V_0(t) \tag{12}$$

to introduce the virtual fragments-Hamiltonian $H_0(t)$, and the corresponding propagator $G_0(t,t')$, satisfying the equation

$$[i\hbar\frac{d}{dt} - H_0(t)]G_0(t,t') = \delta(t-t') \tag{13}$$

For example, for the double ionization problem of present interest, the virtual intermediate fragments can consist of one electron in the Volkov states of all virtual momenta $\{\mathbf{k}\}$, and the singly charged residual ion in its virtual eigenstates; the corresponding fragments-Hamiltonian is, therefore,

$$H_0(t) = [(\mathbf{p}_1 - \frac{e}{c}\mathbf{A}(t)]^2/2m + [\mathbf{p}_2^2/2m - Ze^2/r_2]; (Z=2, \text{He}) \tag{14}$$

The corresponding two-particle fragments-Green's function satisfying Eq. (13) is

$$G_0(t,t') = -\frac{i}{\hbar}\theta(t-t')\sum_j \frac{1}{(2\pi)^3}\int d\mathbf{k}|\mathbf{k}>|\phi_j^+(2)>$$
$$\times e^{-\frac{i}{\hbar}\int_{t'}^{t}\{[(\mathbf{p}_1-e\mathbf{A}(\tau)/c]^2/2m+E_j\}d\tau} < \mathbf{k}| < \phi_j^+(2)| \tag{15}$$

where, $\theta(t-t')$ is the Heaviside theta function, $\{|\mathbf{k}>\}$ is the complete set of plane wave states with wavevectors $\mathbf{k}$, and $\{|\phi_j^+(2)>\}$ is the complete set of residual ionic states.

Once $G_0(t,t')$ is available, we can expand $G(t,t')$ as

$$G(t,t') = G_0(t,t') + \int_{t_i}^{t} dt_1 G_0(t,t_1)V_0(t_1)G_0(t_1,t') + \cdots \tag{16}$$

and substitute it in the expression for the total wavefunction, Eq. (10) above, to obtain,

$$|\Psi(t)> = |\phi_i(t)> + \int_{t_i}^{t} dt_1 G_f^0(t,t_1)V_i(t_1)|\phi_i(t_1)>$$
$$+ \int_{t_i}^{t}\int_{t_i}^{t_2} dt_2 dt_1 G_f^0(t,t_2)V_f(t_2)G_0(t_2,t_1)V_i(t_1)|\phi_i(t_1)> + \cdots \tag{17}$$

(Note that additional intermediate partitions, if desired, can also be introduced exactly in the same way again.)

Last, to obtain the transtion amplitude explicitly, we substitute the above wavefunction, Eq. (17), in Eq. (11), and write the resulting series as

$$(S-1)_{fi}(t) = \sum_{j=1}^{\infty} S_{fi}^{(j)}(t) \tag{18}$$

with

$$S_{fi}^{(1)}(t) = \int_{t_i}^{t} dt_1 <\phi_f(t_1)|V_i(t_1)|\phi_i(t_1)> \tag{19}$$

$$S_{fi}^{(2)}(t) = \int_{t_i}^{t}\int_{t_i}^{t_2} dt_2 dt_1 <\phi_f(t_2)|V_f(t_2)|G_0(t_2,t_1)V_i(t_1)|\phi_i(t_1)> \tag{20}$$

$$S_{fi}^{(3)}(t) = \int_{t_i}^{t} \int_{t_i}^{t_3} \int_{t_i}^{t_2} dt_3 dt_2 dt_1 < \phi_f(t_3) |V_f(t_3)| G_0(t_3,t_2) V_0(t_2) G_0(t_2,t_1) V_i(t_1) |\phi_i(t_1)> \quad (21)$$

....

We note that the first term of IMST is of the same form as the so-called KFR theory [20, 21, 19]. It can be shown that the 'prior', 'post' and IMST series for this first term are equivalent (see, [22], footnote 11) for long interaction times.

RECOIL-MOMENTUM DISTRIBUTION IN DOUBLE IONIZATION OF He

Let us illustrate the above theory (IMST) by applying it to the problem of recoil-momentum distributions in double ionizations in intense femtosecond Ti:sapphire laser pulses. We discuss below the recent results for the doubly charged ions of He, both *parallel* as well as *perpendicular* to the polarization axis of the laser. The momentum of the doubly charged ions is related to that of the two outgoing electrons via the momentum conservation relation. Thus, neglecting the very small contribution from the negligible momentum of the (long wavelength) laser photons, the overall momentum conservation requires that the sum of the electron momenta parallel to the laser plarization axis, (z-axis), satisfies,

(i) $\mathbf{P}_{par.} \approx -[(\hbar \mathbf{k}_a)_{par.} + (\hbar \mathbf{k}_b)_{par.}]$

and that perpendicular to it satisfies,

(ii) $\mathbf{P}_{perp.} = -[(\hbar \mathbf{k}_a)_{perp.} + (\hbar \mathbf{k}_b)_{perp.}]$

where, $\mathbf{P}$ is the momentum of the doubly charged ion, and $\hbar \mathbf{k}_a$ and $\hbar \mathbf{k}_b$ are the momenta of the two outgoing electrons.

Before discussing the results of calculations, we summarise the most prominent features of the observed distributions as found in the experiments [7, 8]. They are:

(a) the component of the recoil momenta *parallel* to the laser polarization direction show a prominent *double*-hump distribution with a central minimum

(b) the perpendicular component of the recoil momentum show rather a single-hump distribution

(c) the parallel component distribution is very broad

(d) the perpendicular distribution is narrow

(e) the maximum momentum transfer of the parallel component show a sharp cut-off at a value $\mathbf{P}_{par.} \approx Re(4\sqrt{mU_p})$ for He [7] and a still lager value for Ne [8], where $U_p = \frac{e^2 A_0^2}{4mc^2}$ is the so-called ponderomotive energy i.e. the mean quiver-energy of the free electron in the laser field.

(f) the distributions are qualitatively different from the corresponding distributions for the $\gamma + 2e$ reaction by (weak field) synchrotron photons, which favours the Warnier (back to back) mechanism near zero momentum.

Previously we had extensively investigated the intensity dependence of the *total* double ionization yields using the present theory [10] and also interms of simplified *models* based on it [11, 12]. These investigations permitted us to identify a so-called correlation mediated energy-sharing diagram, also called the non-sequential (NS) diagram, that automatically incorporates in the quantum domain the well-known classical 'rescattering model' proposed by Corkum [23].

In fact, for the double ionization problem the leading two terms of the IMST series, Eqs. (17) and (18), provide (c.f., e.g., [9]) one first rank, and six second rank Feynman diagrams of which only one is found to be dominant in the intensity *and* wavelength of experimental interest (near infrared, femtosecond Ti:sapphire lasers), in the so-called non-sequential domain (e.g., [5]). (The 'rank' of a diagram corresponds to the number of times the generalised interactions (vertices) appear in the diagram.) The correlation mediated energy-sharing diagram is shown in Fig.1.

The physical picture behind the present double ionization process can in fact be visualized directly by reading the diagram from the bottom to the top (in the indicated direction of the flow of time). Thus, at an initial time t_i the two electrons (say, 1 and 2) are in the ground state of the unperturbed He atom. At a time t_1 (when the field phase is $\phi_1 = \omega t_1$) one of the bound electrons (say, 1) absorbs, via the generalised ATI-like interaction($V_{ATI}(t_1)$), a large amount of field energy by a virtual above-threshold ionization (ATI) process, while the other electron propagates in the virtual states of the residual ion. This intermediate ensemble of virtual states corresponds to the two-particle propagator (Green's function) $G_0(t, t')$, ({Volkov} $\otimes$ {ion states}) [9, 13]. Then, at a time t_2 (field phase $\phi_2 = \omega t_2$ the two electrons interact ('internally collide') via the electron-electron correlation ($V_{corr.}(t_2)$) (plus, in general, any residual interaction with the nucleus) and share the energy until both of them may have enough energy to escape together from the binding force of the atom, into the final two-electron Volkov state (c.f., [16, 17]). Finally, at a large time t_f they proceed to the detector(s) with the final momenta $\hbar \mathbf{k}_a$ and $\hbar \mathbf{k}_b$.

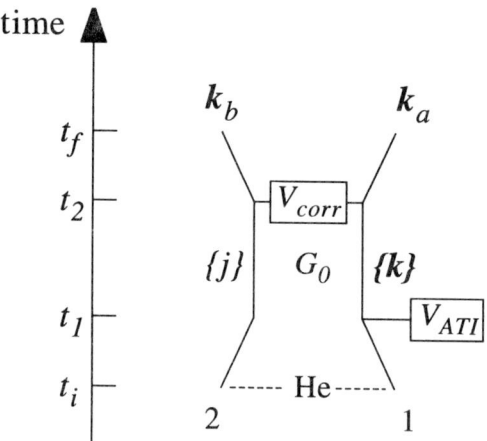

Fig 1. Correlation mediated energy-sharing diagram for laser induced non-sequential double ionization of He. For interpretation see text.

Note that quantum mechanically speaking, since the electrons are not observed in the above intermediate states in the experiment, contributions to the double ionization amplitude, in principle, must be integrated for all values of $\Delta t = t_2 - t_1$ (or the phase difference, $\Delta\phi = \phi_2 - \phi_1$). Under appropriate limiting conditions, however, this integration may be estimated semiclassically. Under the latter condition a significant contribution could then arise from time interval(s) of the order of the classical return time(s) of the first emitted electron to the core region, with a correponding return energy of the order of $3U_p$. This provides the theoretical connection between the present quantum theory and the classical 'rescattering model' of Corkum [23], mentioned above. For the purpose of actual calculations, we write down the NS double ionization amplitude, following the diagram in Fig. 1, as,

$$S_{fi}^{(2)}(t_f, t_i)|_{NS} = -\frac{i}{\hbar} \int_{t_i}^{t_f} dt_2 < \phi_f^V(\mathbf{k}_a, \mathbf{k}_b; \mathbf{r}_1, \mathbf{r}_2; t_2)$$
$$\times |V_{corr}(t_2)|\Psi_i(\mathbf{r}_1, \mathbf{r}_2; t_2) > \quad (22)$$

where, the angular brackets stand for the spatial integrals; $\phi_f^V(\mathbf{k}_a, \mathbf{k}_b; t_2)$ is the two-electron product Volkov final state in the field (c.f. [16]), $V_{corr}(t_2)$ stands for correlation operator and $\Psi_i(t_2)$ is the total wavefunction of the system, at time t_2. In the present approximation, $\Psi_i(t_2)$ is given by (c.f. Fig. 1),

$$\Psi_i(\mathbf{r}_1, \mathbf{r}_2; t_2) = \int_{t_i}^{t_2} dt_1 G_0(\mathbf{r}_1, \mathbf{r}_2, t_2; \mathbf{r}'_1, \mathbf{r}'_2, t_1)$$
$$\times |V_{ATI}(t_1)|\phi_{1S}(\mathbf{r}'_1, \mathbf{r}'_2; t_1) >$$
$$= \int_{t_i}^{t_2} dt_1 \sum_j \frac{1}{(2\pi)^3} \int d\mathbf{k} |\phi^V(\mathbf{k}; \mathbf{r}_1; t_2)\phi_j^+(\mathbf{r}_2; t_2) > \quad (23)$$
$$\times < \phi_j^+(\mathbf{r}'_2; t_1)\phi^V(\mathbf{k}; \mathbf{r}'_1; t_1)|V_{ATI}(t_1)|\phi_{1S}(\mathbf{r}'_1, \mathbf{r}'_2; t_1) >$$

where, $G_0(t_2, t_1)$ is the intermediate two-electron Green's function, $V_{ATI}(t_1)$ is the interaction operator for the virtual above-threshold ionization or ATI-like process at the time t_1, and $\phi_{1S}(\mathbf{r}_1, \mathbf{r}_2; t_1)$ is the ground state wavefunction of the He atom with binding energy $E_B = 2.904$ (a.u.). An exact evaluation of this amplitude, including all orders of correlation and ATI interaction, is practically an impossible task. We have, therefore, restricted ourselves in this work to the lowest significant terms of the theory by replacing $V_{corr}(t_2)$ by $1/r_{12}$ and $V_{ATI}(t_1)$ by $((-\frac{e}{mc}\mathbf{A}(t_1)\cdot\mathbf{p}_1) + \frac{e^2}{2mc^2}A^2(t_1))$ This simplification still requires calculation of a formidable multiple integral which we have evaluated approximately as follows. The Jacobi-Anger formula ([24], p. 7) is used to expand the Volkov wavefunctions, $\phi^V(\mathbf{k}, \mathbf{r}; t)$ in terms of the corresponding Fourier components defined by the generalized Bessel functions of two arguments (e.g., [15]). This allows us to evaluate the two-fold time integrations, over the instants t_1 and t_2 of the two interactions, exactly; for pulses much longer than a laser period (as in the experiments of present interest) this gives,

$$S_{fi}^{(2)}(\infty, -\infty)|_{NS} = -2\pi i \sum_N$$
$$\delta\left(\frac{\hbar k_a^2}{2m} + \frac{\hbar k_b^2}{2m} + E_B + 2U_p - N\hbar\omega\right) T^{(N)}(\mathbf{k}_a, \mathbf{k}_b), \quad (24)$$

where,

$$T^{(N)}(\mathbf{k}_a, \mathbf{k}_b) = \sum_n \sum_j \int \frac{1}{(2\pi)^3} d\mathbf{k}$$
$$\times < \phi^0(\mathbf{k}_a, \mathbf{r}_1)\phi^0(\mathbf{k}_b, \mathbf{r}_2)|\frac{1}{r_{12}}|\phi_j^+(\mathbf{r}_2)\phi^0(\mathbf{k}, \mathbf{r}_1) > \times$$
$$\times \frac{J_{N-n}\left(\boldsymbol{\alpha}_0\cdot(\mathbf{k}_a + \mathbf{k}_b - \mathbf{k}); \frac{U_p}{2\hbar\omega}\right) J_n\left(\boldsymbol{\alpha}_0\cdot\mathbf{k}; \frac{U_p}{2\hbar\omega}\right)}{\frac{\hbar k^2}{2m} - E_j + E_B + U_p - n\hbar\omega + i0}$$
$$\times (E_j - E_B - \frac{\hbar k^2}{2m}) < \phi_j^+(\mathbf{r}_2)\phi^0(\mathbf{k}, \mathbf{r}_1)|\phi_{1S}(\mathbf{r}_1, \mathbf{r}_2) > \quad (25)$$

where $\phi^0(\mathbf{k}, \mathbf{r})$ is a plane wave state.

Next, the six-fold space integrations are carried out analytically, the radial integration in k is performed by pole approximation and the integrals over the angles of $\mathbf{k}$ and the sum over n are performed numerically. Contribution of the lowest term of the sum over j (corresponding to the ground state of the ionic state) is retained only since it is found to dominate over the contribution from any excited state. Finally, Monte-Carlo sampling method is used to evaluate the differential rate of double ionization, $\sum_N \frac{dW^{(N)}}{d\mathbf{k}_a d\mathbf{k}_b}$, as a function of the two outgoing momenta $\mathbf{k}_a$ and $\mathbf{k}_b$, from the formula,

$$\frac{dW^{(N)}}{d\mathbf{k}_a d\mathbf{k}_b} = 2\pi\delta\left(\frac{\hbar k_a^2}{2m} + \frac{\hbar k_b^2}{2m} + E_B + 2U_p - N\hbar\omega\right) \times |T^{(N)}(\mathbf{k}_a, \mathbf{k}_b)|^2 \ . \qquad (26)$$

Computations are carried out for He atom, and the distributions of the parallel and the perpendicular components of the *sum* of the two momenta, $(\hbar\mathbf{k}_a + \hbar\mathbf{k}_b)$, are determined, in each case by integrating over the remaining variables and summing the contributions from all significant Ns. We recall that under the condition of the experiment the He^{2+} recoil momentum, $\mathbf{P}$, satisfies $\mathbf{P} \approx -(\hbar\mathbf{k}_a + \hbar\mathbf{k}_b)$ [7, 8].

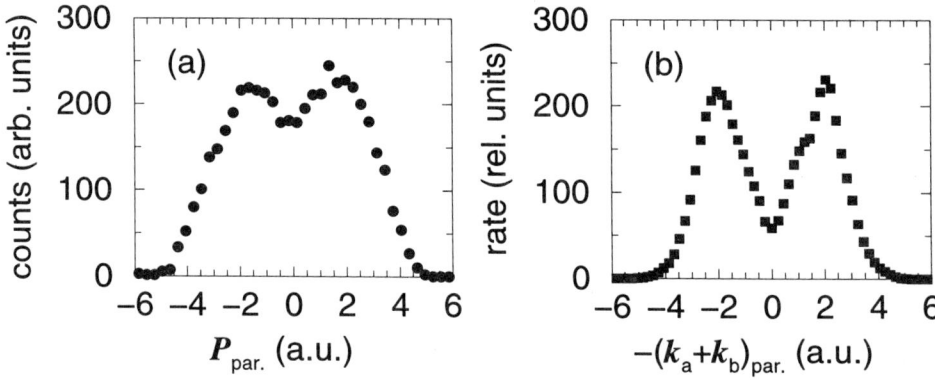

Fig. 2 Recoil ion momentum distribution of He^{2+} parallel to the polarization direction, $P_{par.}$ (experimental data, panel a) and the sum-momentum of the two outgoing electrons in the opposite direction (present theory, panel b).

We show in Fig.2 the experimental result (panel a) obained for the recoil momentum of He^{++}, *parallel* to the (linear) polarization axis, by Weber et al. [7], in the field of a Ti:sapphire laser pulse of $200 fs$, and a peak intensity of $6.6 \times 10^{14} W/cm^2$ and $\lambda = 800 nm$. They are compared with the corresponding results of the present theory [13] in panel b. We note that the experimental data are available only in arbitrary units. One data point is, therefore, fitted to the theory to determine the *relative* scale for the entire set. As it can be seen from the comparison, all the essential features of the experimental distribution of the parallel component of the recoil momentum are reproduced by the theoretical results. Thus, both the experimental and the theoretical distributions show a double-hump structure with a central minimum. The positions and the heights of the of the two maxima are also well reproduced. We observe that the maximum size of the recoil ion momentum (cut-off momentum) of the experiment and the calculation of the sum electron-momenta also agree well with each other and, in fact, in both cases it is as large as $\approx 5 a.u.$. Finally, we note that there is a quantitative difference at the minimum of the distribution. This suggests that either the uncertainty in the intensity measurement and momentum resolution [25] and/or the higher order contribution in the theory might be involved here, which should be investigated in the future.

In Fig.3 we show the distributions for the perpendicular component of the recoil momentum. Again the essential features of the experimental distribution, panel a, are reproduced by the theoretical calculation for the sum momentum of the electrons, panel b. Moreover, comparison of Fig.2 with Fig.3 shows that the theoretical calculations are consistent with the observation with respect to relative widths in the parallel and perpendicular directions, the latter being much narrower than the former.

What is the origin of the double-hump structure in the parallel case *and* its absence in the perpendicular case? To gain a greater insight into their origin we calculated the corresponding distributions by deliberately neglecting the *final*-state interaction of the two electrons with the laser pulse, i.e. dropping the 'Volkov dressing' of the two electron in the final state. This is easily done in the present theory by replacing the final state two-electron Volkov wavefunction, by two free (plane) waves of momenta $\mathbf{k}_a$ and $\mathbf{k}_b$, and keeping everything else the same.

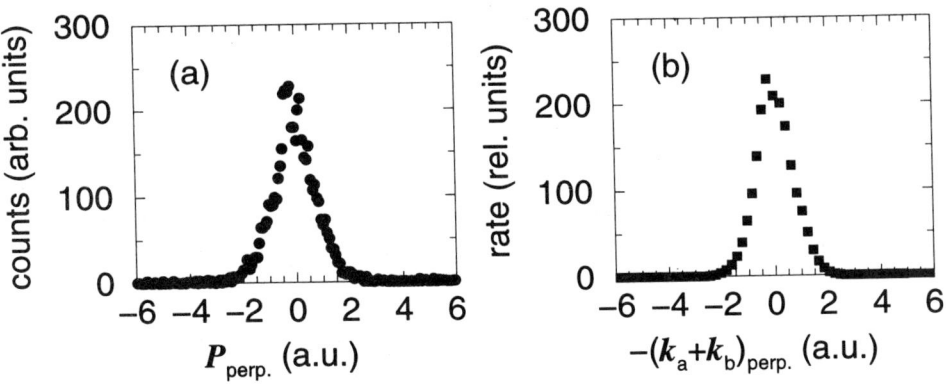

Fig.3 The same as in Fig.2 but perpendicular to the polarization direction.

The result so obtained for the parallel component is shown in Fig.4a. A comparison of the theoretical calculation *with* (Fig.2b) and *without* (Fig.4a) the Volkov dressing in the final state, clearly shows that the double-hump character of the parallel component of the distribution collapses into a single-hump structure in the *absence* of the final state Volkov dressing. This unequivocally suggests that the final-state field interaction of the two electrons is primarily responsible for the double- hump structure of the observed distribution in this case.

In Fig.4b we show the corresponding result of calculation for the perpendicular component in the absence of the final state laser interaction. Comparison of this distribution with that of Fig.3b shows that in this (perpendicular) case the presence of the final state laser interaction does not paly a significant role. This is as might be expected in the present case, since the force due to the electric field is negligible in the direction perpendicular to the polarisation direction. This is also consistent with the observed narrow width of the perpendicular distribution due to the absence of significant laser coupling in the final state. Therefore, in this case we have the interesting situation of observing the double ionization in the laboratory as if the laser field is switched off as soon as the two electrons are freed released from the bound atomic system.

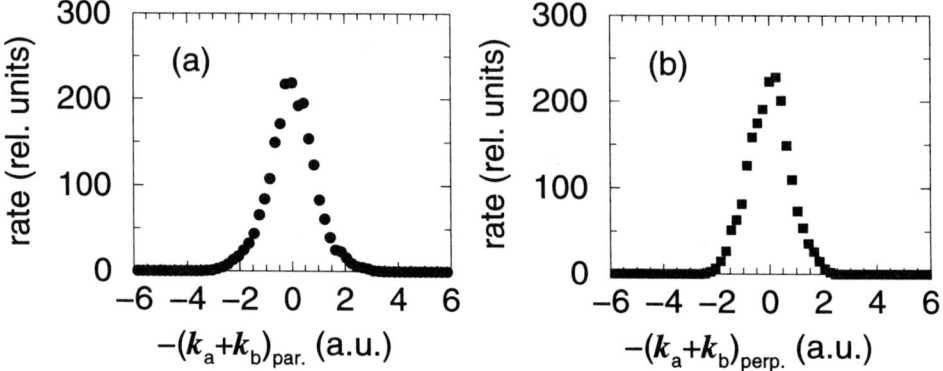

Fig. 4 Sum-momentum distributions calculated *without* the final state 'Volkov dressing' of the two outgoing electrons: (a) parallel case, (b) perpendicular case.

We may finally ask, what determines the observed large width of the parallel distributions in Fig.2? From the explicit expression of the transition matrix element (see, [13], Eqs. (3) and (4)), and using the properties of the (generalised) Bessel functions involved, one could derive [13] the following expression for the cut-off momentum of the distribution,

$|P_{par.}|_{(cut-off)} \approx |(\hbar \mathbf{k_a})_{par.} + (\hbar \mathbf{k_b})_{par.}|_{(cut-off)} = Re(4\sqrt{mU_p} + \sqrt{m(8U_p - E_B)})$, where E_B is the binding energy. For the case of the distribution shown in Fig.2 use of this formula predicts the cut-off momentum ≈ 5 a.u., in good agreement with both the experimental value, and the numerical result. Satisfactory agreement with the above cut-off formula and the available experimental values at other intensities for He and Ne [7, 8] has also been found [13].

ACKNOWLEDGMENTS

The results discussed here have been obtained in collaboration with Dr. Andreas Becker. This work was partially supported under the project number, FA 160/18-2, SPP: "Wechselwirkung intensiver Laserfelder mit Materie", Deutsche Forschungsgemeinschaft (Bonn).

References

[1] D. N. Fittinghoff et al., Phys. Rev. Lett. **69**, 2642 (1992).

[2] B. Walker et al., Phys. Rev. A **48**, R894 (1993).

[3] K. Kondo et al., Phys. Rev. A **48**, R2531 (1993).

[4] D. N. Fittinghoff et al., Phys. Rev. A **49**, 2174 (1994).

[5] B. Walker et al., Phys. Rev. Lett. **73**,1227 (1994).

[6] S. Augst et al., Phys. Rev. A **52**, R917 (1995); A. Talebpour et al., J. Phys. B **30**, 1721 (1997).

[7] Weber et al. Phys.Rev.Lett **84**, 443 (2000).

[8] Moshammer et al. Phys.Rev.Lett **84**, 447 (2000).

[9] F.H.M. Faisal and A. Becker, in *Selected Topics on Electron Physics*, Eds., D.H. Campbell an H. Kleinpoppen, Plenum Press: N.Y. (1996), p. 317.

[10] A. Becker and F.H.M. Faisal, J.Phys. B **29**, L197 (1996);*ibid* **32**,L225 (1999).

[11] F. H. M. Faisal and A. Becker, Laser Phys. **7**, 684 (1997); in *Multiphoton Processes 1996*, ed. by P. Lambropoulos and H. Walther, Int. Nat. Conf. Ser. No. 154 (IOP: Bristol, 1997) p. 118.

[12] A.Becker and F.H.M. Faisal, Phys.Rev. A**59**, R1742 (1999); *ibid* A**59**, R3182 (1999).

[13] A.Becker and F.H.M. Faisal, Phys.Rev.Lett **84**, 3546 (2000).

[14] J.S. Parker, K.T. Taylor, C.W. Clark and S. Blodgett-Ford, J. Phys. B **29**, L33 (1996); E.S. Smyth, J.S. Parker and K.T. Taylor, Comp. Phys. Comm. **144**, 1 (1998).

[15] F.H.M. Faisal, *Theory of Multiphoton Processes*, Plenum Press, N.Y. (1987).

[16] F.H.M. Faisal, Phy. Lett., A **187**, 180 (1994).

[17] A. Becker and F.H.M. Faisal, Phys.Rev. A**50**, 3256 (1994).

[18] C. Joachain, *Qunatum Collision Theory*, 3rd. edn., North-Holland, Amsterdam (1983).

[19] H. R. Reiss, Phys. Rev. A **22**, 1786 (1980).

[20] L. V. Keldysh, Zh. Eksp. Teor. Fiz. **47**, 1945 (1964) [Sov. Phys. JETP **20**, 1307 (1965)];

[21] F. H. M. Faisal, J. Phys. B **6**, L89 (1973);

[22] F.H.M. Faisal, A. Becker and J. Muth-Böhm, Laser Phys. **9**, 115 (1999).

[23] P. Corkum, Phys. Rev. Lett., **71**, 1994 (1993).

[24] A. Erdélyi (Ed.), *Higher Transcendental Functions*, Vol. 2, (New York: McGraw-Hill, 1953).

[25] The uncertainty in the intensity measurement is $\approx$ 15 - 30%, and that of momentum resolution $\approx$ 0.2 - 0.4 a.u. [R. Dörner, and H. Rottke (private communication)].

ELECTRONS AND IONS IN RELATIVISTIC LASER FIELDS

Y. I. Salamin*, M. W. Walser, S. X. Hu[†], and C. H. Keitel

Theoretische Quantendynamik
Fakultät für Physik, Universität Freiburg
Hermann-Herder-Strasse 3
D-79104 Freiburg, Germany

INTRODUCTION

In general, theoretical analysis of the interaction of a single electron with the radiation field, in vacuum, may be carried out at various levels of sophistication, with the transition from one level to the next made on clear physical grounds. Under conditions to be described below, a nonrelativistic classical description is adequate. However, in the presence of fields of sufficiently high intensity the electron gets accelerated to speeds close to that of light, in which case a relativistic treatment becomes essential[1-8]. On the other hand, in situations where the electron motion takes place over small spatial dimensions, rules of the quantum world rather than those of Newtonian mechanics apply[9,10] even the spin degree of freedom becomes important in this regime[11-13]. Still, at a more sophisticated level it becomes necessary to *second* quantize both fields, matter (the electron) and radiation.

The starting point for a classical study of the electron dynamics is the Newton-Lorentz equation

$$\frac{d\bm{p}}{dt} = -e(\bm{E} + \bm{\beta} \times \bm{B}), \qquad (1)$$

where $\bm{p}$ is the momentum of the electron, e is the magnitude of its charge, and $\bm{\beta}$ is its velocity $\bm{v}$ scaled by the speed of light c. Furthermore, $\bm{E}$ and $\bm{B}$ are the electric and magnetic components, respectively, of the radiation field. Electrostatic units will be used throughout, unless noted otherwise. We will also use the terms *radiation field* and

---

*Permanent address: Physics Department, Birzeit University, P. O. Box 14 - Birzeit, Palestine.
[†]Present address: Max-Born Institut, Max-Born-Strasse 2a, 12489 Berlin, Germany.

laser field interchangeably. As usual, the radiation E and B fields will be derived from a vector potential A. Adopting a plane-wave representation for the vector potential and assuming linear polarization, we may write

$$A = \hat{i} A_0 \cos(\omega t - \mathbf{k} \cdot \mathbf{r}). \tag{2}$$

In this equation $\mathbf{k}$ and ω are the propagation vector and frequency, respectively, of the field whose magnitudes are related by $k = \omega/c$, $k = |\mathbf{k}|$. Furthermore, $\mathbf{r}$ and t are the space and time coordinates of the electron. Thus

$$E = -\frac{1}{c}\frac{\partial A}{\partial t} = \hat{i} E_0 \sin(\omega t - \mathbf{k} \cdot \mathbf{r}), \tag{3}$$

$$B = \nabla \times A = \hat{j} B_0 \sin(\omega t - \mathbf{k} \cdot \mathbf{r}), \tag{4}$$

where $\hat{i}, \hat{j}$ and $\hat{k}$ are unit vectors in the x, y and z directions of a Cartesian coordinate system. Note that the field is polarized in the $+x$ direction and propagates along $+z$. Also, $E_0 = B_0 = \omega A_0/c$.

Provided the field intensity is not too high, the electron may not reach *relativistic* speeds and the condition $\beta \ll 1$ holds. Under these conditions the second term on the right hand side of Eq. (1) may be dropped. This may be viewed as essentially equivalent to dropping the magnetic component of the radiation field by comparison with the electric component. Conditions may also be invoked to justify resorting to the long-wavelength approximation in describing the radiation field. So, if the electron moves over spatial dimensions small by comparison to a radiation wavelength, then $kr \ll 1$ and the field phase becomes $\approx \omega t$ in Eq. (3). It becomes immediately evident that, under these conditions, the electron motion is oscillatory and along the polarization direction. With $p = mv$, a single integration of (the approximate form of) Eq. (1) leads to

$$v(t) = \frac{eE_0}{m\omega} \cos \omega t, \tag{5}$$

where a constant term has been dropped. According to Eq. (5) the maximum speed attainable by the electron is $v_m = eE_0/m\omega$. When v_m approaches the speed of light c, a relativistic analysis becomes necessary. Equivalently, this happens when the *dimensionless intensity parameter*

$$q = \frac{eE_0}{m\omega c}, \tag{6}$$

approaches unity. Note that, for radiation of wavelength $\lambda = 1\mu m$, $q^2 = 1$ is equivalent to an intensity $I \approx 1.375 \times 10^{18}$ W/cm^2. Reference will always be to this field intensity whenever the term *relativistic intensity* is invoked in this work.

Transition between the classical and quantum regimes may also be gauged by other dimensionless parameters[14]. The physical quantity playing a decisive role in this respect is the electron's time-averaged kinetic energy of oscillation in the field, or *ponderomotive potential* U_p. From Eq. (5) one has

$$U_p = <\frac{1}{2}mv^2> = \frac{(eE_0)^2}{4m\omega^2}. \tag{7}$$

Field intensities in excess of the relativistic value are now available for laboratory experiments[15]. A few interesting effects and potentially important applications, not accessible with laser systems that deliver fields of nonrelativistic intensities, my now be subjected to the experimental test. For example, the transition from Thomson to Compton scattering[16] and nonlinear Compton scattering[17] have been observed in the laboratory. The effect of radiation reaction, however, has not yet been measured as it has recently been shown that such an effect may be important in electron-field interactions at intensities over ten orders of magnitude greater than the relativistic limit[18].

In a fully relativistic treatment the magnetic component of the radiation field is retained, which in turn affects the ensuing electron dynamics quite dramatically. To begin with, it makes the trajectory two-dimensional (linear-polarization) and three-dimensional (circular-polarization). In Fig. 1, for example, we show the famous figure 8 trajectory followed by an electron in a linearly-polarized laser field, starting initially from rest at the origin and when viewed in the reference frame in which the electron is *on average* at rest. In the laboratory frame of reference, the electron undergoes a forward drift as well (see also Fig. 2 below).

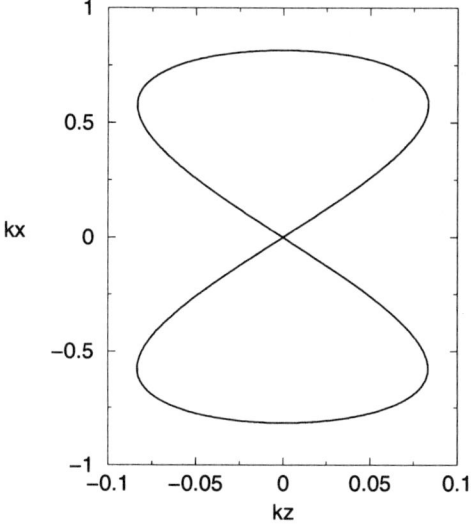

Figure 1. Electron trajectory in the linearly-polarized field of Eq. (2). The dimensionless intensity parameter $q = 1$ and the trajectory is shown in the frame in which the electron is *on average* at rest. See Ref. 5 for the relevant equations.

For high enough intensities the magnetic force becomes so strong that even the electron's spin degree of freedom begins to play an important dynamical role. The latter, a purely quantum property, endows the electron with a magnetic moment, which when properly accounted for in the equation of motion may be shown to affect the dynamics and to result in effects on the harmonic emission spectra that may be observed. There is also a great deal of interest in utilizing the high-intensity laser systems in the design of compact schemes to accelerate electrons to high energies[19-26]. Radiation emitted in such schemes, at multiples of the laser field frequency, also stand a good chance of becoming of practical utility[27].

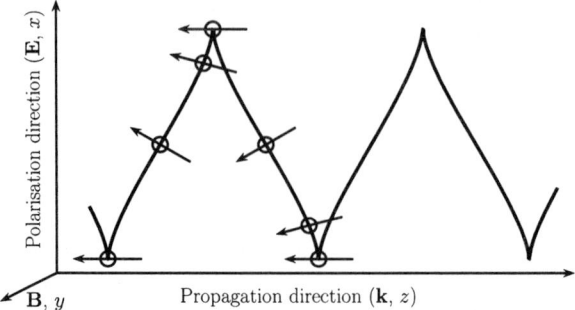

Figure 2. Qualitative picture of the change of the spin orientation while the mean expectation value of the electron's position is evolving in a linearly-polarized intense laser field. (Reproduced from Ref. 12 with permission.)

We will confine attention in this chapter only to two of the above-mentioned issues. In the next section, the spin effects will be discussed on both the motion and emission spectra of free and bound electrons. This will be followed by a brief discussion of one of the recently suggested schemes of particle acceleration, namely the one involving a pair of laser beams crossing at an angle.

SPIN-LASER INTERACTION EFFECTS

The Free Electron Case

The dynamics of a single electron in the presence of a high-intensity laser field gets markedly altered when the spin degree of freedom is taken into account. To see this on sound theoretical grounds, we treat the electron fully relativistically as a classical particle with a magnetic dipole moment m, whose rest-frame magnitude is $|m| = 1/2c$, and an induced electric dipole moment $d = \beta \times m$. In the presence of the plane-wave laser field of Eqs. (3) and (4) the relativistic analogue of the equation of motion (1) of the particle in atomic units ($e = m = 1$) is[11]

$$\frac{d(\gamma v)}{dt} = -\frac{1}{1 + \sigma^{\mu\nu}F_{\mu\nu}}(E + \beta \times B + f_{spin}), \qquad (8)$$

where $\sigma^{\mu\nu}$ is the moment tensor having the property

$$\sigma_{\mu\nu}\sigma^{\mu\nu} = |m|^2 - |d|^2. \qquad (9)$$

Furthermore, $F_{\mu\nu}$ is the electromagnetic field tensor, and $\gamma = (1 - \beta^2)^{-1/2}$ is the Lorentz factor.

Two changes to the equation of motion are brought about by introduction of the spin degree of freedom. First, the denominator of Eq. (8) may be viewed as an effective mass for the *spinning* electron *dressed* by the radiation field. Second, there is also an additional force term which has the following form

$$f_{spin} = -(m_y B_0 + d_x E_0)\frac{\omega}{c} \sin\eta \left[\gamma\beta(1 - \frac{v_z}{c}) - \frac{\hat{k}}{\gamma}\right] + (\dot{m}_y B_0 + \dot{d}_x E_0)\frac{\gamma\beta}{c}\cos\eta$$

$$-\frac{\hat{k}}{\gamma c}\left[m_y E_0 \omega(1 - \frac{v_z}{c})\sin\eta - \dot{m}_y E_0 \cos\eta\right]$$

$$+\frac{\hat{j}}{\gamma c}\left[m_z E_0 \omega(1 - \frac{v_z}{c})\sin\eta - \dot{m}_z E_0 \cos\eta\right] + (m_z \hat{j} - m_y \hat{k})\frac{\gamma}{c^2}(\beta \cdot \dot{\beta})E_0 \cos\eta. \qquad (10)$$

In this equation a dot on a symbol signifies a total time derivative, e.g.,

$$\dot{m}_x = \frac{dm_x}{dt}. \qquad (11)$$

Furthermore, $\eta = \omega t - kz$. On the other hand, the time dependence of the magnetic dipole moment is governed by

$$\frac{d\boldsymbol{m}}{dt} = \frac{1}{\gamma c}(\boldsymbol{m} \times \boldsymbol{B} + \boldsymbol{d} \times \boldsymbol{E}), \qquad (12)$$

with additional acceleration-dependent terms neglected. Equations (8)-(12) have recently been solved numerically[11].

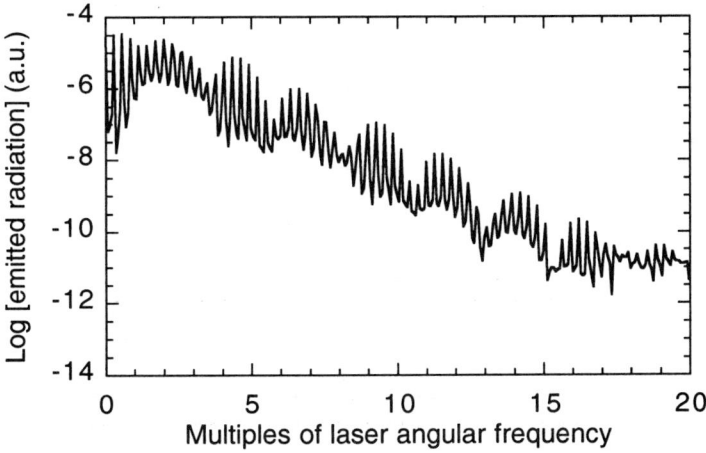

Figure 3. The retarded radiation spectrum as a function of the multiple number n of the fundamental angular frequency. This is the part of the spectrum seen in direction of the electric field component $\boldsymbol{E}_0$ and polarized along the direction of the magnetic field component $\boldsymbol{B}_0$. (Reproduced from Ref. 11 with permission.)

In Fig. 3, a sample emission spectrum is shown which has been calculated using these numerical solutions. In the absence of spin, only harmonics polarized along the incident wave propagation direction $\hat{k}$ may be detected at observation points along the $\boldsymbol{E}$ component of the field. Fig. 3 shows harmonics whose polarization is along $\boldsymbol{B}$. We view this finding as a clear signature of the electron spin effects.

Note that, in addition to terms parallel to the wave polarization and propagation directions, f_{spin} has a component parallel to the direction of the $\boldsymbol{B}$ field of the wave.

The spin dynamics have also recently been shown to induce a force, in the direction of the laser B field, on an electron modeled quantum mechanically by a Gaussian wavepacket[12].

The Bound Electron Case

Bound electrons, on the other hand, have also been the subject of a search for spin signatures in intense laser-ion interactions[13,28]. Effects on the emission spectra and the dynamics of an electronic wavepacket have been studied.

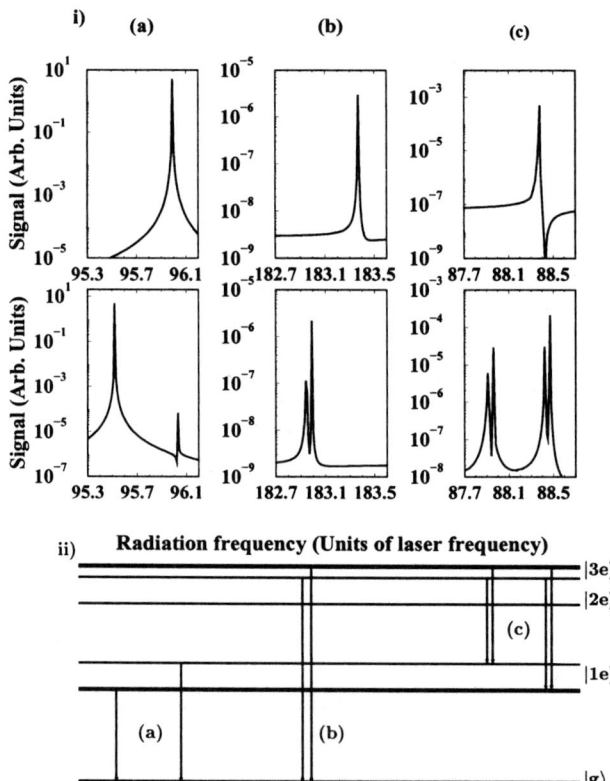

Figure 4. Radiation spectrum of a laser-driven model ion close to its lowest resonances. The top row of i) corresponds to a system modeled by the Pauli equation, and the second row is for the case where spin-orbit coupling including the relativistic mass shift and Zitterbewegung is taken into account. Figures (a), (b) and (c) are associated, respectively, with transitions from $|1e>$ to $|g>$, $|3e>$ to $|g>$, and $|3e>$ to $|1e>$, these being the ground and first- and second-excited states. The spectral lines split into doublets, (a) and (b), and a four-line structure, (c), due to the spin-orbit interaction. (see Ref. 13 for details regarding the parameters used). ii) Schematic diagram of state-splitting induced by the intense laser-enhanced spin-orbit interaction. We note that non-symmetric states split as opposed to symmetric states. Transitions (a), (b) and (c) are associated with the corresponding spectral lines in i). (Reproduced from Ref. 13 with permission.)

It is appropriate, in this case, to solve the Dirac equation (atomic units)

$$i\frac{\partial}{\partial t}\Psi(t) = \left\{c\boldsymbol{\alpha}\cdot\left[\mathbf{p}-\frac{1}{c}\mathbf{A}(t)\right] + c^2\beta + V\right\}\Psi(t). \tag{13}$$

Here, $\Psi(t)$ is a four-component Dirac spinor, V is the binding potential, and $\boldsymbol{\alpha} = (\alpha_x, \alpha_y, \alpha_z)$, and β are the Dirac matrices.

This equation has been solved for one to three dimensional model systems[29-33]. However, in the classical and weakly relativistic regime, an approximate analysis has been shown to suffice[34-45]. Quite a number of interesting reviews of intense laser-atom interactions, which address relativistic aspects, exist[46-52].

Intuitively, the Lorentz force causes enhanced angular motion of the bound electron in the vicinity of the nucleus which gives rise to significant deviations in the spin-orbit coupling as compared to the case in the absence of the laser field. The end result on the radiation spectrum is a significant splitting of the resonance lines. Fig. 4 gives an example.

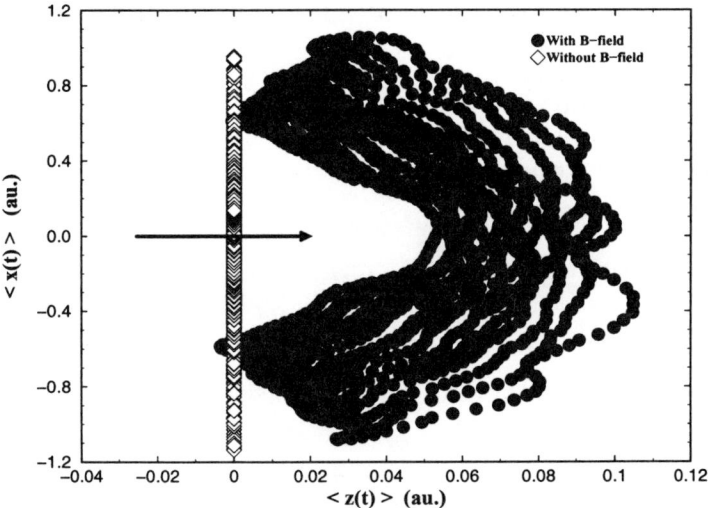

Figure 5. The center-of-mass evolution of the electronic wave packet for the cases with B-field ("full circles") and without B-field ("open diamonds"). The laser parameters involve a wavelength of 248nm, an intensity of $10^{17} W/cm^2$, a three-cycle linear turn-on and a ten-cycle constant amplitude duration. A hole around the ionic core for Be^{3+} (s=1, k=10.7, see Eq. (14)) as indicated by the arrow is clearly observable in the time-evolution of the center-of-mass of the electronic wave packet, when the B-field is considered. (Reproduced from Ref. 28 with permission.)

The calculations leading to Fig. 4 were based on a Foldy-Wouthuysen expansion of Eq. (13) in $1/c$, appropriate to the weakly relativistic regime of intensities up to 10^{17} Wcm^{-2}. Only terms $\mathcal{O}(1/c^2)$ in the expansion were kept and the Coulomb field of the active electron was modeled by the soft-core potential

$$V(x,z) = -\frac{k}{\sqrt{s+x^2+z^2}}, \tag{14}$$

where k and s depend on the *effective nuclear charge* sensed by the active electron.

The resulting equation was first numerically solved for the eigenstate of the bound electron in the ionic core potential. Next the split-operator algorithm was used to investigate the evolution of the system under irradiation with a laser pulse consisting of a 5-cycle linear turn-on followed by 100 cycles at a constant-amplitude. A laser intensity of 7×10^{16} Wcm^{-2}, at the wavelength 527 nm of a frequency-doubled Nd:glass system, was used.

For less charged ions interesting effects have also been noted just arising from first order effects in the ratio of the velocity of the electron to the speed of light. Due to the strong attraction of the nucleus and the magnetic component of the laser field, the Lorentz force causes the center of mass of the electronic wavepacket to move around the inic core. This may be associated with the hollowing out of the single ion[53] (see also Fig. 5), while the X-ray emission after the laser pulse is a signature of population close to the ionic core[28]. The dynamics in the laser propagation direction towards the ionic core has also been studied classically[54].

ELECTRON ACCELERATION BY CROSSED LASER BEAMS

A new scheme has recently been suggested[55-60,24] to accelerate electrons to high energy by means of two laser beams crossing at an angle (see Fig. 6 for a schematic). In this one-particle plane-wave configuration, the magnetic component of the laser field vanishes at all points along the initial injection direction of the electron. On the other hand, the accelerating field along z is[24]

$$E_z(0,0,z) = -2E_0 \sin\theta \sin\eta, \qquad (15)$$

where, in the present context, $\eta = \omega t - kz\cos\theta$, and $k = |\mathbf{k}_1| = |\mathbf{k}_2| = \omega/c$.

To arrive at Eq. (15) the fields have been modeled by $\mathbf{E}_i = \pm E_0 \sin\eta_i(\hat{\imath}\cos\theta \mp \hat{k}\sin\theta)$,

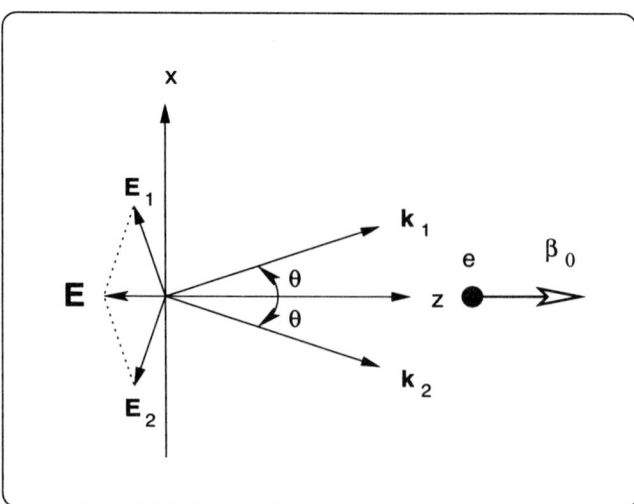

Figure 6. Electron-field interaction: A schematic. (Reproduced from Ref. 24 with permission.)

where $\eta_i = \omega t - \mathbf{k}_i \cdot \mathbf{r}$ and the upper(lower) sign goes with $i = 1(2)$. Solving the equations of motion with the field (15) is straightforward and leads to

$$\beta(\eta) = \frac{\cos\theta + s\sqrt{s^2 + \sin^2\theta}}{1 + s^2}, \tag{16}$$

where

$$s(\eta) \equiv \gamma(\beta - \cos\theta) = \gamma_0(\beta_0 - \cos\theta) + 4q\sin\theta\sin^2(\eta/2). \tag{17}$$

The Lorentz factor (energy scaled by the rest energy mc^2) follows now from Eq. (16)

$$\gamma(\eta) = \frac{s\cos\theta + \sqrt{s^2 + \sin^2\theta}}{\sin^2\theta}. \tag{18}$$

Furthermore, the electron's z coordinate may be found numerically from

$$\frac{dz}{d\eta} = \frac{dz}{dt}\frac{dt}{d\eta} = \frac{c}{\omega}\frac{\beta(\eta)}{[1 - \beta(\eta)\cos\theta]}. \tag{19}$$

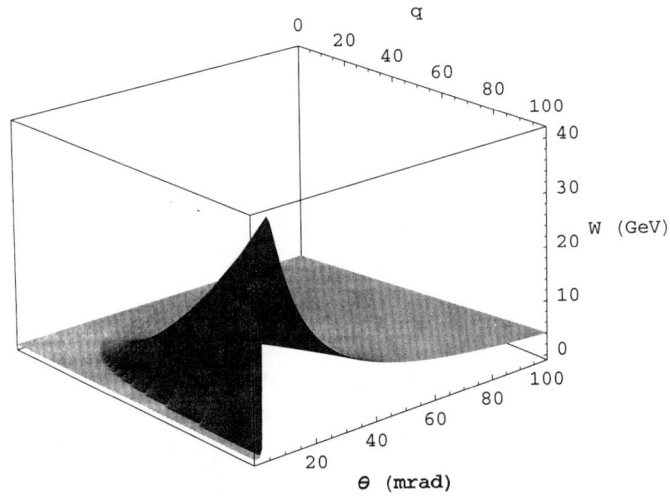

Figure 7. Energy gain vs the crossing angle θ and the intensity parameter q. Electron is accelerated from rest. (Reproduced from Ref. 24 with permission.)

Provided the electron is injected as shown in Fig. 4, we now have a complete picture of the ensuing 1D dynamics. In particular, the electron energy gain $W(\eta)$, as a result of interaction with the laser fields, follows from

$$W(\eta) = mc^2[\gamma(\eta) - \gamma_0]. \tag{20}$$

We show this quantity in Fig. 7 as a function of the crossing angle θ and the dimensionless intensity parameter q. Note from the figure that:

(a) For every choice of q, a preferred crossing angle seems to exist which makes the gain a maximum, apparently due to maximum constructive interference between the two waves being made possible at the given angle.

(b) For the chosen parameters, a maximum gain of about 40 GeV at a crossing angle of 4 mrad seems to be possible when fields of present-day laboratory intensity are employed.

(c) Although the electron is accelerated from rest at the origin of coordinates, yet it may gain up to 4 GeV of energy from such fields crossing at $\theta=100$ mrad.

Note that interaction with only half of a field cycle is considered in the above analysis in order to avoid phase-slippage.

In closing, we mention that this electron-laser accelerator configuration has been the subject of a more recent investigation[61]. In addition to the acceleration *per se*, the emission of Compton harmonics by the accelerated electron is studied. Realistic spatially-confined laser pulses are employed in that work and effects of off-axis injection, injection at an angle to the z-axis (see Fig. 6) and other related issues are discussed.

ACKNOWLEDGMENTS

Support for this work was received from the DAAD Gastdozentenprogramm (YIS), SFB 276 (MWW and CHK), the Austrian Academy of Sciences (MWW), and the Alexander von Humboldt Stiftung (SXH).

REFERENCES

1. J. H. Eberly and A. Sleeper, Phys. Rev. **176**, 1570 (1968).
2. E. S. Sarachik and G. T. Schappert, Phys. Rev. D **1**, 2738 (1970).
3. E. Esarey, S. K. Ride, and P. Sprangle, Phys. Rev. E **48**, 3003 (1993).
4. E. Esarey and P. Sprangle, Phys. Rev. A **45**, 5872 (1992).
5. Y. I. Salamin and F. H. M. Faisal, Phys. Rev. A **54**, 4383 (1996).
6. Y. I. Salamin and F. H. M. Faisal, J. Phys. A **31**, 1319 (1997).
7. Y. I. Salamin and F. H. M. Faisal, Phys. Rev. A **55**, 3964 (1997).
8. F. V. Hartemann, A. L. Troha, and N. C. Luhmann, Jr., Phys. Rev. E **54**, 2956 (1996).
9. L. S. Brown and T. W. B. Kibble, Phys. Rev. **133**, A705, (1964).
10. T. W. B. Kibble, *Lectures in Physics*, ed. M. Levy (Gordon and Breach, New York, 1968).
11. M. W. Walser, C. Szymanowski, and C. H. Keitel, Europhys. Lett. **48**, 533 (1999).
12. M. W. Walser and C. H. Keitel, J. Phys. B **33**, L221 (2000).
13. S. X. Hu and C. H. Keitel, Phys. Rev. Lett. **83**, 4709 (1999).
14. H. R. Reiss, Prog. Quantum Electron. **16**, 1 (1992).
15. D. M. Perry, D. Pennington, and V. Yanovsky, Opt. Lett. **24**, 160 (1999).
16. C. I. Moore, J. P. Knauer, and D. D. Meyerhofer, Phys. Rev. Lett. **74**, 2439 (1995).
17. C. Bula *et al.*, Phys. Rev. Lett. **76**, 3116 (1996).
18. C. H. Keitel, C. Szymanowski, P. L. Knight, and A. Maquet, J. Phys. B **31**, L75 (1998).
19. A. Loeb, L. Friedland, and S. Eliezer, Phys. Rev. A **35**, 1692 (1987).
20. M. S. Hussein, M. P. Pato, Phys. Rev. Lett. **68**, 1136 (1992).
21. M. S. Hussein, M. P. Pato, and A. K. Kerman, Phys. Rev. A **46**, 3562 (1992).

22. Y. I. Salamin and F. H. M. Faisal, Phys. Rev. A **61**, 043801 (2000).
23. Y. I. Salamin, F. H. M. Faisal, and C. H. Keitel, Phys. Rev. A **62**, 053809 (2000).
24. Y. I. Salamin and C. H. Keitel, Appl. Phys. Lett. **77**, 1082 (2000).
25. J. X. Wang *et al.*, Phys. Rev. E **58**, 6575 (1998).
26. J. X. Wang *et al.*, Phys. Rev. E **60**, 7473 (1999).
27. I. V. Pogorelsky, Nucl. Instr. & Meth. in Phys. Res. A **411**, 172 (1998).
28. S. X. Hu and C. H. Keitel, Europhys. Lett **47**, 318 (1999).
29. N. J. Kylstra, A. M. Ermolaev, C. J. Joachain, J. Phys. B **30**, L449 (1997).
30. U. W. Rathe, C. H. Keitel, M. Protopapas and P. L. Knight, J. Phys. B **30**, L531 (1997).
31. C. Szymanowski, *et al.*, Phys. Rev. A **56**, 3846 (1997).
32. R. Taïeb, V. Véniard and A. Maquet, Phys. Rev. Lett. **81**, 2882 (1998).
33. J. W. Braun, Q. Su and R. Grobe, Phys. Rev. A **59**, 604 (1999).
34. J. Grochmalicki, M. Lewenstein, K. Rzazewski, Phys. Rev. Lett. **66** 1038 (1991).
35. C. H. Keitel, P. L. Knight, and K. Burnett, Europhys. Lett. **24**, 539 (1993).
36. T. Katsouleas and W. B. Mori, Phys. Rev. Lett. **70**, 1561 (1993).
37. F. H. M. Faisal and T. Radozycki, Phys. Rev. A **47**, 4464 (1993).
38. O. Latinne, C. J. Joachain, and M. Dörr, Europhys. Lett. **26**, 333 (1994).
39. C. H. Keitel and P. L. Knight, Phys. Rev. A **41**, 1420 (1995).
40. M. Protopapas, C. H. Keitel, and P. L. Knight, J. Phys. B **29**, L591 (1996).
41. D. B. Milosevic and A. F. Starace, Phys. Rev. Lett. **82**, 2653 (1999).
42. R. E. Wagner, Q. Su, and R. Grobe, Phys. Rev. Lett. **84**, 3282 (2000).
43. J. R. V. de Aldana and L. Roso, Phys. Rev. A **61**, 3403 (2000).
44. N. J. Kylstra, *et al.*, Phys. Rev. Lett. **85**, 1835 (2000).
45. J. V. R. de Aldana and L. Roso, J. Phys. B **33**, 3701 (2000).
46. J. San Roman, L. Roso, and H. R. Reiss. J. Phys. B **33**, 1869 (2000).
47. G. H. Rutherford and R. Grobe, J. Phys. A **31**, 9331 (1998).
48. M. W. Walser, C. H. Keitel, A. Scrinzi, and T. Brabec, Phys. Rev. Lett. **85**, in press.
49. T. Brabec T, F. Krausz, Rev. Mod. Phys. **72**, 545 (2000).
50. C. J. Joachain, M. Dörr, N. Kylstra, Adv. At. Mol. Opt. Phys. **42**, 225 (2000).
51. G. A. Mourou, C. P. J. Barty and M. D. Perry, Phys. Tod. **51**, 22 (1998).
52. M. Protopapas, C. H. Keitel, and P. L. Knight, Rep. Progr. Phys. **60**, 389 (1997).
53. S. X. Hu and C. H. Keitel, submitted.
54. D. J. Urbach and C.H. Keitel, Phys. Rev. A **61**, 043409 (2000).
55. C. M. Haaland, Opt. Commun. **114**, 280 (1995).
56. E. Esarey, P. Sprangle, and J. Krall, Phys. Rev. E **52**, 5443 (1995).
57. Y. C. Huang, D. Zheng, W. M. Tulloch, and R. L. Beyer, Appl. Phys. Lett. **68**, 753 (1996).
58. Y. C. Huang and R. L. Beyer, Appl. Phys. Lett. **69**, 2175 (1996).
59. Y. C. Huang and R. L. Beyer, Instr. and Meth. in Phys. Res. A **407**, 316 (1998).
60. Y. C. Huang and R. L. Beyer, Rev. Sci. Instrum. **69**, 2629 (1998).
61. Y. I. Salamin, G. R. Mocken, and C. H. Keitel, submitted.

THE CLASSICAL AND THE QUANTUM FACE OF ABOVE-THRESHOLD IONIZATION

G.G. PAULUS[1] and H.WALTHER[1,2]

[1] Max-Planck-Institut für Quantenoptik, 85748 Garching, Germany
[2] Ludwig-Maximilians-Universität München, Germany

1 Introduction

1.1 What is ATI and why is it interesting?

Above-threshold ionization (ATI) was discovered in 1979 by Agostini et al. [1]. Its discovery marks the beginning of the investigation of extreme highly non-linear phenomena in high-frequency fields, i.e. in the visible spectral region. ATI is photoionization in intense laser fields such that an atom absorbs more photons than necessary for subduing the ionization potential. This can be seen by inspection of the photoelectron kinetic energy spectra which consist of a series of peaks separated by the photon energy. Generally, the peak heights decrease as the order is increased. In this respect, the appearance of ATI spectra is similar to that for the generation of high harmonics in gaseous media (high-harmonic generation, HHG) for comparable laser intensities [2]. For that effect, in 1988, it was discovered that the initial decrease of harmonic yield with increasing harmonic order is followed by a flat annex, i.e. the harmonic intensity is more or less independent of its order [3]. This plateau-like feature ends with a steep cut-off. Years later, an analogous phenomenon was measured for ATI [4, 5].

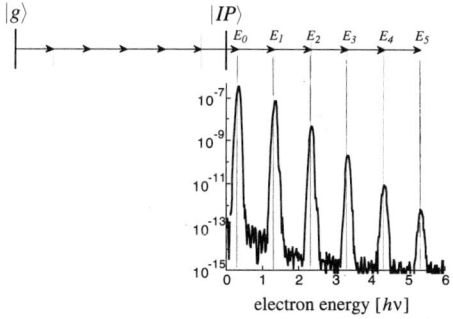

Figure 1: ATI spectra result from the absorption of more photons than necessary for ionization.

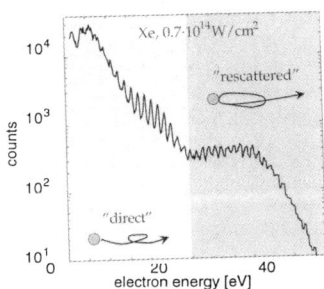

Figure 2: At high laser intensities, ATI spectra display a plateau-like annex to the low-energetic part of the spectrum.

Atoms, Solids, and Plasmas in Super-Intense Laser Fields
Edited by D. Batani *et al.*, Kluwer Academic/Plenum Publishers, 2001

Drawing these parallels should not suggest that both effects are of similar importance from the point of view of applications: the existence of the HHG plateau and its extent up to more than 200 harmonic orders [6, 7] means that there is a surprisingly efficient way to produce coherent radiation in the XUV and even soft X-ray spectral region. However, an HHG spectrum is the product of the atomic response and propagation effects in the medium. Therefore, the interpretation of HHG spectra often is ambiguous. It is now a well-established fact that ATI and HHG are indeed twin effects. Since ATI experiments are done at very low target gas pressures ($< 10^{-6}$ torr vs. > 1 torr for HHG), they give the atomic response alone, which facilitates their interpretation. For the sake of completeness, we would also mention that the future might bring applications also for ATI: Firstly, it could be exploited as a nonlinear medium, in particular for the investigation of the properties of XUV radiation. Secondly, as femtosecond pulses become so short that they consist only of very few optical cycles, the phases of the pulse's carrier frequency vs. the envelope is of significance. Most likely, this will be one of the most important issues in strong-field laser-matter interaction in the future and ATI is a way to investigate and possibly control these effects [8].

1.2 Theoretical models of ATI

The most desirable theory for any physical effect is an analytical solution. Unfortunately, these are known for very few examples, in particular in quantum mechanics. Most well-known is the example of the hydrogen atom, but of similar importance in our field is the analytical solution of the Schrödinger equation of a free electron in an oscillating electrical field, the Volkov eigenfunctions.

If an analytical solution of the Schrödinger equation is not available, the standard method in quantum mechanics is perturbation theory. However, taking the field-free solution as the unperturbed solution means that the first non-vanishing order for ionization of rare gases with Ti:Sapphire lasers is bigger than 8, promising very ugly formulae. In addition, the appearance of ATI spectra, not to mention the plateau, suggests that this approach is unsuitable. The reason for the failure of perturbation theory is that a strong field is by no means a small perturbation of the atom: At 10^{14} W/cm^2 the electrical field of the laser is almost 10% of that of the nucleus on the first Bohrian radius.

The Volkov eigenfunction is exploited by Kedysh-type theories. They approximate the ATI spectrum by the squared modulus of the matrix element projecting the propagated wave-function onto the Volkov function. With suitable approximations the resulting integrals can be simplified such that numerical integration becomes feasible.

Besides these analytical approaches, there are also fully numerical methods, namely Floquet-type ones and brute-force numerical integration of the appropriate Schrödinger equation. It should be pointed out that one variant of the Floquet-type methods, namely the R-matrix-Floquet theory, is a very promising and more or less the only approach to investigate molecules and correlation effects in strong laser fields [9]. While the purely numerical methods are the most accurate ones, the physical interpretation of the results is often cumbersome. Therefore, in particular the numerical integration of the Schrödinger equation is also refered to as a 'computer experiment'.

In contrast, the very simple classical model we are going to discuss in the first part of this paper not only is able to reproduce many effects observed experimentally (although in most cases only qualitatively), but, in addition, provides an intuitive picture of the physical processes involved. Actually, the classical approach relies on an almost trivial analysis of electron trajectories in an oscillating electrical field. The second part of the paper is devoted to the discussion of very recent experimental results. These can only be explained by a quantum theory. In one particular example the influence of quantum interferences makes differences by an order of magnitude, i.e. quantum effects dominate the spectra. Still, also in this section we will try to keep the intuitive picture of electron trajectories as far as possible.

2 The classical model of ATI

2.1 Direct electrons

The classical model of strong-field effects [10, 11, 12, 13, 14] starts by subdividing the ionization process into several steps. In a first step, an electron is assumed to be shifted into the continuum at some phase ωt_0 of the laser field

$$\mathcal{E} = \mathcal{E}_0 \sin \omega t. \quad (1)$$

In a less ad hoc manner one can imagine that the first step is due to optical field ionization. Accordingly, it is reasonable to assume that the trajectory of the electron has the initial conditions $x(t=t_0) = 0$ and $\dot{x}(t=t_0) = 0$. Eq. (1) implies linear laser polarization.

2.1.1 Cutoff for direct electrons

The second step of the classical model is evolution of the electron trajectory in the strong laser field. Hereby, the influence of the atomic potential is neglected due to the large oscillation amplitude of the electron and the strong laser field. In an ATI experiment, it is the drift energy of the photoelectron (i.e. the cycle-averaged kinetic energy) which is measured. The calculation of this quantity is straightforward:[1]

$$\ddot{x} = \mathcal{E}_0 \sin \omega t$$

$$\dot{x} = -\frac{\mathcal{E}_0}{\omega}(\cos \omega t - \cos \omega t_0) = A(t_0) - A(t) \quad (2)$$

$$E = \frac{1}{2}\langle \dot{x}\rangle_t^2 = \frac{1}{2}A(t_0)^2 = 2\frac{\mathcal{E}_0^2}{4\omega^2}\cos^2 \omega t_0 \quad (3)$$

A is that component of the vector potential that is parallel to the polarization. The gauge is chosen such that $\langle A(t)\rangle_t = 0$, where $\langle\ \rangle_t$ indicates averaging over time. The prefactor $U_p := \frac{1}{2}\langle A(t)^2\rangle_t = \mathcal{E}_0^2/(4\omega^2)$ in eq. (3) is the quiver energy of a free electron in an oscillating field and called ponderomotive energy.[2] It is proportional to the intensity and the square of the wavelength and is of great importance as a scaled intensity. Eq. (3) immediately predicts a maximum kinetic energy of $2U_p$ for the photoelectrons. In more detail, electrons born at a phase where the field strength is maximal, will gain no kinetic energy from the pulse in contrast to electrons entering the field at zero field strength which gain $2U_p$. Experiments and numerical solutions of the Schrödinger equation confirm the existence of this cutoff, see fig. 3. The agreement is better the higher the intensity and the lower the frequency. This is not surprising because the oscillation amplitude of the electron is $\mathcal{E}/\omega^2$, i.e. it gets bigger and thus more classical under these assumptions. There have also been experiments in the microwave regime which confirmed these arguments [15].

One might be disturbed by the fact that the most energetic electrons are assumed to come out when the field is zero. While we could list some arguments that this is not

---
[1] We use atomic units throughout this paper, i.e. $e=1$, $m_e = 1$, and $\hbar = 1$

[2] In a laser focus, U_p can also act as a potential: If the laser pulse is long enough, the electrons will surf down the focus because U_p is highest in the center of the focus and goes to zero outside the laser beam. Thus an electron created in the center of the focus gains a kinetic energy of U_p. In this paper we consider short pulses, i.e. the pulse is off before the electron is able to start surfing.

completely unreasonable, we leave it by the remark that this simple theory should not be taken too literally.

Another point is worth mentioning here: The ATI peak structure is due to the periodicity of the ionization process. In every cycle of the laser an electron wavepacket tunnels out of the potential. The wavepackets disperse and interfere with each other eventually. In the frequency domain this gives rise to the ATI peaks.

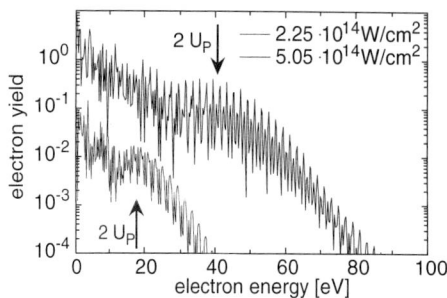

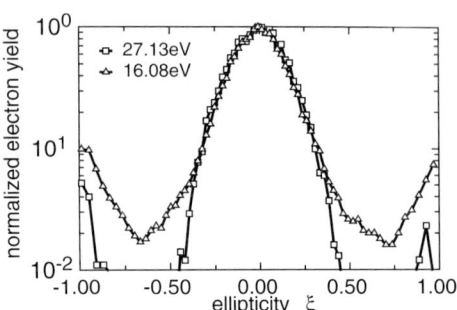

Figure 3: Low-energy part of ATI spectra calculated by numerical integration of the Schrödinger equation for two intensities. The arrows indicate the energy positions of $2U_p$ for the corresponding intensities.

Figure 4: Electron yield as a function of the ellipticity of the laser polarization for two electron energies.

2.1.2 The dodging phenomenon for elliptical polarization

In the preceding section we assumed linear polarization pointing in the direction of the electron detector. While the position of the electron detector is maintained, now we change the ellipticity ξ of the laser polarization. In fig. 4 the result of a measurement is shown, where the ATI electron yields for certain energies have been measured as a function of ξ. The most conspicuous effect is a local maximum in the electron yield for circular polarization. Since the peak electric field strength is lower for circular polarization and since the electron has to overcome the angular momentum barrier, this is counterintuitive. The laser field for elliptical polarization is described by

$$\mathcal{E}(t) = \frac{\mathcal{E}_0}{\sqrt{1+\xi^2}} \begin{pmatrix} \sin\omega t \\ \xi \cos\omega t \end{pmatrix}. \tag{4}$$

The drift velocity can be derived to be

$$\langle \dot{\mathbf{x}} \rangle_t = \frac{\mathcal{E}_0}{\omega\sqrt{1+\xi^2}} \begin{pmatrix} \cos\omega t_0 \\ -\xi \sin\omega t_0 \end{pmatrix}. \tag{5}$$

As shown above, the photoelectrons with the lowest energy are due to ionization near the maximum of the laser field, e.g. for $\omega t_0 \approx \pi/2$. While these electrons linger around the ion core for quite some time in the case of linear polarization, they acquire a rather high energy for elliptical polarization by leaving the ion core in a direction different from the main axis of the polarization, i.e. the electrons dodge the large component of the field. When circular polarization is approached this effect decreases as all directions become more and more equivalent, and the rates rise again [17].

Eq. (5) also predicts a well-defined electron energy for each ξ. This again suggests the classical model not be taken too seriously. Quantum mechanics always softens the sharp predictions of classical mechanics. In addition, the initial conditions for the electron trajectories are not as well justified for circular polarization as for linear polarization.

2.2 Rescattered electrons

The classical model of ATI becomes much more colorful if rescattering effects are taken into account. To this end, we integrate eq. (2) and obtain the trajectory of the electron:

$$x(t) = (t-t_0)A(t_0) - \int_{t_0}^{t} d\tau A(\tau) = -\frac{\mathcal{E}_0}{\omega^2}(\sin \omega t - \sin \omega t_0) + \frac{\mathcal{E}_0}{\omega^2}(\omega t - \omega t_0)\cos \omega t_0 \quad (6)$$

Obviously, $x(t)$ obeys the initial conditions established in section 2.1, in particular $x(t=t_0) = 0$. All the well-known features of atomic strong-field effects like the ATI and the HHG plateau as well as non-sequential double ionization (NSDI) arise from situations like the following: Let us assume that the electron enters the field at $\omega t_0 = 108°$. This electron will revisit the ion core at $\omega t_1 = 342°$, as can be checked by plugging these numbers into eq. (6). The instantaneous electron velocity upon return (as opposed to the drift velocity) corresponds to an (instantaneous) kinetic energy of $3.17 U_p$.[3] Now, there are several possibilities:

i the electron may recombine, which leads to the emission of a photon with an energy equal to the electron's (instantaneous) kinetic energy plus the ionization energy. This process is responsible for the HHG plateau.

ii the electron may kick out another electron and thus ionize the ion. This leads to NSDI.

iii the electron may scatter off the ion core. In the case of rescattering (i.e. $\dot{x}(t)$ changes its sign at $t = t_1$), the initial conditions of the electron's motion are changed such that the electrical field accelerates the electron further instead of decelerating it as it would have been the case without rescattering. This leads to high-energetic photoelectrons, viz. the ATI plateau [4].

Since this paper deals with ATI, we will concentrate on the last possibility in the following.

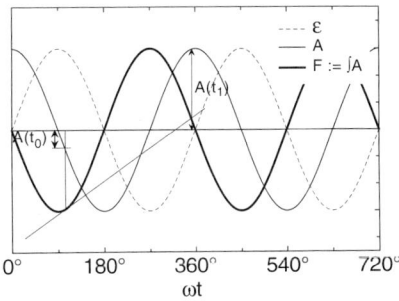

Figure 5: Graphical method to find the return time t_1 and to determine the corresponding instantaneous return energy as well as the drift energy after rescattering.

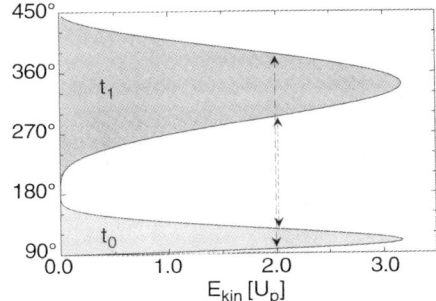

Figure 6: Start times t_0 and return times t_1 for trajectories supposed to return with a given kinetic energy.

---

[3]This, by the way, is the maximum return energy possible. All other start and/or return times lead to lower return energies.

2.2.1 A graphical ATI computer

Many if not all aspects of the classical model of effects due to a revisiting electron can be visualized by a graphical method [14]. The condition for these effects is

$$x(t = t_1) = (t_1 - t_0)A(t_0) - \int_{t_0}^{t_1} d\tau A(\tau) \stackrel{!}{=} 0. \tag{7}$$

Introducing the indefinite integral of the vector potential

$$F(t) := \int^t d\tau A(\tau) \tag{8}$$

we may write eq. (7) as

$$F(t_1) = F(t_0) + (t_1 - t_0)\dot{F}(t_0). \tag{9}$$

This equation determines the return time t_1 and is amenable to a simple graphical solution, see fig. 5. For given t_0, we can find t_1 by intersecting $F(t)$ with its tangent at t_0. An evident consequence is that electrons starting at $t_0 \in [0, \pi/2[$ will never return to the ion core, while those with $t_0 \in [\pi/2, \pi[$ return once or even several times.

The (instantaneous) kinetic energy of the electron upon its return to the origin (just before rescattering) is

$$E_{\text{kin}} = \frac{1}{2}\dot{x}(t_1)^2 \stackrel{(2)}{=} \frac{1}{2}[A(t_1) - A(t_0)]^2. \tag{10}$$

If we are interested in the highest possible harmonic, we should try to make the difference between $A(t_1)$ and $A(t_0)$ as big as possible. For a sinusoidal field this requires t_1 to be as near as possible to 360° and t_0 as far as possible from 90°. The actual solution is, as mentioned above, $t_0 = 108°$ and $t_1 = 342°$ leading to $E_{\text{kin,max}} = 3.17\,U_p$. Obviously, all times somewhat bigger and smaller than t_0 lead to lower return energies. Equivalently, for each return energy $E_{\text{kin}} < E_{\text{kin,max}}$ there exist two trajectories. The graphical method also shows that the trajectory that started later will return earlier. These results, which are also important for the understanding of phase matching effects [18, 19], are summarized in fig. 6. For any return energy the corresponding start and return times for the pair of trajectories can be read off.

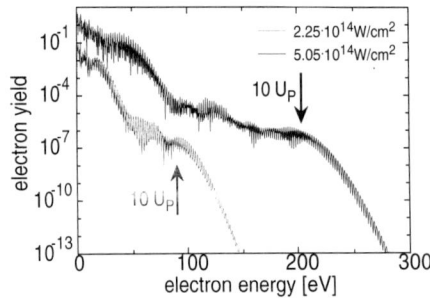

Figure 7: The same spectra as in fig. 3 but with higher dynamic range. The arrows indicate the energy positions of $10\,U_p$ for the corresponding intensities.

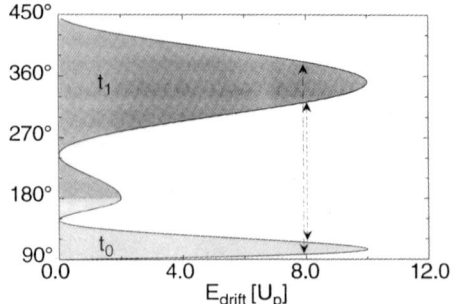

Figure 8: Start times t_0 and return times t_1 for trajectories supposed to have a given drift energy after rescattering.

2.2.2 The cutoff of the ATI plateau

We return to the rescattered electron and look at its drift energy after rescattering. According to eq. (2) the electron velocity is given by $\dot{x}(t) = -A(t) + C$ where the constant C is determined by the initial conditions. At $t = t_1$ the electron velocity changes its sign, i.e. $\dot{x}(t = t_1 - 0) = -A(t_1) + A(t_0)$ and $\dot{x}(t = t_1 + 0) = +A(t_1) - A(t_0) = A(t_1) + C$. This leads to

$$\dot{x}(t > t_1) = 2A(t_1) - A(t_0) - A(t) \tag{11}$$

and to a drift energy of

$$E_{\text{drift}} = \frac{1}{2}\langle \dot{x}\rangle_t^2 = \frac{1}{2}[2A(t_1) - A(t_0)]^2. \tag{12}$$

Eq. (12) is similar to that for the (instantaneous) kinetic energy, see eq. (10). In particular, it means that the start and return times for the maximally possible drift energy $E_{\text{drift,max}}$ are close to those for the maximal return energy $E_{\text{kin,max}}$. The numerical evaluation yields $t_{0,\text{max}} = 105.0°$ and $t_{1,\text{max}} = 351.7°$. The cutoff energy resulting from the corresponding trajectory is much higher due to the additional factor of 2 in eq. (12) as compared to eq. (10). Plugging the numbers in results in $E_{\text{drift,max}} = 10.007\, U_p$. Fig. 7 displays spectra obtained from numerical integration of the Schrödinger equation for two laser intensities. Both exhibit pronounced ATI plateaus which roll off around $10\, U_p$. The predictive power of this simple calculation for such a seemingly complicated problem is really amazing.

With arguments as above we state that for each ATI electron with an energy $E_{\text{drift}} < E_{\text{drift,max}}$ there exists a pair of trajectories, one with longer and one with shorter travel time, see fig. 8.

2.2.3 Rainbow scattering

In the preceding section exact backscattering was considered. This, of course, is not the only possibility, rather the electron can be scattered by any angle. However, as long as the laser polarization points in the direction of the electron detector, only electrons scattered in the forward (0°) and backward (180°) direction will be observed. In order to investigate other scattering angles, ATI spectra corresponding to different angles between the laser polarization and the spectrometer axis have to be recorded. From these spectra it is also possible to extract the electron yield as a function of the angle for any given electron energy (angular distribution, AD).

For a theoretical modeling in the framework of the classical approach, again we are not pedantic. Due to the large oscillation amplitudes we assume that the range of the potential is very small, namely a δ-potential. Then it is consistent to assume that the particle upon its return to the origin at time t_1 scatters by an angle θ_0 with respect to the negative x-axis, where θ_0 is a uniformly distributed random quantity.[4] For short laser pulses (i.e. the electron has no time to surf down the laser focus), the canonical momentum $\mathbf{p} = \dot{\mathbf{x}}(t) + \mathbf{A}(t)$ is conserved. An electron that was scattered by the angle θ_0 arrives at the detector at the angle θ given by

$$\cot \theta = \frac{\langle v_x(t)\rangle_t}{\langle v_y(t)\rangle_t}. \tag{13}$$

Here, the x-axis is assumed to be parallel to the polarization while the y-axis is perpendicular to it and to the laser beam. For $t > t_1$ $\dot{x}_x$ and $\dot{x}_y$ are given by

---

[4]This is the classical description of the quantum-mechanical fact that the scattering amplitude of a zero-range potential does not depend on the scattering angle.

$$\dot{x}_x = [A(t_1) - A(t)] - \cos\theta_0 |A(t_0) - A(t_1)| \tag{14}$$
$$\dot{x}_y = \sin\theta_0 |A(t_0) - A(t_1)|. \tag{15}$$

This leads to

$$\cot\theta = \cot\theta_0 - \frac{A(t_1)}{\sin\theta_0 |A(t_0) - A(t_1)|} \tag{16}$$

and

$$\begin{aligned} E_{\text{drift}} &= \frac{1}{2}\langle \dot{x}_x(t)^2 + \dot{x}_y(t)^2 \rangle_t - U_p \\ &= \frac{1}{2}\left[A(t_0)^2 + 2A(t_1)[A(t_1) - A(t_0)](1 \pm \cos\theta_0) \right] \end{aligned} \tag{17}$$

where the upper (lower) sign holds for $A(t_0) > A(t_1)$ ($A(t_0) < A(t_1)$).
For backscattering eq. (17) reduces to eq. (12) as expected. The largest drift energies occur for backscattering. Fig. 9 shows a measurement of angle- and energy-resolved ATI photoelectrons for linear polarization ($\xi = 0$) in a density plot. High electron yields correspond to dark grey and vice versa. It can clearly be seen that the ATI plateau cutoff retreats to lower energies the bigger the angle between the axis of the spectrometer and that of the polarization. For comparison the result of the classical calculation of the ATI-cutoff as a function of θ is superimposed.

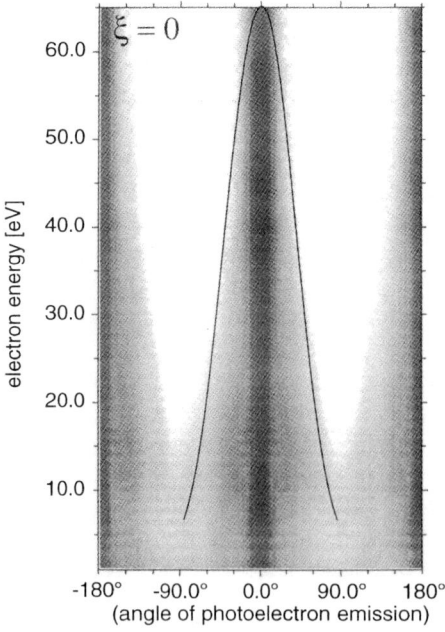

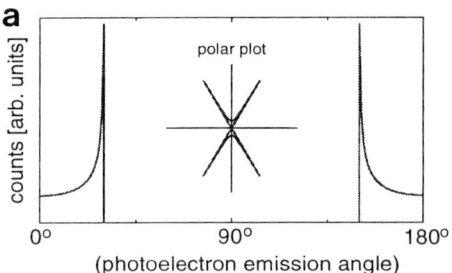

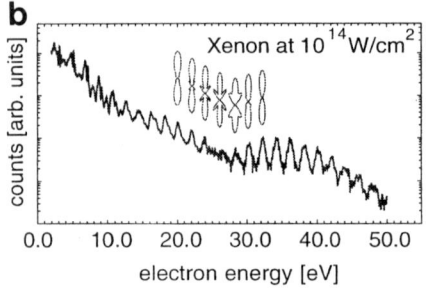

Figure 9: Angle- and energy-resolved ATI spectrum in a density plot. Superimposed is the ATI plateau cutoff as a function of θ. For visual convenience the data was normalized for each ATI peak.

Figure 10: Angular distribution of photoelectrons with an energy of $8U_p$ calculated with the classical model (a). ATI spectrum at $\theta = 0°$ and polar plots of angular distributions corresponding to some electron energies (b).

Rather than looking in the energy direction one can also consider the electron yield as a function of the scattering angle θ with the electron energy as a parameter. The corresponding classical result is shown in fig. 10a for electrons with an energy of $8\,U_p$. The remarkable feature of this figure is that the field bends all initial scattering angles θ_0 into an interval $[180° - \theta_{\max}, 180° + \theta_{\max}]$ where $\theta_{\max}$ is a function of the electron energy. No scattered electrons are seen at $\theta = 90°$. A closer analysis shows that this behavior is quite analogous to rainbow scattering [13, 20].

Under certain circumstances, namely when ionization for $\theta = 0°$ is suppressed for some energy regions due to reasons we will treat in the next section, rainbow scattering in ATI is then particularly nicely revealed. Then the cutoffs in the angular direction are no longer outshone by the spectra close to $0°$ and the cutoffs become visible as side-lobes in polar plots of the angular distribution of the photoelectrons at these energies, see fig. 10b [21, 22].

3 Quantum interference effects in ATI

Notwithstanding the intuitive and successful description the classical theory of strong field effects is able to provide, it has its limitations. This already becomes evident for elliptically polarized light fields, where the initial conditions for the model become questionable, and holds even more for quantum effects of ATI, which we want to treat in the following.

3.1 Keldysh-type theories

Fortunately, there is an analytical and fully quantum-mechanical theory which, interestingly, is much older than ATI and goes back to the work of Keldysh in 1964 [23]. However, the integrals resulting from the theory have to be evaluated numerically and in order to facilitate this, usually approximations are made. We want to introduce some of the key concepts of Keldysh-type theories.

The probability for detecting an ATI electron with momentum $\mathbf{p}$ in a continuum state $|c_\mathbf{p}\rangle$ after interaction with the laser pulse ($t \to \infty$) is given by the squared modulus of the matrix element

$$M_\mathbf{p} = \lim_{\substack{t \to \infty \\ t' \to -\infty}} \langle c_\mathbf{p}(t) | U(t, t') | g(t') \rangle. \tag{18}$$

Here, $|g(t')\rangle$ is the atomic state before interaction with the laser pulse ($t' \to -\infty$), i.e. usually the ground state. $U(t, t')$ is the time-evolution operator. It solves the Schrödinger equation with the Hamiltonian

$$H(t) = \frac{\mathbf{p}^2}{2m} - er\mathcal{E}(t) + V(\mathbf{r}). \tag{19}$$

Further, we introduce the Hamiltonians for the atom and the field alone:

$$H_a(t) := H_{\text{atom}}(t) = \frac{\mathbf{p}^2}{2m} + V(\mathbf{r}) \tag{20}$$

$$H_f(t) := H_{\text{field}}(t) = \frac{\mathbf{p}^2}{2m} - er\mathcal{E}(t). \tag{21}$$

The corresponding Schrödinger equations are solved by the time evolution operators U_a and U_f, respectively. The eigenfunctions of the Schrödinger equation with the Hamiltonian H_f are analytically known and called Volkov states $|v_\mathbf{p}\rangle$. This implies $U_f(t, t')|v_\mathbf{p}(t')\rangle = |v_\mathbf{p}(t)\rangle$.

The Dyson equation enables U to be expanded either in terms of U_a

$$U(t,t') = U_a(t,t') - \frac{i}{\hbar} \int_{\tau=t'}^{\tau=t} d\tau\, U(t,\tau) \left[H(\tau) - H_a(\tau)\right] U_a(\tau,t'), \tag{22}$$

or in terms of U_f

$$U(t,t') = U_f(t,t') - \frac{i}{\hbar} \int_{\tau=t'}^{\tau=t} d\tau\, U(t,\tau) \left[H(\tau) - H_f(\tau)\right] U_f(\tau,t'). \tag{23}$$

With the first version, eq. (18) transforms into

$$M_{\mathbf{p}} = \lim_{\substack{t \to \infty \\ t' \to -\infty}} \Big\{ \langle c_{\mathbf{p}}(t)| \underbrace{U_a(t,t')|g(t')\rangle}_{\substack{= |g(t)\rangle \\ = 0 \text{(orthogonal)}}}$$

$$- \frac{i}{\hbar} \int_{t_0=t'}^{t_0=t} dt_0\, \langle c_{\mathbf{p}}(t)| U(t,t_0)\, \underbrace{[H(t_0) - H_a(t_0)]}_{= -e r \mathcal{E}(t_0) =: D_{WW}(t_0)}\, \underbrace{U_a(t_0,t')|g(t')\rangle}_{= |g(t_0)\rangle} \Big\}. \tag{24}$$

The original Keldysh approximation replaces $\langle c_{\mathbf{p}}(t)| U(t,t_0)$ by $\langle v_{\mathbf{p}}(t_0)|$ and concludes

$$M_{\mathbf{p}} \approx M_{\mathbf{p}}^{(0)} = \frac{i}{\hbar} \int_{-\infty}^{\infty} dt_0\, \langle v_{\mathbf{p}}(t_0)| D_{WW}(t_0) |g(t_0)\rangle. \tag{25}$$

This means that the electron is transfered *directly* into the continuum at $t = t_0$ due to the interaction D_{WW} with the laser field. Numerous varieties of the Keldysh ansatz were developed. Most well known are those by Faisal and Reiss [24, 25]. Therefore, this type of theory is commonly referred to as the Keldysh-Faisal-Reiss (KFR) theory.
Obviously, a better approximation would be to expand $U(t,t_0)$ in eq. (24) once more. This time, we use the expansion with respect to the field:

$$M_{\mathbf{p}} = -\frac{i}{\hbar} \int_{-\infty}^{\infty} dt_0\, \underbrace{\langle c_{\mathbf{p}}(t)| U_f(t,t_0)}_{\langle v_{\mathbf{p}}(t_0)|} D_{WW}(t_0) |g(t_0)\rangle$$

$$- \frac{1}{\hbar^2} \int_{-\infty}^{\infty} dt_1 \int_{-\infty}^{t_1} dt_0\, \underbrace{\langle c_{\mathbf{p}}(t)| U(t,t_1)}_{\approx \langle v_{\mathbf{p}}(t_1)|} \underbrace{[H(t_1) - H_f(t_1)]}_{= V} U_f(t_1,t_0) D_{WW}(t_0) |g(t_0)\rangle$$

$$=: M_{\mathbf{p}}^{(0)} + M_{\mathbf{p}}^{(1)}. \tag{26}$$

The first line of eq. (26) reproduces eq. (25) while the second line gives a correction. An interpretation is straightforward: Due to the interaction D_{WW} with the field, the electron is transfered into the continuum at $t = t_0$. It propagates in the field between t_0 and t_1 and undergoes a further interaction with the potential at $t = t_1$ which transfers it into the continuum. Just as for the direct electrons this sounds like (and in fact is) a literal repetition of the classical model for the rescattered ones. For the first time $M_{\mathbf{p}}^{(1)}$ was used by Becker [26] in response to the discovery of the ATI plateau.

So far, the improved Keldysh theory has been able to reproduce all effects of atoms in strong laser fields. Although the Keldysh-type theories are superior to the classical model with respect to accuracy, the question rises whether there are qualitative effects only accessible to quantum theories. Well, it was known for years that the envelopes of

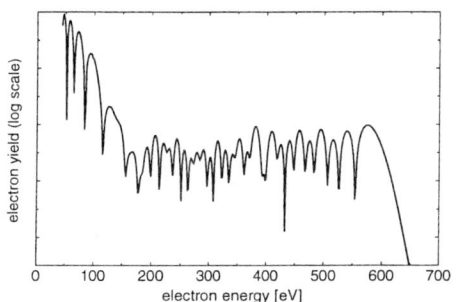
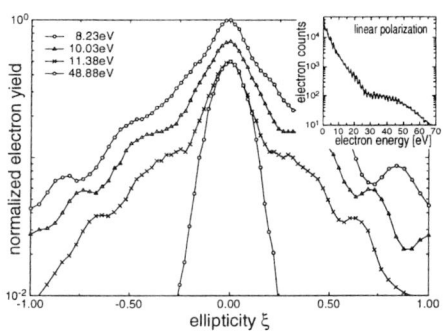

Figure 11: Calculated envelope of an ATI spectrum. The seemingly erratic pattern of sharp minima is due to interference effects [27].

Figure 12: Ellipticity distributions of xenon for some electron energies. The local maxima are interference effects.

ATI (and HHG) spectra calculated with suitable quantum-mechanical models exhibit a seemingly erratic pattern of minima, see fig. 11. It has been a change in the point of view which led to a physical explanation of this feature. The conclusion has been that they are due to interferences between quantum trajectories [28, 29]. However, the experimental verification is quite hard because the energy positions of the minima are very sensitive to intensity fluctuations. A laser focus, however, incorporates all intensities smaller than the peak intensity and thus the minima tend to be washed out. Only recently several other features of ATI spectra were detected and shown to be due to quantum interference effects. In the following, we shall to describe three of them. The first example involves only direct electrons with the consequence that the first term of eq. (26) is sufficient to explain it. The second example will be seen to be due to interference between the direct and rescattered electrons and, finally, it is also possible to have interference effects involving the second term of eq. (26) alone.

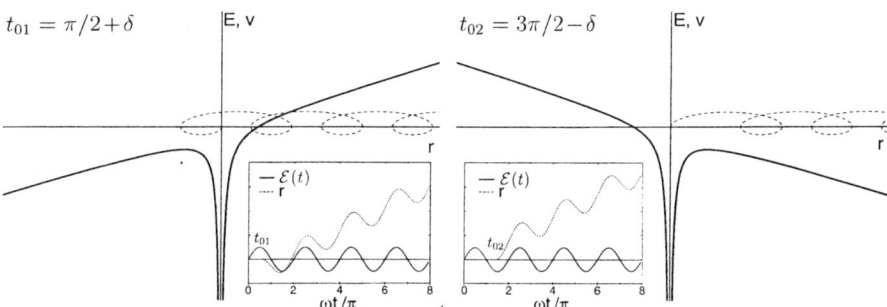

Figure 13: Classical trajectories leading to electrons emitted in the same direction with the same energy. This is a strongly simplified picture of the physics underlying the interferences mentioned in fig. 11.

3.2 Interference of direct electrons

Looking back to eq. (3), we see that in each cycle two times $t_{01} = \pi/2 + \delta$ and $t_{02} = 3\pi/2 - \delta$ exist, which lead to the same drift velocity and thus the same electron energy. Since it is not possible to decide when the electron entered the field, the quantum-mechanical requirements for interference are fulfilled. Interestingly, the two resulting trajectories start out with different directions: Trajectory 2 immediately starts in the direction of the drift velocity, whereas trajectory 1 initially runs the wrong way, see fig. 13. This situation is reminiscent of ballistics and therefore this type of interference is called *ballistic interference*. The situation described here is depicted in fig. 13. Of course, the model presented above is strongly simplified, in particular it neglects the fact that the electrons are born through a tunneling process.

Experimentally, it is instrumental to use elliptical polarized light. Here, we take advantage of the fact that the ellipticity is fairly easy to control by means of a quarter-wave plate. As in 2.1.2, the field of the laser is described by

$$\mathcal{E}(t) = \frac{\mathcal{E}_0}{\sqrt{1+\xi^2}} \begin{pmatrix} \sin\omega t \\ \xi\cos\omega t \end{pmatrix} \quad \text{or} \quad \mathbf{A}(t) = -\frac{\mathcal{E}_0}{\omega\sqrt{1+\xi^2}} \begin{pmatrix} -\cos\omega t \\ \xi\sin\omega t \end{pmatrix}. \tag{27}$$

With the main axis of the polarization ellipse remaining in the direction of the photoelectron detector, 73 ATI spectra corresponding to ellipticities regularly distributed between $\xi = -1$ (left circular) and $\xi = +1$ (right circular) are recorded. From these ATI spectra we extract ellipticity distributions (EDs), e.g. the electron yield as a function of ξ for an energy interval $[E - \Delta E/2, E + \Delta E/2]$, with $\Delta E \approx 100$meV. An example for such a measurement is shown in fig. 12, where the inset displays the ATI spectrum for linear polarization.

As explained in section 2.2, the high energetic photoelectrons in the ATI plateau region are due to rescattering. As the ellipticity is increased, the returning electrons increasingly miss the ion core and let the ATI plateau vanish quickly. The consequence are narrow EDs for plateau electrons. In contrast, the EDs for the direct electrons have a very different appearance. First of all, they are much broader. Most interestingly, however, also structures appear on them. In fig. 12 for both orientations of the polarization ellipse two local maxima are present, where the ones closer to linear polarization have already merged with the global maximum at $\xi = 0$. It is also very typical that the wiggles move towards linear polarization as the electron energy is increased. This structure looks like an interference pattern and, in fact, it is. In order to prove this, we concentrate on the first term of eq. (26) because low-energy electrons are to be analyzed. In addition, it is rewritten as an integral over all possible quantum paths leading to photoelectrons that are allowed by energy conservation:

$$M_n = \sum_n \delta\left(\frac{\mathbf{p}^2}{2} + E_{\text{IP}} + U_p - n\omega\right) \int_0^{2\pi/\omega} dt\, e^{iS_\mathbf{p}(t)} \tag{28}$$

where the phase is just the classical action

$$S_\mathbf{p}(t) = E_{\text{IP}}t + \int^t dt'\, (\mathbf{p} + \mathbf{A}(t'))^2/2. \tag{29}$$

The energy of the corresponding ATI peak is $E = n\hbar\omega - U_p - E_{\text{IP}}$ where the integer n denotes the number of photons absorbed from the ground state. If the process described by the above matrix element would have a classical limit, then the integral in eq. (28) would be dominated by contributions from those times, for which the action (29) is stationary, e.g.

$$dS_\mathbf{p}(t)/dt = (\mathbf{p} + \mathbf{A})^2 + E_{\text{IP}} = 0. \tag{30}$$

Obviously, there are no real times for which eq. (30) holds, i.e. we are confronted with a tunneling process. Nevertheless, it is possible to derive the points t_S of stationary action (saddle points) in the complex t plane by solving eq. (30).

$$\cos\omega t_S = \frac{\sqrt{1+\xi^2}}{1-\xi^2}\left\{-\sqrt{\frac{E}{2U_p}} \pm \underbrace{\sqrt{\frac{E}{2U_p}\xi^2 - \frac{E_{\text{IP}}}{2U_p}(1-\xi^2) - \xi^2\frac{1-\xi^2}{1+\xi^2}}}_{R}\right\}. \tag{31}$$

The fact that only electrons emitted in the direction of the main axis of the polarization ellipse are registered, is taken into account by the term $\mathbf{pA} = |\mathbf{p}||\mathbf{A}|\cos(\angle(\mathbf{p},\mathbf{A}))$. There are four complex solutions for eq. (31). Their symmetry depends on R, which is imaginary for $\xi \in [0\ldots\xi_0]$ and real for $\xi \in [\xi_0\ldots 1]$. The null of R is at

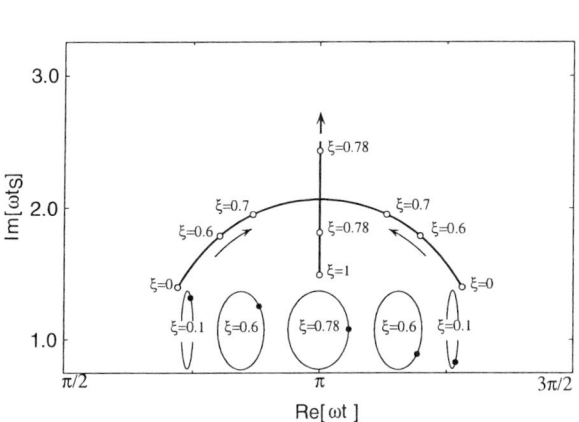

Figure 14: Positions of the saddle points ωt_S in the upper half complex ωt-plane in the interval $\pi/2 \leq \mathrm{Re}\,\omega t \leq 3\pi/2$, calculated from eq.(31) for $E = 17\omega - E_{\mathrm{IP}} - U_p$. The arrows indicate the motion of the saddle points for increasing ξ. For several values of the ellipticity, insets depict the ellipse circumscribed by the electric-field vector, and the positions of the latter at the emission times $\mathrm{Re}\,t_S$ are marked by solid dots.

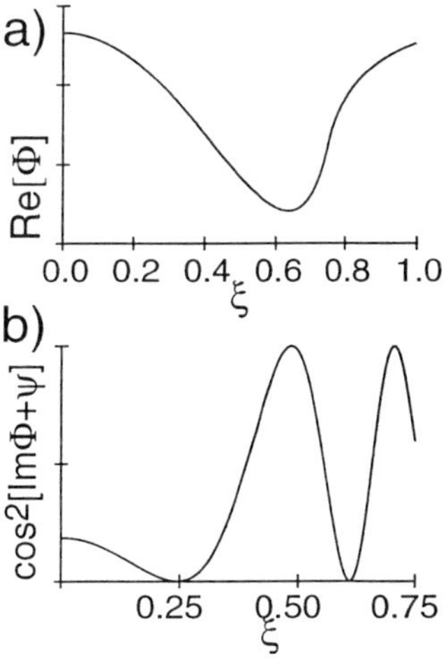

Figure 15: The squared magnitude of the approximation (32) reproducing the dodging phenomenon (a). The corresponding squared cosine of the phase is responsible for the interference pattern (b).

$$\xi_0 = \frac{\sqrt{(1-e) + \sqrt{(1-e^2) + 4e_{\mathrm{IP}}(1+e+e_{\mathrm{IP}})}}}{\sqrt{2\cdot(1+e+e_{\mathrm{IP}})}},$$

where we abbreviated $E/(2U_p)$ with e and $E_{\mathrm{IP}}/(2U_p)$ with e_{IP}. For $\xi \in [0\ldots\xi_0]$ the saddle points are symmetric with respect to a real part of π and the real axis, while for $\xi \in [\xi_0 \ldots 1]$ only the symmetry with respect to the real axis remains and the real part of all four saddle points is π. Fig. 14 displays the behavior of the two saddle points with positive imaginary part as a function of ξ. Now eq. (28) can be calculated by deforming the contour for the integral so that it passes through those saddle point(s) in the upper half plane that are nearest to the real axis. For $\xi \in [\xi_0 \ldots 1]$ this is obviously just one point, whereas for $\xi \in [0\ldots\xi_0]$ two points need to be considered. Accordingly, in the latter case, the contributions from the two saddle points may amplify or cancel each other, or, in other words, there may be constructive or destructive interference of the contributions of the two saddle points. On the other hand, interference is of course not possible for $\xi > \xi_0$.

The calculation yields for $\xi \in [0\ldots\xi_0]$

$$M_n \sim \exp[\mathrm{Re}\Phi(\omega t_{S1})]\cos[\mathrm{Im}\Phi(\omega t_{S1}) + \psi(\omega t_{S1})] \tag{32}$$

with

$$\Phi(\omega t) = i[n\omega t + \sqrt{\frac{8EU_p}{\omega^2(1+\xi^2)}}\sin\omega t + \frac{U_p}{2\omega}\frac{1-\xi^2}{1+\xi^2}\sin 2\omega t]$$

297

and $\psi(\omega t) = -(1/2)\arg\sin\omega t$. For $\xi > \xi_0$ the result is almost the same, only the cosine is absent. This is shown in fig. 15. Part b) shows the interference effect caused by the two saddle points, whereas part a) displays the envelope of the ED.

The saddle points ωt_S have an important physical interpretation as complex tunneling times [16]. The imaginary part sets the scale for the 'tunneling time', i.e. the time the electron spends in the classically forbidden region. Taking typical values for the parameters yields tunneling times of the order of a quarter of an optical cycle. The tunneling barrier, of course, is not constant at all during this interval, rather we deal with a dynamic tunneling problem. The real part can be attributed to the time at which the electron 'leaves the tunnel' (or rather to the phase of the field at this time). Thus the formal interference discussed above has physical significance: in order to reach the detector with a specific energy, for $\xi \leq \xi_0$ the electrons may enter the continuum at one or the other of two specific times during one optical cycle [17].

In fig. 14 these two times are marked by solid dots on the field ellipse; cf. the ellipses for $\xi = 0.1$ and 0.6 in the left-hand and in the right-hand part of the figure. The contributions from emission at these two times interfere. This is in close analogy to the discussion for fig. 13. For $\xi \geq \xi_0$, there is only one emission time contributing and, correspondingly, no interference. In this regime, $\text{Re}\,\omega t_S = \pi$ indicates that emission occurs when the *small* component of the driving field is maximized, cf. the ellipse for $\xi = 0.78$ in fig. 14.

3.3 Interference of direct and rescattered electrons

Mathematically, interference between direct and rescattered electrons is easy to understand. Eq. (26) expresses the matrix element as a sum of the contributions of the corresponding ionization path ways. Clearly, one will interpret differences between the coherent sum $|\langle v|V_f|g\rangle + \langle v|V_a G_f V_f|g\rangle|^2$ and the incoherent sum $|\langle v|V_f|g\rangle|^2 + |\langle v|V_a G_f V_f|g\rangle|^2$ as interference effects. Experimentally, the situation is much more complicated because some minimal contrast is necessary to unambiguously prove an interference effect. Therefore, the amplitudes of the direct and the rescattered electrons should be of comparable strength which is the case just in the tiny energy region where the plateau begins. Even more severe is the averaging of ATI spectra over many intensities inside a laser focus which washes out interference structures.

Again, the ellipticity of the laser polarization is helpful because it weakens the ATI plateau and thus keeps the amplitude of direct and rescattered electrons comparable over an extended energy region. In addition, it was realized that the interference patterns in the angular coordinate are much less sensitive to intensity fluctuations than for the energy coordinate. Fig. 16 shows angle- and energy-resolved ATI spectra as density plots. For linear polarization ($\xi = 0$) we have the Eiffel-tower-like appearance already seen in section 2.2.3. In remarkable qualitative contrast, the dark stripe representing the plateau for $\xi = 0$ has split into two for $\xi = 0.36$. Fig. 17b shows cuts through fig. 16b for a series of electron energies; they are the respective angular distributions. The contrast of the curves reaches 0.5. In fig. 17a the comparison to theory is shown. First of all, the oscillatory behavior is reproduced. In addition, for $s = 17$ the amplitudes for the direct and the rescattered electrons have been drawn. Both display a smooth behavior. This proves that the observed pattern indeed is due to the coherent sum of the amplitudes of direct and rescattered electrons, or in other words, due to interference [30].

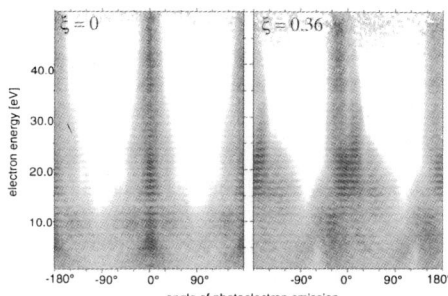
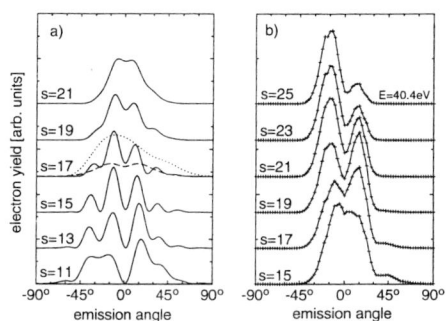

Figure 16: Density plots of the energy-resolved angular distribution for Xe at an intensity of $7.7 \cdot 10^{13}$ W/cm^2 and a wavelength of 800nm for an ellipticity of $\xi = 0$ (a) and $\xi = 0.36$ (b). 0° corresponds to electrons emitted in the direction of the main axis of the polarization ellipse. Dark means high electron yield. For visual convenience the data was normalized for each ATI peak.

Figure 17: Calculated a) and measured b) angular distributions of various ATI peaks in the plateau region. s denotes the order of the ATI peak. In addition, in part a) the amplitudes for the direct electrons (dashed) and the rescattered electrons (dotted) are plotted for $s = 17$. The photon energy is ≈ 1.55eV. The experimental result was obtained by performing cuts of Fig. 16 at the respective energies.

3.4 Interference of rescattered electrons

The same type of interference effect as for direct electrons (see sec. 3.2) is visible for rescattered electrons in the plateau region of the calculated spectrum displayed in fig. 11. The pattern is caused by interference of two trajectories with different traveling times. For rescattered electrons, depending on t_0, it is also possible to have several returns of the electron to the core. To each order of return corresponds a pair of trajectories, one with shorter and one with longer travel time. This can be seen easily by studying the classical theory of rescattered electrons, see 2.2.1. Naturally, under most circumstances, the first return is by far the most important one and the higher order returns cause small corrections only. This holds all the more as pairs of trajectories belonging to different return orders usually interfere in a random way.

However, if they all conspire to interfere constructively, the situation can be completely different. The condition for constructive interference is easy to achieve, in fact, experimentally it can hardly be avoided, as it is the channel-closing condition well known from the early days of ATI: An electron that wants to enter the continuum not only has to acquire the field-free ionization energy, but, in addition, the ponderomotive energy, because, once in the continuum, the electron will wiggle with this energy. Obviously, the ionization order increases if

$$n\hbar\omega = E_{\text{IP}} + U_p. \tag{33}$$

At the corresponding intensity, the lowest-order ATI channel closes [31, 32].

Therefore, the strategy to prove these channel-closing induced interference effects is to study the ATI spectra in dependence of the applied laser intensity. The polarization is kept linear and parallel to the axis of the spectrometer. An example is shown in fig. 18. As the intensity rises, the plateau grows and expands to higher electron energies. At a certain intensity, the low-energy part of the plateau performs a jump and grows again slowly if the intensity is increased further [33, 34, 35, 36, 37]. For an interpretation, it should be kept in mind that a measured ATI spectrum is made up by contributions from all intensities $I \leq I_0$ that are contained within the spatio-temporal pulse profile.

This means that a spectrum for a fixed intensity would show the enhanced group of ATI peaks only at that intensity where it first appears in this measurement, namely at $I \approx 0.85 I_0 = 7 \cdot 10^{13}\,\mathrm{W/cm^2}$, just above the $k = 12$ channel closing.

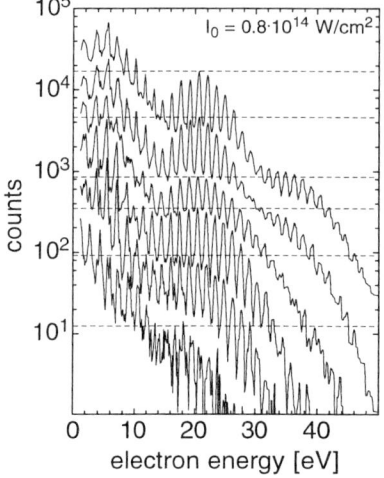 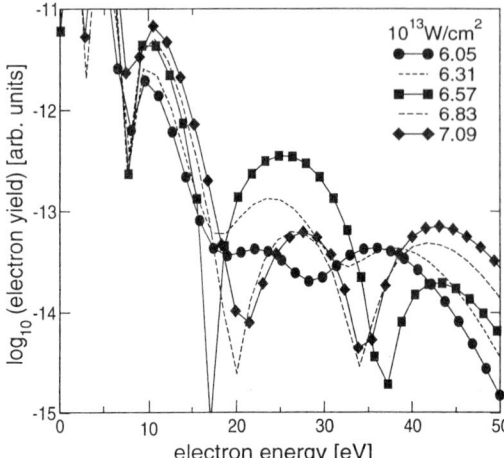

Figure 18: ATI spectra in argon at 800 nm recorded in the direction of the linearly polarized field for various intensities equally distributed between $0.5\,I_0$ and $1.0\,I_0$ with $I_0 \approx 8 \times 10^{13}\,\mathrm{W/cm^2}$. The horizontal lines mark the maxima of the plateaus for each intensity.

Figure 19: Calculated ATI spectra corresponding to fig. 18 for five different intensities, as indicated in $\mathrm{W/cm^2}$ in the figure. The central intensity is at the position of the channel closing with multiphoton order $k = 12$ ($I = 6.57 \times 10^{13}\,\mathrm{W/cm^2}$).

The Keldysh-type theory reproduces that behavior, see fig. 19. By doing a saddle-point analysis for the rescattered electrons (which is much more difficult than for direct ones), the contributions of the trajectories to the ATI spectra can be determined. Off the channel-closing intensity, very few pairs of trajectories are sufficient to approximate the exact Keldysh rate. This is quite different at the channel-closing position, where 20 pairs are hardly able to deliver a good fit.

The amazing point for this type of interference in ATI is that it dominates the appearance of ATI in the plateau region. Here, the quantum effects in strong-field laser matter interaction are not washed out by averaging intensities and they are by no means a small correction [38].

4 Conclusion

We have seen that many effects in ATI can easily be explained by almost trivial classical arguments. In particular, in the plateau region this model has enormous predictive power. However, we have also seen that there are several features which require a quantum model for an explanation. In most cases, these effects are difficult to analyze experimentally, yet, under certain conditions they clearly stand out and may even dominate what is explainable with the classical theory.

It should also be pointed out that throughout this paper we neglected the atomic structure and multi-electron effects. In the literature these effects are classified as being 'simple' (i.e. a single-electron effect) and 'universal' (i.e. present in all atoms with an ionization energy much bigger that of the photon). Interestingly, for atoms in strong laser fields hardly any effect was discovered that was not 'simple' and 'universal' so far.

It is our pleasure to thank all the colleagues from whose collaboration we benefitted during the last years, most notably F. Grasbon, W. Becker, and R. Kopold. We also appreciate the stimulating discussions with C. Joachain during his stays at MPQ.

References

[1] P. Agostini, F. Fabre, G. Mainfray, G. Petite, N. Rahman, Phys. Rev. Lett. **42**, 1127 (1979).
[2] J. Wildenauer, J. Appl. Phys. **62**, 41 (1987).
[3] M. Ferray, A. l'Huillier, X. F. Li, L. A. Lompré, G. Mainfray, C. Manus, J. Phys. B **21**, L31 (1988).
[4] G. G. Paulus, W. Nicklich, Xu Huale, P. Lambropoulos, H. Walther,
 Phys. Rev. Lett. **72**, 2851 (1994).
[5] B. Walker, B. Sheehy, K. C. Kulander, L. F. DiMauro, Phys. Rev. Lett. **77**, 5031 (1996).
[6] Ch. Spielmann, N. H. Burnett, S. Sartania, R. Koppitsch, M. Schnürer, C. Kan, M. Lenzner, P. Wobrauschek, F. Krausz, Science **278**, 661 (1997).
[7] Z. Chang, A. Rundquist, H. Wang, M. M. Murnane, H. C. Kapteyn, Phys. Rev. Lett. **79**, 2967 (1997).
[8] D. J. Jones, S. A. Diddams, J. K. Ranka, A. Stentz, R. S. Windeler, J. L. Hall, S. T. Cundiff, Science **288**, 635 (2000).
[9] C. J. Joachain, M. Dörr, N. Kylstra, Adv. At. Mol. Opt. Phys. **42**, 225 (2000).
[10] H. B. v. Linden v.d. Heuvell and H. G. Muller, in *Multiphoton Processes, Studies in Modern Optics* No. 8, 25, Cambridge (1988).
[11] K. C. Kulander, K. J. Schafer, K. L. Krause, in *Proceedings of the SILAP III Workshop*, Plenum Press, New York (1993).
[12] P. B. Corkum, Phys. Rev. Lett. **71**, 1994 (1993).
[13] G. G. Paulus, W. Becker, W. Nicklich, H. Walther, J. Phys. B **27**, L703 (1994).
[14] G. G. Paulus, W. Becker, H. Walther, Phys. Rev. A **52**, 4043 (1995).
[15] T. F. Gallagher, T. J. Scholz, Phys. Rev. A **40**, 2762 (1989).
[16] N. B. Delone and V. P. Krainov, *Multiphoton Processes in Atoms* (Springer-Verlag, Berlin, 1994); for a review of the concept of the tunneling time in constant fields, see E. H. Hauge, J. A. Støvneng, Rev. Mod. Phys. **59**, 917(1989).
[17] G. G. Paulus, F. Zacher, H. Walther, A. Lohr, W. Becker, M. Kleber,
 Phys. Rev. Lett. **80**, 484 (1998).
[18] M. Bellini, C. Lyngå, A. Tozzi, M. B. Gaarde, T. W. Hänsch, A. L'Huillier, C.-G. Wahlström, Phys. Rev. Lett. **81**, 297 (1998).
[19] A. Rundquist, C. G. Durfee III, Z. Chang, C. Herne, S. Backus, M. M. Murnane, H. C. Kapteyn, Science **280**, 1412 (1998).
[20] M. Lewenstein, K. C. Kulander, K. J. Schafer, P. H. Bucksbaum, Phys. Rev. A **51**, 1495 (1995).
[21] B. Yang, K. J. Schafer, B. Walker, K. C. Kulander, P. Agostini, L. F. DiMauro,
 Phys. Rev. Lett. **71**, 3770 (1993).
[22] G. G. Paulus, W. Nicklich, H. Walther, Europhys. Lett. **27** 267 (1994).
[23] L. V. Keldysh, Zh. Eksp. Teor. Fiz. **47**, 1945 (1964) [Sov. Phys. JETP **20**, 1307 (1965)].
[24] F. H. M. Faisal, J. Phys. B **6**, L89 (1973).
[25] H. R. Reiss, Phys. Rev. A **22**, 1786 (1980).
[26] W. Becker, A. Lohr, M. Kleber,
 J. Phys. B **27**, L325 (1995) and **28**, 1931 (1995) (corrigendum).
[27] A. Lohr, M. Kleber, R. Kopold, W. Becker, Phys. Rev. A **55**, R4003 (1997).
[28] M. Lewenstein, Ph. Balcou, M. Y. Ivanov, A. L'Huillier, P. B. Corkum,
 Phys. Rev. A **49**, 2117 (1994).
[29] M. Lewenstein, P. Salières, A. L'Huillier, Phys. Rev. A **52**, 4747 (1995).
[30] G. G. Paulus, F. Grasbon, A. Dreischuh, H. Walther, R. Kopold, W. Becker,
 Phys. Rev. Lett. **84**, 3791 (2000).
[31] P. Kruit, J. Kimman, H. G. Muller, M. J. van der Wiel, Phys. Rev. A **28**, 248 (1983).
[32] H. G. Muller, A. Tip, M. J. van der Wiel, J. Phys. B **16**, L679 (1983).
[33] M. P. Hertlein, P. H. Bucksbaum, H. G. Muller, J. Phys. B **30**, L197 (1997).
[34] P. Hansch, M. A. Walker, L. D. Van Woerkom, Phys. Rev. A **55**. R2535 (1997).
[35] H. G. Muller and F. C. Kooiman, Phys. Rev. Lett. **81**, 1207 (1998).
[36] M. Nandor, M. A. Walker, L. D. Van Woerkom, H. G. Muller, Phys. Rev. A **60**, R1771 (1999).
[37] H. G. Muller, Phys. Rev. A **60**, 1341 (1999).
[38] G. G. Paulus, F. Grasbon, H. Walther, R. Kopold, W. Becker, (submitted).

RELATIVISTIC EFFECTS IN NON-LINEAR ATOM-LASER INTERACTIONS AT ULTRAHIGH INTENSITIES

Valérie Véniard

Laboratoire de Chimie Physique-Matière et Rayonnement
Université Pierre et Marie Curie
11, rue Pierre et Marie Curie - 75231 Paris
Cedex 05, France

1. INTRODUCTION

Relativity is expected to play an important role in several types of radiative processes in atoms. Its influence on the atomic levels fine structure has been most thoroughly investigated as its signature is easily evidenced in atomic x-ray spectra,[1,2,3]. It manifests itself also in some delicate aspects of the chemical reactivity of the elements[4]. These effects arise from both the standard Dirac-like properties of electrons and from more sophisticated QED corrections. One of the major objective of the present paper is to show that the overall picture has dramatically changed recently, as a consequence of the considerable advances made in the design of ultra intense laser sources operated at intensities well beyond the so-called atomic unit of intensity $I_0 = 3.5 \times 10^{16}$ W/cm^2,[5].

These advances have opened new fields of relativity-related research in two complementary directions. One is related to the advent of laser-based sources of coherent radiation in the X-UV domain, either from high harmonic generation [6,7], from X-ray-laser devices,[8] or from Free Electron Laser (FEL),[9]. The implementation of these sources opens the possibility to perform two-photon bound-bound absorption experiments involving inner shells in heavy atoms or ions. One expects that those kind of experiments will considerably enlarge the scope of traditional (single-photon) x-ray absorption spectroscopy (XAS) and related techniques. Considerable advances are expected in the field, much in the spirit of those made in the IR-visible ranges with standard laser sources. The question then arises of the importance of relativistic and of the (closely related) retardation effects on multiphoton transition amplitudes in high-Z atoms. This question has been the object of active investigations in the field of standard (one-photon) X-ray spectroscopies,[10]. As we will show below, for two-photon bound-bound transitions, see Sec. 2, the magnitudes of the corrections differ significantly from what one could naively infer from the analysis of the one-photon results.

Another kind of processes in which intense lasers have brought relativity into play in atomic physics, results from the fact that even an electron initially at rest can acquire, within the laser field, a ponderomotive energy (i.e. its averaged oscillating energy) comparable to its rest mass energy. In addition, as a result of the dressing by the field, its mass can be significantly shifted from its bare value,[11]. As we shall show below these two combined effects can considerably modify the dynamics of typical atomic processes such as electron-atom collisions, (Sec. 3), and strong field photoionization, (Sec. 4).

The intensity needed to achieve electron-positron pair creation from vacuum in a strong laser field is well beyond the capabilities of currently operated laser devices. This theoretical limit will be briefly addressed in Sec. 5 through a simple model based on the Klein paradox.

2. TWO-PHOTON BOUND-BOUND TRANSITIONS IN HYDROGENIC ATOMS

An intriguing observation made in several areas of X-ray spectroscopy (XAS, XES and even RIXS) is that the Non-Relativistic Dipole (NRD) approximation provides fairly accurate values for the oscillator strengths, well beyond its expected range of validity,[10]. For instance, for the element Sn, ($Z = 50$, $\alpha Z \approx 0.365$), the value of the K-shell photoionisation cross section deduced from a fully relativistic MCDF computation differs from the Non-Relativistic Dipole (NRD) value by less than 5 % at photon energies up to 100 keV,[10]. A commonly advanced explanation for this observation is that two distinct (though deeply entangled) effects may partially compensate each other: one arises from the contribution of retardation which becomes important at higher frequencies, i.e. when the momentum carried by the photon cannot be neglected. The other effect is a consequence of the modifications of the atomic structure under the combined effects of spin and nuclear charge. It seems that in most one-photon processes these two classes of contributions do partially compensate each other. The question is then of determining if whether this situation holds also for two-photon transitions in high-Z elements.

In order to address this question we have considered the simpler case of two-photon bound-bound transitions originating from the ground state of point-nucleus hydrogenic systems. Our treatment is based on the computation of the relevant second-order perturbative amplitudes, the perturbative approach being certainly valid in view of the expected intensities of the currently developed x-ray sources as compared to typical electric field strength in heavier atoms. For such systems it is possible to perform an *exact* calculation of the relevant second-order T-matrix element :

$$T^{(2)}_{f,i} = \langle \psi_f | V G(\vec{r},\vec{r}';\Omega) V | \psi_i \rangle \tag{1}$$

where $V = \vec{\alpha} \cdot \vec{A}$, $\vec{\alpha}$, $\vec{A}$ denoting Dirac matrices and the potential vector associated to the electric field, respectively, and $G(\vec{r},\vec{r}';\Omega)$ is the Dirac Coulomb Green's function. Such second-order matrix elements can be computed conveniently to within a given accuracy with the help of a Sturmian expansion of $G(\vec{r},\vec{r}';\Omega)$,[12,13].

When relativity is taken into account, the Z-dependence of the transition amplitude is intricate, as the nuclear charge enters in an complicated way into the expression of the relevant wave functions and Green's function. In contrast, in the Non-Relativistic Dipole limit, it can be shown that the second-order matrix elements scale as Z^{-4}. Accordingly, in order to clearly display the difference between the NRD approximate results and those deduced from more refined treatments, we have reported the scaled modulus $|T^{(2)}_{f,i}| \times Z^4$ in terms of Z. One of the main outcomes of the analysis, when performed for the most typical two-photon $|1\,^2S_{1/2}\rangle \rightarrow |2\,^2S_{1/2}\rangle$ transition, is that, as compared to the NRD approximation, relativistic contributions amount to well over 10% to the scaled second-order T-matrix element for Z = 50, i. e. for photon energies $\hbar\omega \approx 12.75\,\text{keV}$, see Figure 1. Moreover, one observes that the respective contributions of retardation and of relativistic effects do not compensate each other. We have checked that this observation holds also for other two-photon bound-bound transitions,[12]. To conclude, our model computations, performed for a well defined test-system, clearly indicate that the conclusions drawn from the analysis of one-photon transition amplitudes cannot be transposed to the case of multiphoton processes.

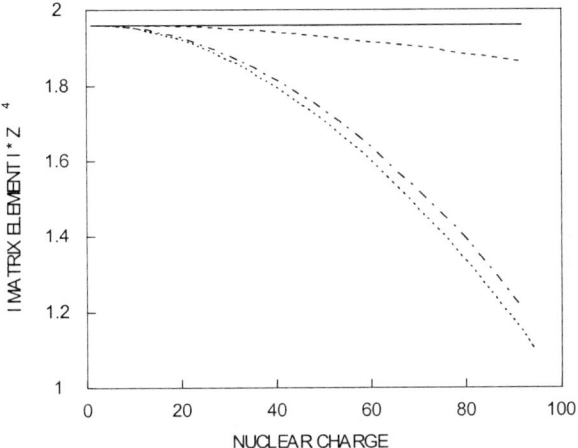

Figure 1. Modulus of the matrix elements M for the transition $|1\,^2S_{1/2}\rangle \rightarrow |2\,^2S_{1/2}\rangle$ in atomic units multiplied by a factor Z^4 as a function of nuclear charge Z. The solid line represents the nonrelativistic dipole (NRD), the dashed line the nonrelativistic retarded (NRR), the dot-dashed line the relativistic dipole (RD), and the dotted line the relativistic retarded (RR) result.

3. LASER-ASSISTED MOTT SCATTERING

When a relatively dense medium (atomic clusters, for instance) is irradiated by a strong IR laser, fast electrons are produced very early during the rise of the pulse. Later in the pulse, these electrons can be scattered by ionic species while the laser has a very high intensity. It is thus of interest to determine to which extent relativistic effects will affect the dynamics of the laser-assisted collisions and, more generally, the exchange of energy between the electrons and the field. In order to address these questions, we have considered the simpler case of Coulomb scattering for a Dirac electron (the so-called Mott scattering) in the presence of a circularly polarized, single-mode, laser field. Our motivation is that the physics of this process can provide a clear distinction between simple kinematics and spin-orbit coupling effects. The main advantage of the model is that, though much simplified, its limitations are well delineated and, as a first-Born treatment of the collision stage can be performed exactlly, it leads to analytical expressions which can be analyzed and discussed,[14].

The solutions of the Dirac equation for a free electron in the presence of a circularly polarized classical field, are the so-called Volkov wave functions:

$$\psi_q = \langle \vec{x} | q \rangle = \left[1 + \frac{kA}{2c(kp)}\right] \frac{u}{\sqrt{2QV}} \exp\left[-i(qx) - i \int_0^{kx} \frac{(pA)}{c(kp)} d\phi\right] \quad (2)$$

where u is the bispinor, solution of the Dirac equation for a free electron with four-momentum p. Here q is the averaged four-momentum of the electron within the field:

$$q^\mu = (Q/c, \vec{q}) = p^\mu - \frac{\overline{A^2}}{2(kp)c^2} k^\mu \qquad (3)$$

where, $\overline{A^2}$ denotes the time-averaged square of the four-vector potential of the laser field. The square of this four-vector is Lorentz invariant.

The first-order Born transition amplitude for the Coulomb scattering of a Volkov electron reads:

$$T_{f,i} = \frac{iZ}{c} \int d^4x \, \overline{\psi}_f \frac{\gamma^0}{|\vec{x}|} \psi_i \qquad (4)$$

Though cumbersome, the calculation can be carried out exactly, the final result being given as a complicated expression in terms of Bessel functions,[14, 15]. We note that the final formula reduces, in the non relativistic limit, to a known expression first derived in the sixties by Bunkin and Fedorov,[16].

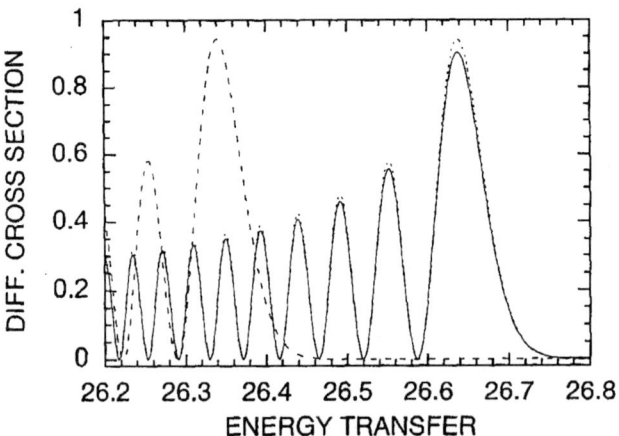

Figure 2. Envelope of the absorption edge part of the differential cross section $d\sigma/d\Omega$ scaled in 10^{-y} a.u. as a function of energy transfer scaled in units of 1000 ω for an electrical field strength of E = 1 a.u The initial electron energy is W = 100 a.u. The solid line denotes the result for electrons, the short-dashed one the differential cross section for spinless particles and the long-dashed one the result for the nonrelativistic limit.

The influence of relativity on the differential cross section is clearly evidenced in the high intensity regime, for scattering events in the course of which large numbers of IR photons are exchanged. For instance, for incoming 2.7 keV electrons, at $I = I_0 = 3.5 \times 10^{16}$ W/cm^2 up to 2.6×10^4 photons from a Nd: Yag laser can be absorbed in a 90° scattering event. Under such circumstances, noticeable differences are already observed between the non-relativistic and relativistic treatments, see Fig. 2. The observed differences can be ascribed to the mass-shift experienced by the electron embedded within the field,[11]:

$$m^* = m\sqrt{1 + \frac{\langle A^2 \rangle}{c^4}} \qquad (5)$$

This is, of course, amplified at I = 1.2×10^{18} W/cm^2 i.e. at an intensity well beyond the atomic unit I$_0$. Then, spin-dependent effects begin to show up, as a result from both the standard spin-orbit coupling and from the coupling with the magnetic component of the field, see Fig. 3.

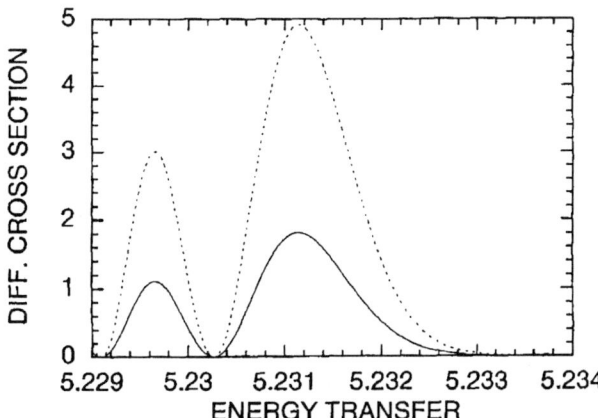

Figure 3. Envelope of the absorption edge part of the differential cross section $d\sigma/d\Omega$ scaled in 10^{-14} a.u. as a function of energy transfer scaled in units of 10^5 ω for an electrical field strength of E = 5.89 a.u. The solid line denotes the result for electrons, the short dashed one the differential cross section for spinless particles.

In conclusion, it appears that relativity can affect significantly the dynamics of laser-assisted collisions at intensities achieved by several laser sources currently operated worldwide,[14].

4. RELATIVISTIC EFFECTS IN PHOTOIONIZATION SPECTRA

A weakly bound electron submitted to an intense pulse of IR radiation can be ionized by a variety of mechanisms, from multiphoton absorption to tunnel or even "over the barrier" ionization, depending of the intensity regime,[17]. At higher intensities tunnel ionization and over the barrier processes are dominant and, as a consequence, the ejected electrons are initially relatively slow. This implies that the changes induced by relativity in the corresponding photoionization spectra are expected to be small. The situation can differ significantly in the presence of a pulse of UV or X-ray radiation. Indeed, a weakly bound electron submitted to a UV pulse is readily ionized, via one-photon absorption, even at low or moderate peak intensities. However, if the peak intensity of the radiation field is increased, several model simulations show that, very counter-intuitively, the ionization yields can *decrease*,[18]. This situation can be pictured as resulting of a *dichotomy* of the electron wave function strongly driven away from the nucleus in the high frequency field. The more intense is the field, the farther from the nucleus is the electron, thus accounting for the "stabilization" of the dressed atom.

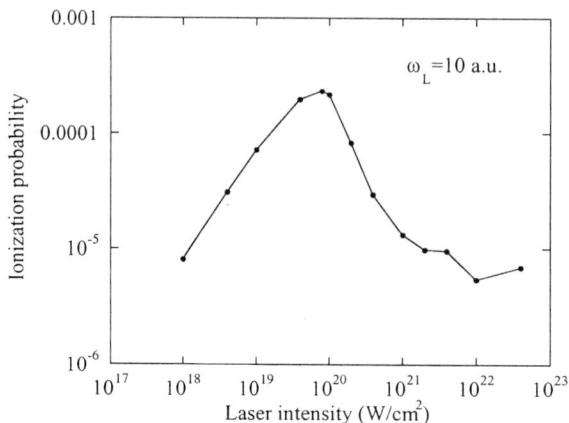

Figure 4. Ionization yield as a function of laser intensity for a radiation pulse with a linear turn-on of 100 T_H. The field frequency is $\omega_H = 10$ a.u. These yields are computed at $t=1100\ T_H$. The line represents the results for the time-dependent 1-D Schrödinger equation treatment while the dots are the results of the relativistic Schrödinger equation treatment.

This is shown in the Figure 4, where ionization yields for a simplified model atom are reported as a function of the peak intensity of the field. Here the data are obtained by solving numerically both the time-dependent Schrödinger equation and the time-dependent relativistic Schrödinger equation:

$$i\hbar \frac{\partial}{\partial t}\Psi(x,t) = \sqrt{m^2c^4 + c^2\left(p - \frac{A(t)}{c}\right)^2}\,\Psi(x,t) + V(x)\Psi(x,t), \qquad (6)$$

for a 1-dimensional "soft-Coulomb" potential,[19]: $V(x) = \frac{-1}{\sqrt{2+x^2}}$ with ground state binding energy $E_0 = -0.5$ a.u. For the sake of illustration, we have chosen the electric field envelope with linear turn-on of 100 high-frequency cycles $T_H = 2\pi/\omega_H$ durations. The dressing high frequency is $\omega_H = 10$ a.u., with field strength intensities around $F_{0H} \approx 10^3$ a.u., the corresponding excursion length being : $\alpha_0 = qF_{0H}/(m\omega^2) \approx 10$ a.u.. These yields are computed at $t=1100\ T_H$. Two interesting comments are in order: first, the stabilization phenomenon, originally demonstrated within the framework of the Floquet formalism, i.e. for constant amplitude fields, is also observed in numerical simulations for an explicitly time-dependent field. Regarding the role of relativity it should be understood that it is a qualified statement, as it has been established in a one-dimensional system which cannot account for spin and retardation effects. Up to know, it is not clear if such effects would affect the general trend observed here,[20]. Another remarkable observation is that the ionization yield is almost the same whether it is computed from a non-relativistic or from a fully relativistic treatment. This holds even at ultra-high intensities at which the dynamics of a free electron is certainly relativistic. Our results clearly indicate that, for a weakly bound electron, relativity does not affect quantitatively the dynamics of the ionization stage even in an ultra-strong field.

However, it is possible to probe such a relativistic dynamics with a second field with a lower frequency, so that multiphoton absorption, leading to ATI, can take place. In order to ionize such a stabilized atom with a significant probability, the second field must force the electron wave function to explore again the vicinity of the nucleus to be able to absorb energy (i.e. photons). This can be achieved by chosing parallel polarizations and field strength intensities and frequencies such that the characteristic excursion lengths α_0 for each field are comparable. We have chosen a "low" frequency $\omega_L = 0.2$ a.u. with a field strength intensity being around $F_{0L} \approx 0.3$ a.u. We note that the contribution of this additional field to the effective mass and to the relativistic parameter $\eta = \dfrac{qF_{OH}}{mc\omega_H}$ are negligible.

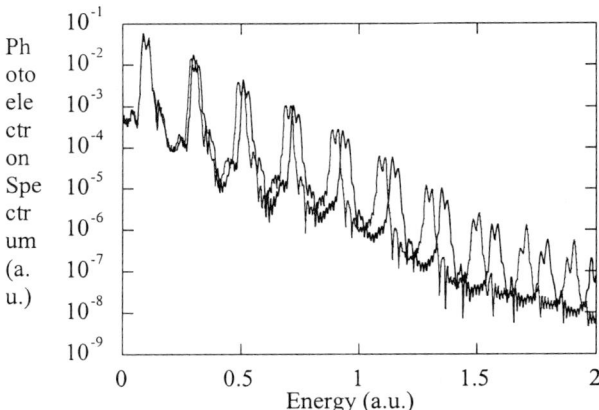

Figure 5. Two-color photoelectron spectrum for a radiation pulse containing two frequencies $\omega_H = 10$ a.u. and $\omega_L = 0.2$ a.u. with intensities $I_H = 5 \times 10^{21}$ W/cm^2 and $I_L = 10^{15}$ W/cm^2. The total duration of the high frequency pulse is $T = 28\, T_L$. The spectrum is obtained by solving the time-dependent relativistic Schrödinger equation (thick line) and for the sake of reference we have shown (thin line) the one obtained by solving the time-dependent Schrödinger equation.

In the figure 5, we have compared the photoelectron energy spectrum lines as obtained from a non-relativistic (time-dependent Schrödinger) and from a relativistic treatment, see Eq. (6),[21]. The fields frequencies and intensities are $\omega_H = 10$ a.u., $I_H = 5 \times 10^{21}$ W/cm^2 ($\eta = 0.28$), $\omega_L = 0.2$ a.u. and $I_L = 10^{15}$ W/cm^2. The most salient feature is the notable difference observed in the locations of the ATI peaks. The non relativistic simulation displays regularly spaced peaks, separated from each other by $\omega_L = 0.2$ a.u., corresponding exactly to the frequency of the probe field. In contrast, the relativistic simulation reveals photoelectron peaks with spacings of $\omega'_L = 0.21$ a.u$=1.05\omega_L$. We mention that a simulation (not shown),[21] for a stronger high frequency field, $I_H = 4 \times 10^{22}$ W/cm^2 ($\eta = 0.8$), exhibits ATI line spacing of $\omega'_L = 0.28$ a.u$=1.24\omega_L$.

The observed spacing cannot be explained using a simple "Doppler shift" argument because the exchange takes place within the high frequency dressing field, i.e. while, not only the electron experiences a periodically accelerated motion but while it has also acquired an effective mass. As the ATI spectrum is recorded after the end of the pulse, one has to analyze how the absorbed energy is transferred into the kinetic energy of the detected "bare" electron. To this end, we model the process with the use of a "free" gaussian wavepacket, "dressed" by the high frequency relativistic field, the multiphoton absorption from the low frequency field being simulated by a "kick" of magnitude p_0 in momentum, taking place at a random time during the pulse. The transferred energy will be,[21]:

$$\left\langle \Delta E^t_{KG} \right\rangle \approx \frac{p_0^2}{2m} \left\langle \left[1 + \frac{q^2 A_H(t_0)^2}{m^2 c^4}\right]^{-3/2} \right\rangle, \qquad (7)$$

while the detected energy is given by : $\left\langle \Delta E^t_{KG} \right\rangle \approx \dfrac{p_0^2}{2m}$.

This implies that, as the transferred energy is a multiple of the low-frequency quantum, the detected electrons have kinetic energies as if they had absorbed photons with shifted frequencies:

$$\omega' \approx \frac{1}{\left\langle \left[1 + \frac{q^2 A_H(t_0)^2}{m^2 c^4} \right]^{-3/2} \right\rangle} \omega \qquad (8)$$

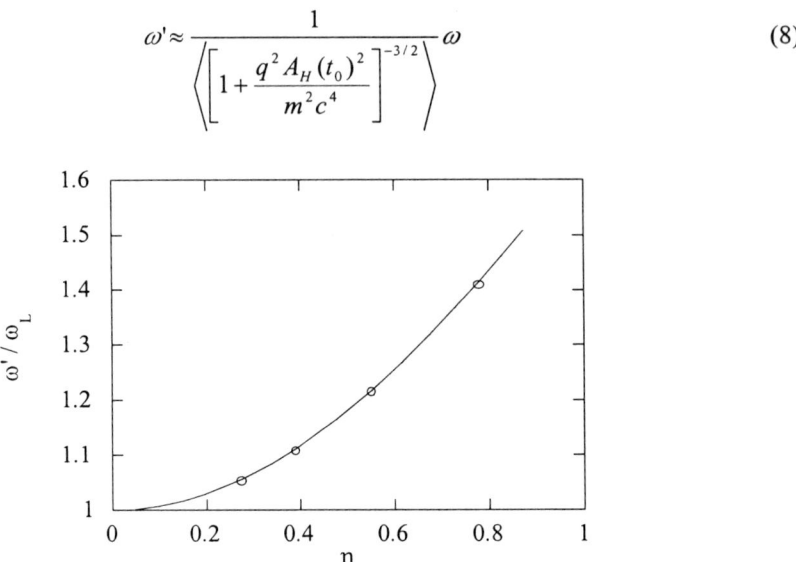

Figure 6. Frequency shift ω'/ω_L, Eq. (8), as a function of the relativistic parameter $\eta=(q\, F_{0H})/(mc\backslash\omega_H)$, with $\omega_H=10$ a.u. The circles indicate the shift deduced from the full time-dependent calculation.

It is worth mentioning that there is no equivalent frequency shift within the framework of the Schrödinger theory. Numerical values of the frequency shift computed from this formula are shown in Fig. 6, where it can be verified that they coincide with the one deduced from the spectral analysis of the final wave function. This effect, which exploits the stabilization of atoms in strong high frequency fields, is in essence different from standard relativistic corrections in high-Z atoms and in laser-plasma interactions. It represents a clear and unambiguous signature of relativistic effects in atom-laser interactions in the regime of ultra-high intensities.

5. PAIR CREATION IN ULTRA-INTENSE LASER FIELDS

The question of the possible electron-positron pair creation (from vacuum) in a strong laser field has arised since the early days of laser physics. The limit often put forward in popular science writings is around $I=10^{30}$ W/cm^2, well beyond the capabilities of currently operated laser sources. In present days, the most advanced "table-top" devices are able to deliver pulses with durations of a few femtosecond and peak intensities (after focusing) around $I=10^{21}$ W/cm^2, [22]. Moreover, as a result of fundamental limitations inherent to the scheme used for the amplification stages and also to the optics elements, there is a limit set around $I=10^{23}$ W/cm^2 imposed upon the maximum laser field strength attainable after focusing. However, alternative schemes are explored which would permit to go beyond the present limitations and to reach $I=10^{28}$ W/cm^2,[22]. This would make feasible to observe significant pair creation in the focus of a strong laser pulse.

The above theoretical limit can be directly derived from Schwinger's Quantum Electrodynamics (QED) treatment, which leads to the following estimated expression for the probability of pair creation,[23]:

$$\Gamma \approx \exp\left(-\frac{\pi m c^3}{F}\right) \qquad (9)$$

where F is the laser field strength. If $F=mc^3=2.57\times 10^6$ a.u., corresponding to a laser intensity $I = \sqrt{\varepsilon_0/\mu_0}|F|^2/2 \approx 2.3\times 10^{29} W/cm^2$, then the probability is $\Gamma \approx \exp(-\pi) = 4.3 \; 10^{-2}$. We wish to point out that this result can be recovered without having recourse to the heavy formalism of QED. In fact it can be derived through the use of a standard semi-classical (WKB) treatment of tunnelling through a potential barrier and can be related to the so-called Klein's paradox, one of the most fascinating effect of relativistic quantum mechanics [24]. Recent discussions of the Klein paradox can be found in references [25,26,27].

The Klein paradox, often presented in textbooks as an illustration of the peculiarities of relativistic quantum mechanics, results from the completely counter-intuive fact that a particle can tunnel through a very high potentiel step, provided the height of the step is larger than twice the rest energy of the particle: $V_0 > 2mc^2$ [2,28]. Under such conditions, shown schematically in Figure 7, the particle can apparently propagate in the Dirac see of negative energy states within the barrier. It has been soon recognized that the component of the wave function, solution of the Dirac (or Klein-Gordon) equation in the region of space within the barrier, could be associated to an anti-particle propagating rightwards, see the reference (25). There is also a component of the wave function which is reflected from the barrier and is associated to an electron created together with the positron and propagating leftwards.

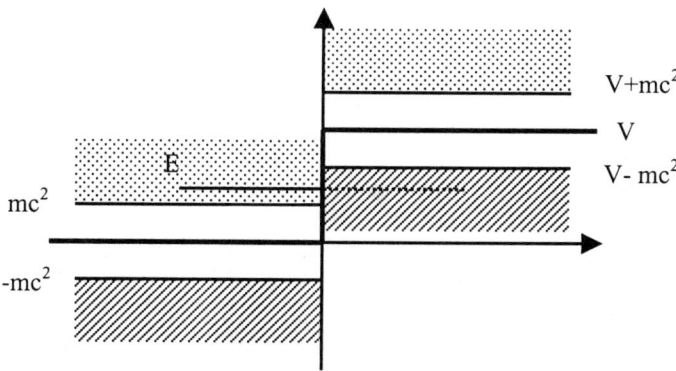

Figure 7. Scattering of an electron, with an incoming energy E, on a potential step of height $V>2mc^2$. The continuum for the particle and the antiparticle overlap when $mc^2<E<V-mc^2$.

For a more physical situation, one can use the potential shown in Fig. 8, representing a constant electric field F_0 in a finite region of space $x\in[0,L]$ requiring that $F_0L>2mc^2$.

The transmission probability, interpreted as the pair creation probability, is obtained through a semiclassical approach, see for instance ref.(29), and is given by:

$$|T|^2 \approx \exp\left[-2\,\mathrm{Im}\int_{x^-}^{x^+} q(x)dx\right] \qquad (10)$$

with $\mathrm{Im}\,q(x) = \sqrt{m^2c^4-(E-F_0x)^2}$. By evaluating Eq(10), one recovers the expression obtained by Schwinger, Eq.(9).

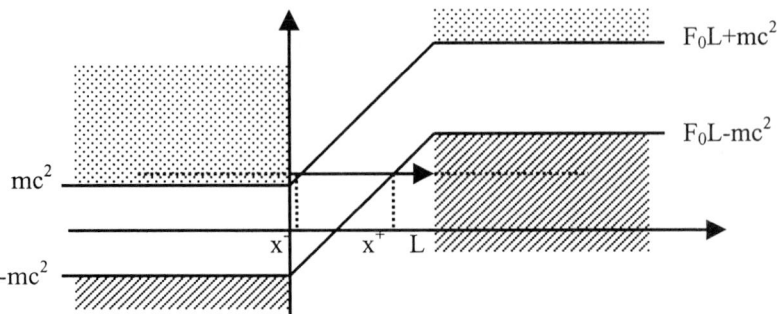

Figure 8. Scattering of an electron, with incoming energy E, on a potential V(x) representing a constant electric field F_0 in the region $0<x<V$. $[x^-,x^+]$ defines the classically forbidden region.

6. CONCLUSIONS

In the present paper, we have reported the results of very recent computations, either analytical or fully numerical, of the transition amplitudes and/or of the cross sections for different phenomena which are expected to be observed when atomic systems experience non-linear radiative processes in intense external fields. In each case, we have addressed the question of the relative importance of the effects of relativity and we have made explicit the conditions in which the signatures of relativistic effects and retardation could be evidenced. We have considered in particular the case of multiphoton transitions, to be observed with the help of intense high frequency fields as produced by X-ray Lasers or Free-Electron Lasers (FEL). As a result of our analysis, we have shown that two-photon bound-bound transition amplitudes in high-Z hydrogenic systems are significantly affected by relativistic corrections, even for low values of the charge of the nucleus. For instance at $Z = 20$, the corrections amount to about 10%, a value much higher than what is observed for standard one-photon transitions in X-ray spectroscopy measurements for which the non-relativistic dipole (NRD) approximation agrees with the exact result to within 99% at comparable frequencies. Relativity affects also significantly the scattering cross section of fast electrons in the Coulomb field of a nucleus, in the presence of an ultra-strong infrared laser. When the laser intensity becomes comparable or larger than the so-called atomic unit of intensity $I_0 = 3.5 \times 10^{16}$ W/cm², notable corrections are observed in addition to the standard influence of the spin-orbit coupling which gives rise to the Mott scattering cross section. For a Dirac electron, the coupling of the spin with the magnetic component of the field contributes also to the changes observed in the energy spectra of the scattered electrons. Eventually, we have shown also that in the presence of ultra-intense laser fields the dynamics and the amount of energy absorbed by an atom can be significantly modified by relativistic effects, when compared to the results of non relativistic simulations. Such effects are expected to be observable in some selected cases clearly identified. Finally we have presented a simple model, based on the tunnelling of an electron through a potential step and related to the Klein paradox, which provides an estimate of the intensity of the field needed to have pair creation from vacuum.

Acknowledgements

The Laboratoire de Chimie Physique-Matière et Rayonnement is a Unité Mixte de Recherche, Associée au CNRS, UMR 7614, and is Laboratoire de Recherche Correspondant du CEA, LRC # DSM-98-16. Parts of the computations have been performed at the Centre de Calcul pour la Recherche (CCR, Jussieu, Paris) and at the Institut du Développement et des Ressources en Informatique Scientifique (IDRIS).

References

[1] I. P. Grant, J. Phys. B: At. Mol. Opt. Phys. **7**, 1458 (1974).
[2] J. P. Desclaux, Comp. Phys. Comm. **9**, 31 (1975).
[3] J. Bruneau, J. Phys. B: At. Mol. Opt. Phys. **16**, 4135 (1983).
[4] a recent reference is : K. Doll, P. Pyykkö and H. Stoll, J. Chem. Phys. **109**, 2339 (1998).
[5] G. A. Mourou, C. P. J.Barty. and M. D. Perry, Physics today **51**, 22 (1998).
[6] Recent references include : R. Zerne, *et al*, Phys. Rev. Lett. **79** 1006 (1997); Z. Chang, *et al*. Phys. Rev. Lett. **79** 2967 (1997); Ch. Spielmann, *et al*. Science **278**, 661 (1997); M. Schnürer, *et al*. Phys. Rev. Lett. **80** 3236 (1998).
[7] M. Bellini, *et al*. Phys. Rev. Lett. **81**, 297 (1998).
[8] Recent references can be found in: *6th International Conference on X-Ray Lasers*, Institute of Physics Conferences Series, to be published.
[9] R. Brinkmann, G. Materlik, J. Rossbach (eds.), DESY 1997-048, ECFA 1997-182.
[10] A. Ron, I. B. Goldberg, J. Stein, S. T. Manson, R. H. Pratt and R. Y. Yin, Phys. Rev. A **50**, 1312 (1994).
[11] L. S. Brown and T. W. B. Kibble, Phys. Rev. **133**, A705 (1964).
[12] C. Szymanowski, V. Véniard, R. Taïeb and A. Maquet, Phys. Rev. A **56**, 700 (1997).
[13] A. Maquet, V. Véniard and T. Marian, J. Phys. B: At. Mol. Opt. Phys **31**, 3743 (1998).
[14] C. Szymanowski, V. Véniard, R. Taïeb, A. Maquet and C. H. Keitel, Phys. Rev. A **56**, 3846 (1997).
[15] M. M. Denisov and M. V. Fedorov, Sov. Phys. JETP, **26**, 779 (1968).
[16] F. V. Bunkin and M. V. Fedorov, Sov. Phys. JETP, **22**, 974 (1966).
[17] M. Protopapas, C. H. Keitel, P. L. Knight, Rep. Prog. Phys **60**, 389 (1997).
[18] M. Gavrila, in *Atoms in Intense Laser Fields*, edited by M. Gavrila, Adv. Atom. Molec. Opt. Phys. Suppl. 1 pp435(Acad. Press, San Diego, 1992).
[19] Q. Su and J. H. Eberly, Phys. Rev. A **44**, 5997 (1991) and references therein.
[20] N. J. Kylstra, R. A. Worthington, A. Patel, P. L. Knight, J. R. Vazquez de Aldana and L. Roso, Phys. Rev. Lett. **85**, 1835 (2000).
[21] R. Taïeb, V. Véniard and A. Maquet, Phys. Rev. Lett. **81**, 2882(1998).
[22] G. A. Mourou, C. P. J. Barty and M. D. Perry, Phys. Today, 22 (1998).
[23] J. Schwinger, Phys. Rev. **82**, 664 (1951).
[24] O. Klein, Z. Physik **53**, 157 (1929).
[25] B.R. Holstein, Am. J. Phys. Rev. **66**, 507 (1998).
[26] N. Dombey and A. Calogeracos, Phys. Rep. **315**, 41 (1999); A. Calogeracos and N. Dombey, Contemporary Physics **40**, 313 (1999).
[27] J.W. Braun, Q. Su and R. Grobe, Phys. Rev. A **59**, 604 (1999).
[28] W. Greiner and J. Reinhardt, *Relativistic Quantum Mechanics*, Springer-Verlag (1992).
[29] L. Landau and E. Lifchitz, *Quantum Mechanics*, Pergamon, Oxford (1971).

PLASMAS AT SOLID STATE DENSITY GENERATED BY ULTRA-SHORT LASER PULSES

K. Eidmann
Max Planck Institut für Quantenoptik
D-85740 Garching, Germany

INTRODUCTION

Ultra-short pulse lasers allow us to heat solid state matter isochorically to temperatures of a few 100 eV at electron densities close to 10^{24} cm^{-3} [1, 2, 3]. These densities are considerable larger than those which can be generated with long laser pulses in the ns-range. In the nature such plasmas occur in stars. Fig. 1 compares laboratory plasmas with the sun plasma. The conditions prevailing in the center of the sun are experimentally achievable in the compressed core of inertial confinement fusion (ICF) plasmas [5], whose realization requires large laser facilities firing a few shots per day only. The atractiveness of using ultrashort pulses results from the high repetition rate provided by the lasers producing these pulses. Thereby the unique possibility is opened up to systematically study the complex behaviour of dense plasmas.

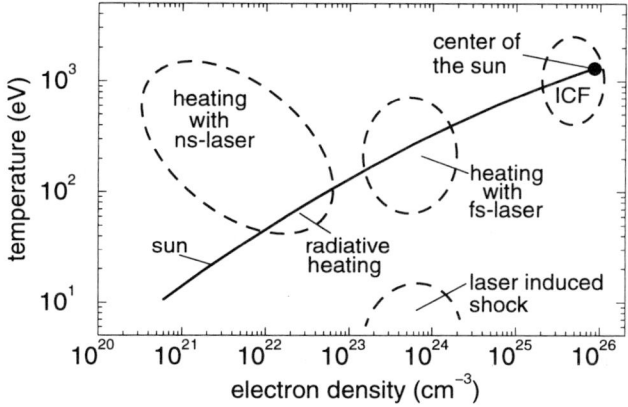

Figure 1: Temperature and density region accessible with laser compared with the sun (for the data of the sun see reference [4]).

In this lecture I discuss the basic processes leading to isochoric heating of solid matter. Then an investigation of the K-shell emission of a dense aluminum plasma is presented. Characteristic features of the spectra emitted by these plasmas are significant high-order satellite line emission, strong line broadening and a shift to the red, which can even be observed in $n = 2$ or $n = 3$ to $n = 1$ transitions [2, 3].

For these experiments, the ATLAS Ti:Sapphire laser of the MPQ was used. This laser system delivers after frequency doubling an energy of ≈ 70 mJ in 150 fs pulses at $\lambda = 395$ nm with a contrast ratio better than $1{:}10^{10}$ in the temporal window 2-30 ns before the main pulse and better than $1{:}10^6$ at 1ps before the pulse maximum.

ISOCHORIC HEATING

Basic Principle

In this section I discuss the mechanism of isochoric heating. The basic points are illustrated in Fig. 2. The short laser pulse irradiates a metal target. Then a thin front layer is heated with a thickness given by the skin depth

$$d_{skin} = \frac{\lambda}{4\pi (Im(\hat{n}))}, \text{ where } \hat{n}^2 = 1 - \frac{\omega_{pe}^2}{\omega_L(\omega_L - i\nu)},$$

with ω_{ei}, ω_L and ν the electron plasma frequency, the laser frequency and the electron-ion collision frequency, respectively. For cold Al d_{skin} is $\approx 7nm$. How fast does this layer expand? The typical expansion time is d_{skin}/s, where $s = [(ZT_e + T_i)k/m_i]^{1/2}$ is the ion sound velocity in a plasma with the electron temperature T_e and the ion temperature T_i. This yields for a temperature of 100 eV, which is reached at $\approx 10^{15}$ W/cm^2, a typical expansion time of 100 fs. Thus at laser intensities exceeding $\approx 10^{15}$ W/cm^2 expansion is no longer negligible!

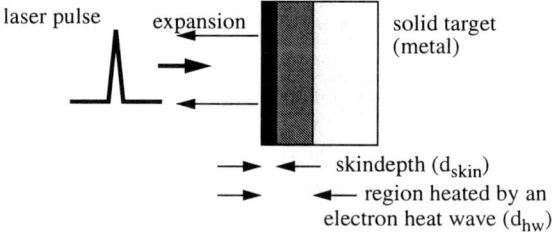

Figure 2: Essential features in the interaction of an ultra short laser pulse with a solid metal target.

Does this mean, we cannot isochorically heat solid matter with ≈ 100 fs to temperatures exceeding ≈ 100 eV? No, because in addition matter is heated by electron heat conduction, which transports the energy deeper into the target in a typical range given by $d_{hw} \approx \sqrt{\chi t}$. χ is the heat diffusivity and given by $\chi \approx \lambda_e v_e$ with λ_e and v_e the electron mean free path and the electron thermal velocity, respectively. For a discussion of the heat conduction in plasmas, confer the book of Zeldovich and Raizer [6]). Calculating d_{skin} and d_{hw} one finds, that for laser pulses longer than ≈ 10 fs the heat wave propagation exceeds the skin depth [7].

Thus heating solid matter by a short laser pulse is a rather complex process and one has to use codes for describing it. Figure 3 shows an example which is characteristic of the experiment presented below. The simulation has been performed with the MULTI-fs hydrodynamic code [7] (the paper [7] contains a detailed description of the physics occurring in the interaction of short laser pulses with solid targets in a large intensity range from 10^{12} to 10^{18} W/cm^2; it can be recommended as an introduction into this subject). Corresponding to the experiment a p-polarized laser beam irradiates an Al target at an angle of $45°$ with an intensity of 10^{17} W/cm^2 and a pulse duration (FWHM)

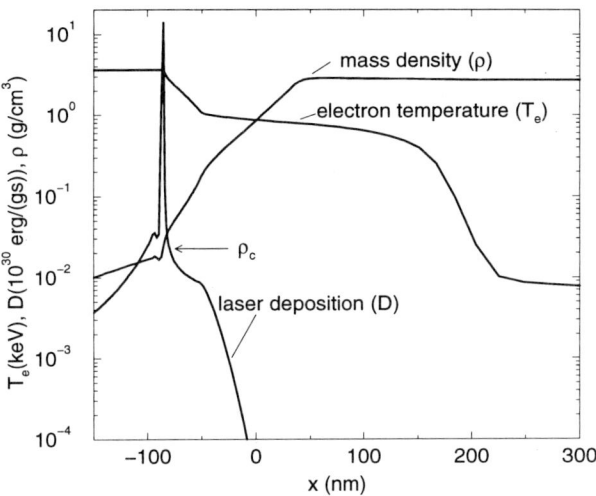

Figure 3: Spatial dependence of the mass density, the electron temperature and the laser energy deposition at 100 fs after the laser pulse maximum. The laser (10^{17} W/cm² with 150 fs FWHM) is incident from the left. Initial target surface is at $x = 0$

of $\tau = 150$ fs. Because at this intensity expansion during the laser irradiation is no longer negligible, a critical density layer develops where the laser energy is absorbed and a hot plasma with temperatures of several keV is generated.

The hot electrons produced at the critical density penetrate into the dense target. MULTI-fs uses a local heat transport approximation, i.e. the heat flow is given by $S_e = min[-\kappa \nabla T_e, f n_e k_B T_e v_e]$. k_B is the Boltzmann constant and κ the electron heat conductivity. The factor f is the flux inhibition parameter. We use for f the free-streaming value 0.6, which means that in very steep gradient plasmas an upper limit for the heat flow exists, namely, when all the Maxwellian electrons, which move into one hemisphere, contribute to the heat flow. In this approximation an electron heat wave is formed which heats the solid matter up to temperatures of a few 100 eV. The existence of a diffusive heat wave depends on the comparison between the mean free path of the hot electrons in solid matter ($\lambda_{e,hot}$) and the depth of the heat wave x_{hw}. For the conditions of Fig. 3, x_{hw} exceeds $\lambda_{e,hot}$. Hence, the hot electrons mostly thermalize in the region heated by the heat wave. Of course, one cannot exclude the fact that hot electrons exist in the dense plasma region resulting in a non-Maxwellian velocity distribution, at least while the laser pulse is on. Figure 3 presents a snapshot 100 fs after the laser pulse maximum. Later, when the pulse is switched off, the hot and low density plasma cools rapidly. Nevertheless, the heat wave propagates further into the target resulting in a lifetime of the dense plasma of a few ps before it finally expands and cools.

It is clear, that the coupling of laser energy into the dense matter depends on the laser absorption and on the energy transport into the dense target. We have addressed both points experimentally.

Absorption

At normal incidence the absorption decreases at higher intensities because the electron ion collision frequency decreases with temperature. At intensities $\geq 10^{17}$ W/cm² the absorption efficiency at normal incidence is therefore small (less than 10%, see [8]). However, efficient absorption can be achieved by using oblique incidence and p-polarized

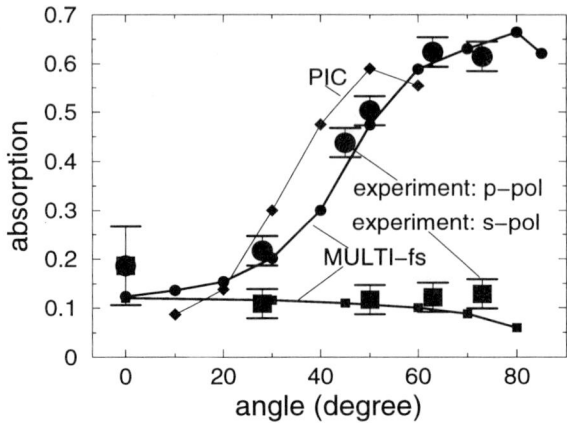

Figure 4: Angular dependence of the absorption for s- and p-polarized laser light. Measured values are compared with hydrodynamic and PIC code simulations.

laser light. This is demonstrated by Fig. 4, which shows the absorption measured as a function of the angle of incidence for s- and p-polarized light. The absorption for p-polarized light increases with the angle of incidence reaching a maximum of $\approx 65\%$ at a rather large angle of $\approx 70°$. In contrast, for s-polarized light the absorption remains on a low level of $\approx 10\%$. The large difference in the absorption between s- and p-polarized light correlates with the observed Al K-shell emission which is much larger for p- than for s-polarized light (by about a factor of 10, see reference [3]). Therefore we have used preferentially p-polarized laser light for investigating the K-shell emission.

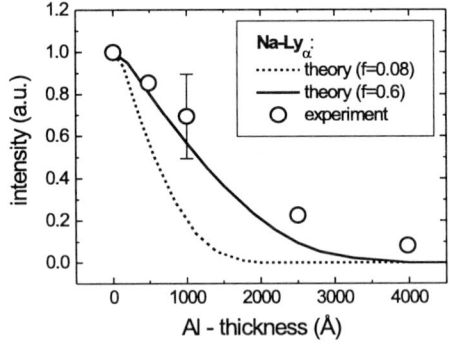

Figure 5: Sodium Ly$_\alpha$ emission from a NaCl substrate covered by an Al layer of variable thickness. The target is irradiated under $45°$ with p-polarized light. The theoretical curves are the result of hydrodynamic code calculations for two different flux inhibition factors f.

The reason for the efficient absorption of p-polarized light at oblique incidence is resonance absorption, which is a well known mechanism in laser generated plasmas [9]. It is caused by an intense longitudinal electric field at the critical layer where plasma oscillations or plasma waves are generated. The optimum angle of incidence for absorption Θ depends on the plasma scale length, $L = n_e/\nabla n_e$, and is given by $\sin\Theta = 0.43(\lambda/L)^{1/3}$ (this formula is valid as long as the plasma scale length L is not much smaller than λ).

Figure 4 shows theoretical results from hydrodynamic and PIC code simulations. The calculations were performed with the hydrodynamic code MULTI-fs in which the interaction of the laser light with the plasma is described by a local refractive index [7]. In this approximation the enhanced absorption for p-polarization is caused by a plasma oscillation at the critical density, as seen in the strongly peaked deposition profile in Fig. 3. The PIC calculations (performed with the PIC code EUTERPE [10]) are more realistic in the sense that they allow for the generation of electron plasma waves and non-Maxwellian electron distributions. However, they completely neglect collisional effects. From both calculations we conclude that indeed resonance absorption is the main absorption mechanism in the case of p-polarized obliquely incident light. Since the maximum absorption occurs at a large angle of incidence, the plasma scale length is rather short ($L \ll \lambda$, $L/\lambda \approx 0.02$). It may be surprising that two rather different models give comparable absorption results. However, this can be attributed to the fact that resonance absorption is rather independent of the detailed dissipation mechanism [9].

Heated Depth

The heated depth in the Al target was measured by overcoating a NaCl substrate with Al layers of different thicknesses and observing the emission of Na K-shell lines as a function of the Al layer thickness. The result for the Na Ly_α-line is shown in Fig. 5. We find typical penetration depths in the range 1000 to 2000 Å. Figure 5 also shows calculations performed with the MULTI-fs hydrodynamic code which were further post-processed with the FLY atomic kinetics code [11]. The FLY code solves the atomic time-dependent rate equations for the different ionization and excitation populations. Good agreement between theory and measurement was achieved by setting the flux inhibition parameter f in MULTI-fs to the free-streaming value f=0.6.

The Al K-shell spectra discussed in the next sections indicate that the density of the heated region is close to the solid density.

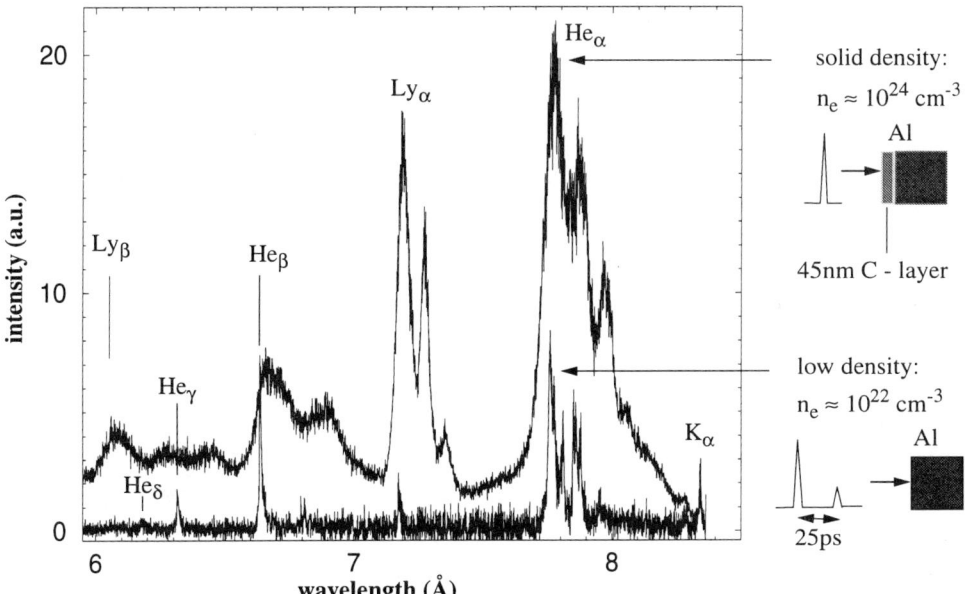

Figure 6: Al K-shell spectra measured at low and high density.

AL-K-SHELL SPECTRA

The K-shell emission of aluminum in the wavelength range 6.0 - 8.4 Å (covering the spectrum from the Ly$_\beta$ to the K$_\alpha$ line) was measured with a von Hàmos crystal spectrometer [12] equipped with a cylindrically bent pentaerythritol (PET) crystal with a lattice constant of 2d = 8.742 Å. The axis of the spectrometer was along the normal of the target surface. The spectra were recorded on photographic x-ray film. The resolving power $\lambda/\Delta\lambda$ was better than 1500. The target was irradiated by the laser at an angle of incidence of 45°.

Figure 6 shows a comparison of two spectra recorded at high and low density. The low density spectrum with narrow lines was generated by adding a prepulse to the main pulse and using a massive Al target. The high density spectrum with broad lines and strong satellite contributions was generated by a single pulse, which irradiated a target that was over coated by a 45 nm thick carbon layer as a tamper. (Satellites are the lines on the red side of the resonance lines. They are caused by an additional so-called bound spectator electron in one of the upper levels, for illustration see Fig. 10). Carbon has no line emission in the region of the Al K-shell lines. Therefore very clear Al spectra are recorded, which have been measured up to the Ly$_\beta$ line.

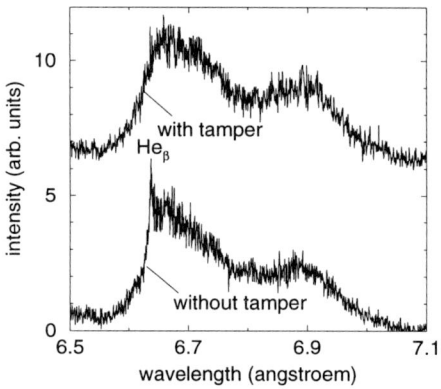

Figure 7: Al He$_\beta$-line with satellites measured with and without a 45nm carbon tamper layer on the Al target.

The tamper layer is very important for obtaining spectra from a more homogeneous high density region. This is demonstrated by Fig. 7, which shows the He$_\beta$-line with its satellites with and without a tamper layer. Without the tamper the He$_\beta$-line exhibits a steep wing on the short wavelength side. We attribute this to the fact, that the Al front layers of the untamped target rapidly expand during the line emission, which results in a low density. Therefore the lines emitted by the untamped target contain contributions from a low density plasma, that add to the contributions from the high density plasma. With a carbon overcoat the expansion occurs in the carbon, and the observed Al emission stems from a deeper more homogeneous high density region.

An interesting line in the spectrum Fig. 6 is the K$_\alpha$ line. Its emission occurs deep in the Al target, where the matter is cold and not ionized. Here an ionization of the inner K-shell electrons can happen either by energetic electrons or by photons propagating into the target. Subsequently a transition of an L-shell electron to the hole in the K-shell gives rise to the K$_\alpha$ emission. This line is important for two reasons. (i) Since the K-shell hole can be created by fast electrons, it gives information on the amount of energetic electrons, as has been for the first time demonstrated by Hares et al.[13]. The fact, that the K$_\alpha$ line is small in our case, indicates a low level of energetic electrons [3]. (ii)

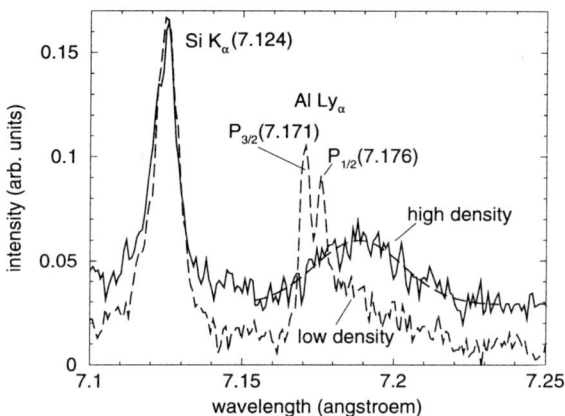

Figure 8: Test of the line shift using the Si K_α as a reference on spectra recorded at low and high densities. The given wavelengths for the Si K_α and the Al Ly_α components are tabulated values, which coincide with the experimental line positions.

Because this line is not shifted and broadened, we can use it for wavelength calibration and determine the exact position of the plasma lines.

Line Shift

Using the K_α-line for wavelength calibration we find that the center of gravity of the lines in the high density spectrum in Fig. 6 is shifted to the red compared to lines in the low density spectrum. The shift of the Ly_α-line is shown in Fig. 8. To confirm the shift further we used in addition to the Al-K_α-line the Si-K_α-line, which is very close to the Al-Ly_α-line. The Si-K_α-line was superimposed on the low and high density spectrum by firing under the same conditions on a glass target after the Al spectrum had been recorded without touching the spectrometer. As a result, the center of gravity of the high density Ly_α is clearly red shifted by $\Delta\lambda = 15 \pm 2$ mÅ (corresponding to $h\nu = -3.5$ eV). The wavelength dispersion was obtained from the two K_α- lines of Si and Al. Using the tabulated values for these lines we find that the low density Al-Ly_α components $P_{1/2}$ and $P_{3/2}$ are at the correct tabulated positions. This result gives good confidence, that our procedure for the wavelength calibration is quite accurate.

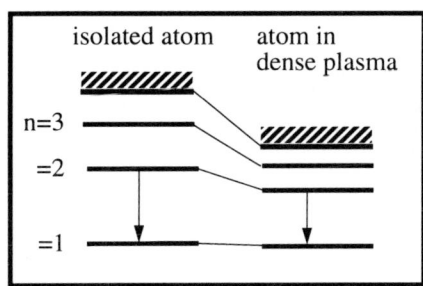

Figure 9: Level shift in dense plasma.

A line shift of this size ($h\nu = -3.5$ eV) can be expected from a simple electrostatic model. Consider an atom embedded in a homogeneous cloud of free electrons with the density n_e within the ion sphere of the radius r_i, which is defined by $(4\pi/3)n_i r_i^3 = 1$. The Poisson equation (in S.I. units) $div E = -e n_e/\epsilon_0$ gives then the electrostatic field

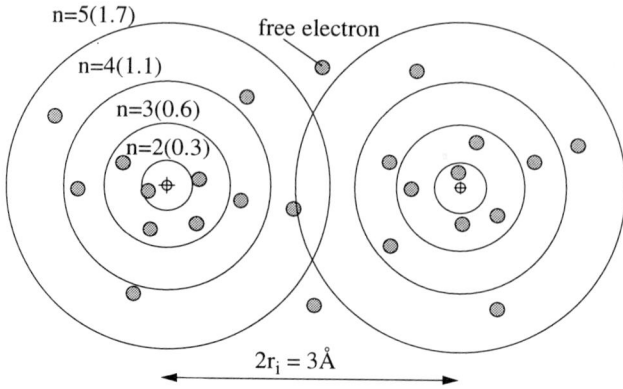

Figure 10: Two ions in isochorically heated Al separated by two ion sphere radii $2r_i$. Numbers in brackets are the radii in Å of the different shells $n = 2$ to $n = 5$.

$E(r) = en_e r/(3\epsilon_0$ or the potential energy

$$P(r) = \frac{e^2 n_e r^2}{6\epsilon_0}$$

due to the free electrons. Thus we expect a level shift increasing with the radius r. Using for r^2 the quadratically averaged value following from the quantum mechanical hydrogen model [14] one finds

$$r^2 = \langle r_{nl}^2 \rangle = r_B^2 \frac{n^4}{Z^2} f_{nl}, \text{ where } f_{nl} = \frac{1}{2}\left(5 + \frac{1 - 3l(l+1)}{n^2}\right).$$

$r_B = 0.53$Å is the Bohr radius. Thus the level shift is given by

$$\Delta \epsilon_{nl} = P(r_{nl}) = Ry \left(\frac{r_B}{r_i}\right)^3 \frac{n^4}{Z} f_{nl}.$$

$Ry = 13.6$ eV is the Rydberg constant. As sketched in Fig. 9, the level shift increases with the quantum number n or with the size of the orbital. For fully ionized Al at solid state ($n_e = 8 \times 10^{23}$ cm^{-3}) the simple model yields a red shift of 1.4 eV for the Ly$_\alpha$-line (with the quantum numbers $(n = 1, l = 0)$ and $(n = 2$ or $3, l = 1)$ for the lower and upper levels, respectively). A more rigorous quantum-mechanical impact theory [15] predicts for this case a larger red shift of 3.2 eV (assuming a temperature of 300 eV), which is close to experimental value.

An illustration of a plasma at the density of solid aluminum is given in Fig. 10. It shows two adjacent Al atoms separated by the ion sphere diameter together with the free electrons. At the density of solid Al ($\rho = 2.7$ g/cm^3) the ion radius is $r_i = 1.5$ Å. The radius of the shells of the different levels is calculated from $< r_{nl} > = \sqrt{f_{nl}}(r_B n^2/Z)$ (using the quadratically averaged radius of the formula given above). Strong overlapping and perturbation of the shells from neighboring atoms is expected for shells with $n > 3$. Therefore these levels will no longer exist. This effect is also called continuum-lowering. It is seen that most of the free electrons are within the shells. Also, they accumulate near the nucleus. This is a result of the rigorous impact theory which predicts that the electron density increases with decreasing distance to the nucleus [15]. It is evident that these free electrons will considerably perturb the levels, resulting in line shift and broadening. The discrepancy between the simple electrostatic model given above and the more rigorous quantum-mechanical impact theory [15] can be attributed to the non-homogeneous distribution of the free electrons, which critically influences the level shift. This can easily be seen by performing the integration of the Poisson equation for inhomogeneous distributions of the electron density.

Spectrum Analysis

A complete theoretical model for the calculation of the total spectrum shown in Fig. 6 or (or at least of sections of the spectrum) is rather complicated, because a huge number of levels is involved. The first step is the calculation of all the energy levels involved and the transition probabilities by solving the Schrödinger equation in an appropriate approximation (for this purpose atomic physics codes exist like the Cowan's code which has been used in our analysis[16]). Then one has to determine the populations of the levels involved by solving the collisional-radiative rate equations. For that purpose one has to take into account all relevant processes, which are electron collisional excitation, deexcitation, ionization and recombination, autoionization and electron capture, photoexcitation and photoionization, spontaneous and stimulated emission and radiative recombination. Very important for the final composite spectrum is the consideration of the Stark broadening due to electrons and ions and the line-shift. In addition, to account for eventual opacity effects, the radiative transfer equation has to be solved. For more details we refer to references [17, 18] or to textbooks on plasma spectroscopy(e. g. [19]).

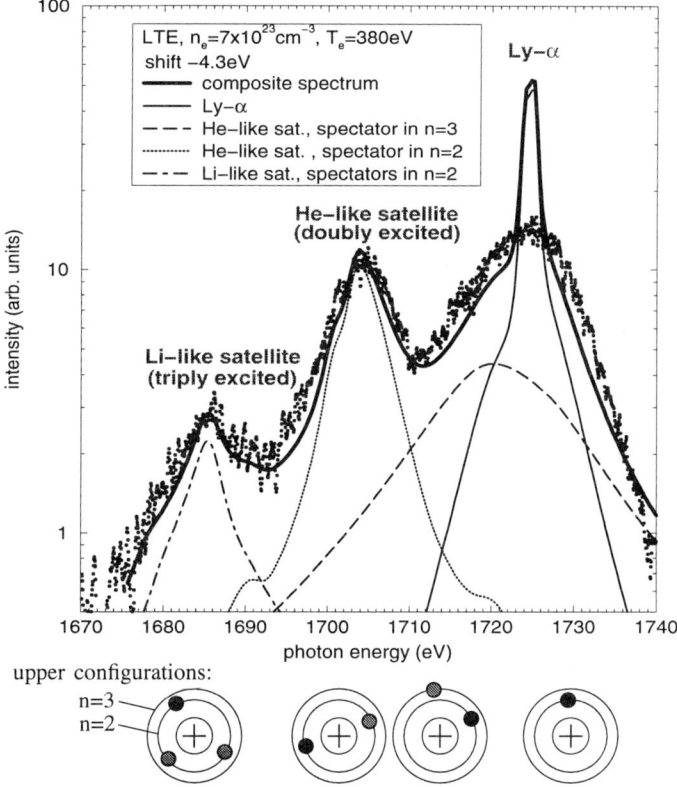

Figure 11: The Ly_α-line with its satellites. Measured data are taken from the high density spectrum of Fig. 6. As indicated at the bottom, the calculations consider the contributions from the Ly_α, the He-like satellites with one spectator electron in the shell $n = 3$ or $n = 2$, and the Li-like satellites with two spectator electrons in the shell $n = 2$. The calculated results refer to an electron density of $n_e = 7 \times 10^{23}$ cm^{-3} and a temperature of 380 eV. An artificial shift of -4.3 eV was used in the calculation.

An example of the analysis of the Ly$_\alpha$-line with its He-like and Li-like satellites is shown in Fig. 11. The upper configurations of the He-like satellites are doubly excited with the active electron in the L-shell and one spectator electron in the L-shell or in higher shells (with n=3,4,...). In addition clearly a second satellite group is observed which is caused by triply excited Li-like ions (also denoted as "hollow atoms"). Here again the active electron is in the L-shell, but now there are two spectator electrons in the L-shell (or higher shells). The theoretical spectrum considers several important configurations (but not all possible configurations) of the spectator electrons. The complexity of the analysis is pointed up by the fact that a large number of levels had to be considered. For example, 15 upper J-energy levels associated with configurations $2l2l'2l''$ and 16 lower J-energy levels associated with configurations $1s2l2l'$, and 67 line transitions were included for the Li-like satellite group. A reasonable fit between experiment and theory has been obtained for an electron density of 7×10^{23}cm^{-3} and an electron temperature of 380 eV. Some approximations have been made in the calculation: (i) The rate equations for determination of the populations have been solved in steady state. (ii) The line shift is not considered in a consistent way, instead the total calculated Ly$_\alpha$+satellite section of the spectrum has been red-shifted. Remembering our simple electrostatic model this can be considered as a reasonable assumption, because it is not expected that one or two additional bound electrons (namely the spectator electrons) significantly affect the level shift due to free electrons. In Fig. 11 the calculated spectrum had been red-shifted by 4.3 eV to achieve good coincidence between experiment and theory. This value is close to the directly measured shift of the Ly$_\alpha$-line (Fig. 8). Also, such a shift is expected from the quantum-mechanical theory of Nguyen et al. [15] for the density and temperature used for the fit in Fig. 11.

However, Fig. 11, also reveals, that the experimental Ly$_\alpha$-line is considerably broader than the theoretical one. We can exclude that optical thickness effects have a major effect on the width of the Ly$_\alpha$-line. This has been concluded by using sandwich targets consisting of an Al layer of finite thickness placed between the tamper front layer and a substrate (consisting of Ti). The Ly$_\alpha$-line width decreased only by some fraction, but not enough to resolve the discrepancy with the theory. Other reasons could be the influence from satellites of higher order than considered in Fig. 11. Also temporal changes and spatial gradients of the density may be important. These questions have to be be addressed in future work.

Another important line in the Al spectrum is the He$_\beta$-line. Its analysis yields similar values for the density and temperature as the Ly$_\alpha$-analysis presented in Fig. 11 [2]. The He$_\beta$-line is of interest, because it serves as diagnostics for the density and temperature in the compressed core if ICF pellets. For this purpose, a small amount of Ar is added as a tracer gas to the hydrogen filling of the pellet. The Ar K-shell spectrum is emitted under similar conditions as the Al K-shell spectrum of the isochorically heated targets. In particular, the Ar He$_\beta$-line has been used in the ICF experiments for diagnostics [20, 21].

CONCLUSIONS

We have demonstrated, that ultra short pulse lasers are a convenient tool to generate hot solid density plasmas. For laser pulses longer than about 10 fs, the dense target is essentially heated by diffusive electron heat flow in a typical depth in the range of 1000 Å. This depth exceeds considerably the skin depth, which is in the range of 100 Å.

Al K-shell spectra emitted from Al targets with an overcoat of magnesium oxide or carbon irradiated by high contrast 150 fs laser pulses show significant satellite emission, large line broadening and red shifts. Analysis of the He-like satellites of the Ly$_\alpha$-line and of the He$_\beta$-line, which is strongly merged with its Li-like satellites, indicates an electron density of $(5-10) \times 10^{23}$ cm^{-3} and electron temperatures of 250 - 400 eV. This corresponds to pressures or energy densities of ≈ 0.4 Gbar and a strong ion coupling parameter of $\Gamma = 3\text{-}4$. (The ion coupling parameter of a plasma is the ratio of the potential to the kinetic energy of the ions: $\Gamma = (Z_{av}e^2/r_i)/(kT_i)$, Z_{av} is the average ionization).

Red-shifts have been accurately measured by using the Al-K$_\alpha$ and Si-K$_\alpha$-lines for wavelength calibration. The measured red shifts are expected from a simple electrostatic model. Quantitatively they are well consistent with results of a quantum mechanical impact approximation. A problem remains, namely the large measured width of the Ly$_\alpha$-line which will be a subject of future studies.

Acknowledgments

The author would like to thank U. Andiel, A. Saemann, I. E. Golovkin, R. Mancini, R. Rix, T. Schlegel and K. Witte for their contributions to the results presented in this paper. The work was supported in part by the commission of the European Communities in the framework of the Euratom-IPP association and by the Deutsche Forschungsgemeinschaft in the framework of the Schwerpunktsprogramm "Wechselwirkung intensiver Laserfelder mit Materie".

References

[1] G. Guethlein, M.E. Foord, and D. Price, *Phys. Rev. Lett.* , **77**, 1055 (1996).

[2] A. Saemann, K. Eidmann, I. E. Golovkin, R. C. Mancini, E. Anderson, E. Förster, and K. Witte, *Phys. Rev. Lett.* , **82**, 4843 (1999).

[3] K. Eidmann, A. Saemann, U. Andiel, I. E. Golovkin, R. C. Mancini, E. Anderson, and E. Förster, *J. Quant. Spectr. Rad. Transfer*, **65**, 173 (2000).

[4] M. Stix, *The Sun*, Springer Verlag Berlin, (1989).

[5] J. Lindl, Phys. Plasmas **2**, 3933 (1995).

[6] Y.B. Zel'dovich and Y.P. Raizer, *Physics of Shock Waves and High-Temperature Hydrodynamic Phenomena*, Academic, New York, (1967).

[7] K. Eidmann, J. Meyer-ter-Vehn, T. Schlegel, and S. Hüller, *Phys. Rev. E* , **62**, 1202 (2000).

[8] D. F. Price, R. M. More, R. S. Walling, G. Guethlein, R. L. Shepherd, R. E. Stewart, and W. E. White, Phys. Rev. Lett. **75**, 252 (1995).

[9] W. L. Kruer, *The Physics of Laser Plasma Interactions*, Addison-Wesley Publishing Company, Inc., pg. 37-43 (1988).

[10] Th. Schlegel, S. Bastiani, L. Gremillet, J.-P. Geindre, P. Audebert, J. C. Gauthier, E. Lefevre, G. Bonnaud, and J. Dellettrez, *Phys. Rev. E* , **60**, 2209 (1999).

[11] R. W. Lee, B. L. Whitten, and R. E. Strout, *J. Quant. Spectr. Rad. Transfer*, **32**, 91 (1984).

[12] L. von Hámos, Ann. Phys. Leipzig, **17**, 716 (1933).

[13] J. D. Hares, J. D. Kilkenny, M. H. Key, and J. G. Lunney, Phys. Rev. Lett., **42**, 1216 (1979).

[14] H. A. Bethe, and E. E. Salpeter, *Quantum Mechanics of One- and Two-Electron Atoms*, Plenum Publishing Corporation, New York, (1977).

[15] H. Nguyen, M. Koenig, D. Benderjen, M. Caby, and C. Coulaud, *Phys. Rev. A* , **33**, 1279 (1986).

[16] R. D. Cowan, *The Theory of Atomic Structure and Spectra*, University of California Press, (1981).

[17] L.A. Woltz, and C. F.Hooper, Jr., Phys. Rev. A, **38**, 4766 (1988).

[18] R. C. Mancini, D. P. Kilcrease, L. A. Woltz, and C. F. Hooper, Jr., Comput. Phys. Commun., **63**, 314 (1991).

[19] H. R. Griem, *Principles of Plasma Spectroscopy*, Camridge University Press, (1997).

[20] N.C. Woolsey, B.A. Hammel, C.J. Keane, C.A. Back, J.C. Moreno, J.K. Nash, A. Calisti, C. Mosse, L. Godbert, L.S. Klein, and R. W. Lee *J. Quant. Spectr. Rad. Transfer*, **58**, 975 (1997).

[21] N. C. Woolsey, C. A. Back, R. W. Lee, A. Calisti, C. Mosse, R. Stamm, B. Talin, A. Asfaw, L. S. Klein, *J. Quant. Spectr. Rad. Transfer*, **65**, 573 (2000).

SHOCK WAVE EXPERIMENTS AND EQUATION OF STATE OF DENSE MATTER

M. Koenig

Laboratoire LULI
Unité Mixte n° 7605 CNRS - CEA - Ecole Polytechnique - Université Pierre et Marie Curie
91128 Palaiseau, FRANCE

INTRODUCTION

The knowledge of Equation Of State (EOS) of dense matter is important in several fields of physics. For example, in astrophysics the star evolution is mainly governed by the thermodynamic properties of matter. Also, planetary physics requires very high pressure parameters as one expects to occur in the planets inner core. Finally, Inertial Confinement Fusion (ICF) success is directly related to the understanding of shell pellet implosion and the final core compression. Both of these processes necessitate a precise knowledge of the microballoon material and the fuel (deuterium) EOS at very high pressures (> 100 Gpa).
However, first of all, one must define what the phrase "dense matter" does mean. From a plasma physicist point of view, it concerns highly coupled matter, i.e., where the coulomb energy between ions is greater than their thermal energy[1]. From an EOS point of view, one can define it roughly where the perfect gas model (or assimilate) does not hold anymore. This implies states around solid density, a few electron-volts temperature and pressure > 10 Gpa.
How can we reach such states in the laboratory? Until the fifties, high pressure physics was mainly governed by Diamond Anvil Cells (DAC). Here, using suitable cut diamond, pressed together with the sample in between (figure 1a), pressures up to 100 Gpa can be reached. Higher pressures can be obtained using dynamic methods, based on shock wave propagation in the material. Among the dynamic methods, gas guns can provide pressures up to several hundreds of Gpa by accelerating a projectile to velocities of several km/s (figure 1b). However in order to reach much higher pressures ($\geq$ Tpa), only nuclear explosions were available until the early seventies. Since, the development of high power lasers provides a new tool to reach extreme conditions of matter in the laboratory[2-4].
Only a few experimental data are available to validate EOS calculations and discriminate between the different theories in this very high pressure regime. In the last few years, direct applications to ICF or planetary physics have been obtained using high power lasers[5, 6].

However, many more experimental data are needed to validate theories and to address uncertainties in modelling the ICF targets or the planets internal structure.

In this paper we shall begin to describe the basic equations (Rankine-Hugoniot) that sustain most of the work done on shock waves and EOS measurements. Then a brief description of the diagnostics will be developed. The key issues to achieve EOS data using lasers will be discussed. and some results obtained at LULI will be shown. Finally we shall consider the most recent applications to planetary physics.

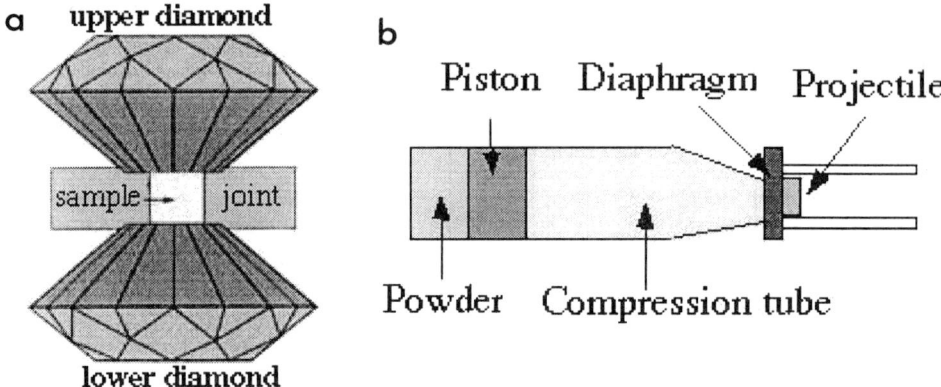

Figure 1: schematic diagrams of a DAC (a) and a gas gun (b).

HUGONIOT RELATIONS, DIAGNOSTICS

When a high power laser is focalized on a target, matter is rapidly vaporized and unloads into vacuum. This ejected hot plasma pushes the solid part of the target inward and compress it. A pressure is then generated (called ablation pressure) which is directly related to the laser intensity, wavelength, …[7].

When a fluid is compressed by a piston moving at a constant velocity U, a shock wave is created which propagates at velocity D[8]. The matter behind the shock front is compressed and is characterized by a density ρ, a pressure P and a fluid velocity equal to the piston velocity U.

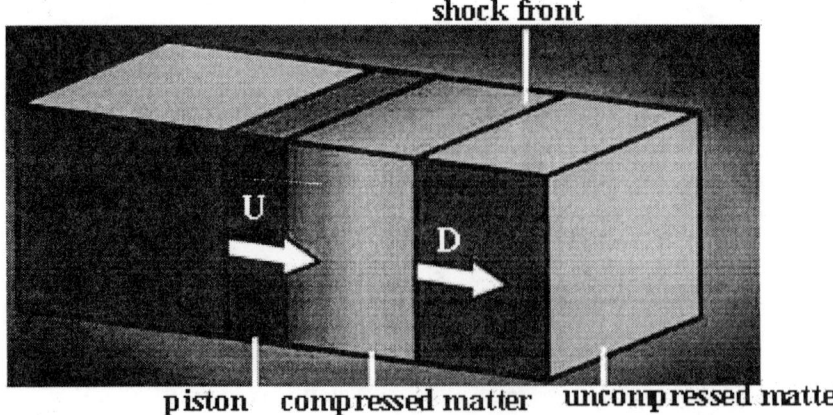

Figure 2: Shock wave propagation scheme.

The Rankine-Hugoniot relations are deduced from conservation laws for the mass, momentum and energy between uncompressed and compressed matter. These equations are the following:

$$\rho_0 D = \rho(D-U) \ ; \quad P - P_0 = \rho_0 DU \ ; \quad PU = \rho D(E - E_0 + \frac{U^2}{2})$$

where the subscript o denotes the parameters of the solid target.

Here we have 3 equations for five unknown parameters. If one can measure two parameters, one point on the equation of state is determined. Therefore all the efforts turn on the determination of two of the 4 parameters: D,U,ρ or T. If the parameters correspond to the same material, one talks of an **absolute** measurement. In that case, it is quite standard to determine D or even U, which are directly obtained through the experiment. For the density ρ or the temperature T, one can only have an indirect measurement. The quantity given by the experiment needs to be include into a model to infer ρ or T. Depending on the target design or other parameters, it is quite often impossible to get an absolute measurement. Therefore we have to adopt another method called the **relative** measurement. Here we still determine two parameters but in two materials, one being a reference[9].

In order to determine all necessary parameters, many diagnostics have been developed around laser experiments. How can we get reliable data? The issues are to be able to reach a high accuracy measurements for a target size of the order of 1 mm within a time $\sim 10^{-9}$ s. Therefore a high instrumental resolution is needed: 10 µm for the spatial one and a few 10^{-12} s for the temporal one.

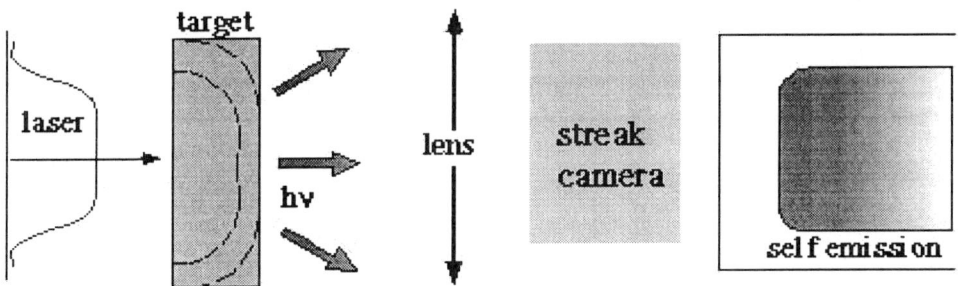

Figure 3: Passive diagnostics scheme.

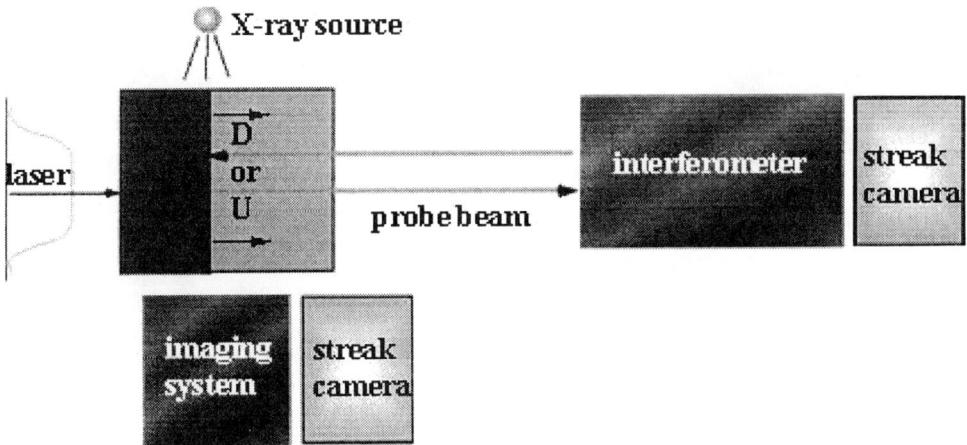

Figure 4: Active diagnostics scheme.

We can distinguish two types of diagnostics: **passive** or **active** ones. The passive diagnostics lie on the fact that for the high pressures, as reached with lasers, when the shock breaks out at the end of the target, the material is very hot (a few electron-volts) and emits suddenly a lot of light in the visible. This light is imaged to the slit of a streak camera which has a high temporal resolution (several picoseconds).

Recently active diagnostics have been developed which all rely on the use of probe beams (figure 4) either in the optical[10] or X-ray domain[11]. With these new techniques, it has been possible to determine, for example, the fluid velocity in deuterium[6] or in water[12].

EOS WITH LASERS: SOME KEY ISSUES

Three mains conditions have to be fulfilled in order to measure EOS with good accuracy using lasers: A **uniform shock** front over a large distance, a **steady state pressure** in the sample during the measurements and a **low level of preheating** in the material ahead of the shock wave.

The problem in realizing high quality shocks is inherent to the use of coherent laser light, which produces lack of uniformity in the irradiation due to beam modulations by interference effects

Recent experiments[13, 14] have proved the possibility of creating spatially very uniform shocks in solids by using two different methods. The first one consists in producing shock waves by direct-laser drive with optically smoothed laser beams; the second one uses X-ray thermal radiation to generate shocks (indirect drive).

The first scheme[15] for optical smoothing (Random Phase Plates) was based on diffraction of laser beams on small rectangular patterns whose phase (0 or π) are randomly distributed. The gaussian shape of the focal spot makes difficult their use to produce shocks for EOS measurements. Hence in our experiments we used a new technique recently developped[16] the Phase Zone Plates beam smoothing. PZPs are made of a Fresnel lens array, each randomly out of phase of 0 or π. In this case we can obtain a top-hat intensity distribution in the focal spot. As a result, 2D effects are almost completely eliminated around the center of the focal spot.

At LULI, the spot was composed of a 200 µm top-hat intensity distribution and gaussian edges so total focal spot FWHM was 400 µm. Figure 5a shows a "flat" shock breakthrough obtained with PZPs.

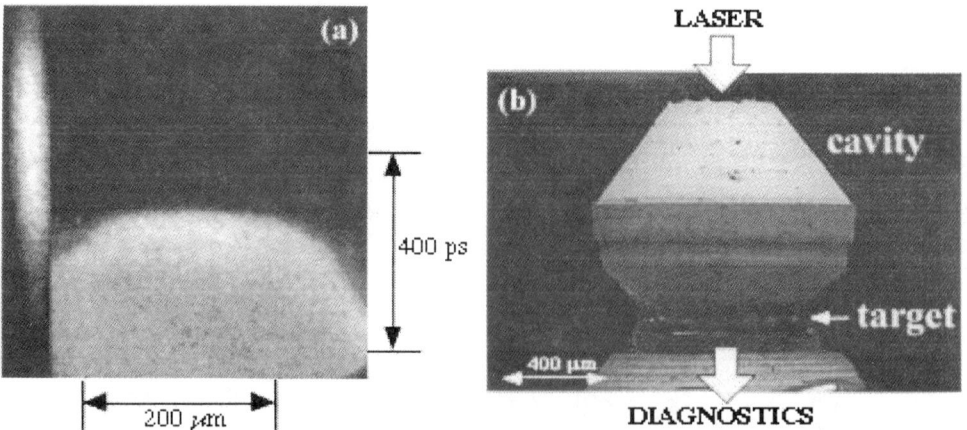

Figure. 5: (a) Rear side shock breakthrough for a 14 µm Al target obtained with PZPs. (b) The labyrinth cavity designed at MPQ Garching.

In order to compare directly and indirectly driven shock waves, we used the ASTERIX laser, which was the only european facility where such experiments can be performed. In the indirect laser drive configuration we focused the laser beam into a 1 mm size gold cavity through a small entrance hole. An isotropic radiation is then created whose temperature depends upon the cavity size and the laser power. It can be determined by observing the velocity of a shock wave generated when radiation is absorbed in low-Z material [17]. In our experiment it has been measured to be in the range of 100-150 eV. This cavity[18] (see figure 5b) has been designed not only to reach such high temperatures, but also to optimize the irradiation uniformity when only one laser beam is used, and to minimize the preheating of the target, produced by direct primary X-rays.

One of the main questions is then: which drive to use to perform EOS measurements? In principle, indirect drive is preferable if one needs a very high uniformity. However, it has been pointed out[19] that both methods can produce similar shock waves. The direct drive approach has a big advantage, that it is much more efficient (factor 3-5).

Regarding the second condition (shock steadiness), some experiments have shown the possibility to obtain a steady shock through a sample[14]. However it is closely related to the target design and the laser parameters. In principle, it is possible to produce much longer steady pressures using direct drive. Indeed, due to wall expansion, the cavity is filled with a hot plasma, which prevent the laser pulse to sustain the pressure due to various plasma instabilities for example.

The control of preheating has important implications in the reliability of EOS measurements. The preheating effect is mainly due to hard X-rays, created either in the hot corona of the target (direct drive) or by the walls of the cavity. The X-rays propagate through the sample before the shock, depositing their energy hence heating it. In this context, a high level of preheating has an unwanted effect since the shock wave propagates then in a medium whose state is unknown. Therefore the quantities of the unperturbed materials into the Rankine-Hugoniot equations are undetermined. In this case the EOS determination becomes difficult or even impossible. The total amount of X-rays increases with the laser intensity, with laser frequency and the charge of the material. Recent experiments[20-22] have shown the way to reduce the preheating effect down to a moderate level.

RELATIVE EOS MEASUREMENTS

The relative EOS method is based on the impedance-matching technique[8] applied to a double-step target, with a target structure sketched in Figure 6.

In general, the target is made of a "base" foil made of a material A, which is irradiated by the laser on one side, and supports, on the opposite side, two steps made, respectively, of the same material A, and of a different material B. Using rear-face, time-resolved imaging it is possible to determine experimentally the velocity of the shock propagating through the two steps, D_A and D_B (corresponding to particle velocities U_A and U_B respectively). Material A is chosen as the reference so it is assumed that its EOS is known. In all our experiments, aluminum is taken as the reference because its principal Hugoniot curve as its release waves are well known up to several hundreds of Gpa. Depending if the tested material has a greater or a lower impedance ($\rho_0 D$) than aluminum, a shock or an unloading wave is reflected at the interface between A and B.

We present here brominated plastic EOS measurements in a pressure range (1 to 10 Mbars) corresponding to the long foot shaped pulse of NIF or Megajoules lasers. In this experiment, the shock emergence from the target was inferred by detection of the emissivity of the target rear face in the visible region. This was imaged by a photographic objective onto the slit of a visible streak camera with 5 ps time resolution.

In general, one of main problems is the error bars of the experimental data, which have to be low in order to distinguish between theoretical models. In the laser experiments,

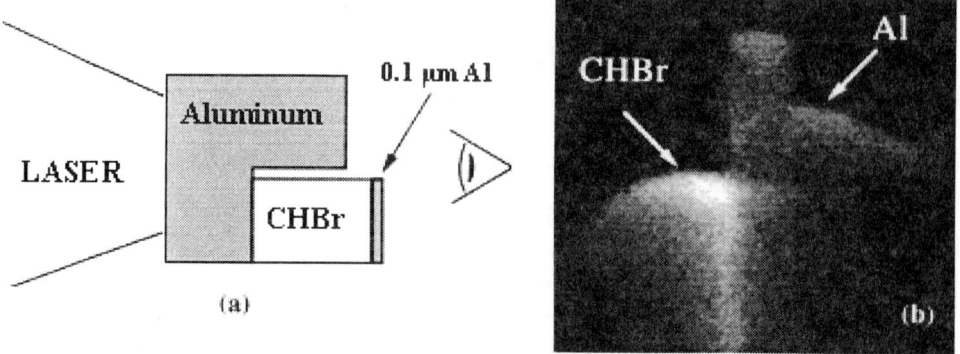

Figure 6: a) Scheme of a Double step targets with a common base and two steps of different materials, Aluminum and CHBr. b) From the shock travelling time in these steps, obtained with a visible streak camera, the shock velocities, D_{Al} and D_{CHBr} are determined.

there are three main source of errors: the quality of the shock itself (requiring flatness over a wide region), the sweep speed (ps/mm) of the streak camera, and the knowledge of the step thicknesses. The calibration of the streak sweep speeds can be made with an etalon giving a series of short laser pulses (FWHM = 240 fs). The relative error with the sweep speed can be reduce to a very low level (< 1%). In order to minimize the second possible source of experimental errors, one needs high quality, well characterized targets with the structure described above. For the experiment presented here, the target was made of an aluminium "base" foil (irradiated by the laser on one side) which supports, on the opposite side, two steps made, of aluminium and of brominated plastic (2.3% of atoms in our case). The plastic step has been coated with a thin layer of aluminium (1000 Å) to avoid shinethrough emission (spurious visible emission of the shock front travelling through the plastic). The step heights were determined with an absolute error smaller than 0.05 μm, i.e., a relative error ≤ 1% in both steps. However in the plastic case, due to the final gluing the error on the plastic step height was ± 0.5 μm giving an error of ± 4.5%. According also to the sweep speed of the streak camera and to the uniformity of the shock produced by the PZP, we estimated a maximum total error of ± 7% and ± 8% on the shock velocities D_{Al} and D_{CHBr} respectively (see figure 7).

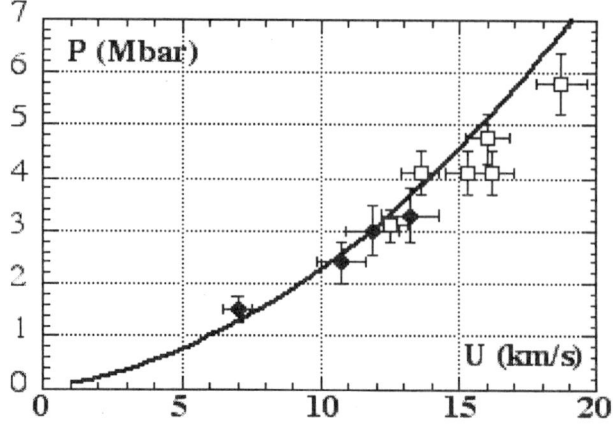

Figure 7: Comparison between experiments (● LULI, ☐ AWE[23] and theory[24] —).

Our results, compared to indirectly driven points obtained at AWE laboratory[23], lies in the 1-5 Mbars part of the Hugoniot. AWE's are above due to the much higher energy used, i.e, 1 kJ. All experimental points are in very good agreement with EOS calculations based on the QEOS model[24]. This experiment was one of the first EOS measurement directly dedicated to ICF.

APPLICATIONS TO PLANETARY PHYSICS

Hydrogen is the simplest and most abundant element in the universe, yet at high pressure, it is one of the most difficult to understand. New experimental data along the principal Hugoniot into the few-100-GPa regime are required to evaluate the competing theoretical models. Indeed, a predicted metallic transition and its effects on the EOS at pressures near 100 GPa are integral to models of many hydrogen-bearing astrophysical objects[25] like Jovian planets[26].

The interiors of giant planets Neptune and Uranus are mostly composed of a thick layer of "hot ices", predominantly water and hydrocarbons. The magnetic field measured by Voyager 2 is assumed to come from the electrical conductivity of this layer. In the "hot ices", pressure and temperature range is from 0.2 Mbar and 2000°K to 6 Mbar and 8000 °K. Recent calculations predict an insulator-metallic phase transition of water in this regime[27]; up to now no measurement has been performed to validate it.

The iron is one of the most abundant component in Earth's interior. Its Equation Of State (EOS) in high pressure range (10 Mbar) has important implications in the description of the Earth's core[28]; indeed the knowledge of the core's thermodynamic properties allows to estimate the heat stocked in the core which is fundamental to model the convection process in the mantle.

In this section, we describe the very first laser shock wave experiment dedicated to the water EOS measurement in the range of pressures 1-7 Mbar and we present iron EOS results. Using the impedance mismatch equation of state measurement method described above, we conducted a study of iron equation of states at pressures in the range 10-35 Mbar. These experiments have been realized at Commissariat à l'Energie Atomique. We used one beam of the PHEBUS laser in the green ($\lambda = 0.526$ µm) smoothed with a KPP[29]. The focal spot was 900x600 µm^2 wide giving a $1.4 \cdot 10^{14}$ W/cm^2 maximum intensity on the target with a square 4ns pulse. A 10 ns probe beam ($\lambda = 1.064$ µm) of a few mJ energy was injected on the rear side of the target, with the back reflection going into two Velocity Interferometer Systems for Any Reflector[30] (VISAR) (see figure 8a).

This interferometer is based on the phase shift due to the Doppler effect associated to a moving surface. To resolve the ambiguity on the initial phase shift[10], we used two VISAR with different sensitivities (one with a 5mm etalon and the other one with a 15 mm) coupled to streak cameras. This gave a velocity, in air, sensitivity of 21 km s^{-1} fringe^{-1} and 7 km s^{-1} fringe^{-1} respectively. Our velocity interferometers employed a Mach-Zehnder configuration. The water target scheme, shown in figure 8b, consists of a plastic ablator (20 µm), an aluminum pusher, with a step (75 µm + 25 µm) to get the Al shock speed, and a cell filled with water (1 mm).

The shock speed D_{H2O} in water can be measured with the VISAR, assuming its metallization, while the aluminum shock speed is given by the transit time in the step. The shock breakout at the Al/H$_2$O interface is detected either by the rear side self-emission and by the changing of reflection (through the interferometers). In the case of no metallization, the probe beam reflects on the Al/H$_2$O interface so the fluid velocity U_{H2O} in water can be, in principle, inferred. From the measurement of D_{H2O} or U_{H2O}, we can determine a point on the water EOS using the impedance mismatch technique.

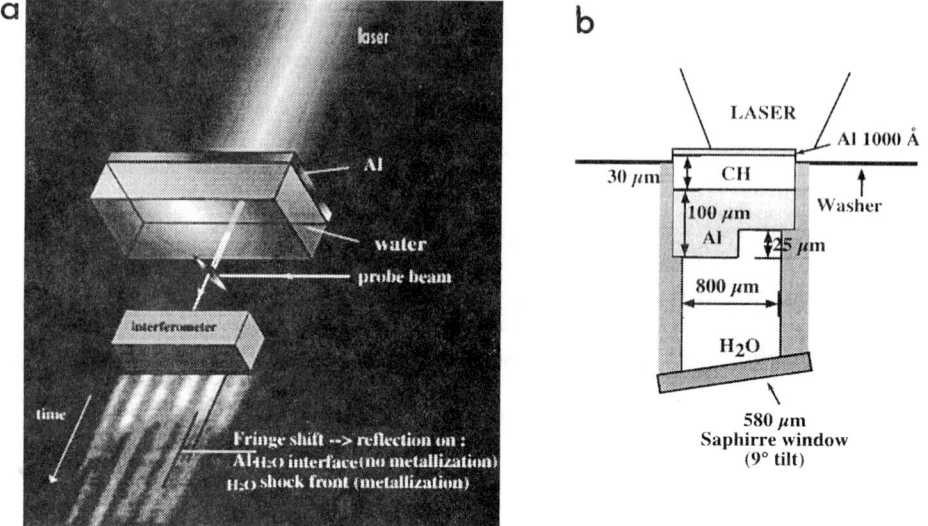

Figure 8: a) VISAR principle; b) Water target scheme.

The water experimental results obtained on the PHEBUS laser are recent and some more experiments are in progress. The most important difficulty encountered on this first experiment was the correct alignment of the target. This was conditioned by the quite complicate imaging system and by the small dimension of the water cell; a few degrees imprecision on the alignment implied a quite important unwanted reflections from the cell. Moreover the best focal position of the aluminum rear side was difficult to find due to the presence of water.

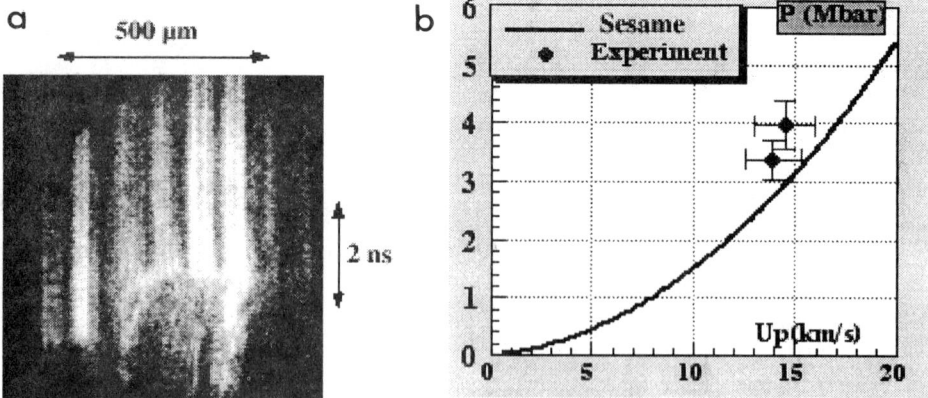

Figure 9: a) Experimental image obtained with a water target. One can see easily the shock breakout from the Al and its propagation through the water. ; b) Comparison between the experimental results and the SESAME table[31] for the water principal Hugoniot.

Despite the encountered problems, we obtained some shots where it is possible to see a fringe shift when the shock crosses the interface Al/H$_2$O (figure 9a). Here we measured the shock speed in the aluminum (knowing the step height) with a high uncertainty (± 10 %) due to target alignment problems. Indeed, we found $D_{Al} = 21 \pm 2$ km/s. By analyzing the two VISAR images, the velocity estimated from the fringe shift is approximately 25 km/s. This value is compatible with the shock speed in water D_{H2O}. From the experiment we were able to deduce only two points on the water principal Hugoniot which are compared (figure 9b) to available theoretical data (SESAME table[31]). These were the highest ones ever obtained with a laser.

In the case of iron, we obtained more quantitative results. In figure 10a one can observe the very high quality of the emission image; indeed the shock was perfectly flat and the transit times of the shock in the two steps can be determined with a good precision. One can notice that the emissivity coming from the iron step is less important than the one from the aluminum step. This is probably due to the fact that the temperature of shocked iron is less than the aluminum one. From the analysis of figure 10b (obtained with the VISAR), one can deduce again the two transit times. The drop of the reflection is fast and then it is easy to determine the shock velocities. Moreover, from this image, one can have quantitative information on the preheating. Before the shock breaks out from the Al base, a fringe motion is observed. This implies that the base rear side moves before the shock breakout, due to the preheating. By using the usual Doppler formula, we deduced that the surface was moving at 2km/s. By assuming that this speed is approximately equal to the sound speed, we can estimate that a temperature of around 0.7 eV corresponds to this speed expansion. Anyway, this preheating could not affect significantly our measurements because it is strongly reduce in the two steps; in fact in this image, no fringe motion is detectable.

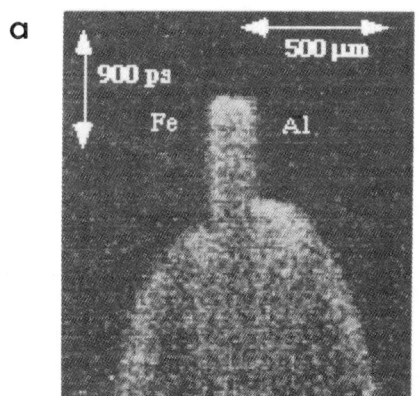

Figure 10: a) Emissivity of the target rear side. b) Image obtained with the VISAR.
Here the laser energy was 1.6 kJ.

Figure 11 shows our experimental results in the plane $(P, \rho/\rho_0)$. These data are compared with a fit by Trunin et al [32] based on experimental data obtained with nuclear explosions. The points in the P<10 Mbar range were obtained with chemical explosions[28, 33].

One can notice that our data are in good agreement with the principal Hugoniot as given by Trunin et al [32], but they show clearly a general trend: most of them lies above this curve. This trend could be explained by the presence of some preheating effects. Indeed due to the X-rays preheating on aluminum, the interface with the iron could expand a little bit. This explains the lower compression given by our results. A deeper analysis is actually under consideration by performing numerical simulations. We would like to point out that our precision on the shock velocity is typically of ± 8 %, hence on the pressure we have ± 15 %. However, for a given shock velocity, the main differences between theoretical models concern density and electronic temperature. It is for this reason that we present the results on the plane $(P, \rho/\rho_0)$. In our case, density can be deduced from the Rankine-Hugoniot equations, but the error propagation is then too high to allow us to discriminate between the various models.

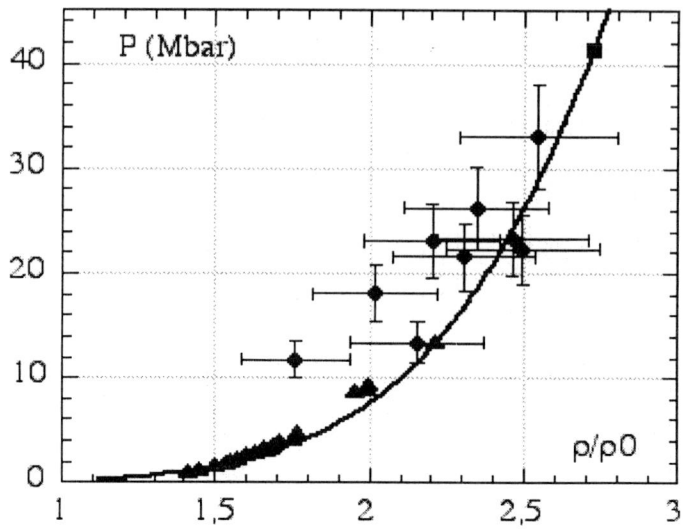

Figure 11: our iron experimental points, the Hugoniot obtained by nuclear explosion[32] and points at lower pressures obtained by conventional explosions[28, 33]

CONCLUSIONS

Laser driven shock waves is now a useful tool for EOS measurement in the range 500 Gpa- 4 Tpa. Many applications dedicate either to ICF[5, 34] or planetary physics[6, 12] have been performed up to now. However, intrinsically a single shock heats the matter to a high level as compared to the temperature existing in the planet cores for example. One of the most exiting challenge will be, in the next few years, to reduce this heating by pre-compressing the target. We know how to hit strongly a material with lasers we have to learn how to do it softly.

ACKNOWLEDGEMENTS

The author would like to thank A. Benuzzi-Mounaix, N. Grandjouan, B. Faral, D. Batani, T. Hall, Th. Löwer, R. Cauble, P. Celliers, G. Collins and L. Da Silva for their collaboration over these last few years.
The work presented here was greatly supported by various E.U. programs under contracts N° CHGE-CT93-0046, CHRX-CT93-0338, FMGE-CT95-0044, ERBFMGE-CT95-0044 ERBFMGE-CT95-0016.

REFERENCES

[1] J. C. Gauthier, in *Matter in super-intense laser fields*, edited by C. Joachain and D. Batani (Plenum), Erice, Italy, (2000), Vol.
[2] F. Cottet, et al., Phys. Rev. Lett. **52**, 1884 (1984).
[3] R. Cauble, et al., Phys. Rev. Lett. **70**, 2102 (1993).
[4] D. Batani, et al., Phys. Rev. B **61**, 9287 (2000).
[5] M. Koenig, et al., Appl. Phys. Letters **72**, 1033 (1998).
[6] G. Collins, et al., Science **281**, 1178 (1998).
[7] S. Atzeni, in *Matter in super-intense laser fields*, edited by C. Joachain and D. Batani (Plenum), Erice, Italy, (2000), Vol.

8. Y. B. Zeldovich and Y. P. Raizer, *Physics of shock waves and high temperature hydrodynamic phenomena* (Academic Press, New York, 1967).
9. M. Koenig, et al., Phys. Rev. Lett. **74**, 2260 (1995).
10. P. M. Celliers, et al., Applied Phys. Lett. **73**, 1320 (1998).
11. L. Da Silva, et al., Phys. Rev. Lett. **78**, 483 (1997).
12. M. Koenig, et al., in *IFSA*, edited by C. Labaune, W. Hogan and K. Tanaka (Elsevier), Vol., p. 1127, Bordeaux, (1999).
13. M. Koenig, et al., Phys. Rev. E **50**, R3314 (1994).
14. T. Löwer, et al., Phys. Rev. Lett. **72**, 3186 (1994).
15. Y. Kato, et al., Phys. Rev. Lett. **53**, 1057 (1984).
16. T. H. Bett, et al., Appl. Opt. **34**, 4025 (1995).
17. R. L. Kauffman, et al., Phys. Rev. Lett. **73**, 2320 (1994).
18. T. Löwer and R. Sigel, in *APS Topical Conference on Shock Waves*, edited by S. C. Schmidt and W. C. Tao (AIP Press), Vol. 2, p. 1261, Seattle, (1995).
19. A. Benuzzi, et al., Phys. Rev. E **54**, 2162 (1996).
20. A. Benuzzi, et al., Phys. of Plasmas **5**, 1 (1998).
21. M. Basko, et al., Phys. Rev. E **56**, 1019 (1997).
22. T. Löwer, et al., Phys. Rev. Lett. **80**, 4000 (1998).
23. S. D. Rothman, et al., in *APS DPP*, Denver, Co., (1996).
24. R. M. More, et al., Phys. Fluids **31**, 3059 (1988).
25. H. M. VanHorn, Science **252**, 384 (1991).
26. G. Chabrier, et al., Astrophys. J. **391**, 817 (1992).
27. C. Cavazzoni, et al., Science **283**, 44 (1999).
28. W. Anderson and T. Ahrens, J. Geophys. Research **99**, 4273 (1994).
29. S. Dixit, ICF Quaterly report **4**, 152 (1994).
30. L. M. Barker and R. E. Hollenbach, J. Appl. Phys. **43**, 4669 (1972).
31. SESAME: The LANL Equation of State database, LA-UR-92-3407, Los Alamos National Laboratory (1992)
32. R. F. Trunin, Phys.-Usp. **37**, 1123 (1994).
33. J. M. Brown and R. G. McQueen, J. Geophys. Research **91**, 7485 (1986).
34. R. Cauble, et al., Phys. Plasmas **4**, 1857 (1997).

LASER PARTICLE ACCELERATION IN PLASMAS

Jean-Raphaël Marquès

Laboratoire d'Utilisation des Lasers Intenses
UMR 7605 du CNRS–CEA–Ecole Polytechnique–Université Paris VI
91128 Palaiseau cedex, France

1. INTRODUCTION

Since the birth of elementary particle research in the thirties, particle accelerators have been the main tool. Each time a particle of higher energy has been available, new phenomena have been observed. The size of the accelerators increases roughly linearly with the particle energy. In "lepton" circular accelerators, this size is governed by synchrotron radiation: the energy loss per turn increases with the fourth power of the particle energy and is inversely proportional to the radius of the trajectory. An electron or a positron in the LEP (101 GeV, 4.2 km radius) radiates around 2 GeV of its energy in one turn. Since a 290 GeV electron would radiates 50 % of its energy per turn, next generation of lepton accelerators should be linear. Because hadrons are heavier, synchrotron radiation is less restrictive, and higher energies are reached. However, for a reasonable accelerator size, large magnetic fields are necessary, which requires supra–conducting magnets. For a given magnetic field, the radius of the ring increases linearly with the particle energy. For example, guiding 7 TeV protons in a 4.2 km diameter ring (LHC) requires 5.5 Tesla.

Linear accelerators do not suffer of these limitations. To reach large energies, one only needs a sufficiently long accelerator. Its length is then given by the accelerating gradient, which is the electric field on the beam axis. This field is created by radio–frequency cavities. For a frequency of the order of the GHz, the breakdown of the cavity wall limits the field to around 10 MV/m. At much higher frequencies, the limitation comes from the surface heating. As a consequence, a 500 GeV linear collider obtained with conventional technology would be 50 km long. Such accelerators seems very difficult to build, not only due to technological problems, but also to their financial costs.

In the last two decades, the generation of large E–fields has been the goal of numerous researches. The more conservative approach is to increase the breakdown threshold by the use of higher frequency cavities. Since the transverse size of the cavity is roughly proportional to the wavelength, increasing the frequency leads to very small cavities, which makes building tolerances more demanding. A more serious problem comes from the wake field created by the accelerated particles. This wake can disturb the particle bunch itself or the following, and increases drastically with the frequency of the EM–wave.

Different solutions have been proposed to take advantage of the very large electric field (> TV/m) associated with high intensity laser pulses. In the following, we describe one type of these possible next generation accelerators, called the laser plasma accelerators. They use a high intensity laser to excite a longitudinal plasma wave that can accelerate charged particles. Since a plasma is already breakdown, plasma waves can support E–fields larger that a few hundreds of GV/m, opening the way to compact high energy accelerators.

2. PLASMA WAVE PROPERTIES

2.1 ELECTRIC FIELD ASSOCIATED WITH A PLASMA WAVE

A hot plasma is made of particles charged negatively, the electrons, and positively, the ions. Obtained by ionisation of a gas or a solid, its global charge is zero. However, if you apply an external force to separate a little bit the electrons from the ions, you create a space charge electric field that tends to recall the particles to their equilibrium position. When you stop this external force, the electrons that are much lighter than the ions start to move back to their initial position. Because their velocity at their equilibrium position is not anymore zero, they start a harmonic oscillation at a characteristic frequency, called the plasma frequency $\omega_p = (ne^2/m_e\varepsilon_0)^{1/2}$, that only depends on the electron density n. The produced electric field is longitudinal (cf. Fig. 1), which means that its direction is parallel to the perturbation wave vector k_p. Such a feature is fundamental for particle acceleration, since the accelerating field has to be oriented along the trajectory of the particle and to follow it.

Figure 1. Left: Charge separation and electric field associated with a plasma wave;
Right: Phase velocity of a plasma wave.

The field amplitude is related to the density perturbation δn by the Poisson equation: $E = (ev_\phi/\omega_p \varepsilon_0) \, \delta n$, where $v_\phi = \omega_p/k_p$ is the phase velocity of the perturbation. Introducing the relativistic factor $\beta_\phi = v_\phi/c$, the maximum electric field of the wave can be expressed as:

$$E_{max} \, (GV/m) \sim 30 \, [n(cm^{-3})/10^{17}]^{1/2} \, \beta_\phi \, [\delta n/n]_{max}$$

For example, at a plasma density of 10^{18} cm^{-3}, a relativistic ($\beta_\phi = 1$) perturbation of only 1 % produces an electric field of the order of 1 GV/m. Recent experiments [cf. chapter 4] on laser plasma acceleration have measured electric fields as high as 100 GV/m.

2.2 ENERGY GAIN IN A PLASMA WAVE

To be able to catch the ocean waves, the surfers know that they need to gain an initial velocity. For a charged particle in a plasma wave, the principle is the same. Three cases can be distinguished [1], depending on the phase velocity and the amplitude of the wave, and on the initial velocity of the particle. A first case corresponds to a particle always slipping backward with respect to the wave. This occurs if its initial velocity is too slow ($\beta = v/c \ll$

β_ϕ), and if the wave amplitude is small. The particle can be accelerated, but the gain is small. If the particle is too fast ($\beta >> \beta_\phi$), it always move forward, and here again it can be accelerated, but the energy gain is small. The optimum case is between these two situations: in the wave frame, the particle moves initially backward. However, because its velocity is not too different from v_ϕ, the E–field acts enough time to reverse the particle motion (in the wave frame), and the particle moves finally forward and faster than v_ϕ. In such a case we say that the particle has been *trapped* by the wave. This optimum acceleration occurs after an optimum length called dephasing length: if the particle continues to follow the wave, it outruns the accelerating region, and starts to see a reversed field and to loose energy. For a trapped particle, this length corresponds to exactly half a wavelength in the wave frame (cf. Fig. 2). Finally, the initial phase of the E–field when the particle enters the wave is also important. Depending on the sign and amplitude of the field at the beginning of the interaction, the final gain can be very different, and even negative (the particle slow down). Let us note that an electron at rest can be trapped if the wave amplitude is very large.

For electron trapped by a relativistic wave of large amplitude ($\gamma_\phi\ \delta n/n >> 1$), the maximum energy gain and the dephasing length are:

$$\Delta W_{max} = 4\ m_e c^2\ \gamma_\phi^2\ \delta n/n \qquad \text{after} \qquad L_d = \gamma_\phi^2\ \lambda_p$$

where $\gamma_\phi = (1 - \beta_\phi)^{-1/2}$ is the relativistic factor associated with the phase velocity of the plasma wave, and $\lambda_p\ (\approx 2\pi c/\omega_p)$ is the plasma wavelength. For a wave with $\gamma_\phi = 100$ and $\lambda_p = 100$ μm, the maximum energy gain is $\Delta W_{max} \approx 20$ GeV, and occurs after $L_d \approx 1$ m.

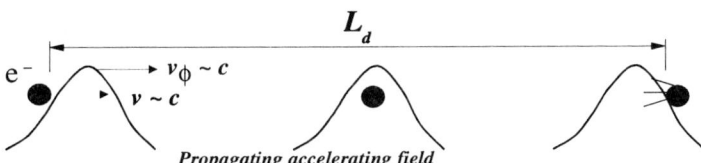

Figure 2. Optimum acceleration of a relativistic electron by a plasma wave of relativistic phase velocity.

2.3 PLASMA WAVE EXCITED BY A LASER

2.3.1 The ponderomotive force: A solution to create a space charge separation in a plasma is to use the radiation pressure associated with a laser pulse. In an EM–field, a charged particle undergoes the Lorentz force $q(E + v \times B)$. If the EM–field is not homogeneous (transverse or longitudinal/temporal profile), the particle not only has a quiver motion at the laser frequency, but also a non–linear slow drift motion toward the region of lower field (cf. Fig. 3).

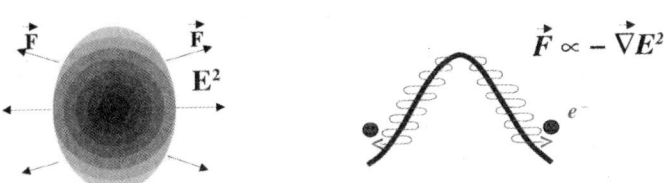

Figure 3. The ponderomotive force pushes the electrons away from the high intensity regions.

To this average motion is associated a force: the ponderomotive force $F_p = -\nabla(qE_0)^2/4m\omega_0^2$, where E_0 is the local amplitude of the E-field of pulsation ω_0. Let us suppose that we apply this force in a time short enough that ion do not have time to move. Because the electrons are much lighter, they move away from the region of high laser intensity before the ions get time to move. The desired space charge separation is created. If the laser force is turned off in a time shorter than the plasma period ($1/\omega_p$), the electrons are called back by the ions, and start to oscillate at the ω_p. However, if we want to use this density oscillation for particle acceleration, we need to give it a non zero phase velocity. This condition is filled if the laser pulse is propagating in the plasma.

2.3.2 Phase velocity of a laser excited plasma wave: If the density oscillation is created by the propagation of a laser pulse, its phase velocity is, by principle, equal to the group velocity of the laser in the plasma: $v_\phi = v_g$. In a one dimensional geometry (laser transverse profile $w_0 \gg \lambda_p$) and for $\omega_p \ll \omega_0$, the laser group velocity is $v_g = c(1 - \omega_p^2/\gamma_\perp\omega_0^2)^{1/2}$, where $\gamma_\perp = (1 + a_0^2/2)^{1/2}$ is the relativistic factor associated with the transverse velocity of the electron in the laser field of normalised potential a_0. This factor is a non-linear correction induced by the mass increase of the plasma electrons when their quiver motion in the laser field becomes relativistic ($a_0 > 1$). A very intense laser pulse propagates faster. The relativistic factor associated with the phase velocity of the plasma wave is then:

$$\gamma_\phi = \gamma_\perp^{1/2} \omega_0/\omega_p = \gamma_\perp^{1/2} \lambda_p/\lambda_0$$

To accelerate particles at very high energy, high phase velocity is required, and so low plasma density. However, the finite size of the laser focal spot induces three dimensional effects that decrease the group velocity of the laser at low density[2] ($\lambda_p = 2\pi/k_p \gg w_0$):

$$\gamma_\phi = \gamma_g = \omega_0/\omega_p \, [\gamma_\perp^{-1} + 2(k_p w_0)^{-2}]^{-1/2}$$

For example, in a plasma of density $n = 10^{16}$ cm^{-3}, a laser of wavelength $\lambda_0 = 1$ µm focused in $w_0 = 10$ µm, will have $\gamma_g = 44$ instead of 330 when focused in a much larger focal spot.

2.3.3 Energy gain in a laser excited plasma wave: From the above expressions, we can get the maximum energy gain and the dephasing length of a relativistic electron trapped by a relativistic wave of large amplitude ($\gamma_\phi \, \delta n/n \gg 1$), excited by a laser pulse (in 1D):

$$\Delta W_{max} = 4 \, m_e c^2 \, \gamma_\perp \, (\lambda_p/\lambda_0)^2 \propto n^{-1} \quad\text{after}\quad L_d = \gamma_\perp \, \lambda_p^3/\lambda_0^2 \propto n^{-3/2}$$

Example: $\lambda_0 = 1$ µm and $n = 10^{19}$ cm^{-3} leads to $\gamma_\phi = 10$, $\Delta W_{max} = 200$ MeV after $L_d = 1$ mm
$\lambda_0 = 1$ µm and $n = 10^{17}$ cm^{-3} leads to $\gamma_\phi = 100$, $\Delta W_{max} = 20$ GeV after $L_d = 1$ m.

High energy gains require low densities but long acceleration lengths.

3. THE MAJOR SCHEMES OF LASER PLASMA ACCELERATORS

Tajima and Dawson[3] proposed two methods to use the ponderomotive force of lasers to excite a plasma wave: the Laser Beat–Wave Accelerator (LBWA) and the Laser Wake–Field Accelerator (LWFA). More recently a third method [4] appeared: the Self–Modulated Laser Wake–Field Accelerator (SMLWFA). A detailed summary can be found in ref. 5.

3.1 THE LASER WAKE–FIELD ACCELERATOR

The laser wake–field mechanism is based on the principle of the swing (cf. Fig. 4). The ponderomotive force associated with the rising edge of a laser pulse pushes forward the electrons of the plasma. The pulse then outruns these electrons and gives them a second kick with its falling edge, but this time in the opposite direction. An oscillation at ω_p is created in the wake of the pulse. To be efficient, this excitation requires a pulse duration τ of the order of half the plasma period: $\omega_p \tau \sim 2$; it is a resonant process.

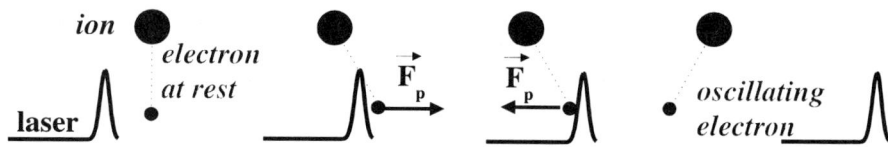

Figure 4. Principle of the LWFA. The ponderomotive force of an ultrashort laser excite a plasma wave.

Even for low plasma densities, this technique requires sub–picosecond pulses. For example, $n = 2 \times 10^{16}$ cm^{-3} implies a duration of the order of 400 fs. The maximum energy that can gain a relativistic electron trapped in a relativistic wave excited by LWF is:

$$\Delta W_{max} [MeV] = eE_{max} \pi z_R \approx 0.8\ E_0(J)\ \lambda_0(mm)\ \tau_{1/2}^{-2}(ps)$$

where $z_R = \pi w_0^2$ is the Rayleigh length of the laser beam and $\tau_{1/2}$ is the full width at half maximum (FWHM) in intensity of the pulse. This result has been obtained by integrating the electric field of the plasma wave excited by a gaussian laser beam. In this case, the wave length is of the order of z_R, which is much shorter than the dephasing length. One can see that the final gain does not depend on the laser focusing, since a smaller focal spot leads to a larger field, but along a shorter length. One can also see that high energy gains require high power lasers ($\lambda_0 = 1$ µm, $E_0 = 10$ J, $\tau_{1/2} = 100$ fs (100 TW) gives $\Delta W_{max} = 1$ GeV).

3.2 THE LASER BEAT–WAVE ACCELERATOR

When, in 1979, Tajima and Dawson proposed the LWFA, ultrashort and intense laser pulses were not available. In order to generate short pulses with the existing lasers, they proposed to mix two identical long pulses of slightly different frequencies. If these two pulses are synchronised and propagate collinear, they beat in time, which creates a sinusoidal modulation in their envelopes. This modulation is at the frequency difference of the two pulses $\Delta\omega = \omega_1 - \omega_2$, and has a wave vector $k = k_1 - k_2$. The resulting temporal pattern is equivalent to a train of micro–pulses, of duration and separation equal to $\Delta\omega^{-1}$. If the two pulses propagates in a plasma of density adjusted so that $\omega_p = \Delta\omega$, the ponderomotive force of the beating pattern is resonant with the natural oscillations of the plasma, and a plasma wave can grow to a large amplitude (cf. Fig. 5). Since the duration of the two pulses is typically of the order of 100 ps, the number of micro–pulses ($\sim$ 100 fs) is very large (few thousand), so that the excitation of a plasma wave by the beat–wave mechanism is a very resonant process, much more than the laser wake–field mechanism. This implies a plasma density very homogeneous and stable in time. In addition, since the excitation is quite slow, the ions can have time to move, which can lead to a saturation and/or a destruction of the plasma wave. For example, in a hydrogen plasma of density $n = 10^{17}$ cm^{-3}, the characteristic time for ion motion is $\omega_{pi}^{-1} = 2.4$ ps.

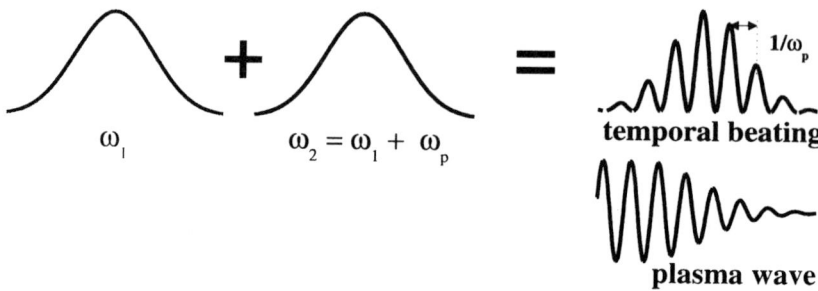

Figure 5. Principle of the LBWA. The temporal beating of two synchronised lasers excite a plasma wave.

3.3 THE SELF–MODULATED LASER WAKE–FIELD ACCELERATOR

This mechanism consists in a self–resonant excitation of a plasma wave by a laser pulse that undergoes an envelope modulation instability during its propagation in the plasma (cf. Fig. 6). The index of refraction seen by a laser pulse in a plasma is $\eta_R \approx (1 - \omega_p^2/\omega^2)^{1/2}$. Let us imagine that the laser pulse excite a very small plasma wave along its propagation. This small plasma wave creates longitudinal and transverse density gradients, and the index of refraction seen by the pulse is not anymore homogeneous. Depending of their signs, the transverse gradients will tends to focus or to diffract the laser beam, while the longitudinal gradients will accelerate or slow down the laser pulse ($v_g = c/\eta_R$). If the pulse length $c\tau$ is longer than the plasma wavelength λ_p, the pulse see periodic gradients that tends to modulate its envelope at λ_p. The ponderomotive force associated with this modulation resonantly enhances the initial plasma wave, which increases again the modulation, and so on. Let us note that in the case of a one dimensional geometry, the modulation is only longitudinal, and the self–modulation instability is the forward Raman instability.

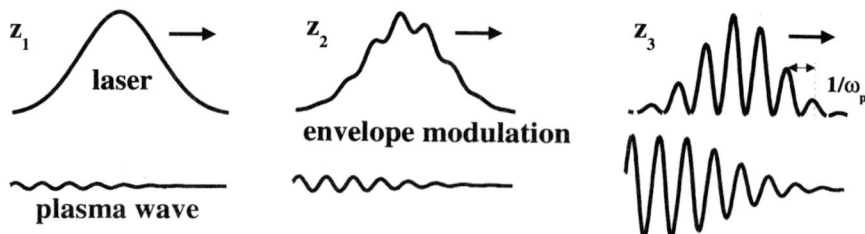

Figure 6. Principle of the SMLWFA. The envelope modulation instability excite a plasma wave.

To be efficient, this envelope instability requires a large intensity along a large interaction length and in a dense plasma. As a result, it usually occurs at laser powers of the order or above the critical power of self–relativistic guiding of the laser pulse. This power is $P_c(GW) = 17\, \omega_0^2/\omega_p^2$, so that a $\lambda_0 = 1$ µm laser pulse propagating in a plasma of density $n = 10^{18}$ cm^{-3} will undergoes self–relativistic guiding if its power is above 20 TW. For this reason, present day lasers impose for SMLWFA the use of ultra–short (~100 fs) lasers.

Because this process is self–resonant, it can excite plasma waves of very large amplitude, up to the wave–breaking limit. In that case, the electrons of the wave have an quiver velocity close to the phase velocity of the wave, so that they can be trapped by it and accelerated to high energies. Even if wake–breaking is not reached, the acceleration of background electrons is still possible. Along its propagation, the laser pulse excites Raman backward scattering. This counter propagating redshifted light of frequency $\omega_0 - \omega_p$ and wave number $-k_0$ beats with the laser (ω_0, k_0) to drive a ponderomotive wave ($\omega_p, 2k_0$) of slow phase velocity ($v_p \approx \omega_p/2k_0$) which can trap and pre–accelerate plasma electrons. These electrons can gain sufficient energy and/or be displaced in phase so that they are trapped and accelerated to high energies by the high phase velocity wave. It has recently been discovered[6] that these pre–accelerated electrons can also be accelerated directly by the laser field, via the inverse of the free electron laser mechanism. This mechanism is explained in this proceeding in the contribution of J. Meyer–ter–Vehn et al.

Compared to the LBWA or the LWFA, the SMLWFA process is the only one that accelerates background electrons, making it attractive as an ultrashort (< 100 fs) and intense (kA) source of energetic (MeV to GeV) electrons. However, this mechanism is very non–linear and based on an instability, so that a precise control of the phase and amplitude of the wave is difficult. In addition, the relatively high plasma densities required leads to a slow phase velocity, which is not acceptable for acceleration at very high energies (> GeV).

3.4 LIMITATIONS ON THE LASER DRIVEN–PLASMA ACCELERATORS

A first intrinsic limitation of plasma accelerators has been discussed above. It is the detuning length $L_d = \gamma_\perp \lambda_p^3/\lambda_0^2$: when the interaction length is longer than L_d, the electron outruns the wave, leaves the accelerating field, and enters the decelerating region. It is for the moment the main limitation in SMLWFA experiments.

A second limitation to the interaction length is the laser diffraction: without guiding mechanisms, the laser spot size increases along the propagation as $w(z) = w_0(1 + z^2/z_R^2)^{1/2}$, so that the high laser intensity region (where the plasma wave is excited) is limited to a few Rayleigh lengths. With typical laser parameters of $\lambda_0 = 1$ μm and $w_0 = 10$ μm we get $z_R = 314$ μm. It is for the moment the main limitation of LBWA or LWFA experiments.

A third limitation is the pump depletion length L_{pd}. After this length, all the available laser energy has been used for the excitation of the wave, and no more wave exists: $E_{Laser}^2 c\tau = E_{wave}^2 L_{pd}$. For LWFA excited by ultra–intense lasers ($a_0 > 1$), L_{pd} can be shorter than L_d.

Another mechanism limits the potential of plasma accelerators. It is a limitation on the number of electrons that the wave can accelerate: if these electrons are too numerous, the electric field they create can compensate the wave field: $E_{bunch} + E_{wave} = 0$. This effect, called *beam loading*, limits the number of electrons to : $N_e = 1.7 \times 10^{-2} (\delta n/n) n(\text{cm}^{-3})^{1/2} w_0(\mu m)^2$. Taking $w_0 = 10$ μm, $n = 10^{19}$ cm^{-3} and $\delta n/n = 1$, one get $N_e = 5 \times 10^9$, or 0.9 nC.

4. EXPERIMENTS ON LASER PLASMA ACCELERATION

Because it does not require very high laser powers, LBWA has been the first demonstrated. Different laboratories[7] have accelerated injected electrons. They measured electric fields of the order of 1 GV/m over a length of the order of 1 mm, and energy gains of 1 to 30 MeV. More recently, SMLWFA has then been demonstrated by different groups[8]. The experimental demonstration of LWFA has been done at LULI[9]. Electric fields of the order of 1 GV/m over a length of the order of 1 mm have been measured. Typical results obtained at LULI on LBWA, LWFA and SMLWFA are presented in the following.

4.1 LBWA

Two synchronised laser pulses of wavelength $\lambda_1 = 1.0530$ μm and $\lambda_2 = 1.0642$ μm were focused in an interaction chamber filled with deuterium gas at the resonant pressure corresponding to $\omega_p = \omega_1 - \omega_2$, close to 2 mbar. Their duration were respectively 90 ps and 160 ps FWHM. They were focused in a diameter of 60 μm FWHM. Their maximum intensities were a few 10^{14} W/cm^2. The two pulses were fully ionising the gas, and their temporal beating was exciting a plasma wave. A Van de Graaf accelerator was delivering a 3 MeV electron beam that was injected and focused in the interaction chamber, at the same place and in the same direction than the laser beams (cf. Fig. 7). About 1000 e⁻/ps were entering the interaction region in a diameter of 100 μm FWHM. The energy spectrum of the accelerated electrons was analysed with a magnetic spectrometer, and measured in ten scintillators coupled to photomultipliers.

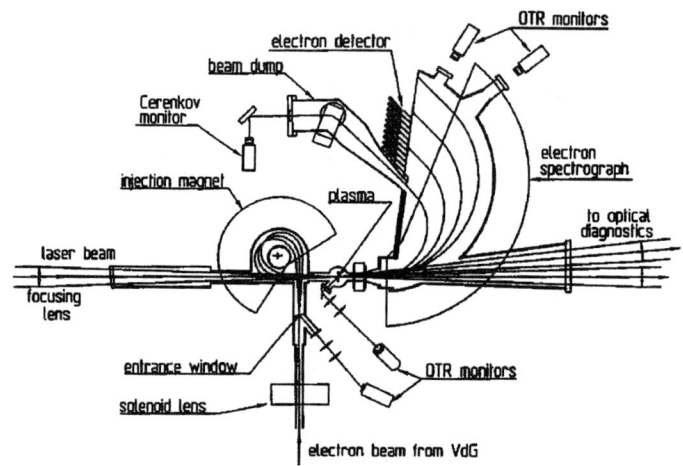

Figure 7. Experimental arrangement of the LBWA experiment at LULI.

The results obtained are summarised in Fig. 8. The energy gain depends on the wave amplitude $\delta n/n$, on the Rayleigh length z_R, and on the initial energy of the injected electrons E_i. The two parameters we have modified were z_R (left figure) and E_i (right figure). In all cases, we observed a large number of accelerated electrons. With $z_R = 1.5$ mm and $E_i = 3$ MeV, the energy gain was $\Delta W = 0.8$ MeV. It increased to 1.3 MeV when $z_R = 1.2$ mm, and reached 1.4 MeV for $E_i = 3.3$ MeV.

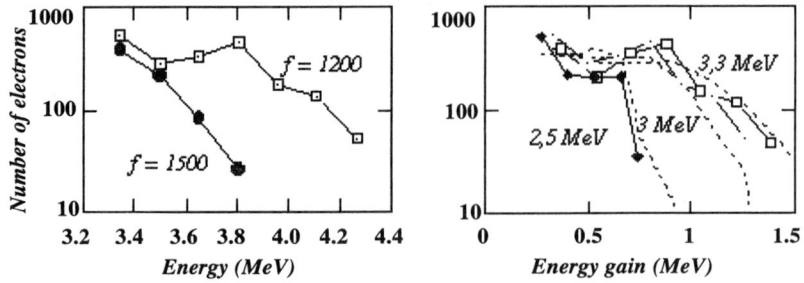

Figure 8. Energy and energy gain spectra of 3 MeV electrons accelerated by LBWA.

These numbers demonstrate the dominant effect in these type of experiments: the dephasing between the accelerated electrons and the plasma wave. The phase velocity of the wave ($\gamma = 94$ in 1D, or 87 taking into account the 3D correction) stay, in our case, much larger than the electron velocity all along the acceleration (γ_e never exceed 9). The dephasing length is then given by $L_d = \gamma_e \lambda_p = \gamma \gamma_e^2 \lambda_0$, where λ_0 is the laser wavelength, and is close to 3.6 mm, a value comparable to the focal length. If the plasma is too long, like in the case of the long focal length, and if the electrons are too slow, they successively undergoes accelerating and decelerating zones, which severely reduces the energy gain.

The characteristic accelerating length L_a and the amplitude of the plasma wave $\delta n/n$ can be obtained from the analysis of the spectra in Fig. 8. The calculation takes into account the measured parameters of the electron beam (divergence, focal spot size) and of the laser beams. Assuming that the wave has a gaussian transverse profile and a Lorentzian longitudinal profile of width L_a (FWHM), the agreement is obtained for $L_a = 2.8$ mm and $\delta n/n = 2.4$ %, giving a maximum electric field of 0.7 GV/m.

Two types of mechanisms can limit the plasma wave amplitude: loss of the resonance condition due to a modification of the plasma density, and/or energy transfer between different eigen modes of plasma oscillations. The contribution of each mechanism depends on the laser parameters and on the plasma density. In our case, the experiment and calculations show that the saturation is mainly due to the excitation of ion acoustic waves of low frequencies that modulate the density profile and kill the resonant coupling between the laser and the accelerating wave. Other groups using a CO_2 laser ($\lambda_0 = 10$ µm) and a resonant density ten times larger have obtained larger amplitudes ($\delta n/n = 30$ %). In their case, the amplitude reached the relativistic detuning limit, and electrons were accelerated up to 30 MeV. However, the wave phase velocity was in their case slower ($\gamma_\phi = 30$), which would limit the maximum energy gain in the case of ultra-relativistic injected electrons.

4.2 LWFA

The LWFA mechanism became possible with the arrival of ultra short (< ps) and intense (> TW) laser pulses. Such a laser has been developed at LULI and allowed us to demonstrate the acceleration of electrons by laser wake-field. Except the laser source, the experimental apparatus was the same than for our PBWA experiments (cf previous chapter).

The laser pulse had a duration of 400 fs (FWHM), an energy up to 10 J (40 TW), at a wavelength of 1.054 µm. It was focused with an off-axis parabola to a 20x40 µm (FWHM) focal spot, leading to a Rayleigh length of 1 to 2 mm. The maximum laser intensity was of the order of 3×10^{17} W/cm^2, much enough to fully ionise the helium gas filling the interaction chamber and to excite a plasma wave in the wake of the laser. The pulse duration implied a resonant density of 2×10^{16} cm^{-3}, corresponding to a helium pressure of 0.5 mbar. Like in our PBWA experiment, the injected electrons were coming from a Van de Graaf accelerator. The energy spectrum of the accelerated electrons has been measured as a function of the gas pressure (0.1 to 4 mbar), and of the laser energy. Typical results are presented in Fig. 9.

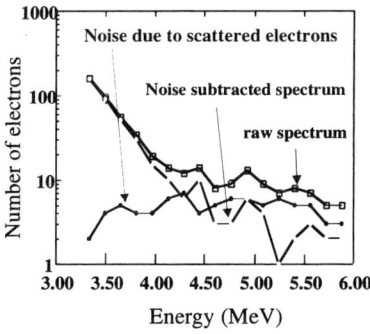
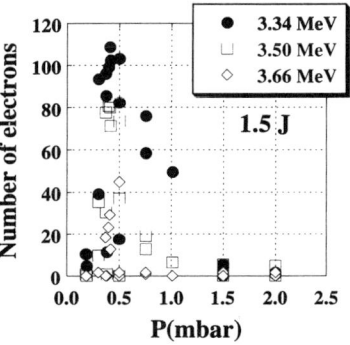

Figure 9. Energy spectrum (right) and resonance curve (left) of 3 MeV electrons accelerated by LWFA.

The 3 MeV injected electrons were accelerated up to 4.5 MeV. Because of the finite size of the laser focal spot, the electric field associated with the plasma wave was not purely longitudinal, but had also an important transverse component. The observed energy spectra are in good agreement with theory (simulations) if we take into account the effect of focusing and defocusing of the electron beam induced by this transverse field. The comparison between the experiment and simulations indicates that the accelerating (longitudinal) field is of the order of 1 GV/m.

4.3 SMLWFA

If the mechanism is very complex, an experiment of SMLWFA is quite simple. It consists in focusing an ultra short and intense laser pulse in a high density gas jet. If the laser power P and the plasma density are large enough to get $P > P_c$ (cf chapter 3.3), a high amplitude wave is created, and electrons from the plasma itself are accelerated. An electron beam collinear with the outgoing laser beam is then ejected from the interaction region.

One important phenomenon that can prevent the plasma wave excitation is the laser beam refraction induced by the gas ionisation: in the presence of an inhomogeneous transverse intensity profile, the ionisation first occurs in the high intensity regions, and creates density gradients that tend to refract the laser. In the case of LBWA or LWFA, the electron density is relatively small ($10^{15} - 10^{17}$ cm^{-3}), so that refraction is often negligible. In SMLWF accelerators, the density is much larger ($10^{18} - 10^{20}$ cm^{-3}), and refraction can completely destroy the laser beam before it has reached its focus. To avoid such phenomenon, the laser beam is usually focused under vacuum and its focus is longitudinally adjust on the edge of a gas jet. In order to have a very well defined edge, supersonic gas jet are used. Experiments have shown that the position of the laser focal plane relative to the gas jet is a very sensitive parameter for the SMLWF excitation.

The first SMLWFA experiments have been done with high energy (1 – 50 J), glass ($\lambda_0 \approx 1$ µm) lasers. Such lasers have a very low repetition rate, less than one shot per minute. Very recently, we have obtained the same kind of results, but with a low energy (600 mJ) high repetition rate (10 Hz), Ti:Sapphire ($\lambda_0 \approx 0.8$ µm) laser. The minimum pulse duration was 35 fs (FWHM), giving a maximum power of 20 TW. The experimental set–up is presented in Fig. 10.

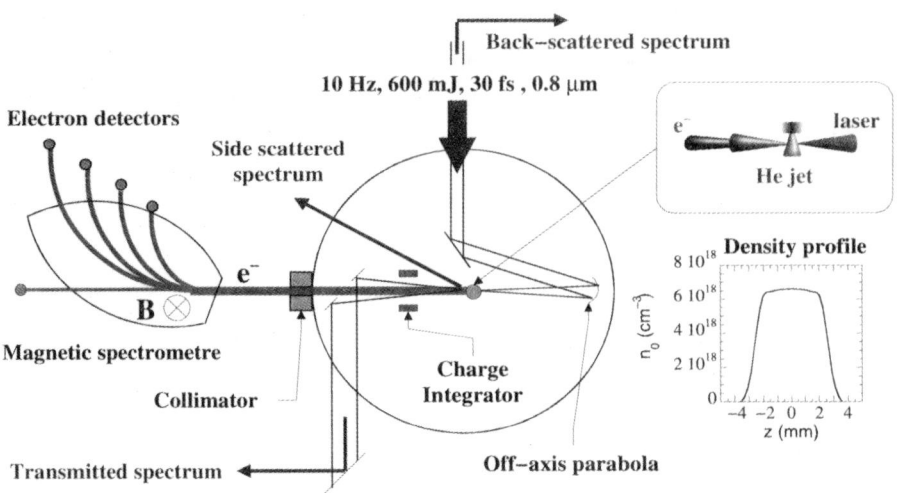

Figure 10. Typical set–up of a SMLWFA experiment.

The laser beam was focused by an off-axis parabola to a focal spot of $w_0 = 6$ µm, leading to a maximum intensity of 2×10^{19} W/cm^2. The focal plane was placed at the edge of a 2 mm diameter laminar plume of helium gas produced by a pulsed supersonic gas jet. The energy spectrum of the ejected electron beam was analysed with a magnetic spectrometer coupled to diodes. The total beam charge was measured with an integrated charge transformer. This device converts the charge passing through a coil to a DC electric signal. The transverse profile of the electron beam was measured with a scintillator placed perpendicular to the beam, and covered on its front face with black paper to cut the laser light. The electron beam divergence was deduced by measuring this transverse profile at different longitudinal positions.

By changing the backing pressure of the gas jet, the plasma density was adjust from 10^{18} to 10^{20} cm^{-3}, allowing to study the transition between LWFA and SMLWFA. The left curve of Fig. 11 shows that as expected, at a plasma density close to or below the LWFA resonance, no electron beam was observed (plasma electrons are too cold to be trapped by the high phase velocity wave). When the density was increased so that the laser power became larger than the critical power of self-focusing ($P/P_c > 1$), and so that the plasma wavelength was a few times shorter than the pulse length ($\lambda_p < c\tau$), an electron beam was created. Its total charge reached a few nC. Its transverse profile was gaussian, and the divergence of the order of 100 mrad, a little bit more than the laser divergence. Since the electrons are accelerated in the focal spot, the beam emittance was of the order of 0.1 mm.mrad, competitive with the conventional electron injectors. Typical energy spectra are presented in the right curve of Fig. 11 for two plasma densities. They are Maxwellian and one can associate them a temperature. The maximum energy was of the order of 70 MeV. When the density is increased, the phase velocity of the wave is slower, which should lead to lower energies. This is what we observed.

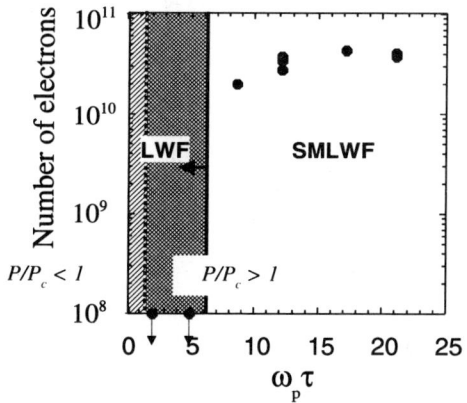
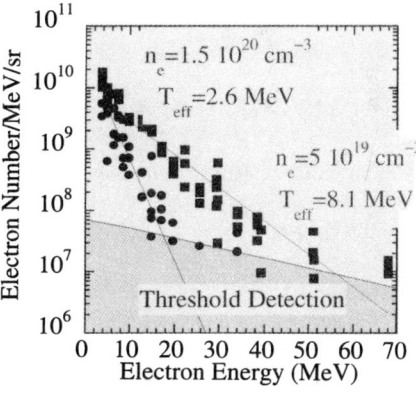

Figure 11. Left: Transition between the LWF (no acceleration of background electrons) and the SMLWFA regimes. Right: Typical electron energy spectra obtained at plasma densities: $n = 1.5 \times 10^{20}$ cm^{-3} (squares) and $n = 5 \times 10^{19}$ cm^{-3} (disks). The spectra are Maxwellian of temperatures $T_{\it eff} = 2.6$ and 8.1 MeV respectively.

5. PERSPECTIVES

Up to now, the goal of the described acceleration experiments was simply to demonstrate the different schemes of laser–plasma accelerators. The obtained energy gains were far from the limits of the mechanisms. The future experiments will try to increase the energy gain and to characterise in detail the accelerated beam (in order to know if it can be injected in conventional accelerators).

The main limitation is at the moment the length of the acceleration region which, in the case of LBWA and LWFA, is limited to the Rayleigh length of the laser. Much larger energy gains are possible if the laser pulse stay focused over centimetres. The SMLWFA could also deliver electrons of much higher energy if much lower plasma densities are used. In that case, the critical power for relativistic guiding becomes much larger, and exceed the present day lasers capabilities. The envelope modulation instability could however be excited if the interaction length is long enough. Calculations and simulations show that energy gains of the order of 1 GeV are possible with the present day lasers if the laser beam is guided on a length of the order of 1 cm.

Different schemes for laser guiding have been proposed and already tested. A possible solution is to send the laser beam in a plasma channel: the plasma density has a parabolic transverse profile that is minimum on the laser axis. Such channels can be produced by ionising and rapidly heating a gas with a long laser pulse[10], or by using a capillary discharge[11]. Plasma guides of the order of 1 cm long have been produced. However, such techniques present one default: the density profile of the plasma guide is determined by the guiding condition of the laser and can be quite different from what is required for LBW, LWF or SMLWF acceleration. Another solution is to use dielectric micro–capillary tubes filled with gas[12]. In that case, the plasma density can be adjust independently of the guiding structure. Glass tubes of internal radius as small as $a = 5$ microns are available, and can be meters long. Mono–mode guiding of TW lasers in glass capillary tubes of 25 µm radius over lengths up to 12 cm has been demonstrated recently at LULI at intensities up to a few 10^{17} W/cm^2, and in helium gas at a pressure of a few mbar. These experiments have shown that a good temporal and spatial contrast of the laser beam is crucial for high intensity guiding ($> 10^{18}$ W/cm^2): the damage threshold of glass with nanoseconde lasers fall to a few 10^{10} W/cm^2. The entrance of the capillary tube can thus be ionised by a nanoseconde pre–pulse, and blocked before the arrival of the sub–picoseconde part of the pulse.

Guided LWFA and SMLWFA experiments are planned, and energy gains of the order or larger than the GeV should be obtained in the next few years.

ACKNOWLEDGEMENTS

Thanks to F. Amiranoff and V. Malka for their contribution to this manuscript.

REFERENCES

1. F. Amiranoff and P. Mora, *J. Appl. Phys.* **66**, 3476 (1989).
2. E. Esarey et al., *J. Opt. Soc. Amer. B* **12**, 1695 (1995).
3. T. Tajima and J. M. Dawson, *Phys. Rev. Lett.* **43**, 267 (1979).
4. N. E. Andreev et al., *JETP Lett*, **55**, 571 (1992); P. Sprangle et al., *Phys. Rev. Lett.* **69**, 2200 (1992).
5. E. Esarey, *IEEE Trans. Plasma. Sci.* **24**, 252 (1996).
6. A. Pukhov et al., *Phys. Plasmas* **6**, 2847 (1999).
7. C. E. Clayton et al., *Phys. Rev. Lett.* **70**, 37 (1993); N. A. Ebrahim, *J. Appl. Phys.* **76**, 7645 (1994); F. Amiranoff et al., *AIP* **335**, 612 (1995).
8. C. A. Coverdale et al., *Phys. Rev. Lett.* **74**, 4659 (1995); A. Modena et al., *Nature* **337**, 606 (1995); D. Umstadter et al., *Science* **273**, 472 (1996); A. Ting et al., *Phys. Rev. Lett.* **77**, 5377 (1996); C. Gahn et al., *Phys. Rev. Lett.* **83**, 4772 (1999); J. Faure et al., *Phys. Plasmas* **7**, 3009 (2000).
9. F. Amiranoff et al., *Phys. Rev. Lett.* **81**, 995 (1998).
10. C. G. Durfee et al., *Phys. Rev. E* **51**, 2368 (1995), V. Malka et al., *Phys. Rev. Lett.* **79**, 2979 (1997).
11. Y. Ehrlich et al., *Phys. Rev. Lett.* **77**, 4186 (1996).
12. F. Dorchies et al., *Phys. Rev. Lett.* **82**, 4655 (1999).

ELLIPTIC DICHROISM IN ANGULAR DISTRIBUTIONS IN FREE-FREE TRANSITIONS IN HYDROGEN

Aurelia Cionga[1], Fritz Ehlotzky[2], and Gabriela Zloh[1]

[1]Institute for Space Science
P.O. Box MG-23, R-76900 Bucharest, Romania
[2]Institute for Theoretical Physics, University of Innsbruck
Technikerstrasse 25, A-6020 Innsbruck, Austria

INTRODUCTION

Dichroism is a well known concept in classical optics where it denotes the property shown by certain materials of having absorption coefficients which depend on the state of polarization of the incident light[1]. This concept has been extended to the case of atomic or molecularinteractions with a radiation field. In particular, the notion of ellipticdichroism in angular distribution (EDAD) refers to the difference between the differential cross sections (DCS) of laser assisted signals for *left* and *right* elliptically polarized (*EP*) light[2].

We discuss here the effect of the photon helicity in laser induced and inverse bremsstrahlung for *high energy scattering* of electrons by hydrogen atoms. We demonstrate that it is possible to find EDAD for high scattering energies of the electrons if the *dressing of the atomic target* by the laser field is taken into account. We consider higher optical frequencies of the laser field and we restrict ourselves to the use of moderate field intensities. In this case we can employ a hybrid treatment ofthe problem[3]: the interaction between the scattered electronand the laser field is described by Volkov solutions, while thelaser-dressing of the atomic electron is evaluated within the framework oftime-dependent perturbation theory (TDPT). Using this approximation, we candemonstrate that EDAD becomes a non-vanishing effect provided second orderTDPT is used to describe the dressing of the atomic electron by the ellipticlaser field. Moreover, we analyze the role of the virtual transitions betweenthe bound and continuum states and we show that these transitions areessential in order to be able to predict the existence of EDAD effects.

The basic equations will be presented in section II. In section III we shall consider in some detail the case of two photon transitions in the weak field limit. The helicity dependence of the DCS for two photon absorption/emission by the colliding system in interaction with the *EP* laser field will be discussed in section IV.

Atoms, Solids, and Plasmas in Super-Intense Laser Fields
Edited by D. Batani *et al.*, Kluwer Academic/Plenum Publishers, 2001

BASIC EQUATIONS

We consider free-free transitions in electron-hydrogen scattering in the presence of an EP laser field of polarization vector $\vec{\varepsilon}$ given by

$$\vec{\varepsilon} = \cos(\xi/2)\left[\vec{e}_i + i\vec{e}_j \tan(\xi/2)\right], \qquad (1)$$

where ξ is the ellipticity, $-\pi/2 \leq \xi \leq \pi/2$, and $\vec{e}_{i,j}$ are orthogonal unit vectors in the polarization plane. We are particularly interested to know whether the DCS are sensitive to the *helicity* of the EP photons, defined by $\eta = i\vec{n} \cdot (\vec{\varepsilon} \times \vec{\varepsilon}^*) \equiv \sin\xi$, with $\vec{n}$ the direction of propagation of the EP laser beam. Right hand EP has $\eta < 0$, it corresponds to $-\pi/2 \leq \xi < 0$. Left hand EP that has opposite helicity, $\eta > 0$, corresponds to $0 < \xi \leq \pi/2$. For optical frequencies we adopt the dipole approximation, thus the resulting electric field can be described by

$$\vec{\mathcal{E}}(t) = i\frac{\mathcal{E}_0}{2}\vec{\varepsilon}\exp(-i\omega t) + \text{c.c.}, \qquad (2)$$

where the intensity of the laser field is given by $I = \mathcal{E}_0^2$.

We assume that at moderate laser field intensities the interaction between the laser field and the atomic electron can be described[3] by TDPT. We find it necessary to use *second order perturbation theory* and, following Florescu et al[4], the approximate solution for the ground state of an electron bound to a Coulomb potential in the presence of an EP laser field can be written in the form

$$|\Psi_1(t)\rangle = e^{-iE_1 t}\left[|\psi_{1s}\rangle + |\psi_{1s}^{(1)}\rangle + |\psi_{1s}^{(2)}\rangle\right]. \qquad (3)$$

Here $|\psi_{1s}\rangle$ is the unperturbed ground state of the hydrogen atom, of energy E_1 and $|\psi_{1s}^{(1),(2)}\rangle$ denote the first and second order laser field dependent corrections, respectively. We made use of the published expressions[4,5] of these corrections.

The scattering electron of kinetic energy E_k and momentum $\vec{k}$ in interaction with the field (2) can be described by the well known Gordon-Volkov solution

$$\chi_{\vec{k}}(\vec{r},t) = \frac{1}{(2\pi)^{3/2}}\exp\left\{-iE_k t + i\vec{k}\cdot\vec{r} - i\vec{k}\cdot\vec{\alpha}(t)\right\}, \qquad (4)$$

where $\vec{\alpha}(t)$ describes the classical oscillation of the electron in the electric field $\vec{\mathcal{E}}(t)$. The amplitude of this oscillation is given by $\alpha_0 = \sqrt{I}/\omega^2$. Using Graf's addition theorem[6] of Bessel functions, the Fourier expansion of the Gordon-Volkov solution (4) leads to a series in terms of ordinary Bessel functions J_N since one has

$$\exp\left[-i\vec{k}\cdot\vec{\alpha}(t)\right] = \exp\left\{-i\mathcal{R}_k\sin(\omega t - \phi_k)\right\} = \sum_{N=-\infty}^{N=\infty} J_N(\mathcal{R}_k)\exp\left[-iN(\omega t - \phi_k)\right]. \qquad (5)$$

Following the definitions of the arguments and phases given in Watson's book,[6] we can write

$$\mathcal{R}_k = \alpha_0 \cos(\xi/2)\sqrt{(\vec{k}\cdot\vec{e}_i)^2 + (\vec{k}\cdot\vec{e}_j)^2 \tan^2(\xi/2)} \equiv \alpha_0|\vec{k}\cdot\vec{\varepsilon}| \qquad (6)$$

and

$$\exp(i\phi_k) = \frac{\vec{k}\cdot\vec{\varepsilon}}{|\vec{k}\cdot\vec{\varepsilon}|}. \qquad (7)$$

We recognize that a change of sign of the helicity of the EP photons, corresponding to the replacement $\vec{\varepsilon} \to \vec{\varepsilon}^*$, will lead to a change in sign of the dynamical phase ϕ_k.

Therefore, by searching for the signature of helicity in the angular distributions of the scattered electrons, it will be crucial to look for the presence of the dynamical phase in the expressions of the DCS.

For high scattering energies, the first order Born approximation in terms of the interaction potential is reliable. Neglecting exchange effects, this potential is given by $V(r,R) = -1/r + 1/|\vec{r}+\vec{R}|$, where $\vec{R}$ refers to the atomic coordinates. Then, the S-matrix element reads

$$S_{fi}^{B1} = -i \int_{-\infty}^{+\infty} dt < \chi_{\vec{k}_f}(t)\Psi_1(t)|V|\chi_{\vec{k}_i}(t)\Psi_1(t) >, \qquad (8)$$

where $\Psi_1(t)$ and $\chi_{\vec{k}_{i,f}}(t)$ are given by the dressed atomic state (3) and by the Gordon-Volkov states (4), respectively. $\vec{k}_{i(f)}$ represent the initial(final) momenta of the scattered electron. After Fourier decomposition of the S-matrix element (8), the DCS for a scattering process involving N laser photons can be written in the standard form

$$\frac{d\sigma_N}{d\Omega} = (2\pi)^4 \frac{k_f(N)}{k_i}|T_N|^2. \qquad (9)$$

N is the net number of photons exchanged between the colliding system and the laser field (2), thus the scattered electrons have the final energy $E_f = E_i + N\omega$. ($N \geq 1$ refers to the absorption and $N \leq -1$ to the emission of laser quanta, while $N = 0$ corresponds to the elastic scattering process.)

In the foregoing equation (9), the nonlinear transition matrix elements T_N, obtained from the S-matrix element (8), have the following general structure

$$T_N = \exp(iN\phi_q) \left[T_N^{(0)} + T_N^{(1)} + T_N^{(2)}\right]. \qquad (10)$$

ϕ_q is the dynamical phase defined in (7), referring here to the momentum transfer $\vec{q} = \vec{k}_i - \vec{k}_f$ of the scattered electron. The first term in equation (10),

$$T_N^{(0)} = -(2\pi)^{-2} f_{el}^{B1} J_N(\mathcal{R}_q), \qquad (11)$$

would yield the well-known Bunkin-Fedorov formula[7]. f_{el}^{B1} is the amplitude of elastic electron scattering in the firstorder Born approximation: $f_{el}^{B1} = 2(q^2+8)/(q^2+4)^2$. The other two terms in the transition matrix element(10) are related to the atomic dressing by the laser field. Theseterms were discussed in considerable detail in our precedingwork[8]. The second term, $T_N^{(1)}$, refers to first orderdressing of the atom in which case *one* of the N photons exchangedbetween the colliding system and the radiation field is interacting with thebound electron, while the third term $T_N^{(2)}$ refers to second orderdressing and here *two* of the N photons exchanged during thescattering interact with the atomic electron. For an EP field, thesedressing terms are given by

$$T_N^{(1)} = \frac{\alpha_0 \omega}{4\pi^2 q^2} \frac{|\vec{q}\cdot\vec{\varepsilon}|}{q} \left[J_{N-1}(\mathcal{R}_q) - J_{N+1}(\mathcal{R}_q)\right] \mathcal{J}_{1,0,1}(q;\omega) \qquad (12)$$

and

$$T_N^{(2)} = \frac{\alpha_0^2 \omega^2}{8\pi^2 q^2} \{ J_{N-2}(\mathcal{R}_q)\left[|\vec{q}\cdot\vec{\varepsilon}|^2 q^{-2}\mathcal{T}_1(q;\omega) + \mathcal{T}_2(q;\omega)\cos\xi\, e^{-2i\phi_q}\right]$$
$$+ J_{N+2}(\mathcal{R}_q)\left[|\vec{q}\cdot\vec{\varepsilon}|^2 q^{-2}\mathcal{T}_1(q;\omega) + \mathcal{T}_2(q;\omega)\cos\xi\, e^{2i\phi_q}\right]$$
$$+ J_N(\mathcal{R}_q)\left[|\vec{q}\cdot\vec{\varepsilon}|^2 q^{-2}\widetilde{\mathcal{T}}_1(q;\omega) + \widetilde{\mathcal{T}}_2(q;\omega)\right] \}. \qquad (13)$$

The five radial integrals, denoted by $\mathcal{J}_{1,0,1}$, $\mathcal{T}_1$, $\mathcal{T}_2$, $\tilde{\mathcal{T}}_1$ and $\tilde{\mathcal{T}}_2$ in the foregoing equations (12) and (13), depend not only on the absolute value of the momentum transfer q but also on the photon frequency. For the numerical evaluations performed in the present work, we used the analytic expressions for the above five radial integrals which are presented explicitly elsewhere[8-10].

The transition matrix elements $T_N^{(1)}$ and $T_N^{(2)}$ in (12)-(13) are written in a form which evidently permits to analyze their dependence on the dynamical phase ϕ_q. We recognize immediately that $T_N^{(0)}$ and $T_N^{(1)}$ do not depend on the helicity of the photon. On the contrary, $T_N^{(2)}$ exhibits such an explicit dependence. This dependence is determined by the phase factors $e^{\pm 2i\phi_q}$ by which $\mathcal{T}_2$ is multiplied in (13). This demonstrates the necessity to describe target dressing in second order TDPT. In order to stress the important role of the virtual transitions to the continuum, we shall analyze small scattering angles. Here the dressing of the target is considerable and the EDAD effect can be large.

WEAK FIELD LIMIT

For small arguments of the Bessel functions, i.e. either for weak fields at any scattering angle or for moderate fields at small scattering angles, we can keep the leading terms in (10) only. We discuss in some detail the case of two photon absorption, $N = 2$. The corresponding matrix element is

$$T_2 = \frac{\alpha_0^2}{8\pi^2 q^2} \left[(\vec{q} \cdot \vec{\varepsilon})^2 \mathcal{A}(q;\omega) + \cos\xi \, \mathcal{B}(q;\omega) \right], \tag{14}$$

where the amplitudes $\mathcal{A}$ and $\mathcal{B}$ depend on q and on ω

$$\mathcal{A}(q;\omega) = -\frac{q^2}{2^2}\left[f_{el}^{B1} - \frac{4\omega}{q^3}\mathcal{J}_{1,0,1} - \frac{4\omega^2}{q^4}\mathcal{T}_1 \right],$$
$$\mathcal{B}(q;\omega) = \omega^2 \mathcal{T}_2. \tag{15}$$

If we consider $N = -2$ (i.e. two photon emission), then the complex conjugate of the EP polarization vector $\vec{\varepsilon}$ in (14) has to be used and the invariant amplitudes for the appropriate value of q have to be evaluated since $q = |\vec{k}_i - \vec{k}_f|$ is a function of N.

The DCS formula, derived from the transition matrix element (14), turns out to be

$$\frac{d\sigma_2}{d\Omega} = \alpha_0^4 \frac{k_f}{k_i} \frac{1}{2^2 q^4} \left\{ |\vec{q}\cdot\vec{\varepsilon}|^4 |\mathcal{A}|^2 + \cos^2\xi \, |\mathcal{B}|^2 + 2\cos\xi \, \mathrm{Re}\left[(\vec{q}\cdot\vec{\varepsilon})^2 \mathcal{A}\mathcal{B}^* \right] \right\}. \tag{16}$$

This formula is sensitive to a change of helicity only if $\mathrm{Im}\,\mathcal{A} \neq 0$ and $\mathrm{Im}\,\mathcal{B} \neq 0$. This happens if virtual transitions to continuum states are energetically allowed[8-9].

EDAD, defined as the difference between the DCS for left hand (LH) and right hand (RH) elliptic polarizations, follows from (16) as

$$\Delta_E = -\alpha_0^4 \frac{k_f}{k_i} \frac{q_i q_j}{2q^4} \sin(2\xi) \mathrm{Im}\,(\mathcal{A}^*\mathcal{B}) \quad \text{where} \quad q_{i;j} = \vec{q}\cdot\vec{e}_{i;j}. \tag{17}$$

Δ_E depends on the ellipticity ξ, its maximum value corresponds to $\xi = \pi/4$. Δ_E is also symmetric with respect to the replacement $\xi \to \pi/2 - \xi$.

RESULTS AND DISCUSSION

We discuss numerical results for EDAD in laser-assisted electron-hydrogen scattering at high energies of the ingoing particle. We concentrate our analysis on the cases in which the number of exchanged photons between the scattering system and the laser field is $N = \pm 2$ since here the effects turn out to be large enough to be accessible to observation. On the basis of the formalism developed above, we present the DCS evaluated from (9) for a fixed scattering angle θ as a function of the azimuth φ. We also present the dichroism Δ_E in the same azimuthal plane. As initial energy of the scattered electrons we have taken $E_i = 100$eV and we have chosen a laser frequency that is close to an atomic resonance, namely $\omega = 10$ eV. We show numerical results for the moderate fieldintensity $I = 3.51 \times 10^{12}$ Wcm^{-2}. The initial electron momentum $\vec{k}_i$ is taken to point along the z-axes; the EP laser beampropagates along the same axes.

In figure 1 we plot the DCS at $\theta = 20°$ for two photon emission ($N=-2$) in panel (a) and two photon absorption ($N=2$) in panel(b). The signals obtained for LH polarization ($\xi = \pi/4$) are represented by full lines and those obtained for RH polarization ($\xi = -\pi/4$) by dotted lines. None of the two axes of the ellipse are symmetryaxes for the DCS, but a change of helicity $LH \leftrightarrow RH$ is equivalent to a reflection in each of the planes xOz and yOz. The elliptic dichroism is shown in panel (c). In this geometry

$$q_i q_j = k_f^2 \sin^2 \theta \sin \varphi \cos \varphi \qquad (18)$$

and Δ_E has an overall $\sin(2\varphi)$ dependence that determines its four-leaved clover pattern. The outer clover leaves correspond to absorption, while the inner ones are obtained for emission. The signs of the leaves aredifferent for emission and absorption, respectively.

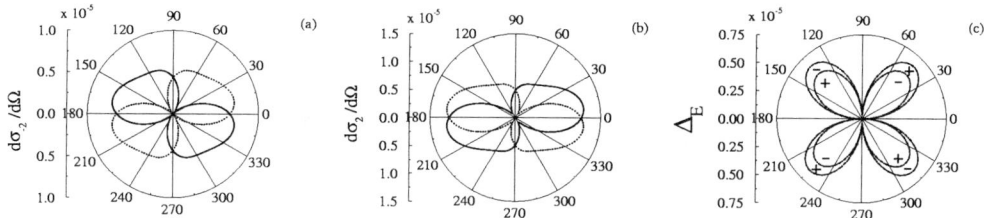

Figure 1: (a) DCS for two photon emission as a function of the azimuthal angle φ at the scattering angle $\theta = 20°$. Full line is used for LH elliptic polarization and dotted line for RHelliptic polarization. (b) Same as panel (a) but for two photon absorption. In panel (c) the dichroism Δ_E is shown. The outer clover correspondto $N=2$, while the inner one is obtained for $N=-2$.

The nonlinear signals exhibit also a strong ellipticity dependence, shown in figure 2, where the DCS for two photon absorption are plotted for LH elliptic polarization using three values of the ellipticity, namely $\xi = 10°$, $\xi = 45°$, and $\xi = 80°$. It is interesting to note that although the DCS may be so different, in this geometry they will always lead to a four-leaved clover pattern. According to (17) onlythe magnitude of the clover leaves will be modulated by the ellipticity.

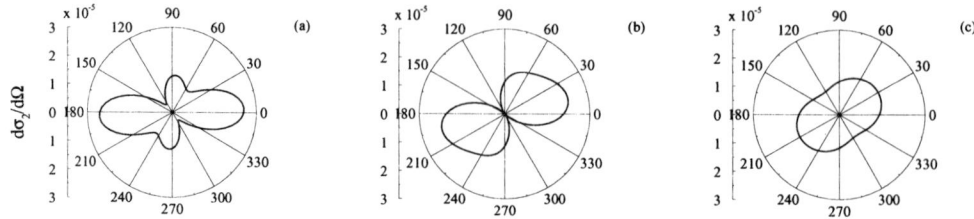

Figure 2: (a) DCS for two photon absorption ($N=2$) as a function of the azimuthal angle φ for LH helicity and ellipticity $\xi = 10°$. The scattering angle is $\theta = 10°$. The rest of the parameters are the same as in figure 1(b). (b) same as in panel 2(a) but for $\xi = 45°$. (c) same as in panel 2(a) but for $\xi = 80°$.

Summarizing we find that EDAD in free-free transitions at high projectile energies in a EP laser field can only be predicted if laser dressing of the target atom is taken into account in second order TDPT including virtual bound-continuum transitions irrespective of the scattering configuration considered.

Acknowledgments

This work has been supported by the Jubilee Foundation of the Austrian National Bank under project number 6211 and by a special research project for 2000/1 of the Austrian Ministry of Education, Science and Culture. We also acknowledge financial support by the University of Innsbruck under reference number 17011/68-00.

REFERENCES

1. M. Born, *Optik* (Springer, Berlin, 1981).
2. N. L. Manakov, A. Maquet, S. I. Marmo, V. Veniard and G. Ferrante, Elliptic dichroism and angular distribution of electrons in two-photon ionization of atoms *J. Phys. B* **32**, 3747 (1999).
3. F. W. Byron Jr. and C. J. Joachain, Electron-atom collisions in a strong laser field *J. Phys. B* **17**, L295 (1984).
4. V. Florescu, A. Halasz, and M. Marinescu, Second-order corrections to the wave functions of a Coulomb-field electron in a weak uniform harmonic electric field *Phys. Rev. A***47** , 394 (1993).
5. V. Florescu and T. Marian, First-order perturbed wave functions for the hydrogen atom in a harmonic uniform external electric field, *Phys. Rev. A* **34**, 4641 (1986).
6. G. N. Watson, *Theory of Bessel Functions*, 2nd. Ed. (University Press, Cambridge, 1962), p. 359.
7. F. V. Bunkin and M. V. Fedorov, Bremsstrahlung in a strong radiation field, *Zh. Eksp. Theor. Fiz.* **49** , 1215 (1965) [*Sov. Phys. JETP*, **22**, 884 (1966)].
8. A. Cionga, F. Ehlotzky, and G. Zloh, Electron-atom scattering in a circularly polarized laser field, *Phys. Rev. A* **61**, 063417 (2000).
9. A. Cionga and V. Florescu, One-photon excitation in the e-H collision in the presence of a laser field, *Phys. Rev. A* **45**, 5282 (1992).
10. A. Cionga and G. Zloh, unpublished.

SHOCK ELECTROMAGNETIC WAVES RESULTING FROM HIGHER HARMONICS GENERATION IN TRANSPARENT SOLIDS

Vitali E.Gruzdev and Anastasia S.Gruzdeva

State Research Centre "S.I.Vavilov State Optical Institute"
12 Birzhevaya Liniya
St.Petersburg, 199034
Russia

INTRODUCTION

Fast developing of femtosecond lasers and general trend in increasing of damage-threshold intensity with decreasing of pulse duration have resulted in new situation in area of high-power nonlinear optics. Many nonlinear optical phenomena can be studied at laser intensity $10^{13} - 10^{14}$ W/cm^2 in transparent materials that cannot be reached for nano- and picosecond laser pulses. On the other hand, characteristic intensities of laser-induced damage and ablation have exceeded 10^{14} W/cm^2 what, being combined with femtosecond pulse duration, gives possibility of new insight into process of laser-matter interaction.

Nonlinear processes induced by femtosecond laser pulses differ very much from similar processes induced by longer pulses because of very high laser intensity and very small pulse duration. Laser-induced variations of refractive index are large at laser intensity close to damage threshold that is a reason to come back to concept of electromagnetic shock waves (SHEW) which appeared in high-frequency electrodynamics some 50 years ago[1-4] and did not found applications in nonlinear optics. New understanding and models of nature of nonlinear material response[5, 6] allow now more correct consideration of factors influencing SHEW formation, in particular, dispersion of linear and nonlinear parts of refractive index. Important reason to consider high-power nonlinear propagation of femtosecond pulses and possibility of SHEW formation is connected with one of main problems of femtosecond laser-matter interactions that is mechanism of energy transfer from radiation to matter. That is especially true for the case of laser ablation and damage of transparent materials by femtosecond pulses. Any study of femtosecond laser interaction with transparent materials should start with consideration of nonlinear pulse propagation and accompanying optical phenomena because they determine conditions of energy deposition to electron and phonon systems and, thus, conditions of material modification.

We start with consideration of formation of optical SHEW using modelling of that process. After consideration of SHEW interaction with transparent solids we conclude with discussion of obtained results.

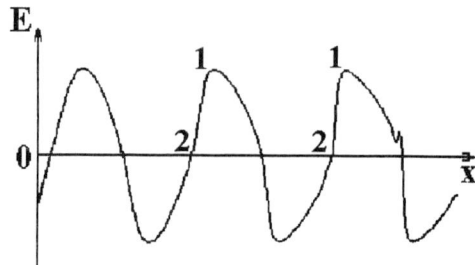

Figure 1. Sketch of deformation of plane-wave profile during formation of shock electromagnetic wave in nonlinear medium with positive nonlinear addition to refractive index.

FORMATION OF ELECTROMAGNETIC SHOCKS

Qualitative picture of SHEW formation on optical cycle due to generation of higher harmonics is as follows for the case of positive nonlinear addition to refractive index. Laser-induced nonlinear polarisation of the medium increases refractive index and results in slowing down the part of optical cycle corresponding to large electric field strength (point 1 in Fig. 1). Bottom part of wave profile (2 in Fig. 1) moves with unperturbed speed of light in the material, i. e., faster than top part of wave profile. That results in distortion of wave profile in such a way that top of the profile "falls down" onto the region of the profile where electric field is small and which moves behind the top. That leads to appearing of very abrupt field variation within length much smaller than wavelength what looks like a step of field strength that is referred to as SHEW front. In general, there is no exact disruption, and structure of SHEW front should be studied with taking into account complicated mechanism of dispersion and energy dissipation near SHEW front[1-6].

Mathematical description of SHEW formation is based on the following model. Nonlinear wave propagation is described by full wave equation:

$$\frac{\partial^2 E}{\partial t^2} - c_0^2 \frac{\partial^2 E}{\partial z^2} = -4\pi \frac{\partial^2 P}{\partial t^2} - \frac{4\pi\sigma}{\varepsilon_0} \cdot \frac{\partial E}{\partial t}. \tag{1}$$

Wave equation (1) should be coupled with a system of nonlinear equations describing material response P consisting of fast P_e (electronic) and slow P_i (ionic) components:

$$P = P_e + P_i. \tag{2}$$

Each of them includes linear and nonlinear contributions. Considering input pulse to be far from any absorption band, one can neglect absorption contributions to polarisation.

Description of polarisation variations in time must give correct dispersion law for both linear and nonlinear parts of refractive index in all considered spectrum range. One of the simplest models for that is phenomenological model of two parametrically coupled oscillators[5, 6] giving the same form of dispersion as quantum-mechanical three-level model. According to it, variations of polarisation are described by the following equations:

$$\frac{\partial^2 P_e}{\partial t^2} + \frac{2}{T_e} \cdot \frac{\partial P_e}{\partial t} + \omega_e^2 P_e = \alpha_e E + \beta_e (R_e + R_i) E \tag{3}$$

$$\frac{\partial^2 P_i}{\partial t^2} + \frac{2}{T_i} \cdot \frac{\partial P_i}{\partial t} + \omega_i^2 P_i = \alpha_i E \tag{4}$$

$$\frac{\partial^2 R_e}{\partial t^2} + \frac{2}{T_{el}} \cdot \frac{\partial R_e}{\partial t} + \omega_{el}^2 R_e = \gamma_e (P_e + P_i) E \tag{5}$$

$$\frac{\partial^2 R_v}{\partial t^2} + \frac{2}{T_v} \cdot \frac{\partial R_v}{\partial t} + \omega_v^2 R_v = \gamma_v (P_e + P_i) E. \tag{6}$$

These equations describe polarisation contributions from vibrations of optical electrons (3) and ions (4) excited straightway by laser electric field as well as contributions from secondary vibrations of valence electrons (5) and ions (6) excited by vibrations of electrons and ions. That allows to take into account electronic and electron-vibrational contributions to nonlinear optical response and correct description of Raman and *inertia-free* Kerr nonlinear response. Two-photon absorption is also included into polarization response deduced from (3) – (6). With suitable parameters, dispersion law obtained from (3) - (6) describes real dispersion of linear part of refractive index of transparent materials with accuracy of 0.01% within full transparency band[6]. We do not touch details of two-oscillator model for description of polarization dispersion, which can be found in papers[5, 6].

In general case, formation and propagation of SHEW is a self-consistent problem. To start, we reduce self-consistent problem to a couple of independent problems:
1) electrodynamical propagation problem - formation and propagation of SHEW in dielectric with constant properties[7];
2) interaction problem including processes induced by unperturbed SHEW (which space distribution and time evolution are results of solving of the first problem)[7, 8].

Several comments should appear in connection with this reducing. First, it can be applied to describe initial stage of SHEW formation near small leading part of laser pulse where laser-induced variation of material parameters are small enough to be neglected. Duration of the leading part to be considered is determined by characteristic time of laser-induced plasma excitation up to the level when it gives sufficient distortions to pulse propagation. That time is about 15-20 fs. Plasma excitation is assumed to be the fastest, other laser-induced processes of material modification have longer excitation time and their influence on nonlinear processes at leading edge of laser pulse can be neglected. Thus, throughout this paper we consider only few (about 10) first cycles of femtosecond pulse.

Second, natural result of reducing to independent problems is breaking of energy conservation law. Two points should be mentioned in this connection: 1) energy dissipating from SHEW to material is assumed to be small enough at leading part of laser pulse not to influence significantly on SHEW propagation; 2) presented consideration is the first approximation to exact solution while absorption rate calculated in the next section can be used to estimate energy dissipation in further approximations.

All processes of SHEW formation and evolution are well known[1-4] to be divided into two groups – slow processes and fast processes. Characteristic space and time scales for slow processes are much larger than laser wavelength λ_0 in vacuum and laser period T_0. Time and space scales of fast processes are much less than laser period and wavelength. Example of slow process is accumulation of nonlinear distortions connected with higher harmonics generation resulting in formation of abrupt SHEW front. That process takes about 10 T_0 and develops within distances of (10 – 15) λ_0. Fast processes are connected with SHEW front evolution due to energy dissipation and interplay between higher harmonics which characteristic time is less than $T_0/6$ and space scale is less than $\lambda_0/6$.

Making use of (3)-(6), one can estimate what contributions to polarization play key role in developing of slow and fast processes. For both processes let us assume all frequencies ω giving sufficient contribution to SHEW formation to be between ionic and electronic absorption bands, i.e., the following condition must be satisfied:

$$\omega_i << \omega < <\omega_e, \ \omega_v << \omega < <\omega_{e1}. \qquad (7)$$

Applying Fourier integral transformation to (3) – (6), then applying perturbation technique with respect to small parameters ω/ω_e, ω_i/ω, ω/ω_{e1}, ω_v/ω to obtained expressions and then applying inverse Fourier transformation to obtained perturbation series, one can obtain expression for material polarization[6, 7]. Their analysis described in[7] shows that slow processes are dominated by *inertia*-free Kerr nonlinearity while SHEW front evolution is dominated by both nonlinearity and dispersion. Bearing that in mind, one can obtain simple estimation of threshold intensity from geometrical consideration of evolution of nonlinear distortions of optical cycle[7]. The threshold intensity of input laser pulse is as follows:

$$I_{th} \approx \frac{n_0}{\gamma} \cdot \left(\exp\left\{ \frac{\Delta x}{s} \right\} - 1 \right), \qquad (8)$$

where n_0 is constant linear part and γ is nonlinear coefficient of refractive index having the form $n = n_0 + \gamma \cdot I$, s is length of path passed by laser pulse in the medium required for SHEW formation, Δx is length of optical cycle mashed during SHEW formation. For example, threshold of appearing of field disruption is $I_{th} \leq 9.46 \cdot 10^{13}$ W/cm^2 for $n_0 = 1.41$, $\gamma = 2 \cdot 10^{-16}$ cm^2/W (fused silica), $s = z_c = 15\lambda_0 \ll L_{DISP}$, $\Delta x = 0.2\lambda_0$. Confocal parameter corresponds to focal spot radius $d_f = 2.185\lambda_0$ ($=6.555$ μm for $\lambda_0 = 3.0$ μm).

More detailed investigation of SHEW can be fulfilled numerically only. The authors applied FDTD technique[7] to investigate SHEW formation and evolution. Fig. 2 depicts one of obtained results - a snapshot of instant field profile in non-absorbing medium. Appearing and evolution of SHEW is clearly seen from that figure.

INTERACTION OF SHEW WITH TRANSPARENT SOLIDS

As it follows from experimental results[9-13], initial stages of femtosecond laser-matter interactions should be managed by fast mechanisms with very small *inertia* because typical pulse duration is less than time of energy transfer from electrons to phonons[9]. The mechanisms must be rather universal, without strong dependence on wavelength if the latter is within transparent band and must act for pulse duration from several hundreds to few tens of femtosecond. We suppose SHEW to be the most suitable nonlinear electrodynamical process for being fast mechanism of initiating of femtosecond damage and desorption. There are two reasons for that connected with *main features of SHEW*[7, 8]:

➢ generation of higher harmonics (odd harmonics in isotropic medium) – can result in effective ionization;
➢ formation of abrupt shock front (can result in straight action on crystal lattice).

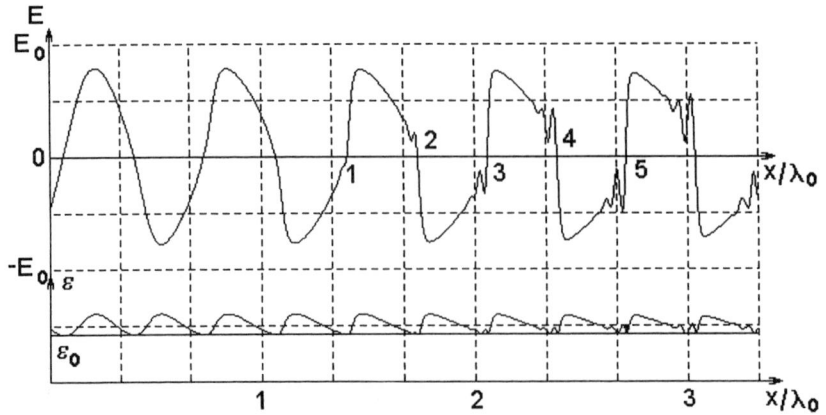

Figure 2. Evolution of SHEW propagating in non-absorbing dielectric (fused silica, $n_0 = 1.41$). Points of interest: 1 – appearing of disruption of space derivation of electric-field strength on wave profile (weak SHEW); 2 – formation of disruption of electric-field strength (strong SHEW) and formation of field variations near SHEW front; 3, 4, 5 – developing of SHEW front and formation of high-frequency pulsation near it resulting from higher harmonics generation. Laser-induced variation of dielectric response is shown in the lower part of this figure. Input field amplitude is above damage threshold in order to obtain good illustrative picture with fast SHEW formation (high-frequency pulsations can not be made out in the picture otherwise).

The first reason for considering SHEW as the most suitable candidate is possibility of effective one-photon ionization within SHEW model by HH[8, 14]. Really, laser-induced ionization goes through generation of higher harmonics and single-photon absorption. Let compare density of SHEW-induced free electrons with density of electrons appearing due to multiphoton ionization.

According to estimations and numerical results[1-4, 7, 14] intensity of higher harmonics I_{HH} can reach about 10% of radiation intensity I_0 at initial laser wavelength. On the other hand, cross-section for one-photon absorption[15] $\sigma_{1\text{-}photon}$ is about 10^{-16} cm^2 what gives absorption $\alpha_{1\text{-}photon} \cong 10^6$ cm^{-1}. That allows estimating of ionization rates for one-photon absorption of N-th order higher harmonic and N-photon absorption. The most optimistic estimations for multiphoton absorption can be obtained for the case N=3 because absorption cross-section σ_N decreases fast with increasing of N. On the other hand, it follows from experimental data[9-14] that three-photon absorption is the lowest-order multiphoton absorption that can take place in wide band-gap materials. Three-photon absorption is given as[15] $\alpha_{3\text{-}photon} = \sigma_3 \, n' = \delta_3 E^4 n' = \beta E^4$ where $\delta_3 \cong 10^{-54}$ 1/(cm^2 s). Thus, for speed of free-electron generation at laser intensity $5\cdot 10^{13}$ W/cm^2 (corresponding to $E=2\cdot 10^8$ V/cm) one obtains[14]

$$\frac{G_{3\text{-}photon}}{G_{3\text{-}harmonic}} = \frac{\alpha_{3\text{-}photon} I_0}{\alpha_{1\text{-}photon} I_{3\text{-}harmonic}} = \frac{\delta_3 E^4 I_0}{\sigma_{1\text{-}photon} 0.1 I_0} \approx 0.0001 \quad (9)$$

Thus, in presence of SHEW in isotropic dielectric, the lowest order odd harmonic along produces 10000 times more electrons than 3-photon ionization, i.e., multiphoton ionization gives less than 0.01% of total amount of free electrons. Ratio of ionization speeds decreases with increasing of order of harmonics and multiphoton absorption because probability of multiphoton absorption decreases fast while absorption of HH increases. Thus, ionization by HH can be much more effective than widely accepted multiphoton ionization.

The second reason for consideration of SHEW as the most suitable candidate for femtosecond interactions is possibility of effective straight action on atoms and ions on crystal lattice. Rough estimation can be obtained in the following way. Interaction of electric field of strength E with a dipole d results in a mechanical force F

$$\vec{F} = \vec{\nabla}(\vec{E}\vec{d}) \quad (10)$$

that is determined by spatial derivations of electric-field strength. Considering plane harmonic wave of amplitude E_{PHW} resulting in a force F_{PHW} and a SHEW resulting in a force F_{SHEW} one can see that $F_{SHEW} \gg F_{PHW}$ even for $E_{SHEW}=E_{PHW}$, because

$$\frac{F_{SEW}}{F_{PW}} = \frac{(\partial E/\partial x)_{SEW}}{(\partial E/\partial x)_{PW}} \gg 1 \quad (11)$$

near SHEW front. Thus, energy of field-dipole interaction stays the same for both waves if their amplitudes are equal while power of interaction is much more for SHEW than for harmonic wave. This can result in specific processes considered in[8, 14].

More correct consideration of SHEW interaction with a dipole in potential well can be found in [8, 14] where interaction process is treated within classical approach: Newton's equations described motion of the particles forming the dipole, and Maxwell's equations were used to describe SHEW formation. In spite of obvious roughness of the approach, the authors obtained interesting qualitative results. In particular, there were obtained different values of ionization and damage threshold with the latter to be higher, there were deduced linear dependence of ionization threshold on material band gap and linear dependence of damage threshold on specific heat, no dependence of both thresholds on laser wavelength and pulse duration were also obtained. Those facts are in good agreement with experimental data[9-14]. Further considerations of the problem will be based on quantum mechanical approach for description of collective motion of a chain of dipoles interacting with SHEW. Mutual influence of the dipoles is being expected to lead to formation of feedback resulting in variation of dipole potential wells and damage of the chain.

Formation of abrupt SHEW front can also result in appearing of large gradient forces proportional to gradient of laser intensity[16]. They can be an important factor of femtosecond laser-matter interaction because they can result in large pressure of about 10^8 Pa (or 1000 atmospheres).

CONCLUSIONS

Thus, shock electromagnetic waves can appear near leading edge of femtosecond laser pulse of mid-IR wavelength range propagating in transparent solids in case of tight focusing. Threshold intensity required for appearing of strong SHEW is close to damage threshold, i.e. $10^{13} - 10^{14}$ W/cm^2.

SHEWs are connected with intensive higher harmonics generation and formation of abrupt front which width is much smaller than input laser wavelength. That makes SHEW a good candidate for being a mechanism of initiating of non-thermal damage and ablation by femtosecond pulses. Possible mechanisms of SHEW action on transparent materials are connected with ionisation, action of large gradient forces[16] and bulk pressure, and straight action of electric field near SHEW front on structural units of material[8, 14] – ions, atoms, molecules resulting in generation of point defects.

REFERENCES

1. A.M.Beliantsev, A.V.Gaponov, G.I.Freidman, "Structure of shock electromagnetic wave front in transmission lines with nonlinear parameters", in *Electromagnetic Wave Theory*, URSI Symposium, Pergamon Press, New York (1967).
2. A.V.Gaponov, L.A.Ostrovsky, M.I.Rabinovich, "Electromagnetic waves in non-linear transmission lines with active parameters", in *Electromagnetic Wave Theory*, URSI Symposium, Pergamon Press, New York (1967).
3. G.B.Whiteham, "Nonlinear dispersive waves", *Proc. Roy. Soc.*, Ser. A, **283**, 238 (1965)
4. L.A.Ostrovsky, "Propagation of modulated waves in nonlinear dispersive media", in *Electromagnetic Wave Theory*, URSI Symposium, Pergamon Press, New York (1967).
5. S.A.Kozlov, «On classical dispersion theory of high-power light», *Optics and Spectroscopy*, **79** (2), 198 (1995).
6. S.A.Kozlov, S.V.Sazonov, "Nonlinear propagation of optical pulses with a few light oscillation duration in dielectric media", *JETP (Sov. Physics - JETP)*, **111** (2), 404 (1997).
7. V.E. Gruzdev, A.S. Gruzdeva, " Formation of shock electromagnetic waves during femtosecond pulse propagation in transparent solids", in *Optical Pulse and Beam Propagation*, Proc. of SPIE, v. 3927, 2000 (to appear).
8. A.S. Gruzdeva, V.E. Gruzdev, "Interaction of shock electromagnetic waves with transparent materials: classical approach", in *Laser Applications in Microelectronic and Optoelectronic Manufacturing – VI*, Proc. of SPIE, v. 3933, 2000 (to appear).
9. R.F.Haglund, Jr., "Mechanisms of Laser-Induced Desorption and Ablation", in *Laser Ablation and Desorption*, J.C.Miller and R.F.Haglund, Jr., eds., Academic Press, Boston (1998).
10. T.V.Kononenko, V.I.Konov, S.V.Garnov, et.al., "Comparative study of material ablation by femtosecond and pico/nanosecond laser pulses", *Quantum Electronics*, **28** (2), 167 (1999).
11. C.B.Schaffer, E.N.Glezer, N.Nishimura, and E.Mazur, "Ultrafast laser induced microexplosions: explosive dynamics and sub-micrometer structures", in *Commercial Applications of Ultrafast Lasers*, Proc. SPIE, v. **3269**, 36 (1998).
12. E.E.B.Campbell, D Ashkenasi, and A.Rosenfeld, "Ultra-short-Pulse Laser Irradiation and Ablation of Dielectrics", in *Lasers in Materials*, R.P.Agarwal, ed., Trans Tech Publ., (1998).
13. H.K.Park, R.F.Haglund Jr., "Laser Ablation and Desorption from Calcite from Ultraviolet to Mid-Infrared Wavelengths", *Appl. Phys. A*, **64**, 431 (1997).
14. V. E. Gruzdev, A. S. Gruzdeva, "Thermal and non-thermal effects in femtosecond laser ablation and damage of transparent materials", in *High-Power Laser Ablation-III*, Proc. of SPIE, v. 4065, 62 (2000).
15. R.H.Pantell, H.E.Puthoff, *Fundamentals of Quantum Electronics*, John Wiley and Sons, Inc., New York (1969).
16. L.D.Landau, E.M.Lifshits, *Electrodynamics of Continuous Media*, Academic Press, New York, any edition.

STUDY OF FAST ELECTRON PROPAGATION IN ULTRA-INTENSE LASER PULSE INTERACTION WITH SOLID TARGETS USING REAR SIDE OPTICAL SELF-RADIATION AND REFLECTIVITY-BASED DIAGNOSTICS

J. J. Santos[1] and E. Martinolli[1],
F. Amiranoff[1], D. Batani[2], S. D. Baton[1],
A. Bernardinello[2], G. Greison[2], L. Gremillet[1],
T. Hall[3], M. Koenig[1], F. Pisani[1],
M. Rabec Le Gloahec[4], and C. Rousseaux[4]

[1] *Laboratoire pour l'Utilisation des Lasers Intenses*
UMR7605, CNRS-CEA-Université Paris VI-
Ecole Polytechnique
91128 Palaiseau, France
[2] *Dipartimento di Fisica G. Occhialini*
Università degli Studi di Milano-Bicocca and INFM
Via Emanueli 15, 20126 Milan, Italy
[3] *Department of Physics, University of Essex*
Colchester CO4 3SQ, United Kingdom
[4] *Commissariat à l'Energie Atomique*
91680 Bruyères-le-Châtel, France

INTRODUCTION

Experimental results are reported on transport through a solid target of fast-electrons created by an ultra-intense laser pulse interaction. In particular, the goal was to determine the heating induced in the material by the fast electrons. Such a study is of great interest within the context of the Fast Igniter [1] approach to Inertial Confinement Fusion, where the heating needed to ignite nuclear reactions is supposed to be achieved by a sub-ps fast electron bunch. Experimentally, the main point is therefore to observe the propagation geometry of the fast electron beam and to estimate the amount of energy which can be carried and deposited in dense matter by a given electron source. So far, theory and simulations have not yet provided a complete picture of the propagation phenomena. Therefore, experimental work is required in order to understand and discriminate the basic processes involved.

The experiment was performed at the *Laboratoire pour l'Utilisation des Lasers Intenses* (LULI) with its $100TW$ laser based on chirped pulse amplification technique. A $350fs$, $1.057\mu m$ laser pulse with an energy up to $10J$ was focused by a $f/3$ off-axis parabola at normal incidence onto Al flat targets with thickness ranging from $17\mu m$ to $400\mu m$. The laser focal spot is estimated to be less than $20\mu m$ $FWHM$ corresponding to an incident laser intensity of $10^{18} - 10^{19} W cm^{-2}$. Two main diagnostics were implemented which results and discussion are presented below.

TARGET REAR SIDE OPTICAL SELF-RADIATION

We present results from time-integrated and time-resolved diagnostics of flat targets rear side self-emission. Ultra-intense ultra-short laser pulse generated fast electron jets have been observed at the back of the targets. A very short duration ($\sim ps$), narrow and bright structure becomes progressively weaker and larger with target thickness, but still visible at least up to $400\mu m$.

The observation of very collimated high intensity laser-produced electron jets traveling through solid targets has already been reported previously [2, 3] but several questions concerning fast electron propagation are still poorly understood. Foreseen aspects are the measurement of the material induced heating, critical to electron propagation mechanisms and to ignition efficiency, and a precise characterization of the electron jet during its propagation, its transverse and angular geometric properties and energetic distribution. In this section we mostly discuss the electronic monitoring aspect.

The target rear side was image relayed using three achromat doublets, with total $f\# \simeq 2$, on both CCD camera and streak camera, as shown in Fig. 1. A spatial resolution of a few μm has been obtained. These cameras were adequately filtered to supress the undesirable $1.06\mu m$ and $0.53\mu m$ wavelengths from the probe beam used simultaneously on the target rear side reflectivity diagnostic (see reflectivity below).

The time-integrated images were obtained with a 1024x1024 CCD $16bits$ camera and time-resolved ones with a Streak camera Hadland S20, coupled with a 512x512 CCD $16bits$ camera. The time resolution depended on the chosen sweep speed and on the slit width. Taking into

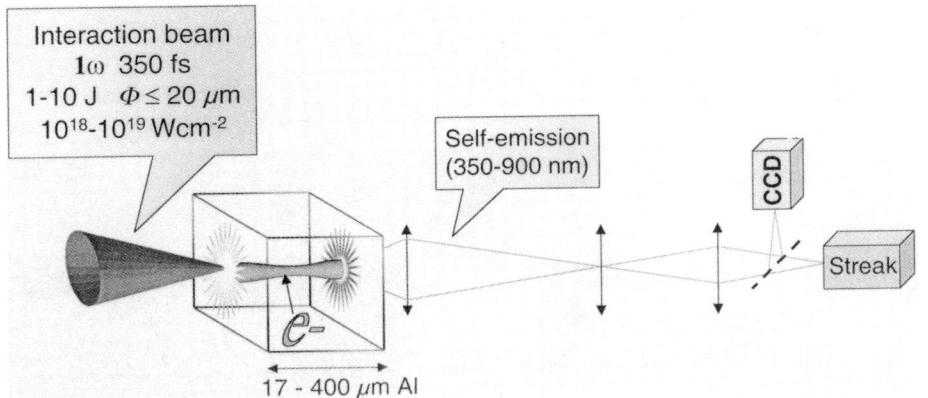

Figure 1: Self-emission diagnostic experimental setup.

account the optics, all the filters and camera responses, acquisition was possible in the spectral waveband between 350 and 900nm.

In the Fig. 2 we present the images obtained from a medium energy configuration shot onto a 75μm Al target. In the time-integrated image, on the left, we see a central narrow bright spot surrounded by a less intense region. The corresponding time-resolved image, presented on the right, allowed to distinguish two types of radiation: a first very short-duration (inferior to 10ps) and very intense signal, being a signature of the hot electronic current flux out from the target, followed by a much less intense and long-duration signal, being an equilibrium thermal radiation from the cooling expanding plasma heated by the fast electron energy deposition. Its intensity slowly decreases for 4.5ns. The signal coming later is interpreted as the arrival to the rear side of a radiative thermal wave propagating in the target.

Using a larger slit and a greater sweep speed we have managed to froze the first brief and intense signal, that is we have obtained a snapshot of the signal. This allows to conclude that most of the radiation is emitted on the first picoseconds, so connected with the fast electron jet crossing the target rear side. The Fig. 3 shows target rear side snapshots for different target thicknesses, ranging from 35μm up to 400μm. The short signal intensity decreases and becomes progressively larger with target thickness, showing a small electron beam spread within the target of about $1/2\theta \approx 17°$ (Fig. 4).

Numerical simulations of electron transport into dense matter have been done with the 3D hybrid code PâRIS. This code models both collision effects (multiple scattering and slowing down) and self-generated magnetic fields [2, 4]. Simulation conditions for the initial electronic

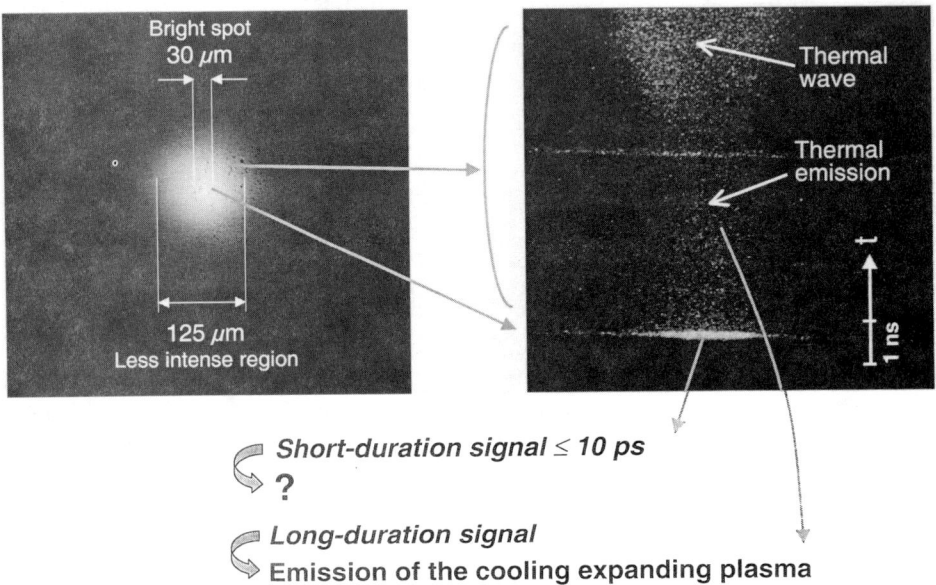

Figure 2: Time-integrated and time-resolved images self-radiation from the rear side of a $75\mu m$ thickness Al target.

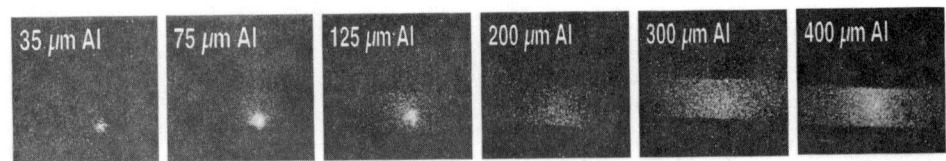

Figure 3: Time-resolved snapshots of the rear side of Al targets self-emission with thicknesses ranging from $35\mu m$ up to $400\mu m$.

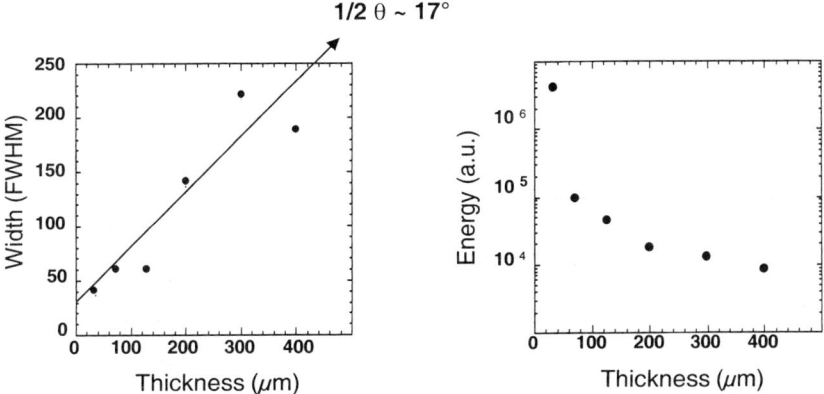

Figure 4: Rear side self-emission evolution with target thickness: emission intensity decreases and emission region increases quasi linearly with a small angle.

population were chosen in order to be as close as possible to experimental conditions. The initial hot electrons temperature is given by the rule [5]

$$T_h(r,t) = 100(I_{17}(r,t))^{1/3}[keV] \qquad (1)$$

where $I_{17}(r,t)$ is the laser intensity in units of $10^{17} W/cm^2$. In Fig. 5 are shown simulated target heating results for an electronic population with $1.6J$ initial energy, spread over a $20\mu m$ focal spot and injected over $300fs$ into a $50\mu m$ thickness Al target. The experimental small electron spreading is confirmed by the simulations (two left figures). On the other hand, if we neglect the electromagnetic effects in the transport code, we obtained (right figure) a target heating about 7 times weaker and much more spreaded ($50\mu m$ FWHM to be compared to the $20\mu m$ with electromagnetic fields). This points out the importance of a magnetically assisted regime electron transport.

Coming back to the time-resolved experimental results, there are different possibilities to explain the origin of the initial short radiation signal:

- The first is Optical Transition Radiation (OTR), which is light emitted when a charge, even in uniform rectilinear motion, crosses the boundary between two media of different dielectric properties [6].

- The second is Synchrotron Radiation: the fast electrons leaving the material create a positive potential that pulls them back within

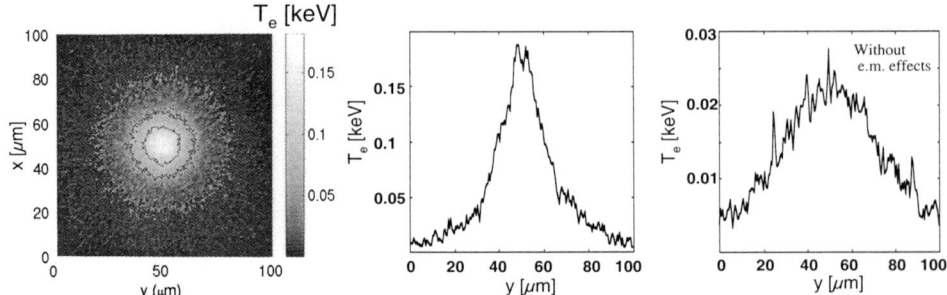

Figure 5: Numerical electron transport simulations (PâRIS) material heating results: target rear side view and profile. The resulting profile if electromagnetic effects are neglected is also shown.

- a distance the order of the Debye length. Their "rotation" to reenter the target would be responsible for a radiative emission type synchrotron, mostly in the direction of its trajectory.

- And finally, a third possibility could be a very localized target heating by the fast electrons, which could emit an out of equilibrium thermal radiation lasting no more then $10ps$.

With the aim of accuratly characterize the electron jet geometry it is then important to identify the relative contribution of each radiating phenomena mentioned above.

In order to accomplish our goal of precisely characterize the electron jet during its propagation, its transverse and angular geometric properties and energetic distribution, a development of the target rear side self-emission diagnostics are scheduled for the next experimental campaign. Measurements will be implemented also in the X-UV region as well as a spectral time-integrated and time-resolved analysis of the emitted light. Exploring experimentally the spectral and angular dependence of OTR and synchotron emission, we expect to be able to quantify the relative contributions to detected signals of each type of radiation.

TARGET REAR SIDE REFLECTIVITY-BASED DIAGNOSTICS

A $\lambda = 0.53\mu m$ probe beam has been used, in order to measure backside surface reflectivity at 45 degrees. The aim was to study geometry and dynamics of the heated spot in the first $200ps$ after the main pulse with a *ps* time resolution. Reflectivity is used as a diagnostic for temperature. For this purpose, $17 - 400\mu m$ massive aluminum targets have been used. Rearside surface unperturbated reflectivity was measured for each set of targets. Two configurations have been set up as in figure 6, in order to study respectively geometry and temporal evolution.

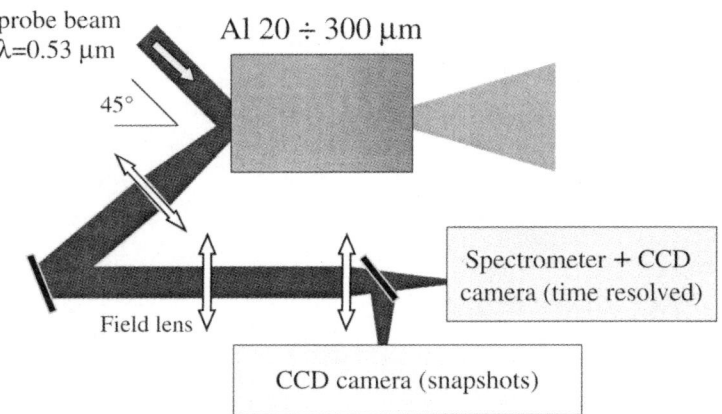

Figure 6: Reflectivity diagnostic setup.

- In the first one, the probe beam is used with a maximum temporal compression of 300 fs, in order to take snapshots of the target at different times $(0-40ps)$ after the main laser pulse interaction. A 3-lenses optical system was used to forme the target image on a 12-bits visible CCD camera. The system had a spatial resolution of several micrometers.

- In the second one, a $2-3ps$ time resolution over a $100ps$ temporal window was obtained, using a $50ps FWHM$ *chirped* probe beam [7]. After reflecting on the target, the probe was focused onto a visible spectrometer entrance slit. The linearity between wavelength and time in the chirped probe allows to consider the spectrometer spectral axis as a temporal axis. On the other axis, the spatial resolution is preserved.

Snapshots and chirped-pulse images, at a given laser energy and for a given target thickness, are shown in Figures 7 and 8. Snapshots were obtained in the same conditions (same laser energy and target type).

- Snapshots show a central reflectivity spot, which expands at a radial speed of about $2-3 \bullet 10^6 m/s$. At the center of the spot, reflectivity drops to values in the range $0.05 - 0.3$. For thin targets ($17 - 25 \mu m$), spot diameter values range from $40 - 50$ to $180 - 200 \mu m$ in the first $30 - 40 ps$. For thick targets ($25 - 400 \mu m$), diameters are greater and increase with target thickness. Comparing laser focal spot ($20 - 25 \mu m\ FWHM$) with the heated spot width at early times, a weak angular divergence can be estimated.

- Chirped pulse images show a strong reflectivity perturbation coming up some ps after the main pulse interaction. It then starts expanding with an initial radial speed of the order of $5 - 7 10^6 m/s$, which then decreases to a few $10^4 m/s$ after $30 - 50 ps$. The high time resolution of chirped pulse method allows to see in more detail the speed variation vs time. Snapshots give only an average speed, which is compatible with the chirped pulse one. In the case of very thin targets ($17 \mu m$), a perturbation is already visible (on the left in figure 9) before $t = 0$, indicating that an ASE-induced shock could breakout *before* the main fs pulse.

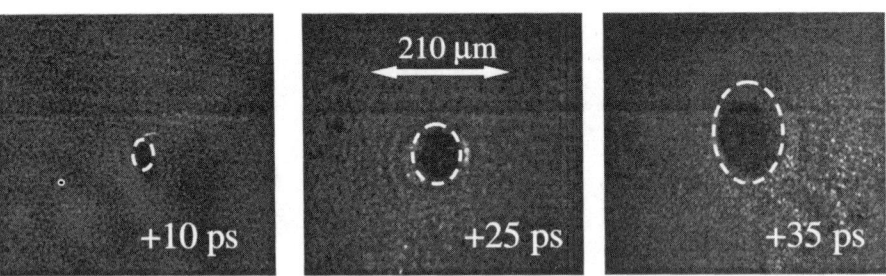

Figure 7: Rear side snapshots of a $17 \mu m$ Al target. Intensity: $8.5 10^{17} W/cm^2$. Probe beam: $300 fs$, $\lambda = 0.53 \mu m$, $45\ degrees$.

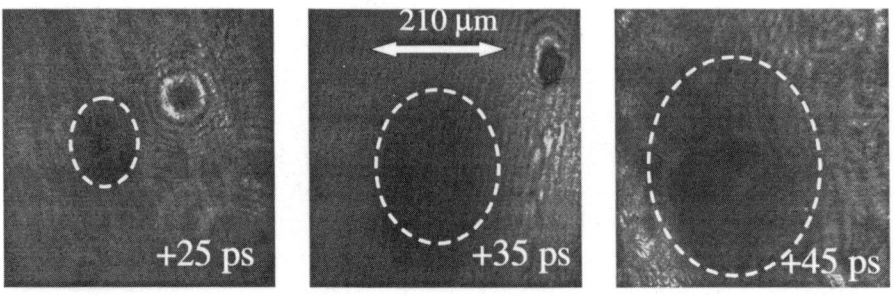

Figure 8: Rear side snapshots of a $80 \mu m$ Al target. Intensity: $5 10^{18} W/cm^2$. Probe beam: $300 fs$, $\lambda = 0.53 \mu m$, $45\ degrees$.

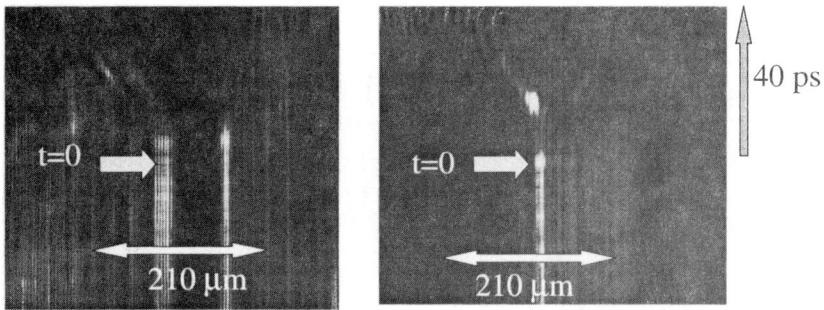

Figure 9: Time-resolved images of 17 and $35\mu m$ Al targets. Intensity: $8.5 10^{17} W/cm^2$.

In general, reflectivity drop could be due to ASE, X-ray or fast electron heating. We discussed the first two possible sources of heating as follows:

- ASE threshold intensity, above which target back is already heated *before* the arrival of the fs beam, was estimated doing hydrodynamic simulations. For this purpose, MULTI code [14] and measured LULI laser ASE parameters were used. ASE can affect the interpretation of thin target ($< 35\mu m$) at maximum intensity ($10^{19} W/cm^2$) shots. These shots have been neglected in the further analysis.

- Simple absorption estimations and energy conservation show that no significant rear side heating can be produced by an hard X-ray burst, coming from target front side.

Therefore, the increased absorption can only be explained by fast electron heating of the target. In fact, a fast($< 1ps$) energy deposition at the rear surface can provide the strong reflectivity drop observed in the first few *ps* after the shot. Two approaches are possible:

- If we neglect the hydrodynamic expansion or during the first $2ps$, when the gradient is steep with respect to probe wavelength, a Fresnel reflection model can be used. When temperature falls between 1 and $100 eV$, no precise determination is possible, because reflectivity does not vary much in this interval, as reported in figure 10.

- After the first $2ps$, this approximation no longer holds and a more detailed calculation is necessary, by taking into account hydrodynamic expansion of the rear plasma and calculating the propagation of the probe beam through it. The reflectivity was then calculated following Celliers & Ng [9] and Basko [10].

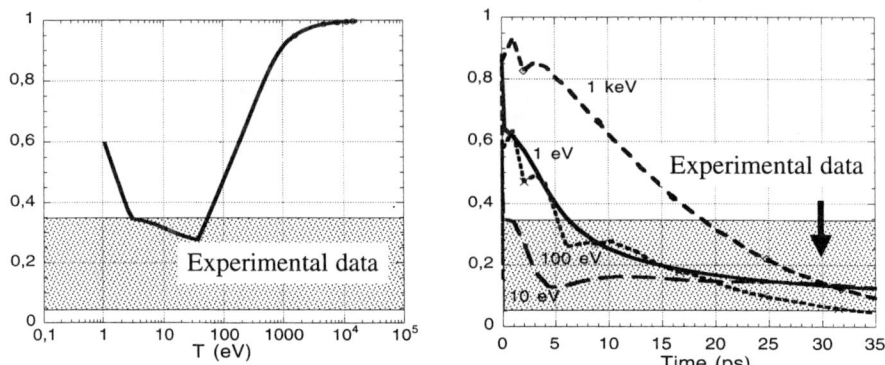

Figure 10: Steep gradient reflectivity model(1) and plasma expansion model (2). Comparison with experimental data (in error bar)

For the plasma dielectric constant, the Drude's frequency-dependent formula was used with Lee and More model for collision frequency[13]. The plasma expansion was calculated, running MULTI 1D hydrodynamic code.

Results give a good agreement with experimental data for a temperature range of $10 - 100 eV$, as in figure 10 (right). Self-emission diagnostic data give, in a preliminary analysis, temperatures of the order of 100 eV, which are coherent with reflectivity. More experimental work is needed in order to estimate more precisely fast electron heating: X-ray emission geometry and spectrum will be measured in the next experiment.

This work was supported by the E.U. TMR Laser Facility Access Program (Contract No. ERBFMGECT950044).

References

[1] M. Tabak et al. Ignition and high gain with ultrapowerful lasers. *Phys. Plasmas*, 1(5):1626, 1994.

[2] L. Gremillet et al. Time-resolved observation of ultrahigh intensity laser-produced electron jets propagating through transparent solid targets. *Phys. Rev. Lett.*, 83(24):5015–5018, 1999.

[3] M. Borghesi et al. Observations of collimated ionization channels in aluminium-coated glass targets irradiated by ultraintense laser pulses. *Phys. Rev. Lett.*, 83(21):4309–4312, 1999.

[4] L. Gremillet. PhD thesis, Ecole Polytechnique, *forthcoming in* 2001.

[5] F. N. Beg et al. A study of picosecond laser-solid interactions up to $10^{19} wcm^{-2}$. *Phys. Plasmas*, 4(2):447–457, 1997.

[6] I. Frank and V. Ginzburg. *J. Phys. USSR*, 9:353, 1945.

[7] A.Benuzzi-Mounaix et al. Ignition and high gain with ultrapowerful lasers. *Phys. Plasmas*, 1(5):3547, 1994.

[8] Y. P. Raizer Y. B. Zeldovich. Physics of shock waves and high temperature hydrodynamic phenomena. *Academic Press, New York,* 1967.

[9] P.Celliers et al. Optical probing of hot expanded states produced by shock release. *Phys. Rev. E*, 47(5):3547, 1993.

[10] M. Basko et al. Optical probing of laser induced indirectly driven shock waves in aluminum. *Phys. Rev. E*, 56(1):1019, 1997.

[11] E. Wolf M. Born. Principles of optics. vi edition. *Cambridge*, 1980.

[12] A. Benuzzi. *Generation de hautes pressions par choc laser: application a la mesure d'equations d'etat.* PhD thesis, Ecole Polytechnique, 1997.

[13] Y. T. Lee et al. An electron conductivity model for dense plasmas. *Phys. Fluids*, 27:273, 1984.

[14] R. Ramis et al. *Comp.Phys.Comm*, 49:475, 1988.

[15] F. Pisani. *Etude experimentale de la propagation et du depot d'energie d'electrons rapides dans une cible solide ou comprimee par choc laser: application a l'allumeur rapide.* PhD thesis, Ecole Polytechnique, 2000.

DEMONSTRATION OF A HYBRIDLY PUMPED SOFT X-RAY LASER

Fulvia Bortolotto

Max-Born-Institut für Nichtlineare Optik und Kurzzeitspektroskopie
Max-Born Str. 2a
D-12489 Berlin-Adlershof, GERMANY

INTRODUCTION

The experiment here presented, has been described in more detail in other papers[1,2] to which can be referred for a more complete understanding. The idea behind it is to combine Transient Collisional Excitation (TCE) X-Ray Lasers (XRLs) schemes, which exhibit high gain coefficients and short pulse duration[3,4,5,6,7], together with plasma production through fast electric capillary discharge. The latter has proved[8,9] to be extremely efficient in producing lasing at short wavelengths in a Quasi-Steady-State (QSS) regime, i.e., for our purposes, in producing plasmas with a suitable abundance of lasant ions.

In our „hybrid" scheme the capillary discharge plasma is longitudinally irradiated by a short (~1-2 ps) heating pulse allowing for the onset of a transient population inversion. Through the discharge, the electric energy from the socket is directly delivered to the material to be ionised. Thus this pumping scheme, when proved to work in a repetible and reliable way, promises to be far more efficient than the „traditional" TCE schemes, where the plasma is pre-formed by means of a long (~1 ns or sub-ns)[3,4] laser pulse focused in a line onto a slab target.

The new scheme would also prevent the use of the complicated optical arrangements needed in the „traditional" TCE schemes: the large aspect ratio of the plasma necessary in Amplified Spontaneous Emission (ASE) regimes being due to the capillary geometry and a quasi-Travelling Wave (TW) pumping being a consequence of the finite velocity of the pump pulse through the plasma. Additionally, the capillary plasma could act as a waveguide allowing it to overcome the strong refraction effects affecting the „traditional" TCE schemes[10].

Two principal requirements have to be fulfilled to produce lasing: the electron density of the pre-formed plasma has to be high enough to allow an efficient energy deposition through Inverse Bremsstrahlung (IB) by the pump laser and, at the same time, the pulse has to be guided along the whole plasma column to make use of the entire capillary length as active medium.

The „hybridly" pumped XRL, proposed in 1997[11], was supported by experiments[12] and numerical simulations[13] of guiding intense laser pulses over long distances in capillary discharge plasmas. The experiment, performed at the MBI laser facility in Berlin in collaboration with Prof. Rocca's group at Colorado State University, has given results that demonstrate the feasibility of the new scheme.

EXPERIMETAL SET-UP

The original design of the capillary discharge system[9,14] has been modified in order to achieve the plasma irradiation by the pump laser beam. Both electrodes are hollow to allow for the focusing of the pump laser at the entrance of the capillary and to record the emitted radiation at its output, i.e. in the inherent quasi-TW direction. In order to obtain the concave electron density profile necessary to guide the pump laser, the plasma is created by direct ablation of the capillary wall instead of being injected into the discharge capillary channel as in Rocca's experiment[9].

A peak current of ~ 3 kA and ~ 150 ns half cycle is obtained by discharging through the evacuated capillary a 0.08 µF capacitor bank charged up to 3-10 kV depending on the capillary length. The current flow in the capillary is measured by a Rogowsky coil to monitor the relative laser-discharge timing and to measure the discharge parameters.

Sulphur capillaries with a diameter of 0.5 mm and lengths of 10, 20 and 30 mm were used during the experiment, sulphur being to date the only solid material to show lasing when pumped in a QSS regime by pure capillary discharge[9]. The lasing transition we expect to observe is the 3p-3s J = 0-1 transition in Ne-like S at a wavelength $\lambda = 60.8$ nm.

The experimental set-up is sketched in fig.1: the CPA laser beam from the MBI glass laser system[15], compressed to a ~ 2 ps pulse with energies ranging from ~ 300 to ~ 450 mJ, was focused by a $f = 3$ m lens inside the capillary a few mm from its entrance. The focal spot, measured at a low signal level has a dimension of 170 µm (FWHM).

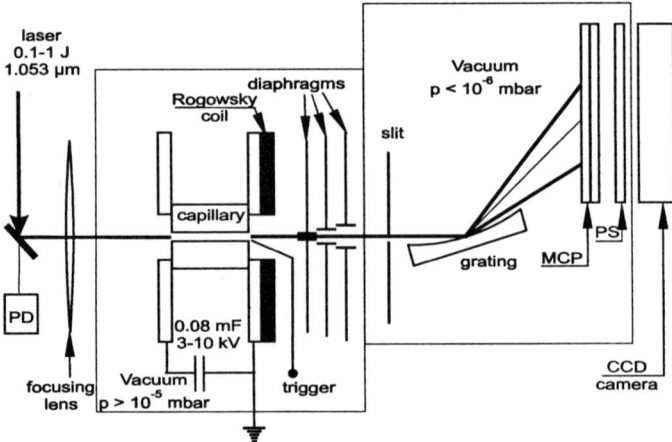

Figure 1. Experimental set-up: sketch of the capillary discharge unit and of the detection system.

The radiation emitted from the capillary output, passing through a 100 µm slit to allow for spectral resolution, is dispersed in wavelength by a flat-field, 1200 lines/mm, Al-coated, Harada-type[16] concave diffraction grating. As this grating is designed for the 5-30 nm spectral range, hence its spectral resolution at the wavelength of interest (~ 60 nm) is not very high (it has been estimated to be ~ 0.1 nm). The dispersed radiation is imaged onto a 16-bit CCD camera via a double stage MCP coupled together with a phosphorus screen. The MCP was gated within a time window of 40 ns at the flat-top in order to have a temporal resolution. The arrival time of the laser pulse was measured by means of a photodiode (marked as PD in fig.1). The timing of the incoming laser pulse, the MCP gate and the discharge current relative to one of the shots that gave lasing, is shown in fig.2.

A set of diaphragms was interposed between the capillary stage and the vacuum chamber to maintain a differential pressure between them, the MCP working pressure being below 10^{-6} mbar while the pressure in the capillary stage rises up to p ~ 10^{-4} mbar during the discharge.

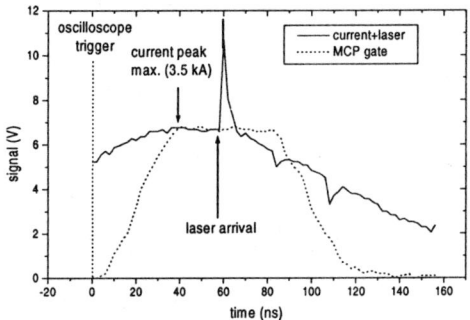

Figure 2. Timing between the discharge current, the MCP gate and the laser pulse in one of the shots showing lasing.

RESULTS

In order to define the temporal window in which the conditions for an efficient heating by the drive laser are met, we implemented a set-up that differs from that sketched in fig.1 in that the system slit-spectrograph-detector was substituted by an imaging system to monitor the transmitted laser beam. A glass wedge attenuates the transmitted beam by reflecting it (R ~ 10%) to a $f = 1$ m lens that produces a magnified image of a plane situated just behind the capillary output on a 8-bit CCD camera. Between the lens and the CCD a second glass wedge was inserted to partially reflect the beam onto a piezoelectric calorimeter to simultaneously monitor the transmitted beam shape together with a measurement of its energy. In front of the CCD a series of neutral filters were inserted that could be changed according to the expected intensity of the transmitted beam. This strongly depends on the incoming laser time relative to the discharge.

This arrangement allowed us to identify the time at which the pump laser energy begins to be strongly absorbed by the plasma as well as the time at which the plasma becomes opaque to the incident radiation (cut-off). These two instants define the temporal window in which the electron density has a distribution allowing an efficient coupling between the plasma and the heating pulse. They were found to be approximately the same for all capillary lengths. Indeed what is playing a role in the plasma behaviour is the current intensity, the current being responsible for the wall ablation processes and for the plasma dynamics.

The temporal window of interest was found to begin just after the current peak and to last 80-90 ns up to the cut-off instant. Indeed we observed lasing for all capillary lengths at laser arrival times ranging from 10 up to 60 ns relative to the current peak.

As for the main experiment, for each capillary length we recorded spectra as those presented in fig.3, where the spectrum emitted by the plasma heated by the laser pulse is presented together with a pure discharge spectrum gated at the same instant.

Fig.3 shows clearly how the pump laser strongly affects the plasma emission at 60.8 nm while it hardly changes the emission at other wavelengths. This effect increases with the capillary length as is shown in fig.4.

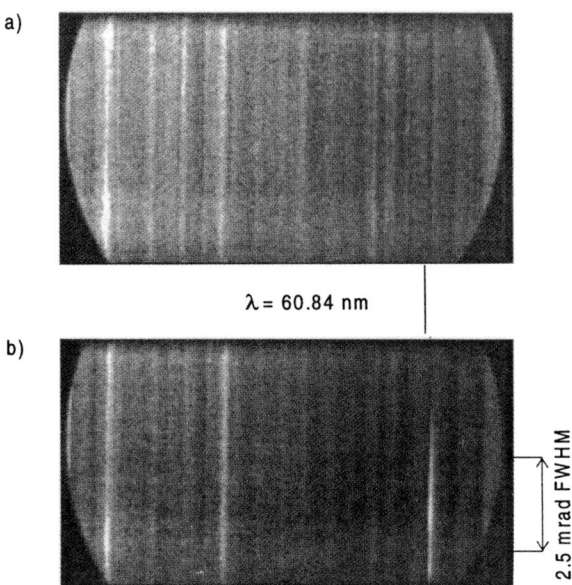

Figure 3. Image of the spectra relative to a) pure discharge and b) discharge irradiated by the laser pulse gated at the same time relative to the current peak.

Fig.3 undoubtedly shows an evidence of an additional plasma heating by the pump laser. Indeed no emission at the lasing wavelength can be observed from pure discharge excited plasma nor from the plasma produced in the capillary by the pump beam alone. Moreover, the reduced divergence of the line at 60.8 nm, that is clearly visible in fig.3, confirms the fact that we observed lasing.

Due to the longitudinal irradiation, the energy deposited by the pump laser along the plasma column and, consequently, the population inversion, will be non-uniform. This implies that the observed output would not increase exponentially with the active medium length. Hence the Linford formula[17] is no longer applicable to estimate the small signal gain unless we introduce a length dependence in the expression of the gain coefficient. A realistic model describing such a dependence would have to take into account the radial electron density profile, the plasma dynamics during the transit time of the pump pulse and the beam profile. This is, at present, beyond the scope of this paper.

In addition, due to the impossibility of controlling precisely the delay between the trigger signal and the discharge current as well as the laser energy, the experimental results for the different capillary lengths are relative to different heating conditions, making difficult a quantitative estimation of the gain coefficient[2].

The poor yield of streak-camera photocathodes at the wavelength of interest did not allow us to record time resolved spectra that could directly prove the transient character of the gain. This conclusion is anyway supported by the argument that the short duration of the heating processes induced by the pump pulse should not sustain the onset of a QSS population inversion[11,18].

It has to be remarked that in our experiment lasing has been obtained by using a pump laser energy below 0.5 J for a total energy of ~ 2 J delivered to the material.

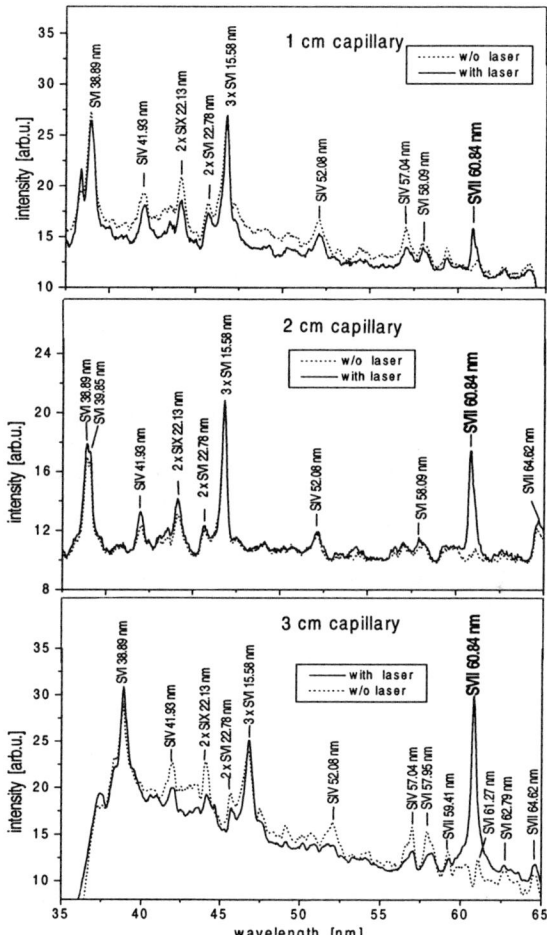

Figure 4. Spectra emitted by the S plasma for different capillary length. Emission from the plasma irradiated by the laser pulse (continuous line) is compared with that from pure discharge plasma (dotted line) gated at the same time relative to the current peak for each capillary length.

CONCLUSIONS AND OUTLOOKS

We have experimentally demonstrated that it is possible to obtain lasing at EUV wavelengths by combining a capillary discharge plasma together with the irradiation by a short laser pulse. We obtained lasing using currents one order of magnitude lower than those necessary for pure capillary discharge pumped XRLs: this may open the possibility to use the „hybrid" scheme to pump very efficiently XRLs at shorter wavelengths, not reachable at present with pure capillary discharge devices.

Future work on this pumping scheme at the MBI will be aimed to collect new data relative to Ne-like sulphur as well as to obtain lasing on higher Z elements, i.e. to scale to shorter wavelengths. For this last purpose one candidate is titanium that has already been shown[19] to reach a suitable Ne-like ions abundance. The sulphur experiment is intended to allow the comparison between the amplified outputs from capillaries of different lengths to give a quantitative, reliable estimation of the gain coefficient. The experiments at short wavelengths could also allow us to get time-resolved measurements to directly prove our idea of its transient character.

This work was supported by the European X-ray laser Network and the DFG Schwerpunkt „Wechselwirkung intensiver Laserfelder mit Materie", SPP 1053.

ACKNOWLEDGEMENTS

I would like here to thank all the researchers and the technicians of the MBI XRL group, who are actively working on this project.

REFERENCES

1. P.V. Nickles et al., Hybridly pumped collisional XRL on Ne-like Sulphur, in: *Proc. 7th Int. Conf. on X-Ray Lasers*, G. Jamelot, C. Möller and A. Klisnick eds., Editions de Physique, (in press.)
2. K.A. Janulewicz et al., Demonstration of a hybrid collisional soft X-ray laser, submitted to: *Phys. Rev. E*
3. P.V. Nickles et al., Short pulse X-ray laser at 32.6 nm based on transient gain in Ne-like titanium, *Phys. Rev. Lett.* 78:2748 (1997)
4. P.J. Warwick et al., Observation of high transient gain in the germanium X-ray laser at 19.6 nm, *J. Opt. Soc. Am. B* 15:1808 (1998)
5. M.P. Kalachnikov et al., Saturated operation of a transient collisional X-ray laser, *Phys. Rev. A*, 57:4778 (1998)
6. J. Dunn et al., Tabletop transient collisional excitation X-ray lasers, in: *SPIE Proc. Soft X-Ray Lasers and Applications III* 3776, J.J. Rocca and L.B. Da Silva eds., SPIE, Denver (1999)
7. A. Klisnick et al., Generation of a transient short pulse X-ray laser using a travelling-wave sub-ps pump pulse, in: *Proc. 7th Int. Conf. on X-Ray Lasers*, G. Jamelot, C. Möller and A. Klisnick eds., Editions de Physique, (in press.)
8. J.J. Rocca, D.P. Clark, J.L.A. Chilla and V.N. Shlyapstev, Energy extraction and achievement of the saturation limit in a discharge-pumped table-top soft x-ray amplifier, *Phys. Rev. Lett.* 77:1476 (1996)
9. J.J. Rocca et al., Lasing in Ne-like S and other new developments in capillary discharge ultrashort wavelength lasers, in: *SPIE Proc. Soft X-Ray Lasers and Applications II* 3156, J.J. Rocca and L.B. Da Silva eds., SPIE, S. Diego (1997)
10. J.L.A. Chilla and J.J. Rocca, Beam optics of gain-guided soft-x-ray lasers in cylindrical plasmas, *J. Opt. Soc. Am. B* 13:2481 (1996)
11. V.N. Shlyaptsev et al., Modelling of table-top transient inversion and capillary x-ray lasers, in: *SPIE Proc. Soft X-Ray Lasers and Applications II* 3156, J.J. Rocca and L.B. Da Silva eds., SPIE, S. Diego (1997)
12. Y. Ehrlich et al., Guiding of high intensity laser pulses in straight and curved plasma channel experiments, *Phys. Rev. Lett.* 77:4186 (1996)
13. N.A. Bobrova, S.V. Bulanov, A.A. Esaulov and P.V. Sasorov, Capillary discharge for guiding of laser pulses, *Plasma Phys. Rep.* 26:10 (2000) translated from *Fizika Plazmy* 26:12 (2000)
14. J.J. Rocca et al., Demonstration of a discharge pumped table-top soft X-ray laser, *Phys. Rev. Lett.* 73:2192 (1994)
15. M.P. Kalashnikov et al., Multi-terawatt Ti:Sa-Nd:glass dual-beam laser:a novel XUV laser driver, *Opt. Comm.* 133:216 (1997)
16. T. Kita, T. Harada, N. Nakano and H. Kuroda, Mechanically ruled aberration-corrected concave gratings for a flat-field grazing incidence spectrograph, *Appl. Opt.* 22:512 (1983)
17. G.J. Linford, E.R. Peressini, W.R. Sooy and M.L. Spaeth, Very long lasers, *Appl. Opt.* 13:379 (1974)
18. J.J. Rocca, Table-top soft x-ray lasers, *Rev. Sci. Instr.* 70:3799 (1999)
19. J.J. Rocca et al., Efficient generation of highly ionised calcium and titanium plasma columns for collisionally excited soft x-ray lasers in a fast capillary discharge, *Phys. Rev. E* 48:R2378 (1993)

X-RAY EMISSION FROM LASER IRRADIATED STRUCTURED GOLD TARGETS

T.Desai[1,2], H.Daido[1], M.Suzuk[1], N.Sakaya[1], Ariel R.Guerreiro[2] and K.Mima[1]

[1]Institute of Laser Engineering, Osaka University, 2-6 Yamada-oka, Suita Osaka 565-0871, Japan
[2]GoLP/Centro de Fisica de Plasmas, Instituto Superior Tecnico, 049-001 Lisboa Codex, Portugal

INTRODUCTION

Laser produced plasma is a potential source of X-rays and there has been a substantial work on the generation of such sources[1-4] using various laser parameters like wavelength, pulse duration, laser intensity and also target atomic numbers etc.. X-ray enhancement by micro structuring the target surface[5-9] and target configurationlike spherical cavity[10] and half cylindrical grooves[11] have been reported. A continuum X-ray spectra in 1-100 nm range can be easily produced by using a laser beam of wavelength ~1 µm, pulse duration ~10 ns, power density ~10^{12} W/cm^2 on a high "z" target materials[3] like gold (z =79). Such spectra is obtained due to various transition to N and O shells ie.on the dynamics of the populations of electron energy levels. The purpose of the experiment reported here was to study the possibility of obtaining higher X-ray yield for a given laser parameter and target material. Present results show that we can obtain almost twice the X-ray flux as compared to planar gold slab target by simply modifying the target surface. ie. the size and cost of the laser system to produce a given X-ray flux can be reduced.

EXPERIMENTS

Experiments were performed using 1.06 µm laser beam (p-polarised) of 10ns (FWHM) pulse duration with an optical energy $E_L \leq 700$ mJ for the study of X-ray emission spectra. Fig.1 shows the schematic diagram of the experimental set up.Laser radiation was incident normal to the target surface and focused to a planar spot of 100 µm in diameter on the surface of both the targets. Energy on the target surface was varied by changing the laser energy. Two types of targets were used Viz. 1) Planar gold target: Optically polished (flatness ~3000 A°) gold target of thickness ~100μm glued to aluminum base. 2) Structured gold target: Fine grooves were made on the aluminum slab surface and were coated with high purity (~99.9%) gold. The final groove dia~35μm and depth ~16μm. Photographs of the structured gold target surface are presented in Fig.2. Thickness of the gold coating was ~1.5μm and the inner surface of the groove was very smooth.

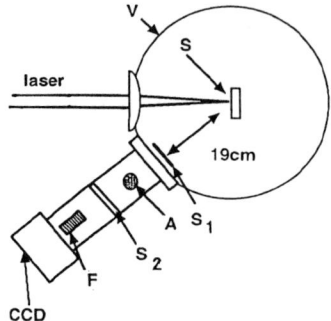

Fig.1. Schematic of the experimental set up. V-Vacuum chamber, S-Solid target, S1-External slit (100 μm width), A-Gold coated spherical mirror, S2- Slit 300 (μm width), F-Flat field grating (1200 lines/mm) and CCD- Back illumination type CCD camera.

X-ray emission spectra were recorded using grazing incidence spectrometer (GIS) in 5-22 nm spectral range and the details are described in our earlier work[12]. X-ray spectra were recorded for both the targets for an observation angle of 45° and 90° to the laser axis. We have also estimated the X-ray source size from the spatially resolved intensity profile along the vertical axis as shown in fig. 3(b) for planar target and 4(b) for structured gold target. Source size is defined as the spatial width at half the peak intensity ($I_{max}/2$) as shown in these figures.

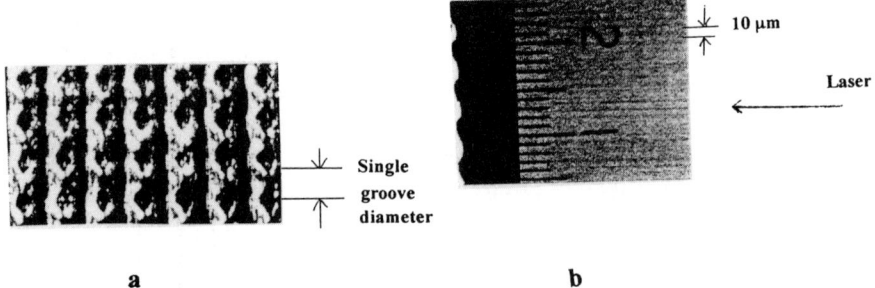

Fig.2. Photograph of the structured gold target; a) Front view and b) Transverse view.

RESULTS

X-ray yield increases with laser energy for planar and structured target but the magnitude is higher from structured target for E_L=250-700 mJ. and for both the detectors. In the following text we discuss the results for E_L=700 mJ on the target surface.

Fig. 3(a) shows X-ray emission spectra as a function of wavelength for a planar gold target and the corresponding space resolved X-ray intensity profile along the vertical axis is presented in fig. 3(b). Similarly fig. 4(a,b) describe similar values respectively for structured gold target surface. X-ray intensity (AU) as a function of X-ray wavelength is presented in fig.5 for planar 'A' and structured 'B' gold targets for an observation angle of 45° to the laser axis for $E_L \sim 700$ mJ. It is clear from the magnitude of intensity that the structured target 'B' shows higher X-ray yield as compared to planar target 'A'.Similarly an enhanced X-ray yield was also observed from structured gold target at 90° to laser axis. This indicates, there is an over all increase of X-ray yield in the structured target at all wavelengths in 5-22 nm range.

Vertical size of the X-ray source for planar gold and mesh structured groove surface show that the source size is smaller for planar target and large for structured gold target under our experimental condition. Fig 3b and 4b depict X-ray source size and X-ray intensities are in arbitrary units. The ratio of source size of mesh groove target to planar target surface varies up to ~1.3 at the incident laser energy $E_L \sim 700$ mJ. This can be visualised by comparing the spatial width AB from fig. 4b for structured (AB~ 206 µm) and fig.3b for planar (AB ~ 156 µm) targets

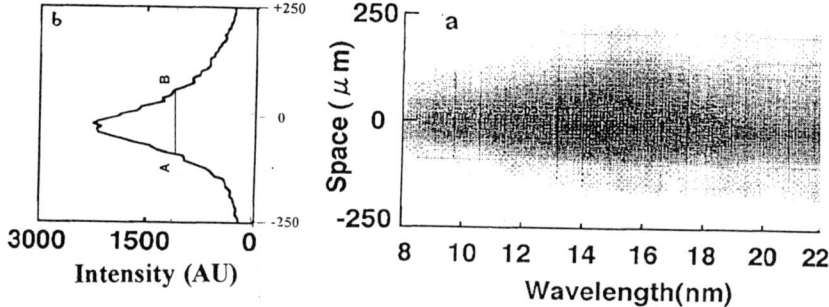

Fig.3 a) X-ray emission spectra as a function of X-ray wavelength from planar gold target surface and the corresponding b) X-ray intensity (AU) along the space resolved vertical axis.

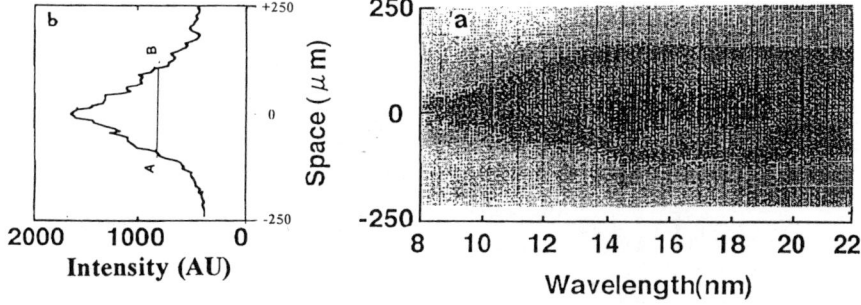

Fig.4. a) X-ray emission spectra as a function of X-ray wavelength from structured gold target surface and the corresponding b) X-ray intensity (AU) along the space resolved vertical axis.

DISCUSSION

Higher spectral intensity in the case of structured gold targets as shown in fig.5, indicates the efficiency of grooved surface as compared to planar target. Maximum X-ray yield enhancement from groove target surface was approximately 2 times at the peak of the spectra as compared to planar target and there is an over all increase at all the spectral range as seen in fig.5. X-ray source size is also larger in structured target compared to planar target. The possible explanation for the increased X-ray yield is discussed on the following aspects. 1. Absorption of laser radiation in the grooved surface due to the Classical absorption (Resonance absorption is neglected as laser energy is low). 2. A brief stagnation of colliding plasma (emanating from the curved surface of the groove) and its impact on the plasma. Enlarged source size in structured target is explained on the basis of the shape of the groove and the subsequent plasma expansion.

1. Absorption of laser radiation in structured targets due Classical mechanism

Grooved structure increases surface area. Total surface area of all the grooves (groove radius r ~17.5 µm) enveloped under the laser focal spot area (=$2\pi R^2$, Radius R is 50 µm) is the sum of the area of all the grooves (each of area ~$2\pi r^2$). Here laser irradiance with radius R=50 µm covers few full and few partial hemispheres. The sum of the area of all the grooves was found to be ≈$2\pi R^2$. This decreases the laser intensity ($I_g = I_p/2$) and plasma temperature T_g scales according to self regulating model[13-15] as per the simple calculations given below. $T_g \sim (I_g)^{4/9} \sim (I_p/2)^{4/9} \sim (T_p/2^{4/9})$. Plasma collision frequency[16] due to electrons and ions is, $\upsilon \approx ZN/T^{3/2}$, where I= laser intensity (W/cm^2), T= plasma temperature (eV), N=plasma density/cc, subscripts g,p are groove and planar targets. The charge state of the plasma Z is calculated[17] using the eqn. $Z = 2(AT)^{1/3}/3$. where A is atomic mass no.of the target. Thus for $Z_g/Z_p \sim 0.9$ and $N_g/N_p \sim 2$, the ratio of plasma collision frequency of structured to planar target is, $(\upsilon_g/\upsilon_p) \sim 2.8$. ie; high collision frequency in groove targets

Higher collision rate effectively leads to better damping (γ) of the laser beam according[16] to γ≈υ. Inverse Bremsstrahlung absorption is efficient in high z material[18] at I~10^{12}W/cm^2 and it is experimentally observed that at I ~ 10^{12}W/cm^2, reflection losses from laser produced plasma are ignorable[19]. Therefore we assume a complete absorption of laser beam by classical mechanism in the present case. Absorbed laser energy is distributed to other forms of plasma energy like ionization, thermal, kinetic, X-ray etc.. X-ray conversion efficiency[20] for I ~ 10^{12}W/cm^2 is ~18% for 10 ns laser pulse interaction with gold targets.

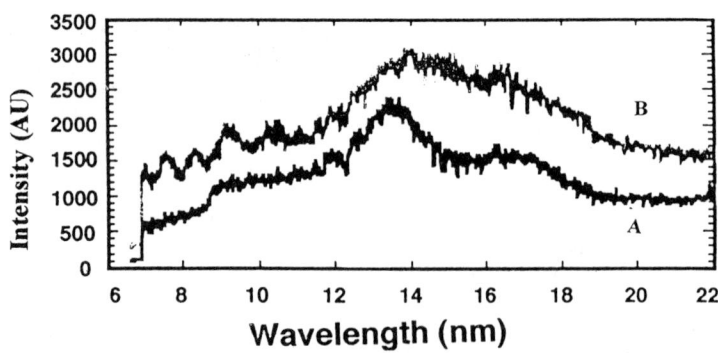

Fig.5. X-ray intensity (AU) Vs. X-ray wavelength for A) planar and B) structured gold target.

Our above calculations show that collision frequency in grooved target is significantly increased. But this will not have any effect on laser energy absorption which is~100% in the present intensity regime and hence can not explain the increase in X-ray yield. Therefore, we have tried to explain the observed ~2 fold X-ray yield in the structured target as compared to planar target in the following way by considering the plasma frequency and stagnation inside the groove structure.

2. A brief stagnation and interpenetration of the colliding plasma

Considering the curved geometry ray tracing at the groove structure, plasma expanding perpendicular to the curvature of the groove surface, from all the direction will converge in the vicinity of 'O' which is equidistant from all the points of the curved surface. To visualise this phenomenon, a simple model (with certain limitation) is presented in fig. 6. For simplicity we consider laser interaction with a single groove which can be extended to experimental condition where laser interaction with many grooves take place under the incident laser focal spot area. Fig. 6 (a) represents a single groove on which laser beam is incident. Plasma is generated along the curved surface with the on-set of laser beam at $t_p=0$ where t_p corresponds to the time of plasma production with respect to laser-target interaction which is instantaneous. Plasma expands normal to the curvature at every point as in fig 6 (b) with a velocity V_p which is assumed to be constant till it collides in the vicinity of the centre of the groove as in fig. 6(c). This will increase the plasma density at the center. We also assume that the hydrodynamic expansion of the plasma in the lateral direction is negligible inside the groove. Therefore, when plasma wave front converges near the centre, it appears like a point as in fig. 6 (c) but in reality, the colliding plasma may form a disc like structure and not a point due to hydrodynamic expansion of the plasma.

The time t_c required for a plasma to collide and to interpenetrate at the center of the groove ~ radius of the hemisphere/V_p~ nano second where V_p is plasma expansion velocity ~2×10^6 cm/sec for gold plasma[21] at T~ 80 eV and the colliding region will exists as long as the laser pulse shines on the groove surface. Plasma generated along the inner periphery of the groove moves radialy inward till it collides at the centre. Such collisions can be treated as head-on collisions. Contrary to this, plasma from the other part of the hemisphere will collide at all angles up to 90° and suffer a greater inhibition in the forward direction. Thus the plasma inside the groove gets confined. Plasma density just before the collision at the centre for an uninterrupted flow (like the plasma emanating from the edge of the groove), can be estimated considering an isothermal expansion (neglecting 2D effects), using the eqn. $n_x \sim n_c \exp(-x/L)$; at x~17 μm~radius of the groove and L= 100 μm, $n_x \sim 8 \times 10^{20}$/cc where L is the density scalelength ~ 2R, n_c is critical density. However, when the entire ablated mass coalesces at the centre, it results in an increase of the plasma density. This high density plasma expands due to its internal pressure at later times $t > t_c$.

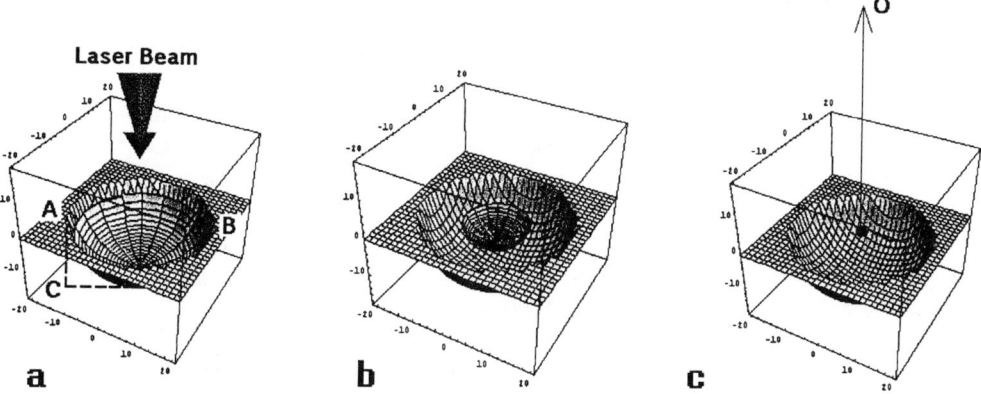

Fig.6. Schematic of the plasma expansion from a single groove of a structured gold target surface at different times. a) Initial plasma formation inside the groove surface at t_p~ 0; corresponds to the onset of laser beam on the target surface. AB= Depth and BC= Diameter of a single groove. b) Plasma expansion along the normal to the curved surface at times $t_p < t < t_c$. c) Plasma collapse in the vicinity of the groove center at t ~ t_c

The average kinetic energy $(m<V^2_p>/2)$ associated with the expanding plasma from the groove surface is calculated which is of the order of ~70 mJ where $m=2\pi R^2\rho\delta t$ where $<V>$ is average plasma velocity, ρ is mass density of gold and δt is ablation depth. Incident laser energy of 700 mJ corresponds to $I=8\times10^{11}$ W/cm^2 on planar target surface. Therefore for structured gold target surface, ablation depth[22] $\delta t\sim1.2$ μm for $I=4\times10^{11}$ W/cm^2 (as $I_g=I_p/2$). Ablation depth was estimated experimentally by measuring the hydrodynamic efficiency of a thin gold foil target using rear side calorimetric technique described in our earlier work[23]. Ablation depth is higher for gold target than low z material due to X-radiation transport.

Inter penetrating of colliding plasma in the vicinity of the centre of the groove can result in the conversion of kinetic energy in to localised thermal energy of the colliding plasma. In the first approximation we assume a complete conversion of kinetic energy in to thermal energy of the colliding plasma. As the plasma from all the sides of the groove surface converges in the vicinity of the centre at 'O', we assume that the entire groove plasma is now confined within a spot of ~4-5 μm diameter and thus the local plasma density will be ~10^{22}/cc. At the inter penetrating region electron-ion collision frequency (Kruer 1988) will be ~10^{15}/s and hence the thermalisation of the plasma can be attained within sub picosecond order. Thus the enhanced plasma density and temperature at the centre of the groove can contribute to increase X-ray yield. The shape of the spectra from structured target is slightly modified than the spectra from planar target; implies plasma ionization level is different than the plasma of planar target surface. This can occur either due to in-elastic collisions at the colliding plasma region, plasma confinement or both.

Reasons for enlarged X-ray source size

The spatial bandwidth provides information on the maximum width of the source size. Large X-ray emitting source in structured target implies larger plasma size. This can be explained on the basis of plasma expansion. Plasma expansion specially from the rim of the groove is responsible for a large plasma plume compared to the plasma expansion from a planar target. This analogy can be drawn from fig.6. It appears from fig.6 (b,c), each groove can independently generate an X-ray emitting region in two phases; firstly during the plasma formation along the target surface and at later stage, during the collapse of the expanding plasma at the center of the groove as described above and schematically shown in fig. 6 (c). Plasma from each groove can merge to form a large plasma plume resulting in to an elongated X-ray emitting source. This can also happen due to various reasons like the hydrodynamic expansion of the collided plasma at $t>t_c$, a small fluctuation in groove structure, laser energy, angle of laser incidence etc. In general, the source size for mesh grooved surface has increased up to ~30% than planar target in the present experiment for $E_L=700$ mJ as shown in fig. 4b for structured target and fig. 3b for planar targets.

Further study to estimate the distribution of absorbed laser energy into ionization, thermal, kinetic energy etc. and the possibility of some of these energy conversion into X-ray energy are proposed to carry out in the future for various groove dimensions. Present results can be advantageous as an intense X-ray source. Study of the colliding plasma from the groove has a direct relevance to the collision and interaction of astrophysical plasmas, X-ray lasers, photon acceleration, opacity calculation, hohlraum target studies where plasma expands in the opposite direction from the inner wall of the target and X-ray yield inside the hohlraum is the deciding factor of the efficacy of the hohlraum etc.

CONCLUSION

We conclude that the X-ray spectra from structured gold targets show higher X-ray yield than planar targets under identical experimental conditions. Increased X-ray yield can be attributed to plasma confinement and conversion of kinetic energy of the plasma in to the localised plasma thermal energy. Either or both phenomenon together may contribute to higher X- ray yield in the structured target. X-ray yield increases with laser energy for planar and structured target but the magnitude is higher from structured target for all intensities and at both the angles of the detector. Source size of the X-ray emitting zone is broader for structured gold target surface due to the curvature of the groove as compared to planar gold target and also due to the possibility of each groove behaving as an independent X-ray source and their subsequent merging.

REFERENCES

1. Zeng G. M,et al., J.Appl. Phys 69, 7460, 1991
2. Zeng G. M,et al., J. Appl. Phys 72, 3355, 1992
3. Zeng G. M,et al., J. Appl. Phys 75, 1923, 1994
4. Turcu C.E.et al,"Application of Laser Plasma Radiation", Proc.SPIE Vol. 2015, 243. 1994
5. Murnane M.M.et al, Appl. Phys. Letts. 62(10), 1068, 1993
6. Gordon S.P et al, Optics Letts. 19(7) 484, 1994
7. Walker T.et al, Appl.Phys. Latts, 62(10), 1068,1996
8. Nishikawa T et al, Appl. Phys. Letts., 70(13),1653, 1997
9. Nishikawa T et al, Appl. Phys. Letts., 75(26), 4079, 1999
10. Eidmann K and Kishimoto T, Appl.Phys.Lett.,49,377,1986
11. Yao-lin Li et al., Phy.Rev.A, 41,4528, 1990
12. Desai T. and Pant H.C Laser and Particle beams 17(1), 2000
13. Puell H., Z. Naturforsch, 25,1807, 1970
14. Mora P., Phys.Fluids, 25,1051, 1982
15. Shirsat (Desai)et al, Laser Part. Beams 7, 795, 1989
16. Kruer W.L,'The physics of Laser Plasma Interaction'Addison–Wesley Publ.Co.55,1988
17. Colombant D.G and Tonon G.F, J Appl. Phy. 44, 3524, 1977
18. Max C.E, Physics of Laser Fusion Vol.1 UCRL-53107, 8, 1981
19. Pant H. C.et al., Appl. Phys., 23,183, 1980
20. Babonneau D. et al. "Laser interaction and related plasma phenomenon"
    Ed. by H.Schwartz and H.Hora Vol.VI, 817, 1983
21. Pant H. C. Private communication. To be published.
22. Desai T. Ph.D thesis " Hydrodynamics of laser irradiated thin solid targets"
    Univ. of Bombay, Ch.5 X-radiation transport. 1992
23. Godwal B. K, Shirsat(Desai) T and Pant H C, J.Appl.Phys. 65(12),4608, 1989

MEASUREMENT OF SPECTRAL AND ANGULAR DISTRIBUTION OF HARD X-RAYS FROM LASER PRODUCED PLASMAS AND THEIR APPLICATION

S. Düsterer[1], H. Schwoerer[1], R. Behrens[1,2], C. Ziener[1], C. Reich[1], P. Gibbon[1], R. Sauerbrey[1]

[1] Institut für Optik und Quantenelektronik, Friedrich-Schiller-Universität
Max-Wien-Platz 1, D-07743 Jena, Germany
[2] Physikalisch-Technische Bundesanstalt, Bundesallee 100
38116 Braunschweig, Germany

INTRODUCTION

In the last few years it has become possible to compress a moderate light energy on the order of one joule into some tens of femtoseconds. This light pulse, focussed down to several µm^2, results in a peak intensity exceeding 10^{18} W/cm^2. This enormous intensity focussed onto a solid target creates relativistically moving electrons in a hot and dense plasma spot which radiates in the entire spectral range up to several MeV. In addition to the radiation, electrons and ions are emitted with rather high energies and even nuclear reactions (e.g. fusion, photofission) can take place [1-7].

We demonstrate measurements of the spectral composition and the angular distribution of the emitted hard x-rays (10 keV to 2 MeV). The absolutely calibrated x-ray spectrometers used allow us to measure absolute photon numbers in 16 channels. As a first application of these laser produced MeV x-rays neutrons via a ^{9}Be+$\gamma \rightarrow \alpha + \alpha + n$ reaction were created, demonstrating for the first time laser fission with a table top laser system.

EXPERIMENTAL SETUP AND X-RAY MEASUREMENTS

The experiments described here were performed with the multi Terawatt Ti:Sapphire laser system in Jena. The laser generated pulses of 60 fs duration, center wavelength of 800 nm, an energy of 250 mJ and a repetition rate of 10 Hz. The pulses were focused with a f/2 parabola to 4×7 µm^2, containing 50% of the energy. The p-polarized laser light impinged under 45° onto the 1.0 mm thick smoothed tantalum (Z=73) target. A peak intensity of 5×10^{18} W/cm^2 was reached. The target was moved between the shots to a new position to provide equal conditions for each shot. To get a significant reading from the thermoluminescence detector (TLD)[8] stacks, the radiation of 20,000 laser shots was accumulated in each experiment.

The hard x-ray spectrum was measured using 12 spectrometers based on TLDs (see Fig.1). These spectrometers allowed the absolute measurement of the photon fluence for photon energies between 10 keV and 2 MeV with a spectral resolution of 20%. It was carefully verified that no scattered x-rays excite the TLDs. No evidence of electrons

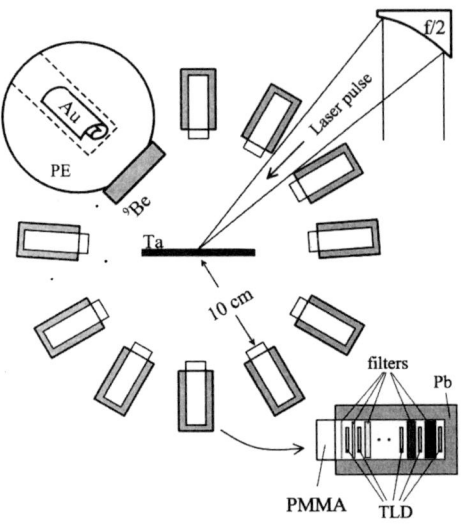

Figure 1: Experimental setup, located in a vacuum chamber, for the x-ray and the photoneutron measurements. The inset shows the alternating arrangement of filters and TLDs in the x-ray spectrometers.

penetrating through the 2 cm thick PMMA (Polymethylmetacrylate) cover into the TLD stacks was found. It was reported[9] that this can significantly change the deconvolution of the data. Contributions to the x-ray spectrum by Bremsstrahlung generated in the PMMA were determined to be less than 15% by comparing the results from several TLD stacks equipped with various front filters. The filters consisted of either low or high Z material, owing different radiation conversion coefficients (i.e. 1 MeV electron converts about 0.3% of its energy in radiation when stopped in PMMA, whereas 7% are emitted as radiation when stopped in Pb.)

Fig. 2 displays two x-ray spectra, detected by TLD spectrometers in the direction of specular reflection (solid circles) and in forward direction (open squares) with respect to the incident laser pulse. The absolute photon yield is given in number of photons per laser shot, keV and sr. From the exponential decrease of the x-ray spectrum at high energies (solid line), we calculated a hot electron temperature T_e supposing Bremsstrahlung comes from electrons with a Maxwellian energy distribution. Deconvolution of the TLD readings using a SAND II algorithm[8] yields $T_e \approx 700$ keV in specular direction and $T_e \approx 300$ keV for the forward laser direction on the back side of the tantalum. Spectra similar to Fig. 2 were recorded in twelve directions within the plane of incident laser and target normal. Fig. 3 shows the angular distribution of the electron temperatures.

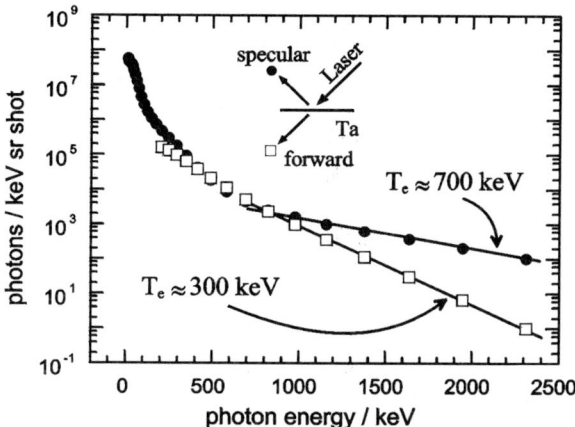

Figure 2. Measured x-ray spectra in front and behind the target are shown. The difference in the hot electron temperatures for the two directions can clearly be seen.

DISCUSSION OF THE RESULTS AND THEORETICAL CONSIDERATIONS

The transverse electric field of the laser pulse forces the electrons to a wiggling motion. Therefore the electron gains the so called quiver energy. PIC simulations show that this quiver energy equals – derived for perpendicular incidence - the hot electron temperature $k_B T_e$ [1]

$$k_b T_e \approx m_e c^2 \left(\sqrt{1 + 7.28 \times 10^{-19} (I\lambda^2)} - 1 \right). \tag{1}$$

For 45° angle of incidence and p-polarized light a similar scaling law ($k_b T_e \propto (I\lambda^2)^{1/3}$) is in good agreement with several experiments[10]. Equation (1) predicts an electron temperature of 420 keV for a laser intensity of 5×10^{18} W/cm^2 which is consistent with our measurement, where the hot electron temperature varies between 200 keV and 700 keV (see Fig. 3). However, the total yield of electrons, photons or secondary created particles depends on various experimental parameters such as target element, target surface properties, density scale-length of plasma generated by laser prepulses and finally laser polarization and direction of emission.

The angular distribution of the x-rays points to some interesting features. Bremsstrahlung from weakly relativistic electrons ($E_\gamma \leq 0.5$ MeV) shows no significant deviation from isotropic emission (circles in fig. 4). This may be attributed to multiple scattering of the electrons in the target and the dipolar emission character of the Bremsstrahlung. Therefore no direction is favored. This behavior changes distinctly for x-ray energies greater than the electron rest mass ($E_\gamma \gg 0.5$ MeV). Bremsstrahlung emission of electrons in this energy range is peaked in the propagation direction of the electrons (triangles in fig. 4).

This suggests that the detected angular distribution of MeV radiation resembles the distribution of the highly relativistic electrons. Our data therefore imply the existence a jet of highly energetic electrons close to the specular reflection direction of the laser. Similar effects – but for other conditions and due to different mechanisms - were seen by other groups using laser powers as low as 0.5 TW (50mJ, 120fs) [13] and as powerful as

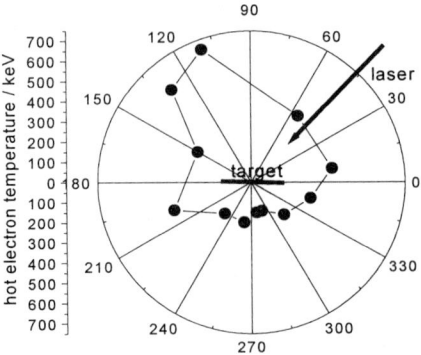

Figure 3 Angular dependence of the hot electron temperature derived from the measured x-ray spectra.

100TW (50J, 500fs)[14]. As experimentally verified by ref. 1 and 13 we assume 0.5-1% energy transfer from the laser pulse to suprathermal electrons. Yu et al.[15] performed PIC simulations with parameters similar to our experiment, but with perpendicular incidence. They found more than half of the suprathermal electrons that are created in the interaction were accelerated in back direction by the reflected laser light. Assuming a similar fraction of electrons was accelerated in the specular direction for our geometry and taking into account the measured T_e, one finds that the plasma density in front of the target, required to generate the measured x-ray yield in the specular direction is $n_i \approx (10^{18}\text{-}10^{19})$ cm^{-3}, for a plasma volume of $(10 \text{ µm})^3$ (laser focus × the preplasma expansion within a few ps). This is a realistic preplasma density generated by the unavoidable prepulse of TW Ti:sapphire laser systems.

APPLICATION OF THE MEV X-RAYS: PHOTONEUTRONS

We used the MeV x-rays to initiate photonuclear reactions. Neutrons were produced by a (γ, n)-reaction in Beryllium via $^9\text{Be}+\gamma \to \alpha + \alpha + n$. The Beryllium was split by γ (i.e. x-rays from the laser produced plasma) exceeding the threshold energy of 1.67 MeV.

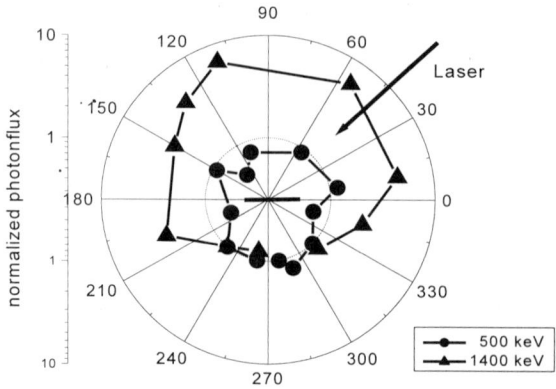

Figure 4: Logarithmic plot of the photon flux (arbitrarily normalized to the flux at 235°) for weakly (500 keV) and stronger relativistic (1.4 MeV) photon energies. The asymmetry for higher photon energies can clearly be seen.

We placed a Beryllium disk (12 mm thick, 70 mm diameter) 4.5 cm away from the laser plasma in specular direction of the incident laser, covering a solid angle of 1.3 sr (Fig. 1). The Be disk was mounted on the surface of a polyethylene sphere (Bonner sphere) of 12.5 cm diameter. About 40% of the neutrons generated in the Be disk entered the Bonner sphere and were thermalized by inelastic scattering with hydrogen and carbon nuclei. The neutron thermalization efficiency of the Bonner sphere was absolutely calibrated for the relevant neutron energies. A gold disk (^{197}Au, 16 g) was placed in the center of the Bonner sphere, capturing thermal neutrons ^{197}Au + n → ^{198}Au* with a cross section of 99 barn. ^{198}Au* decays by emission of a 411.8 keV γ with a half-live of 2.7 d. The γ-spectrum was collected by a Ge-detector located 1 km below ground level for reduction of Au excitation by cosmic neutrons and low γ background.[11] After 20,000 laser shots the activity of the ^{198}Au* due to thermalized photoneutrons exceeded the natural background activation of gold by a factor of 2.5. This corresponds to about 100 photoneutrons per shot after correction for solid angles, thermalization probability in the Bonner sphere, capturing cross section of Au and Ge-detector response. To our knowledge this was the first observation of a (γ,n)-disintegration with a laboratory size laser system.

From the number of neutrons generated in the Beryllium disk, one can infer the total number N_P of photons with energies larger than 1.67 MeV emitted from the relativistic plasma: Using an averaged photoneutron cross section of 0.5 mbarn above 1.67 MeV, we calculated $N_{photoneutron} = 2.5 \times 10^4$ (sr^{-1}), which corresponds well to the integrated photon yield of $N_{TLD} = 1 \times 10^5$ in the hot electron wing above 1.67 MeV achieved in the TLD measurement in specular direction.

In our measurement the neutron yield was used to cross check the TLD readings, whereas in interactions with more energetic laser pulses (resulting in a higher photon flux) and higher intensities (increased hot electron temperature) the photon fluxes above 2 MeV will dramatically increase. Since the detection range of the TLDs ends at about 2 MeV one has to think of a different method. Photonuclear reactions with well defined thresholds and well known reaction cross-sections are an interesting substitute. At the Lawrence Livermore National Laboratory[16] the γ-spectrum of the interaction of a PW laser pulse, focussed to 10^{20}W/cm^2 was characterized up to 20 MeV by the photonuclear activation of Ni and Cu slabs placed next to the target.

Another way to apply (γ,n) reactions is to activate nuclei not to use them as detectors but as a source for medical applications. So far, isotopes for medical x-ray examinations (e.g. radioactive iodine is used for thyroid gland diagnosis) have to be produced in large scale facilities (accelerators or reactors) and transported to the individual hospital. Therefore isotopes with long half-lives have to be used. Since the x-ray examinations require a certain activity during the measurement. The total dose absorbed by the patient increases exponentially with the half-live of the isotope. Therefore it seems to be favorable to use fast decaying isotopes instead, but they would have to be produced in the hospital. This could be done using powerful lasers. Current lasers cannot produce enough neutrons, but extrapolating the progress in laser development it should be possible to have turn-key table-top lasers producing enough isotopes for everyday application.

CONCLUSION

In conclusion we have performed the first measurement of the complete angular distribution in the plane of incidence of the hard x-ray spectrum of a relativistic, laser produced plasma, generated on a tantalum surface with a laser intensity of 5×10^{18} W/cm^2. We showed the strongly anisotropic x-ray emission characteristics at relativistic photon energies by Bremsstrahlung generation of hot electrons reflected on the target surface.

Utilizing the detailed knowledge of the x-ray spectra we induced the photonuclear reaction ^{9}Be+$\gamma \rightarrow \alpha + \alpha + n$ by MeV photons from the laser produced tantalum plasma and demonstrated the neutron capture reaction ^{197}Au + n $\rightarrow$ ^{198}Au* with our compact laser neutron source.

The authors thank H. Langhoff and R. Nolte for stimulating discussions and gratefully acknowledge technical assistance by F. Ronneberger. The work was funded by the Deutsche Forschungsgemeinschaft (Schw 766/2-2 and Gi 300/1-1).

REFERENCES

1. T. Feurer et al., Onset of diffuse reflectivity and fast electron flux inhibition in 528- nm-laser-solid interactions at ultrahigh intensity, *Phys. Rev. E* 65:4608 (1997)
2. E. Krushelnick et al., Multi-MeV ion production from high-intensity laser interactions with underdense plasmas, *Phys. Rev. Lett.*, 83:737 (1999).
3. T. Ditmire et al. , Nuclear fusion in gases of deuterium clusters heated with a femtosecond laser, *Phys. Plasmas*, 7:1993 (2000).
4. J.D. Kmetec et al., MeV x-ray generation with a femtosecond laser, *Phys. Rev. Lett.*, 68:1527 (1996).
5. K. W. D. Ledingham et. al., Photonuclear physics when a multiterawatt laser pulse interacts with Solid targets, *Phys. Rev. Lett.*, 84:899 (2000)
6. T. E. Cowan et. al., Photonuclear fission from high energy electrons from ultra intense laser solid interactions, *Phys. Rev. Lett.*, 84:903 (2000)
7. M. H. Key et. al.,Hot electron production and heating by hot electronsin fast ignitor research, *Phys. Plasmas*, 5:1966 (1998)
8. R. Nolte et al. , A TLD-based few-channel spectrometer for x-ray Fields with high fluence rates, *Rad. Prot. Dos.*, 84:367 (1999).
9. M. Schnürer et al. , Dosimetric measurements of electrons and photon yields from solid targets irradiated with 30 fs pulses from a 14 TW laser, *Phys. Rev. E,* 61:4394 (2000)
10. P. Gibbon and E. Förster, Short-pulse laser-plasma interactions, *Plasma Phys. Control. Fusion*, 38:769 (1996).
11. S. Neumaier et al., The PTB underground laboratory for dosimetry and spectroscopy , *Appl. Rad. and Isotopes,* 53:173 (2000).
12. M.J. Jakobbson, Photodisintegration of Be^9 from threshold to 5MeV, *Phys. Rev.*, 123:229(1961)
13. S. Bastani et al., Experimental study of the interaction of subpicosecond laser pulses with solid targets of varying initial scale length, *Phys. Rev. E*, 56:7179 (1997)
14. R. Kodama et al., Long-scale jet formation with specularly reflected light in ultraintense laser-plasma interactions, *Phys. Rev. Lett.*, 84:674 (2000)
15. W. Yu et al., Electron acceleration by a short relativistic laser pulse at the front of solid targets, *Phys. Rev. Lett.*, 85:570 (2000)
16. T. W. Phillips et al., Diagnosing hot electron production by short pulse, high intensity lasers using photonuclear reactions, *Rev.Sci. Inst.*, 70:1213 (1999)

EQUATION OF STATE OF WATER IN THE MEGABAR RANGE

[1] Dipartimento di Fisica G. Occhialini
Universita' degli Studi di Milano-Bicocca
Via Emanueli 15, 20126 Milano, Italy
[2] L.U.L.I., C.N.R.S., Ecole Polytechnique
91128 Palaiseau, France
[3] C.E.A. Limeil, France
[4] Lawrence Livermore National Laboratory, U.S.A.
[5] University of Essex, U.K.
[6] C.I.N.E.C.A., Bologna, Italy

E. Henry[1,2]*, D. Batani[1],
M. Koenig, A. Benuzzi[2],
I. Masclet, B. Marchet, M. Rebec, Ch. Reverdin[3],
P. Celliers, L. Da Silva, R. Cauble, G. Collins[4],
T. Hall[5],
C. Cavazzoni[6]

INTRODUCTION

Astrophysical context

The mantle of Neptune and Uranus is mainly constituted by ice layers containing water, methane and ammonia. The magnetic field of both planets, as measured by the probe Voyager 2, is larger than what was expected and asymmetrical, hence originates from the conductivity of those layers. The range of temperature and pressure in the ice layers is of 0.2 to 6 Mbar and 2,000 K to 8,000 K. Estimations of the minimum conductivity capable of sustaining the magnetic field by dynamo effect give about 200 $(\Omega.cm)^{-1}$. Recent calculations predict a transition from electrolyte to metal in this regime for water and ammonia [1]. Yet no measure has so far confirmed its existence.

Goals of our experiment

We describe here the first experiments measuring the equation of state (EOS) of water with laser driven shock waves in the range of pressure 1-10 Mbar. This technique of measurement has been much improved in recent years [2] and has become a reliable tool for high-pressure physics [3]. The first experiments were conducted at the Commissariat a l'Energie Atomique in Limeil, France, and completed later on by experiments on the LULI laser, at the Ecole Polytechnique, France. Both were funded by the European Union in the framework of the *Access to Large Scale Facilities* programme.

Our goals were to obtain new experimental points for the EOS of water in the Megabar range, as well as to estimate the conductivity in this range of pressures.

EXPERIMENTAL SET-UP

The experiments are based on the impedance mismatch method, where the shock velocity is simultaneously measured in two different materials, one of which used as a reference. We chose Aluminium, as its EOS is well known up to 40 Mbar [4]. As both experiments are very similar, we only describe the set-up for the experiment performed at LULI. The parameters for each set-up are compared in table 1.

Table 1. Experimental parameters for the Limeil and LULI set-ups

		Limeil	LULI
	Wavelength λ	0.532 µm	0.532 µm
	Square pulse duration	4 ns	600 ps
High power	Maximum energy	2000 J	100 J
laser	Optical smoothing	Kinoform Phase Plate	Phase Zone Plate
	Focal spot	900 x 600 µm^2	400 x 400 µm^2
	Maximum intensity	$1.4 \cdot 10^{14}$ W.cm^{-2}	$8 \cdot 10^{13}$ W.cm^{-2}
Probe	Wavelength λ	1.064 µm	0.537 µm
beam	Square pulse duration	10 ns	8 ns

We used 3 beams of the LULI laser (cf. figure 1a) in the green, optically smoothed with Phase Zone Plates (PZP), and focused on the target (cf. figure 1b).

A probe beam with a longer duration and very little energy (a few mJ) was reflected on the rear-side of the target. The reflected beam was sent to two velocity interferometers VISAR (Velocity Interferometer Systems for Any Reflector) [5] which measure the velocity of the rear-side of the target: the Doppler effect at the reflecting surface induces a shift in wavelength for the probe beam, hence modifies the interference pattern. The sideways shift of the fringes is proportional to the velocity of the reflecting surface. To resolve the ambiguity on the initial shift of the fringes at the shock arrival [6] we used two VISAR with different sensitivities (16.7 km.s^{-1} and 3.4 km.s^{-1} per fringe respectively) coupled with streak cameras. Assuming that water is metallised, the probe beam crosses cold water and gets reflected on the shock front. Evaluating the Doppler effect at the interface brings to a modification of the usual VISAR formula [7] involving the refractive index of "cold" water n at the wavelength λ of the probe beam:

$$V(t) = F(t) \frac{\lambda}{2\tau n},$$

where F(t) is the fringe shift, τ the delay introduced by the interferometer, and V(t) the velocity measured.

To image the rear-side of the target, we used an image relaying system to avoid vignetting (i.e. a luminosity drop on the borders of the image). The two streak cameras coupled with the VISAR had a spatial resolution of 4 µm and a temporal resolution of 10 ps. A third streak camera with a temporal resolution of 4 ps was used to record the target self-emission (VDC).

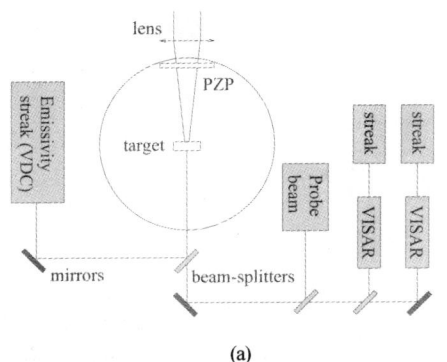 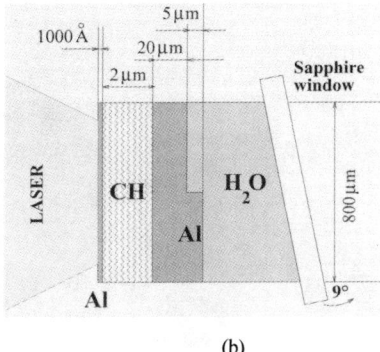

(a) (b)

Figure 1. Experimental set-up and water targets for the LULI experiment.
a. Set-up: the high-power laser beam is optically smoothed by PZP and focused on the target; a probe beam is reflected on its rear side and analysed by two velocity interferometers (VISAR) with different sensitivities; the emissivity of the target is recorded by a third diagnostic.
b. Target: a layer of CH is used to minimise preheating ahead of the shock, and a thin Al foil avoids laser shine-through at early times. The Al layer includes a step of thickness h = 5 μm.

The targets consist of an Aluminium step of height h (typically a few microns) and a cell filled with water (cf. figure 1b). To minimise preheating of the target, a layer of plastic (CH) was added on the front-side; finally, a very thin Al foil was placed to avoid laser shine-through at early times.

EXPERIMENTAL RESULTS

The transit time Δt of the shock in the Aluminium step is measured with the VDC: when the shock breaks through, Al begins to emit, and a visible signal is detected (fig. 2a). This measurement is confirmed with the Visars (fig. 2b & c). Hence, knowing the thickness h of the step (and assuming that the shock is stationary - cf. figure 3a), the shock velocity in Al can be simply calculated as $D_{Al} = h / \Delta t$.

The use of the Visars diagnostics can only be done if we know where the probe beam gets reflected. If water is not metallised, the probe beam crosses compressed, ionic water and is reflected by the Al layer. In this case, the Visars measure the fluid velocity. If on the other hand water is metallised, it reflects the probe beam, and we measure the velocity of the shock front.

For all our shots, we could say that the probe beam was reflected on metallised water. Indeed, calculations by Cavazzoni [8] predict that the Hugoniot curve reaches the metallic state above approximately 1 Megabar. Moreover, the high reflectivity measured indicates the metallisation of water. Finally, assuming that water was *not* metallised leads to EOS points very far from the Sesame table [4] or even in regions that are not physically possible (such as negative density).

The shock velocity in water is thus measured by the fringe shift on the Visars (cf. figure 2b & c). We used an image processing code to reconstruct the interference pattern from the images.

Once that we have measured two parameters - the shock velocity in Al and Water - we can use the impedance mismatch method: when the shock arrives at the interface between Al and Water, a shock is transmitted in Water and a relaxation wave propagates backwards in Al. As the EOS of Al is well known, we can compute the parameters of the relaxation wave in Al. Then, stating that pressure and fluid velocity have to be equal on each side of the interface, we obtain a point on the EOS of Water.

The results for the EOS are shown on table 2 and figure 4; the agreement with the Sesame table [4] is good.

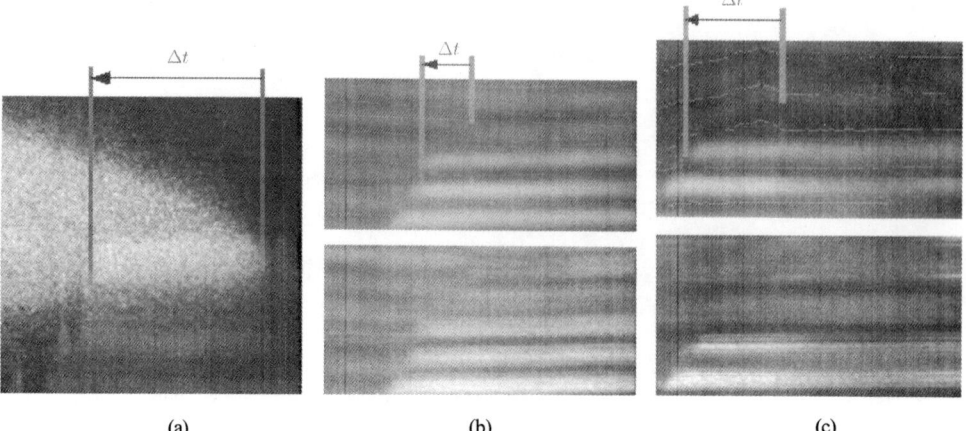

(a) (b) (c)

Figure 2. Experimental images from shot # 60:
a. Target self-emission. We measure the transit time Δt of the shock in the Al step.
b & c. Images from the two Visars and the processed images: we measure the fringe shift, get an independent measurement of the shock transit time in the Al step, and measure the reflectivity of the target.

In our experiment we also measured the reflectivity of the interface between metallised and cold water by comparing the signal from shocked and unperturbed regions of the target (cf. figure 3b). High reflectivities of the order of 50% are observed.
From the experimentally measured reflectivity, electromagnetism allows calculating the conductivity of metallised water. This conductivity is of electronic nature, as it is the response to an optical signal (i.e. of typical frequency 10^{15} Hz). Results are shown on figure 5a & b. Conductivities of the order of 10^3 $(\Omega.cm)^{-1}$ are observed, well in agreement with the hypothesis of water metallisation.
We also observe a decrease in conductivity as pressure increases, and this can be fairly well reproduced by a semi-classical formula for conductivity:

$$\sigma \propto \frac{\rho^{2/3} Z^*}{\sqrt{T}},$$

where the ionisation Z^* has been computed with the Thomas-Fermi model (cf. fig. 5b).

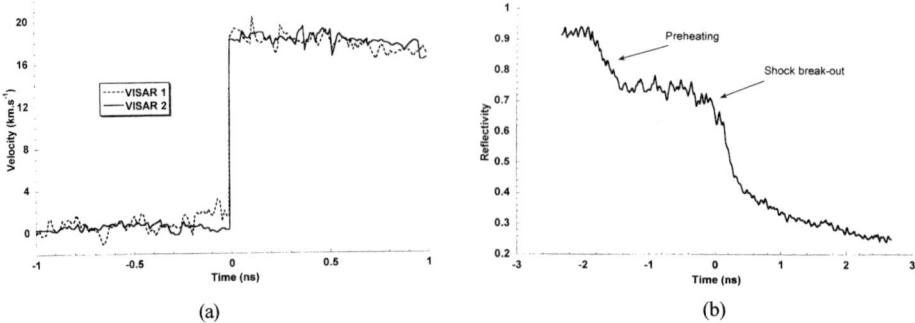

(a) (b)

Figure 3. Measurements from shot # 60:
a. Shock velocity profile measured with the VISARS from the images of figure 2b. The shock remains stationary for about 1 nanosecond.
b. Reflectivity of the target measured from the images of figure 2b. Initially, the reflectivity decreases because of a small preheating of the target. Immediately after the shock breakout, its value is of about 0.45.

Table 2. Experimental results

Shot #	D Al (km.s^{-1})	D water (km.s^{-1})	P water (Mbar)	Conductivity σ (Ω.cm)$^{-1}$
59	18.7	19.2	2.7	
60	18.15	18.5	2.4	
61	14.4	15.3	1.4	
63	12.5	10.7	0.8	7 10^3
65	19.6	20.9	3.1	2.5 10^3
66	17.2	19.1	2.3	2.7 10^3
67	11.8	11.3	0.8	2 10^4
L1	20.3	26.1	3.8	210^3
L2	19.4	25.9	3.6	2 10^3

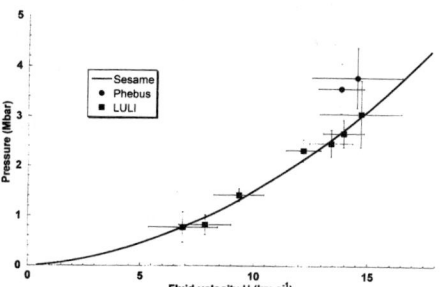

Figure 4. Experimental points for the EOS of Water and comparison with the Sesame table [4].

CONCLUSIONS

The results presented here are the first measurement of the EOS of water with laser driven shock waves. This method proved to be a reliable tool for this measurement, as new points have been obtained in good agreement with previous data.

We evidenced a high electronic conductivity of Water at P $\geq$ 0.8 Mbar and T $\geq$ 0.4 eV, in good agreement with calculations by Cavazzoni [8] which predict the transition from ionic to metallic conductor above about 1 Megabar along the Hugoniot curve.

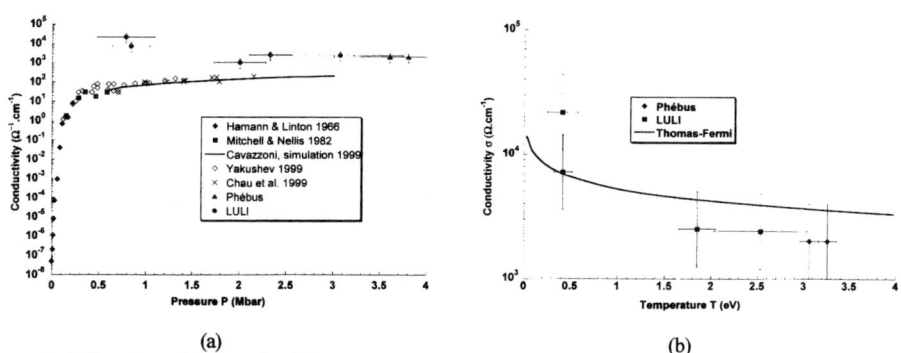

Figure 5. Estimation of the conductivity:
a. Comparison with measurements of the conductivity along an isentrope.
b. Comparison with a semi-classical formula for conductivity, where ionisation is computed with the Thomas-Fermi model.

Acknowledgements

This work was supported by European Programmes E.U.TMR, under the contract ERBFMGECT 950016.

References

[1] C. Cavazzoni, G. Chiarotti, S. Scandolo, M. Tosatti, E. Bernasconi, and M. Parrinello, Superionic and metallic states of water and ammonia at giant planet conditions, *Science* **283**, pp.44-46, 1999.
[2] M. Koenig, B. Faral, J. Boudenne, D. Batani, A. Benuzzi, S. Bossi, C. Rémond, J. Perrine, M. Temporal, and S. Atzeni, Relative consistency of equations of state by laser driven shock waves, *Phys. Rev. Lett.* **74**(12), pp. 2260-2263, 1995.
[3] Y. Zeldovich and Y. Raizer, *Physics of shock waves and high temperature hydrodynamic phenomena*, Academic Press, New York, 1967.
[4] SESAME : Report LA-UR-92-3407. Los Alamos National Laboratory,1992.
[5] P. Celliers, G. Collins, L. Da Silva, D. Gold, and R. Cauble, Accurate measurement of laser-driven shock trajectories with velocity interferometry, *Applied Phys. Lett.* **73**(10), pp.1320-1322,1998.
[6] R. Trunin, Shock compressibility of condensed materials in strong shock waves generated by underground nuclear explosions, *Phys.-Usp.* **37**(11), pp.1123--45, 1994. Uspekhi Fizicheskii Nauk **164** (11) 1215-1237.
[7] L. Barker, Laser interferometer for measuring high velocities of any reflecting surface, *J. Appl. Phys.* **43**(11), pp.4669-4675, 1972.
[8] C. Cavazzoni. Unpublished. Calculation of Hugoniot points for Water.

XUV INTERFEROMETRY USING HIGH ORDER HARMONICS: APPLICATION TO PLASMA DIAGNOSTICS

J.-F. Hergott

CEA/DSM/DRECAM, Service des Photons, Atomes, et Molécules
Centre d'Etudes de Saclay
91191 Gif-sur-Yvette, France
hergott@drecam.cea.fr

High order harmonic generation in gases is now a widely studied non linear process. The research in this domain shows two main directions. On the one hand, researchers are working on the different fundamental aspects involved in high order generation. On the other hand, they have developed several applications of this unique XUV source. High order harmonics (HOH) present a good beam quality, which allows to focus the harmonic radiation to very tight spots[1-3], a high brightness and a high repetition rate (up to 1 kHz)[4-6]. Since high order harmonic generation is a field driven process, the properties of the laser that is used to generate the harmonics are partially transferred to the XUV radiation. As a result, the harmonic pulse presents a short duration, equal to or shorter than the generating laser pulse duration[7,8] depending on the harmonic order. The harmonic radiation is also naturally synchronized with the fundamental driving IR laser. These two last properties are very important for all experiments of the pump probe type. Up to now high order harmonics have already been used in atomic and molecular spectroscopy[9,10]. This type of experiments can also be applied in solid state physics for instance to probe the relaxation time of electrons that have been excited in the conduction band of a dielectric material[11]. Another important property of the harmonic source is the good temporal[12] and spatial[13-15] coherence that can be obtained in some generation conditions. Moreover, a very unique property in this XUV range is the possibility to produce two mutually coherent sources either separated in space or in time. In the following, we discuss the generation of these phase-locked harmonic sources and their applications to plasma diagnostic.

In 1997, Zerne et al.[16] have demonstrated the possibility to produce two phase locked harmonic sources, separated *in space* by splitting the generating fundamental laser beam into two parts focused at two different places in a gas jet. The two harmonic beams that are produced interfere coherently when they overlap in the far field due to diffraction. Let us recall briefly the basics of interference[17]. Consider two point sources S_1 and S_2 emitting monochromatic spherical waves of same frequency. The separation a between this sources should be greater than their wavelength λ. The two waves can be written as

$E_i(\vec{r},t) = \frac{E_{0i}}{r}\cos(\vec{k}\cdot\vec{r}_i - \omega t + \varphi_i)$ with i=1 and 2. If the observation point P is far enough, the irradiance at this point is proportional to the time average of the magnitude of the electric field intensity squared: $I = \langle(\vec{E}_1 + \vec{E}_2)^2\rangle = I_1 + I_2 + 2\sqrt{I_1 I_2}\langle\cos\delta\rangle$ where $\delta = k(r_1 - r_2) + (\varphi_1 - \varphi_2)$. If the relative phase $(\varphi_1 - \varphi_2)$ of the two waves varies quickly in time (uncorrelated sources), then $\langle\cos\delta\rangle = 0$: the interference term vanishes and the irradiance is constant in the observation plane. If the relative phase $(\varphi_1 - \varphi_2)$ of the two waves remains constant in time (phase locked sources), $\langle\cos\delta\rangle = \cos\delta$ and a fringe pattern is observed in the far field. In a Young's two slit experiment, a monochromatic spherical wave illuminates two parallel narrow slits. These two slits produce two phase locked secondary waves that can interfere in the far field. The irradiance can then be expressed as $I = 4I_0 \cos^2 ya\pi/d\lambda$, where d is the observation distance, y the dimension perpendicular to the fringes and a the separation between the two slits. When splitting the laser beam into two spatially separated beams, we in fact perform a Young experiment at the laser frequency. Since high order harmonic generation is a field driven process, the phase locking is transferred to the generated harmonic beams giving rise to fringes in the harmonic far field pattern

Figure 1a shows the experimental setup that has been used to produce the interference of the two spatially separated harmonic sources and the two application experiments. The IR laser pulse is split into two identical pulses with a Michelson interferometer. One of the arms is slightly tilted so that the laser beams are focused in the krypton gas jet at two spatially separated foci, where two harmonic sources are generated. A monochromator (spherical grating coupled to exit slits in the image plane) is used to select a given harmonic order. After the slit, the two harmonic beams diverge and interfere in the far field. The fringe pattern can be recorded either with a charge coupled device (CCD) camera or with micro channel plates (MCP) coupled to a phosphor screen and a CCD. The fringe pattern resulting from the far field interference exhibits a fringe period $\Delta i = \lambda d/a$ (see above).

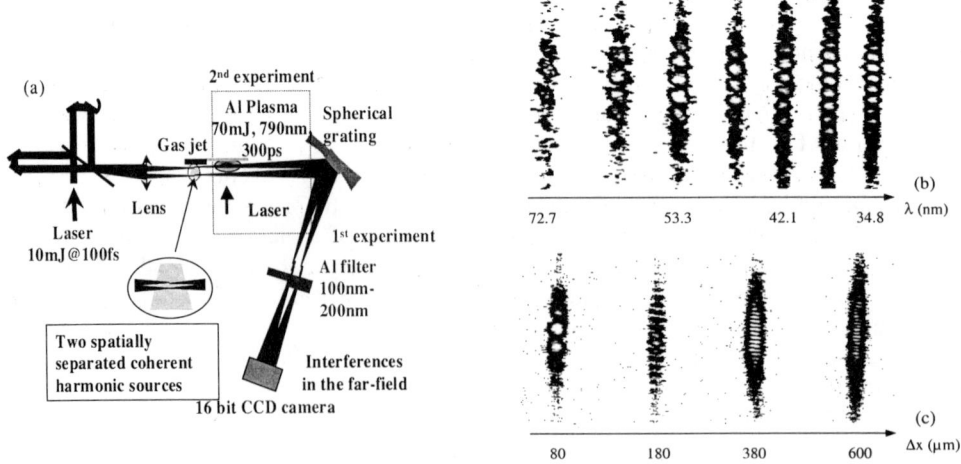

Figure 1 a) Experimental setup to produce two phase locked harmonic sources and application experiments. Fringe pattern variation: b) as a function of the harmonic order for a 100 μm separation between the sources c) of the 17th harmonic as a function of the separation between the sources.

In Figure 1b, one can see the linear variation of the fringe period as a function of the harmonic wavelength for two harmonic sources that are separated by a fixed distance a=100μm. The fringe contrast is better than 50% over a large wavelength range and the limitation is given by the MCP resolution. Note that the recorded cross section of the beam in the vertical dimension is very large because we have tilted the MCP to a grazing incidence of 8° in order to increase the effective spatial resolution. We also observed the inverse linear variation of the fringe period as a function of the separation distance a between the two harmonic sources for the 17th harmonic as shown in Fig. 1c. One can observe a good fringe contrast even when the two sources are quite far from each other (a=600µm). This gives evidence of a good mutual coherence of the two harmonic sources, even for a large spacing that can be very useful for applications.

The *spatial* interferometry should be a powerful tool for probing dephasing objects[18]. Indeed since the two harmonic beams are clearly separated by some 100µm we can put a dephasing object in one of the arms without perturbing the other one. That was the aim of the first experiment we performed at the Lund Laser Center. We have probed the thickness of a free standing aluminum (Al) step filter using harmonics 9 to 15. We chose aluminum because of its good transmission in this spectral range and its refractive index different from unity which made it possible to see a fringe shift. The filter is composed of a 100-nm layer coupled with a second one, of same thickness, that is covering half part of the first one, in order to build a step. To study the step region we use one harmonic beam which section covers the step (probe beam) whereas the second one passes through one layer (reference beam). As shown on Fig. 1a, the filter is placed slightly after the image plane in order to record a pure projection of the filter. A typical interference pattern obtained with the 13th harmonic is shown in Figure 2a. The top part of the image is the reference fringe pattern and the bottom part includes the phase shift that is introduced by the step of Al layer. There is a clear shift of $\Phi=0.4$ fringe between the two fringe patterns. The fringe shift is related to the Al step size ΔL by: $\Delta L=\Phi\lambda/(1-n_{Al})$. This gives a value of $\Delta L=72\pm8$-nm when taking into account the Al refractive index. Note that these interferometry measurements are not affected by the presence of the thin oxide layer that always recovers Al, whereas absorption measurements would strongly be affected by this oxide, that absorbs a lot in this spectral range[19].

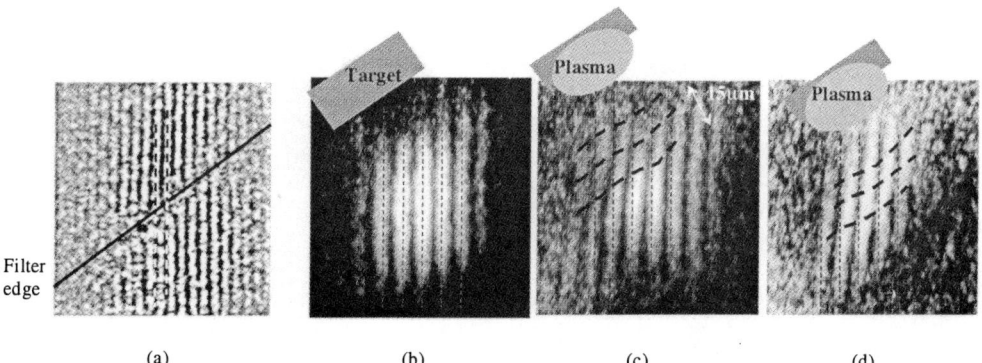

Figure 2: a) Interference pattern with the 13th harmonic through an Al step filter Single shot fringe patterns of the 11th harmonic b) without plasma c), d) with plasma. Initial fringe position (short dash line) and isodensity lines (dashed curves)

The second experiment was to demonstrate that our method can be used to probe a laser produced plasma. Using the tunability of the HOH we are able to choose a wavelength for the probe which minimizes refraction (by the density gradient), reflection or absorption due to the plasma. Indeed the plasma refraction index is $n^2=1-n_e/n_{cr}$ where n_e is the electronic density and n_{cr} the critical density. When $n_e>n_{cr}$ the probe beam can not penetrate in the plasma and is reflected. For IR light at 800-nm $n_{cr}=1.710^{21}$ e$^-$/cm^3 whereas for XUV light at 72-nm $n_{cr}=2.10^{23}$ e$^-$/cm^3. Therefore we should be able to probe high density plasmas using harmonics. The plasma we probed is generated by focusing a 50mJ 300ps IR laser on an Al plate placed a few centimeters after the gas jet (see Figure 1a). The plasma expands in the normal direction and perturbs the first harmonic beam that is close to the target and not the second one. The delay between the harmonic pulses and the pump pulse is held constant to 1.2ns. In Figure 2b is shown the fringe pattern without plasma. When the plasma is generated, there is a clear deviation of the fringes that increases when the target probe distance decreases (Figure 2c and 2d). To estimate the electron density of the plasma, we can approximate the observed fringe shift Φ at wavelength λ as: $\Phi = dn_e/2\lambda n_{cr}$ where d is the distance traveled by the probe beam in the plasma. This distance is experimentally estimated to 0.1mm. In Figure 2c only the upper part of the fringes is shifted the lower part giving the reference position: this allows us to determine the electron density. In Figure 2c and 2d we have plotted isodensity curves corresponding, from bottom to top, to n_e of $0,5.10^{20}$, 1.10^{20} and 2.10^{20} e$^-$/cm^3. If we assume that the bottom part of the fringes is not perturbed in Figure 2d, we estimate an electron density $\geq 2,5.10^{20}$ e$^-$/cm^3 close to the Al plate. A simulation of the plasma in the conditions of the experiment has been performed and confirms the measured value.

Similarly to the generation of two phase locked harmonic sources separated in space, we have demonstrated that it is also possible to produce two phase-locked harmonic sources separated *in time* by focusing a double laser pulse in the generating medium[20]. After dispersion by a grating the two harmonic pulses can overlap and interfere in the spectral domain giving rise to a highly contrasted spectral fringe pattern. This frequency domain interferometry is the temporal analog of the Young's slits experiment. The double laser pulse was created by using the group velocity difference on the two axes of a birefringent plate rotated at 45° from the laser polarization. A polarizer placed after the plate projects both components on the same axis. Both birefringent plate and polarizer were placed before the laser compressor in order to avoid self-phase modulation by the intense compressed beam. The calibrated thickness of the plates fixed the time delay between the pulses (120fs and 450fs). This set-up offers a high stability making it possible acquisition over thousands of shots. The fringe period in the wavelength domain is now given by $\Delta\lambda=\lambda^2/c\Delta t$ and a high resolution spectrometer is necessary. The experimental spectra of harmonics 11, 15, 19, and 23 generated by two phase-locked laser pulses delayed by 120 fs and focused at 2.10^{14} W/cm^2 in argon are shown in Figure 3. Regular contrasted fringes are measured for all

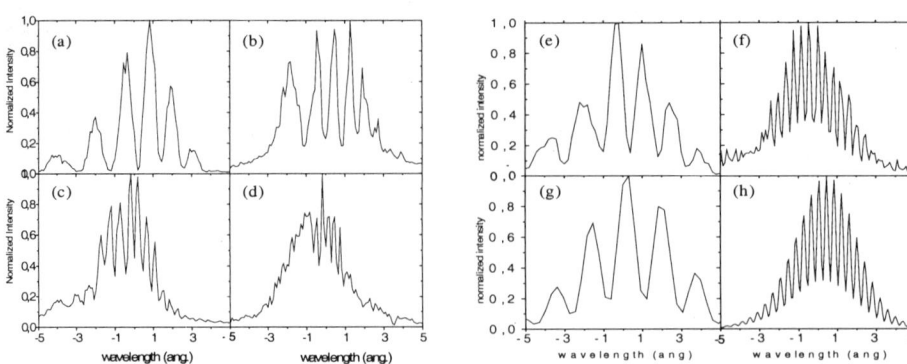

Figure 3: Experimental spectra of harmonics 11 (a), 15 (b), 19 (c) and 23 (d) generated by two laser pulses delayed by 120 fs and focused at 2.10^{14} W/cm^2 Experimental [(e) and (f)] and theoretical [(g) and (h)] spectra of harmonic 11. (e) and (g) generated at 2.10^{14}W/cm^2 with a delay of 120fs. (f) and (h) obtained at 3.10^{14}W/cm^2 with a delay of 450fs.

orders, with a reduced contrast when the order increases: from 90% for harmonic 11 to about 25% for the 23rd. The period of the fringes varies quadratically with the wavelength, as expected from the above formula. One can note that the high energy part of the spectrum is not modulated. This asymmetry in the spectra is also observed when increasing the IR pulse intensity up to the point where the fringes totally disappear. We also observed the dependence of the fringe period with the delay: when increasing the delay from 120 fs to 450fs the fringe period decreases as expected by more than a factor 3 (Figure 3e,f). The numerical simulations performed with the Saclay 3D propagation code[21] reproduce very well the experimental curves (Figure 3g,h). We can explain the reduced contrast in Figure 3f compared to that of Figure 3a by an increase of ionization due to the higher intensity. The intensity of the second harmonic pulse generated in the partially ionized medium is reduced and so is the contrast. A more detailed study of the fringe patterns as a function of intensity should give interesting information on the respective dynamics of HHG and ionization.

Besides its fundamental interest, the above study opens up new perspectives for plasma diagnostics using XUV frequency-domain interferometry. As a demonstration, we have probed the temporal evolution of the electron density in an He high density gas jet, ionized by an intense ultra short laser pulse (50fs, 10^{18}W/cm^2). The experimental setup is shown in Figure 4a. The two 11th harmonic pulses are generated in a xenon gas jet with a 300 fs delay and refocused without magnification with a toroidal mirror into a helium jet. After analysis by an XUV flat-field spectrometer, the fringes are detected by MCP coupled to a phosphor screen and a CCD camera. To achieve the best spectral resolution, the MCP were tilted to a grazing incidence of 8°. Note that harmonic 11 is below the first excited state of helium (21 eV), thus preventing absorption of the harmonic by the neutral gas surrounding the plasma. We have recorded the harmonic fringe pattern as a function of the delay between the TW pulse generating the plasma and the harmonic probes (zero delay for pump in between the two probes). When the plasma is generated before (positive delays) or in between the two pulses, the fringes are shifted. In Figure 4b, we show the spatially integrated fringe pattern from single-shot images as a function of the pump-probe delay.

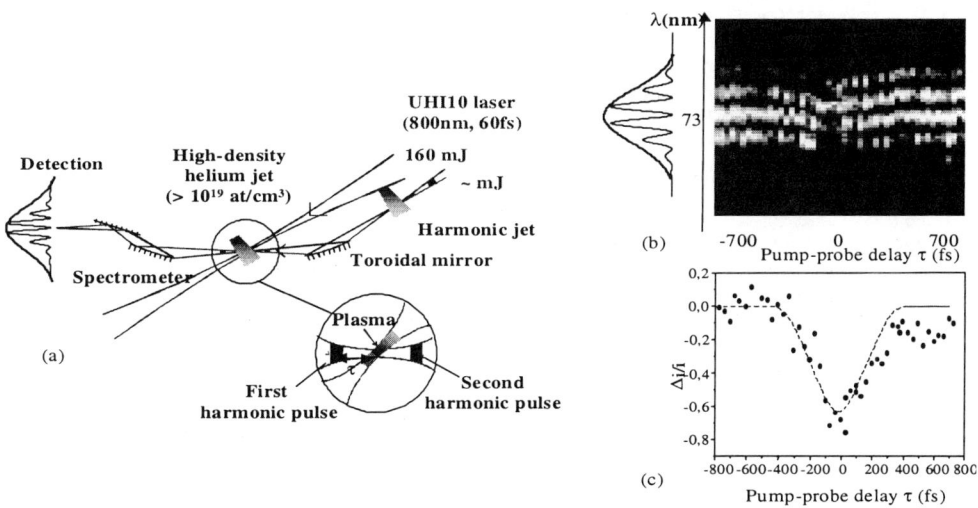

Figure 4: a) Experimental setup for the spectral interferometry with harmonics Variation of the: b) spatially integrated fringe pattern c) average fringe shift with the delay between the plasma generating laser beam and the 11th harmonic beams

A clear shift can be seen in the middle of the image, when the plasma is created in between the two harmonic pulses. The average shift is plotted in Figure 4c. The dashed curve is the theoretical shift obtained for the relevant experimental parameters. The calculated curve fits well the experimental one for an electron density of 6.10^{19} cm^{-3} which is in quite good agreement with the expected density ($4,8.10^{19}$ cm^{-3}) assuming fully stripped ions. The slower shift of the fringes compared to the expected very fast optical field ionization front is, in fact, due to the geometry (probe at 45°) that lowers the time resolution to about 200 fs. One notices that a nonzero phase shift remains when both harmonic pulses propagate through the plasma. This behavior was systematically observed and is not understood yet. It can be due to an in-crease of the plasma size due to postpulse ionization of the surrounding gas by energetic electrons. Further studies are underway in order to clarify this point.

In conclusion we have demonstrated the possibility to perform interferometry in the XUV range with high order harmonics in the *spatial* as well as in the *frequency* domain. This is made possible by using the mutual coherence (in space or in time) of two separated harmonic sources. This results are not only interesting from a fundamental point of view, they also open the possibility of new applications of the harmonic light. High order harmonics are becoming a powerful source to perform time resolved interferometric experiments, e.g. measurements of high electron densities in ultra short transient plasmas with a femtosecond time scale resolution. The matter of this paper is the result of a collective work, namely of P. Salières, B. Carré, L. Le Déroff, H. Merdji, P. Monot, T. Auguste, P. d'Oliveira in CEA Saclay and D. Descamps, C. Lyngå, J. Norin, A. L'Huillier, C.-G. Wahlström in Lund Laser Center.

References

1. L. Le Deroff *et al*, Opt. Lett. **23**, 043802 (2000)
2. S. Le Pape *et al.*, J. Opt. Soc. Am. Applied Optics (submitted)
3. M. Schnürer *et al.*, Appl. Phys. B **70** (Suppl.), 227 (2000)
4. Hergott *et al*, in preparation
5. Constant *et al*, Phys. Rev. Lett. **82**, 1668 (1999)
6. M. Schnürer *et al.*, Phys. Rev. Lett. **83**, 722 (1999)
7. A. Bouhal *et al*, J. Opt. Soc. Am. B **14**, 950 (1997)
8. Y. Kobayashi *et al*, Opt. Lett. **23**, 64 (1995)
9. J. Larsson *et al.*, J. Phys. B **28**, L53 (1995)
10. M. Gisselbrecht *et al.*, Phys. Rev. Lett. **82**, 4607 (1999)
11. F. Quéré *et al.*, Phys. Rev. B **61**, 15 (2000)
12. M. Bellini *et al*, Phys. Rev. Lett. **81**, 297 (1998)
13. P. Salières *et al.*, Phys. Rev. Lett. **74**, 3776 (1995)
14. T. Ditmire *et al.*, Phys. Rev. Lett. **77**, 4756 (1996)
15. L. Le Deroff *et al*, Phys. Rev. A **61**, 043802 (2000)
16. R. Zerne *et al.*, Phys. Rev. Lett. **79**, 1006 (1997)
17. E. Hecht *Optics*, Addison-Wesley publishing company, Reading, 334 (1974)
18. D Descamps *et al.*, Opt. Lett. **25**, 135 (2000)
19. W. J. Tropf *et al*, *Handbook of Optical Constants III*, Academic Press, San Diego, 653 (1998)
20. P. Salières *et al.*, Phys. Rev. Lett. **83**, 5483 (1999)
21. A. L'Huillier *et al*, Phys. Rev. A **46**, 2778 (1992).

INDEX

Ablation
   front, 124
   pressure, 126128, 134, 135
Above Threshold Ionisation, *see* ATI
Absorption processes, 204, 206
   classical, 105, 121
   collisional, 105, 193, 206
   non-collisional, 204, 206, 215, 223
   Resonant, 208-209, 212, 215
Adiabatic approximation, 251
Airy function, 147, 208
Alfvén
   current, 169
   limit, 213, 238
Amplified Spontaneous Emission, *see* ASE
Antihermitian matrix, 254
ASE, 148, 202
Astrophysics, 193, 199, 327, 333, 315
ATI, 16, 28, 59, 105, 266, 285, 299, 309
   perturbative limit, 61
   plateau, 72
   plateau cut-off, 291, 292, 296
   simpleman's theory, 62
ATI-like interaction, 266267
Atomic unit of intensity, 303, 312
Attosecond pulses, 94
Auger
   decay, 216
   electrons, 63
Autoionizing states, 251, 255
Average atom model, 218

Ballistic interference, 296
Baroclynic effect, 235
Barrier ionisation 18
Be atom, 276
Beat-wave, 210
Bessel functions, 175, 267, 270, 306
Betatron
   frequency, 177-178, 185
   oscillation, 180-186
   resonance, 186-187
Bethe stopping power, 216
Binding energy, 255
Bispinor, 306
Black body, 196
Bloch operator, 253
Boltzmann, 199, 213
Bohr radius, 286, 322
Born-Oppenheimer
   approximation, 197, 200

Born-Oppenheimer (*cont'd*)
   electronic energies, 40
Born First-order transition, 306
Bound-bound transitions, 194, 217
Bound-free transitions, 194, 217, 220
Bremsstrahlung 149, 194, 216-217, 384, 375
Brewster angle, 207
B-spline, 257
Bunching, 182

Capillary
   discharge, 375
   tube, 350
Cavitation, 175
Channel
   virtual-fragments, 263
   closure, 74, 299-300
Channelling, 176, 181, 213
Chemical potential, 195, 199
Chirped Pulse Amplification, *see* CPA
Circular dichroism, 70
Circular polarization, 168-174, 182, 189
Close-coupling, 254
Cluster, 305
Cold relativistic limit, 191
Collision frequency, 204-205, 209-215, 221
Collisional broadening, 219
Collisional phenomena, 218, 233, 234
Collisional-Radiative equations, 218
COLTRIM Spectroscopy, 50
Complex-energy degeneracy, 255
Compton
   wavelength, 9
   harmonics, 282
   scattering, 274
Condensation, 101
Conductivity, 398
Continuity equation, 187, 241
Continuum lowering, 197, 220
Correlation, 49
Correlation mediated energy-sharing diagram, 265
Coulomb
   collisions, 214, 218
   correlation interaction, 261, 263
   deflections, 194
   explosion, 109, 117, 240
   field, 279, 312
   gauge, 173, 241
   potential, 21, 37, 108, 214, 261, 308
   pressure, 108
   logarithm, 122, 214-215

Coulomb (*cont'd*)
   scattering, 215-216, 305-306
   soft potential, 308
Coupling parameter, 196
CPA, 1, 3, 102, 119, 140, 146-149, 152-159, 167, 193, 201-202
Critical density, 107
Cyclotron
   rotation, 180
   frequency, 240

Damping period, 121
Debye
   length, 106, 194, 198-199, 215
   temperature, 195, 215
Debye-Hückel, 199, 220
Degeneracy, 197
Dense matter, 167
Density effects, 220
Diamond anvil cell, 327
Dichroism, 351
Dielectric constant, 204-209, 212-215, 221- 224
Dirac
   equation, 200, 218, 279, 305-306, 311
   matrices, 279, 304
   electron, 305, 312
Dirac-like properties, 303
Dirac Coulomb Green's function, 304
Direct acceleration, 167, 180
Direct drive, 136
Dispersion relation, 121-122, 168, 173-174
Dissociation, 42
Distribution function, 214-216, 233-236
Dodging phenomenon, 288, 297
Doppler
   broadening, 219
   shift, 211-212, 309
Drude
   approximation, 221
   dielectric constant, 108
   model, 209-214, 221
Drift
   energy, 287, 290-292
   velocity, 295
Dynamics
   perturbative, 249
   non-perturbative, 249-250
   correlated electrons, 257
   spin, 278
Dyson
   series, 263
   equation, 294

ECFI, 236, 237, 240
Effective nuclear charge, 279
Einstein coefficients, 219
Electrical conductivity, 194-195, 204-205, 210-216
Electromagnetic current filamentation instability, *see* ECFI
Electromagnetic field tensor, 276
Electron
   acceleration, 167, 177-179, 182, 187, 280-282, 309
   atom scattering, 149, 304
   beams, 167, 175
   canonical momentum, 233, 240, 250, 265, 273, 291

Electron (*cont'd*)
   conduction, 221
   correlation, 255, 263, 266
   degeneracy, 215
   dipole moment, 276
   hydrogen problem, 257
   jets, 210,
   thermal , 211-213
   thermal wave, 211
   wavepacket, 278, 280, 288
   waves, 215
Electronic energy gain, 281
Electronic fluid, 235, 240-241
Electronic pressure tensor, 233-236
Electronic temperature gradient, 235
Electronic vortices, 244
Elliptical polarization, 288
Ellipticity distribution, 296-298
Emissivities, 217
Energetic electrons, *see* fast electrons
Energy conservation law, 296
EOS, 193, 196-199, 216, 220, 395, 327, 329
Equation of motion, 169-173, 178, 183, 187-190, 200
Equation of State, *see* EOS
Euler-Lagrange equation, 169
Evolution operator, 262-264, 293

Faraday's law, 233-235
Fast electron, 175, 201, 213, 215-216, 222-223, 237, 305, 363,
   penetration range, 237
   transport, 213-216
Fast ignitor 120, 139-141, 158, 210, 237-239, 363
FEL, 167, 177-178, 182, 303, 312
Fermi
   distribution, 204
   energy, 196, 215
   integral, 199
   temperature, 195
Fermi-Dirac, 195, 199, 200, 213
Feynman diagrams, 265
FDI, 288, 404
Filamentation, 193, 201
   relativistic, 237
Filamentary instability, 175
Fission, 389
Floquet
   blocks, 249, 258
   Hamiltonian, 24, 249, 258
   theory, 15, 24, 46, 286, 308
   type numerical method, 286
Floquet-Fourier expansion, 24, 30
Fluid
   closure, 234
   equations, 188, 220
   singularity, 239
   theory, 175
   turbulence, 241
Fluorescence efficiencies, 216
Foldy-Wouthuysen expansion, 278
Fourier mode, 121, 237
Free Electron Laser, *see* FEL
Free-free transitions, 217
Frequency domain interferometry, *see* FDI
Frequency doubling, 280

Fresnel
  formula, 207
  limit, 207, 209

Gamma ray, 6
Gamow-Siegert boundary conditions, 254
Gaussian, 171-175, 178-179, 219
Gaussian wavepacket, 278, 309
Gas jet, 67, 84, 100
Geophysics, 193
Gouy phase, 87
Gordon-Volkov, 324, 327, 352
Green's operator, 23, 262, 267, 304

Hagena parameter 100
Hamiltonian 21, 169, 197
  motion, 241
  non relativistic, 250-251
  time-independent atomic, 250-251
  total, 262-264
  virtual-fragments, 263-264
Harmonic emission spectra, 275-278, 290
Hasegawa-Mima equation, 240
He atom, 249-258, 261-270
Heating
  Impulsive, 224
  J x B, 215
  Joule, 205
Heaviside theta function, 264
Helmoltz equations, 205, 211, 221
Hermitian eigenvalues problem, 253, 254
HFFT, 30
HHG, 223-224, 249, 255, 285, 295, 303, 401
  plateau, 286-289
High-frequency Floquet-theory, see HFFT
High harmonic generation, see HHG
High order harmonic, see HOH
HOH, 401
Hole boring, 210
Hot electrons, see fast electrons
Hot spots, 193
H-shell emission, 315
Hugoniot relations, 328
Hydrocodes, 220-223
Hydrodynamic
  approximation, 189
  equations, 221
  expansion, 203, 213
  instabilities, 193
  pressure, 108
Hydrogen atom, 286
Hydrogen-like, 193
Hydrogenic system, 261, 304

ICF, 120, 132, 138, 193, 237, 363, 315
IFAR, 136
IFEL, 177, 182-185
Impedance mismatch, 397
Inertial Confinement Fusion, see ICF
Indirect drive, 136
Inglis-Teller
  effect, 220
  merging, 218
Interference, 282, 288, 295, 300, 401-402
Interferometry, 211, 212
Instability
  azimuthally symmetric, 239

Instability (*cont'd*)
  electrostatic two-stream, 237
  Weibel-type, 236
Inverse bremsstrahlung, 105, 121, 193
Inverse free electron laser, *see* IFEL
Ion
  acceleration, 167
  correlation, 215
Ion sphere, 196-200, 220
Ionic-core potential, 280
Ionisation, 38, 107, 286, 299
  double, 250, 257, 261-266, 270, 289
  energy, 197, 220, 285
  multiphoton, 249, 258
Isotope Enrichment 6

Jacobi-Anger formula, 267

Keldysh
  approximation, 294
  parameter, 18, 39, 62
  theory, 286, 293-294, 300
Keldysh-Faisal Reiss approximation, *see* KFR
Keldysh Klein-Gordon equation, 311
KFR approximation, 61, 265, 294
Kinetic
  description, 189, 213
  equation, 204, 213
Klein's paradox, 311-312
K-matrix, 254
Kramers-Henneberger frame, 21, 73, 254
Krook approximation, 204, 213
K-shell photoionisation, 304

Lagrange function, 169
Langmuir
  frequency, 237
  wave, 237
  time, 240
Laser
  acceleration, 167, 191
  accelerator
    beat-wave, 343-345
    wake-field, 343, 347
    self-modulated wake-field, 344, 348
  channel, 173, 182
  CO2, 193
  Excimer, 193
  infrared, 305, 401
  KrF, 208
  Megajoule, *see* LMJ
  Nd:glass, 193, 201-202, 280
  Nd:YAG, 202
  Ti:Sapphire, 1, 3, 52, 102, 201-203, 265, 268, 286
Laser-assisted single photon ionisation, *see* LSPI
Laser-driven model ion, 278
Laser-Induced Degenerate States, *see* LIDS
LASPI, 28
Lee and More model, 212-215
LICS, 27, 255
LIDS, 255
Light-induced continuum structure, *see* LICS
Line broadening, 218
Linear polarization, 168-172, 178-182
Lippmann-Schwinger equation, 262
LMJ, 139

409

Local Thermal Equilibrium, *see* LTE
LOH, 249, 255
Lorentz
    model, 215
    profile, 219
    force, 233, 279, 280
    relativistic factor, 241, 276, 281
    invariant, 306
Low order harmonics, *see* LOH
LTE, 213-218

Magnetic coalescence, 239
Magnetic field
    annihilation, 234
    freezing, 235
    generation, 233-234
    quasi-static, 233-234, 240
    quasi-stationary structure, 240
    self generated, 240
    spatial structure, 238
    superstrong 233-234
Magnetic flux conservation law, 235
Magnetic guiding, 239
Magnetic moment, 275
Magnetic quantum numbers, 253
Magnetic vortice, 234
Magneto hydrodynamic equations, *see* MHD
Many body system, 261-262
Mass-shift, 306, 310
Master equation, 222
Matrix method, 207
Maxwell
    distribution, 195, 204, 215
    equations, 172, 205-206, 236, 241
MCP, 67, 103, 402
Metallisation, 397
MHD, 234, 241
Micro Channel Plate, *see* MCP
Micromachining, 5
Microwave regime, 287
Molecular fragmentation 49
Momentum
    virtual, 264
    conservation, 265
    photon, 304
Monte Carlo
    sampling method, 268
    simulations, 31, 218
Mott scattering, 305, 312
Multilayer mirrors, 67
Multiphoton process, 249, 256, 305-308
Multipole potential, 254

National Ignition Facility, *see* NIF
Ne atom, 270
N-electron system, 251, 253
Newton-Lorentz equation, 273
Newtonian mechanics, 273
NIF 139,147
Non-linear atom-laser interaction, 303
Non-linear optics, 357, 349
Non-linear pendulum equation, 185
Non-linear phenomena, 241, 285, 312
Non-linear plasma kinetics, 167
Non-linear saturation, 237
Non-linear Thomson Scattering, 6
Non-relativistic dipole approximation, *see* NRD

Non-relativistic electrons, 168, 174
Non-relativistic limit, 121
Non-relativistic retard, *see* NRR
Non-relativistic wave breaking, 191
Non sequential diagram, 265
Non sequential double ionisation, *see* NSDI
NRD, 304, 305, 312
NRR, 305
NSDI, 289
Nuclear Coulomb potential, 254
Nuclear Fusion, 116

Ohm's law, 205
One-electron problem, 254, 261
One-particle plane wave configuration, 280
Opacities, 213, 217- 218
OPA, 1, 92
Optical depth, 216
Optical Parametric Amplifier, *see* OPA
Optical smoothing, 396
Oscillating mirror, 223
Oscillator strength, 218
OTB, 18
Over the barrier ionisation, *see* OTB

Pair creation, 149, 304, 310
Parametric instabilities, 193, 210
Particle acceleration, 120, 178, 187, 339
Particle in cell code, *see* PIC
Pauli
    equation, 278
    exclusion principle, 200
Petawatt, 119,145-147, 152
Phase
    locking, 401, 402
    matching, 290
    slippage, 282
Phonons, 215
Photoionization, 285, 304, 307
Photoelectron, 285-299
    kinetic energy spectra, 285
    drift energy, 287
Photo nuclear process, 149
Photon source, 120
PIC, 233, 242-245
    2D PIC, 179-182
    3D PIC, 175, 179, 187
Planck formula, 217
Plasma
    astrophysical, 233-234
    channel, 167, 174, 177, 186-187, 350
    cold, 167, 172
    critical density, 238, 404
    degenerate, 195, 198, 218
    dense, 193-194, 197-201, 210, 213-219, 223, 404
    frequency, 107, 196-197, 201, 236, 242
    fluid velocity, 234
    highly correlated, 195-196
    ideal, 194, 197, 199
    laboratory, 234
    magnetised, 234
    mirror, 210
    non-degenerate, 200
    non-ideal, 194-195, 215
    non-relativistic, 190
    oscillation, 198

Plasma (*cont'd*)
   overdense, 239
   sound speed, 106, 203
   relativistic, 130-131, 241, 322
   relativistic warm, 191
   strongly coupled, 195, 196, 220
   underdense, 233, 237-240, 244
   transient, 406
   wave, 167, 177, 180, 187, 191, 209, 216, 340
   weakly coupled, 220
   warm, 188, 191
Poisson equation, 174, 187, 199
Ponderomotive bucket, 185
Ponderomotive energy, 167, 175, 196, 265, 287, 304
Ponderomotive expulsion, 175, 182
Ponderomotive force, 173-174, 187, 210, 235
Ponderomotive phase, 185
Ponderomotive potential, 65, 274
Ponderomotive pressure, 175, 210, 221, 241
Positron, 120,149
Poynting vector, 168
Pressure ionisation, 217-218, 220
Propagator, 262-263, 266
Proton beam, 152, 163
Pump and probe, 207, 210-211, 224, 405
Polarizability, 249

Quantum electrodynamics, *see* QED
Quantum interference, 286, 293-295
Quasi static approximation, 219
Quasi-molecular models, 200
QED, 303, 310, 311
Quiver energy, 103
Quiver velocity, 120-121, 168, 201

Radiative phenomena, 218
Radiative transfer equation, 216
Rainbow scattering, 291-293
Rarefaction wave, 210
Rayleigh
   scattering, 101
   length, 174
Rayleigh Taylor instability 137-139, 210
Recoil-momentum, 261, 265, 268-269
Relativistic effects, 182, 190, 303-312
Relativistic electrons, 120, 149, 160, 177-178, 237
Relativistic dipole, 305
Relativistic dynamics, 308
Relativistic intensity, 168, 201, 279
Relativistic laser field, 273
Relativistic mass shift, 278
Relativistic plasma interaction, 167, 172-173, 193, 196
Relativistic regime, 170, 221
Relativistic retard, 305
Relativistic self-focusing, 148, 167, 173-175, 237
Relativistic speed, 274
Relaxation time, 401
Reflectometry, 211
Rescattering, 265-266, 289-290, 295-300
Resistivity, 209-210
Resonance acceleration, 180
Resonance condition, 184
Resonance mechanism, 167
Resonant drive, 180
Resonant states, 248-249, 255- 256

Rest-mass energy, 304
Return current, 180, 213, 216, 237
R-matrix method, 25, 286
R-matrix-Floquet method, 249-258
Runge-Kutta, 207
Rutherford cross section, 216
Rydberg states, 16, 40, 220

SAE, 27, 85, 250, 257, 261
Saha equilibrium, 197
Saha-Boltzmann equation, 221
Scattering, 251, 257
Schrödinger equation, 200, 218, 244, 252, 261, 286-287, 291, 293
Schwinger field, 9
Second harmonic emission, 175, 211
Second quantization, 273
Self-Focusing, 6, 174-178, 187, 201, 210, 239, 242
Self-trapping, 241-242, 244
Sesame table, 397
SFA, 86
Shadowgraphy, 211
Sheath inverse bremsstrahlung, 205, 206
Shock compressed matter, 216
Shock wave, 124, 193, 196, 327
   Bow, 210
   Electromagnetic, 357
Single active electron, *see* SAE
Skimmer, 103
Skin depth, 107, 201, 203, 205, 206, 207
   electronic, 235, 238-241
Slater
   exchange potential, 200
   integrals, 218
S-matrix theory, 261-264
Smith-Purcell effect, 7
SRS, 105
Soft-core potential, 279
Soliton, 233, 240-245
   s-polarised, 241-242
   p-polarized, 241-242
   circularly polarized, 242
   toroidal, 243
   subcycle relativistic, 240-241
   subcycle low frequency, 241
Soliton-like structure, 242
Soliton energy burst, 244
Space charge, 167, 172
Spatially-confined laser pulse, 282
Spin, 251-253, 275-277, 304
Spin-laser interaction, 276
Spin-orbit coupling, 251, 278, 305-307, 312
Spitzer transport, 204, 214
Split-operator, 279
Stark
   broadening, 210, 219
   effect, 218, 219, 220
State-splitting, 278
Stewart-Pyatt, 220
Stimulated Raman Scattering, *see* SRS
Stopping power, 216
Strain waves, 224
Streak camera, 211, 223
Strong field approximation, *see* SFA
Strong-field effects, 287
Sturmian expansion, 304

Sturmian-Floquet, 25
Super-channelling, 175
Suprathermal electrons, *see* Fast electron
Surgery, 5
Synchrotron
    oscillation, 185
    rotation, 185

TDSE, 27, 85, 250-258, 262, 308, 309
TDRSE, 308
Thermal conduction, 203-204
Thermal conductivity, 194-195, 213-215, 221
Thermal electron transport, 201, 206, 211, 215
Thermal transport, 213, 215
Thermodynamic equilibrium, 196, 197
Thomas-Fermi model, 199
Thomas-Fermi-Dirac model, 200
Thomson scattering, 275
Three body coulomb problem, 257
Time dependent Schrodinger Equation, *see* TDSE
Time dependent relativistic Schrodinger Equation,
    see TDRSE
Time-of-flight spectrometer, *see* TOF spectrometer
TOF spectrometer, 38, 43, 67, 103
Tunnel ionisation, 18, 85, 307
Transport coefficients, 194, 213-214
Transport inhibition, 213, 216
T-matrix, 304
Tunnelling, 296, 298, 311
    time, 298
    barrier, 298, 311
    dynamic problem, 298
Two-color above threshold ionisation, 59
Two-electron fluid approximation, 236
Two-electron system, 249-258, 261-262
Two-photon bound-bound absorption, 303-305

Ultrafast melting, 224
Ultrarelativistic electron, 191
Ultrarelativistic laser amplitude, 171
Ultrarelativistic limit, 184
Ultraviolet Photoelectron Spectroscopy, *see* UPS
UPS, 93
UV, 307

Vacuum heating, 206, 215

Vector potential, 274, 287
Virtual-fragments, 263-264
Visar, 396, 333
Vlasov's equation, 233, 234
Volkov
    dressing, 269
    eigenfunctions, 286
    electron, 306
    hamiltonian, 263
    propagator, 262, 266
    solution, 262
    state, 264-267, 293
    wave function, 267, 286, 305
Von Karmàn row, 240
Vortex, 233, 236, 241, 245

Warnier mechanism, 265
Wakefield, 167, 177, 185, 189-191, 210
Warm dense matter, 195-196
Wave
    breaking, 167, 187,-191
    collapse, 242
    equation, 173
Wenttel-Kramer-Brillouin, *see* WKB
WKB, 207, 208, 212, 311

XAS, 303
Xe atom, 299
XES, 304
X-ray, 193-194, 207, 210-211, 216-224, 381, 389
    absorption, *see* XAS
    emission, *see* XES
    diffraction, 194
    laser, 193, 303, 312
    lithography, 193
X-ray (*cont'd*)
    sources, 120, 130, 193
    spectroscopy, 280, 286, 303-304, 307, 312
XUV
    interferometry, 93
    radiation, 60, 67, 286, 303, 401

Young experiment, 401

Ziman formula, 215
Zitterbewegung, 278